SURVEY METHODS
IN MULTINATIONAL,
MULTIREGIONAL, AND
MULTICULTURAL CONTEXTS

SURVEY METHODS IN MULTINATIONAL, MULTIREGIONAL, AND MULTICULTURAL CONTEXTS

Edited by

**Janet A. Harkness • Michael Braun • Brad Edwards
Timothy P. Johnson • Lars Lyberg
Peter Ph. Mohler • Beth-Ellen Pennell • Tom W. Smith**

A JOHN WILEY & SONS, INC., PUBLICATION

Published by John Wiley & Sons, Inc., Hoboken, New Jersey.
Published simultaneously in Canada.

For general information on our other products and services or for technical support, please contact our Customer Care Department within the United States at (800) 762-2974, outside the United States at (317) 572-3993 or fax (317) 572-4002.

Wiley also publishes its books in a variety of electronic formats. Some content that appears in print may not be available in electronic format. For information about Wiley products, visit our web site at www.wiley.com.

Library of Congress Cataloging-in-Publication Data is available.

ISBN 978-0470-17799-0

Printed in the United States of America.

10 9 8 7 6 5 4 3 2 1

CONTENTS

PART VIII GLOBAL SURVEY PROGRAMS

CONTRIBUTORS

Alexandra Achen, University of Michigan, Ann Arbor, Michigan
Jennifer S. Barber, University of Michigan, Ann Arbor, Michigan
David Berrigan, National Cancer Institute, Bethesda, Maryland
Georgina Binstock, Cenep-Conicet, Argentina
Annelies G. Blom, Mannheim Research Institute for the Economics of Aging, Mannheim, Germany
Axel Börsch-Supan, Mannheim Research Institute for the Economics of Aging, Mannheim, Germany
Michael Braun, GESIS–Leibniz Institute for the Social Sciences, Mannheim, Germany
Pat Dean Brick, Westat, Rockville, Maryland
Matthieu J. S. Brinkhuis, Utrecht University, Utrecht, The Netherlands
Rachel A. Caspar, RTI, Research Triangle Park, North Carolina
Athanasios Chasiotis, Tilburg University, Tilburg, The Netherlands
Tzu-Yun Chin, University of Nebraska-Lincoln, Lincoln, Nebraska
Julie de Jong, University of Michigan, Ann Arbor, Michigan
Edith de Leeuw, Utrecht University, Utrecht, The Netherlands
Steve Dept, cApStAn Linguistic Quality Control, Brussels, Belgium
Brad Edwards, Westat, Rockville, Maryland
Neli Esipova, Gallup, Inc., Princeton, New Jersey
Andrea Ferrari, cApStAn Linguistic Quality Control, Brussels, Belgium
Rory Fitzgerald, City University, London, United Kingdom
Barbara H. Forsyth, University of Maryland Center for the Advanced Study of Language, Maryland
Joachim R. Frick, DIW Berlin, University of Technology Berlin, IZA Bonn, Germany
Wade M. Garrison, University of Kansas, Lawrence, Kansas
Dirgha J. Ghimire, University of Michigan, Ann Arbor, Michigan
Patricia L. Goerman, U.S. Bureau of the Census, Washington, DC
Markus Grabka, DIW Berlin, University of Technology Berlin, Germany
Peter Granda, ICPSR, University of Michigan, Ann Arbor, Michigan
Reto Hadorn, SIDOS, Neuchatel, Switzerland
Jacques A. Hagenaars, Tilburg University, Tilburg, The Netherlands
Karsten Hank, Mannheim Research Institute for the Economics of Aging, Mannheim, Germany
Sue Ellen Hansen, University of Michigan, Ann Arbor, Michigan
Janet A. Harkness, University of Nebraska-Lincoln, Lincoln, Nebraska
Anne M. Hartman, National Cancer Institute, Bethesda, Maryland
Steven G. Heeringa, University of Michigan, Ann Arbor, Michigan
Marjorie Hinsdale-Shouse, RTI, Research Triangle Park, North Carolina
David A. Howell, University of Michigan, Ann Arbor, Michigan
Joop J. Hox, Utrecht University, Utrecht, The Netherlands
Frost Hubbard, University of Michigan, Ann Arbor, Michigan
Ronald Inglehart, University of Michigan, Ann Arbor, Michigan

Annette Jäckle, Institute for Social and Economic Research, University of Essex, Colchester, United Kingdom

Rukmalie Jayakody, Pennsylvania State University, University Park, Pennsylvania

Yang Jiang, University of Michigan, Ann Arbor, Michigan

Timothy P. Johnson, University of Illinois at Chicago, Chicago, Illinois

Roger Jowell, City University, London, United Kingdom

Hendrik Jürges, Mannheim Research Institute for the Economics of Aging, Mannheim, Germany

Knut Kalgraff Skjåk, Norwegian Social Science Data Services, Bergen, Norway

Katherine King, University of Michigan, Ann Arbor, Michigan

Ashley Landreth, U.S. Bureau of Census, Washington, DC

Deirdre Lawrence, National Cancer Institute, Bethesda, Maryland

Ron J. Lesthaeghe, Belgian Royal Academy of Sciences, Brussels, Belgium

Rachel Levenstein, University of Michigan, Ann Arbor, Michigan

Kerry Levin, Westat, Rockville, Maryland

Lars Lyberg, Statistics Sweden and Stockholm University, Stockholm, Sweden

Peter Lynn, Institute for Social and Economic Research, University of Essex, Colchester, United Kingdom

Michael O. Martin, Lynch School of Education, Boston College, Chestnut Hill, Massachusetts

Sohair Mehanna, American University in Cairo, Cairo, Egypt

John P. Meriac, University of Missouri-St. Louis, St. Louis, Missouri

Debra R. Miller, University of Nebraska-Lincoln, Lincoln, Nebraska

Colter Mitchell, Princeton University, Princeton, New Jersey

Mansoor Moaddel, Eastern Michigan University, Ypsilanti, Michigan

Peter Ph. Mohler, University of Mannheim, Mannheim, Germany

Ina V. S. Mullis, Lynch School of Education, Boston College, Chestnut Hill, Massachusetts

Alicia Norberg, Westat, Rockville, Maryland

Colm O'Muircheartaigh, National Opinion Research Center and University of Chicago, Chicago, Illinois

Daniel Oberski, Tilburg University, Tilburg, The Netherlands and Universitat Pompeu Fabra, Barcelona, Spain

Mary Beth Ofstedal, University of Michigan, Ann Arbor, Michigan

Daphna Oyserman, University of Michigan, Ann Arbor, Michigan

Yuling Pan, U.S. Bureau of Census, Washington, DC

Hyunjoo Park, RTI, Research Triangle Park, North Carolina

Beth-Ellen Pennell, University of Michigan, Ann Arbor, Michigan

Emilia Peytcheva, RTI, Research Triangle Park, North Carolina

Martine Quaglia, Institut National d'Etudes Démographiques, Paris, France

Willem E. Saris, ESADE, Universtat Pompeu Fabra, Barcelona, Spain

Alisú Schoua-Glusberg, Research Support Services, Evanston, Illinois

Mathis Schröder, Mannheim Research Institute for the Economics of Aging, Mannheim, Germany

Norbert Schwarz, University of Michigan, Ann Arbor, Michigan

Tom W. Smith, National Opinion Research Center, University of Chicago, Chicago, Illinois

Rajesh Srinivasan, Gallup, Inc., Princeton, New Jersey

Martha Stapleton Kudela, Westat, Rockville, Maryland

Debra S. Stark, Westat, Rockville, Maryland

Diana Maria Stukel, Westat, Rockville, Maryland

Wendy Thomas, University of Minnesota, Minneapolis, Minnesota

Frances E. Thompson, National Cancer Institute, Bethesda, Maryland

Arland Thornton, University of Michigan, Ann Arbor, Michigan

Robert D. Tortora, Gallup, Inc., London, United Kingdom

Ayşe K. Uskul, Institute for Social and Economic Research, University of Essex, Colchester, United Kingdom

Fons J.R. van de Vijver, Tilburg University, Tilburg, The Netherlands, and North-West University, South Africa

Ana Villar, University of Nebraska-Lincoln, Lincoln, Nebraska

Joachim Wackerow, GESIS–Leibniz Institute for the Social Sciences, Mannheim, Germany

Guangzhou Wang, Chinese Academy of Social Sciences, Beijing, China

Laura Wäyrynen, cApStAn Linguistic Quality Control, Brussels, Belgium

Gordon Willis, National Cancer Institute, Bethesda, Maryland

David J. Woehr, University of Tennessee-Knoxville, Knoxville, Tennessee

Christof Wolf, GESIS–Leibniz Institute for the Social Sciences, Mannheim, Germany

Yu Xie, University of Michigan, Ann Arbor, Michigan

Li-Shou Yang, University of Michigan, Ann Arbor, Michigan

Yongwei Yang, Gallup, Inc., Omaha, Nebraska

Linda C. Young-DeMarco, University of Michigan, Ann Arbor, Michigan

Kathryn Yount, Emory University, Atlanta, Georgia

PREFACE

This book aims to provide up-to-date insight into key aspects of methodological research for comparative surveys. In conveying information about cross-national and cross-cultural research methods, we have often had to assume our readers are comfortable with the essentials of basic survey methodology. Most chapters emphasize multinational research projects. We hope that in dealing with more complex and larger studies, we address many of the needs of researchers engaged in within-country research.

The book both precedes and follows a conference on Multinational, Multicultural, and Multiregional Survey Methods (3MC) held at the Berlin Brandenburg Academy of Sciences and Humanities in Berlin, June 25–28, 2008. The conference was the first international conference to focus explicitly on survey methodology for comparative research (http://www.3mc2008.de/). With the conference and monograph, we seek to draw attention to important recent changes in the comparative methodology landscape, to identify new methodological research and to help point the way forward in areas where research needs identified in earlier literature have still not been adequately addressed.

The members of the Editorial Committee for the book, chaired by Janet Harkness, were: Michael Braun, Brad Edwards, Timothy P. Johnson, Lars Lyberg, Peter Ph. Mohler, Beth-Ellen Pennell, and Tom W. Smith. Fons van de Vijver joined this team to complete the conference Organizing Committee. The German Federal Minister of Education and Research, Frau Dr. Annette Schavan, was patron of the conference, and we were fortunate to have Denise Lievesley, Lars Lyberg, and Sidney Verba as keynote speakers for the opening session in the splendid Konzerthaus Berlin.

The conference and book would not have been possible without funding from sponsors and donors we acknowledge here and on the conference website.

Five organizations sponsored the conference and the preparation of the monograph. They are, in alphabetical order: the American Association for Public Opinion Research; the American Statistical Association (Survey Research Methods Section); Deutsche Forschungsgemeinschaft [German Science Foundation]; what was then Gesis-ZUMA (The Centre for Survey Research and Methodology, Mannheim); and Survey Research Operations, Institute for Social Research, at the University of Michigan. These organizations variously provided funds, personnel, and services to support the development of the conference and the monograph.

Additional financial support for the conference and book was provided by, in alphabetical order:

Eurostat
GfK Group
Interuniversity Consortium for Political
and Social Research (ICPSR)
(U.S.) National Agricultural Statistics
Services

Radio Regenbogen Hörfunk in Baden
The Roper Center
SAP
Statistics Sweden
TNS Political & Social
U.S. Census Bureau

Nielsen   University of Nebraska at Lincoln
NORC at the University of Chicago   Westat
PEW Research Center

We must also thank the Weierstraß Institute, Berlin, for donating the use of its lecture room for conference sessions, the Studienstiftung des Deutschen Volkes [German National Academic Foundation] for providing a room within the Berlin Brandenburg Academy for the conference logistics, and colleagues from the then Gesis-Aussenstelle, Berlin, for their support at the conference.

Planning for the book and the conference began in 2006. Calls for papers for consideration for the monograph were circulated early in 2007. A large number of proposals were received; some 60% of the book is derived from these. At the same time, submissions in some topic areas were sparse. We subsequently sought contributions for these underrepresented areas, also drawing from the ranks of the editors. In this way, the book came to consist both of stand-alone chapters which aim to treat important topics in a global fashion and sets of chapters that illustrate different facets of a topic area. The result, we believe, reflects current developments in multinational, multilingual, and multicultural research.

The book is divided into eight parts:

    I Setting the Stage
    II Questionnaire Development
    III Translation, Adaptation, and Assessment
    IV Culture, Cognition, and Response
    V Key Process Components and Quality
    VI Nonresponse
    VII Analyzing Data
    VIII Global Survey Programs

Chapter 1 looks at comparative survey methodology for today and tomorrow, and Chapter 2 discusses fundamental aspects of this comparative methodology. Chapters 3 and 4 consider question and questionnaire design, Chapter 3 from a global perspective and Chapter 4 from a study-specific standpoint. Chapters 5 and 6 discuss developments in pretesting translated materials. Chapter 5 moves toward some guidelines on the basis of lessons learned, and Chapter 6 applies discourse analysis techniques in pretesting. Chapters 7, 8, and 9 focus on producing and evaluating survey translations and instrument adaptations. Chapter 7 is on survey translation and adaptation; Chapter 8 presents a multistep procedure for survey translation assessment; and Chapter 9 describes translation verification strategies for international educational testing instruments. Chapters 10 and 11 consider cultural differences in how information in questions and in response categories is perceived and processed and the relevance for survey response. Together with Chapter 12, on response styles in cultural contexts, they complement and expand on points made in earlier parts.

Part V presents a series of chapters that deal with cornerstone components of the survey process. Chapter 13 outlines the quality framework needed for multipopulation research; Chapter 14 discusses cross-national sampling in terms of design and implementation; Chapter 15 is a comprehensive overview of data collection challenges; and Chapter 16 discusses the role of documentation in multipopulation surveys and emerging documentation standards and tools for such surveys. Chapter 17 treats input and output variable harmonization. Each of these chapters emphasizes issues of survey quality from their particular perspective. Part VI consists of two chapters on nonresponse in comparative contexts: Chapter 18 is on unit nonresponse in cross-national research and Chapter 19 is on item nonresponse in longitudinal panel studies. Both these contribute further to the discussion of comparative data quality emphasized in contributions in Part V.

Chapter 20 introduces Part VII, which contains five chapters on analysis in comparative contexts. Chapter 20 demonstrates the potential of various techniques by applying them to a single multipopulation dataset; Chapter 21 treats multigroup and multilevel structural equation modeling and multilevel latent class analysis; Chapter 22 discusses polytomous item response theory; Chapter 23 explores categorization problems and a Multitrait Multimethods (MTMM) design, and the last chapter in the section, Chapter 24, discusses mixed methods designs that combine quantitative and qualitative components.

Part VIII is on global survey research and programs. It opens in Chapter 25 with an overview of developments in global survey research. Chapters 26–31 present profiles and achievements in a variety of global research programs. Chapter 26 is on the European Social Survey; Chapter 27 presents the International Social Survey Programme; Chapter 28 deals with the Survey of Health, Ageing, and Retirement in Europe. Chapter 29 discusses developments in two international education assessment studies, the Trends in International Mathematics and Science Study and the Progress in International Reading Literacy Study. Chapter 30 profiles the Comparative Study of Electoral Systems, and the concluding chapter in the volume, Chapter 31, describes the Gallup World Poll.

Pairs of editors or individual editors served as the primary editors for invited chapters: Edwards and Harkness for Chapters 4, 5, and 6; Harkness and Edwards for Chapters 8 and 9; Braun and Harkness for Chapters 10 and 11; Johnson for Chapter 12; Lyberg for Chapters 14, 18, and 19; Lyberg, Pennell, and Harkness for Chapter 17; Braun and Johnson for Chapters 21–24; Smith for Chapters 26–31. The editorial team also served as the primary reviewers of chapters in which editors are only or first author (Chapters 2, 3, 7, 13, 15, 16, 20, and 25).

The editors have many people to thank and acknowledge. First, we thank our authors for their contributions, their perseverance, and their patience. We also thank those who helped produce the technical aspects of the book, in particular Gail Arnold at ISR, University of Michigan, who formatted the book, Linda Beatty at Westat who designed the cover, Peter Mohler, who produced the subject index, and An Lui, Mathew Stange, Clarissa Steele, and, last but not least, Ana Villar, all at the University of Nebraska-Lincoln, who aided Harkness in the last phases of completing the volume. In addition, we would like to thank Fons van de Vijver, University of Tilburg, Netherlands, for reviewing Chapter 12 and students from Harkness' UNL spring 2009 course for their input on Chapters 13 and 15. We are grateful to our home organizations and institutions for enabling us to work on the volume and, as relevant, to host or attend editorial meetings. Finally, we thank

Steven Quigley and Jacqueline Palmieri at Wiley for their support throughout the production process and their always prompt attendance to our numerous requests.

Janet A. Harkness
Michael Braun
Timothy P. Johnson
Lars Lyberg
Peter Ph. Mohler
Beth-Ellen Pennell
Tom W. Smith

PART I

SETTING THE STAGE

1

Comparative Survey Methodology

*Janet A. Harkness, Michael Braun, Brad Edwards,
Timothy P. Johnson, Lars Lyberg, Peter Ph. Mohler,
Beth-Ellen Pennell, and Tom W. Smith*

1.1 INTRODUCTION

This volume discusses methodological considerations for surveys that are deliberately designed for comparative research such as multinational surveys. As explained below, such surveys set out to develop instruments and possibly a number of the other components of the study specifically in order to collect data and compare findings from two or more populations.

As a number of chapters in this volume demonstrate, multinational survey research is typically (though not always) more complex and more complicated to undertake successfully than are within-country cross-cultural surveys. Many chapters focus on this more complicated case, discussing multinational projects such as the annual International Social Survey Programme (ISSP), the epidemiologic World Mental Health Initiative survey (WMH), the 41-country World Fertility Survey (WFS), or the triennial and worldwide scholastic assessment Programme for International Student Assessment (PISA). Examples of challenges and solutions presented in the volume are often drawn from such large projects.

At the same time, we expect many of the methodological features discussed here also to apply for within-country comparative research as well. Thus we envisage chapters discussing question design, pretesting, translation, adaptation, data collection, documentation, harmonization, quality frameworks, and analysis to provide much of importance for within-country comparative researchers as well as for those involved in cross-national studies.

This introductory chapter is organized as follows. Section 1.2 briefly treats the growth and standing of comparative surveys. Section 1.3 indicates overlaps between multinational, multilingual, multicultural, and multiregional survey

---

[1] Survey Methods in Multinational, Multiregional, and Multicultural Contexts, edited by Harkness et al.
Copyright © 2010 John Wiley & Sons, Inc.

research and distinguishes between comparative research and surveys deliberately designed for comparative purposes. Section 1.4 considers the special nature of comparative surveys, and Section 1.5 how comparability may drive design decisions. Section 1.6 considers recent changes in comparative survey research methods and practice. The final section, 1.7, considers ongoing challenges and the current outlook.

1.2 COMPARATIVE SURVEY RESEARCH: GROWTH AND STANDING

Almost without exception, those writing about comparative survey research—whether from the perspective of marketing, the social, economic and behavioral sciences, policy-making, educational testing, or health research—remark upon its "rapid," "ongoing," or "burgeoning" growth. And in each decade since World War II, a marked "wave" of interest in conducting cross-national and cross-cultural survey research can be noted in one discipline or another (see contributions in Bulmer, 1998; Bulmer & Warwick, 1983/1993; Gauthier, 2002; Hantrais, 2009; Hantrais & Mangen, 2007; Øyen, 1990; and Chapters 2 and 25, this volume).

Within the short span of some 50 years, multipopulation survey research has become accepted as not only useful and desirable but, indeed, as indispensable. In as much as international institutions and organizations—such as the European Commission, the Organization for Economic Co-operation and Development (OECD), the United Nations (UN), the United Nations Educational, Scientific and Cultural Organization (UNESCO), the World Bank, the International Monetary Fund (IMF), and the World Health Organization (WHO)—depend on multinational data to inform numerous activities, it has become ubiquitous and, in some senses, also commonplace.

1.3 TERMINOLOGY AND TYPES OF RESEARCH

In this section we make a distinction which is useful for the special methodological focus of many chapters in this volume—between comparative research in general and deliberately designed comparative surveys.

1.3.1 Multipopulation Surveys: Multilingual, Multicultural, Multinational, and Multiregional

Multipopulation studies can be conducted in one language; but most multipopulation research is nonetheless also multilingual. At the same time, cultural differences exist between groups that share a first language both within a country (e.g., the Welsh, Scots, Northern Irish, and English in the United Kingdom) and across countries (e.g., French-speaking nations/populations). Language difference (Czech versus Slovakian, Russian versus Ukrainian) is, therefore, not a necessary prerequisite for cultural difference, but it is a likely indicator of cultural difference.

Within-country research can be multilingual, as reflected in national research conducted in countries as different as the Philippines, the United States, Switzerland, Nigeria, or in French-speaking countries in Africa. Cross-national projects may thus often need to address within-country differences in language and culture in addition to across-country differences, both with respect to instrument versions and norms of communication.

Multiregional research may be either within- or across-country research and the term is used flexibly. Cross-national multiregional research may group countries considered to "belong together" in some respect, such as geographical location (the countries of Meso and Latin America), in demographical features (high or low birth or death rates, rural or urban populations), in terms of developmental theory (see Chapter 4, this volume) or in terms of income variability. Other multiregional research might be intent on covering a variety of specific populations in different locations or on ensuring application in a multitude of regions and countries. Within-country multiregional research might compare differences among populations in terms of north-south, east-west or urban-rural divisions.

1.3.2 Comparative by Design

This volume focuses on methodological considerations for surveys that are deliberately planned for comparative research. These are to be understood as projects that deliberately design their instruments and possibly other components of the survey in order to compare different populations and that collect data from two or more different populations. In 1969, Stein Rokkan commented on the rarity of "deliberately designed cross-national surveys" (p. 20). Comparative survey research has grown tremendously over the last four decades and is ubiquitous rather than rare. However, Rokkan's warning that these surveys are not "surefire investments" still holds true; the success of any comparative survey requires to be demonstrated and cannot be assumed simply on the basis of protocols or specifications followed. Numerous chapters in this volume address how best to construct and assess different aspects of surveys designed for comparative research.

Comparative instruments are manifold in format and purpose: educational or psychological tests, diagnostic instruments for health, sports performance, needs or usability assessment tools; social science attitudinal, opinion and behavioral questionnaires; and market research instruments to investigate preferences in such things as size, shape, color, or texture. Several chapters also present comparative methodological studies.

Comparative surveys are conducted in a wide variety of modes, can be longitudinal, can compare different populations across countries or within countries, and can be any mix of these. Some of the studies referred to in the volume are longitudinal in terms of populations studied (panels) or in terms of the contents of the research project (programs of replication). Most of the methodological discussion here, however, focuses on synchronic, across-population research rather than on across-time perspectives (but see Lynn, 2009; Duncan, Kalton, Kasprzyk, & Singh, 1989; Smith, 2005).

Comparative surveys may differ considerably in the extent to which the deliberate design includes such aspects as sampling, the data collection process, documentation, or harmonization. In some cases, the instrument is the main component "designed" to result in comparable data, while many other aspects are decided at the local level (e.g., mode, sample design, interviewer assignment, and contact protocols). Even when much is decided at the local level, those involved in the project must implicitly consider these decisions compatible with the comparative goals of the study.

If we examine a range of large-scale cross-national studies conducted in the last few decades (see, for example, Chapters 25–31, this volume), marked differences can also be found in study design and implementation. Studies vary greatly in the level of coordination and standardization across the phases of the survey life cycle, for example, in their transparency and documentation of methods, and in their data collection requirements and approaches.

1.3.3 Comparative Uses of National Data

Comparative research (of populations and locations) need not be based on data derived from surveys deliberately designed for that purpose.

A large body of comparative research in official statistics, for instance, is carried out using data from national studies designed for domestic purposes which are then also used in analyses across samples/populations/countries. Early cross-national social science research often consisted of such comparisons (cf. Gauthier, 2002; Mohler & Johnson, this volume; Rokkan, 1969; Scheuch, 1973; Verba, 1969). Official statistics agencies working at national and international levels (UNESCO Statistics; the European statistical agency, Eurostat; and national statistical agencies such as the German Statistisches Bundesamt and Statistics Canada) often utilize such national data for comparative purposes, as do agencies producing international data on labor force statistics (International Labour Organization; ILO), on income, wealth, and poverty (Luxembourg Income Study; LIS), and on employment status (Luxembourg Employment Study; LES). Such agencies harmonize data from national studies and other sources because adequately rich and reliable data from surveys that were deliberately designed to produce cross-national datasets are not available for many topics. The harmonization strategies used to render outputs from national data comparable (ex-post output harmonization) are deliberately designed for that purpose (see, for example, Ehling, 2003); it is the national surveys themselves which are not comparative by design. A partnership between Eurostat and many national statistical offices has resulted in the European Statistical System, an initiative which aims to provide reliable and comparable statistics for all the European Union and the European Free Trade Association Member States on the basis of national data.

Instruments designed for a given population are also frequently translated and fielded with other populations. Such translated versions can be tested for suitability with the populations requiring the translations (see Chapters 5, 6, and 7, this volume) and may produce data that permit comparison. Nonetheless, the original (source) instrument was not comparative by design. Publications arguing the validity and reliability of "translated/adapted" instruments abound, particularly in health, opinion, and psychological research. While the suitability of these procedures and instruments is sometimes contested (e.g., Greenfield, 1997), such instruments may be translated into many languages and used extensively worldwide. In some cases, feedback from implementations in other languages can lead to adjustments to the original instrument. One prominent example is the development of the SF-36 Health Survey, a short (36-question) survey that has been translated and adapted in over 50 languages. The development of translated versions led to related modifications in the original English questionnaire (cf. Ware, undated, at http://www.sf-36.org/tools/SF36.shtml/).

Finally, we note that the distinction between comparative research and research that is comparative by design used here is not one always made. Lynn, Japec, and Lyberg (2006), for example, use the term "cross-national surveys" to refer to "all types of surveys where efforts are made to achieve comparability across countries. Efforts to achieve comparability vary on a wide spectrum from opportunistic adjustment of data after they have been collected to deliberate design of each step in the survey process to achieve functional equivalence" (p. 7). The latter of these would fall under our definition of "surveys comparative by design;" those based on "opportunistic adjustment of data after they have been collected" would not.

1.4 WHAT IS (SO) SPECIAL ABOUT COMPARATIVE SURVEY RESEARCH?

Many discussions of comparative survey research note at some point that all social science research is comparative (cf. Armer, 1973; Jowell, 1998; Lipset, 1986; Smith, forthcoming).

Some also suggest that there is nothing really special about comparative (survey) research. Verba (1971 and 1969) and Armer (1973) seem to take this position—but simultaneously also document difference. Verba (1969) states, for example: "The problems of design for within-nation studies apply for across-nation studies. If the above sentence seems to say there is nothing unique about cross-cultural studies, it is intended. The difference is that the problems are more severe and more easily recognizable" (p. 313). Armer (1973) goes a step further: "My argument is that while the problems involved are no different in kind from those involved in domestic research, they are of such great magnitude as to constitute an almost qualitative difference for comparative as compared to noncomparative research" (p. 4).

Later researchers, focusing more on the design and organization of comparative surveys, point to what they consider to be unique aspects. Lynn, Japec, and Lyberg (2006) suggest "Cross-national surveys can be considered to have an extra layer of survey design, in addition to the aspects that must be considered for any survey carried out in a single country" (p. 17). Harkness, Mohler, and van de Vijver (2003) suggest that different kinds of surveys call for different tools and strategies. Certain design strategies, such as decentering, certainly have their origin and purpose in the context of developing comparative instruments (cf. Werner & Campbell, 1970). The distinction between comparative research and surveys that are comparative by design accommodates the view that all social science research is comparative and that national data can be used in comparative research, while also allowing for the need for special strategies and procedures in designing and implementing surveys directly intended for comparative research.

There is considerable consensus that multinational research is valuable and also more complex than single-country research (Kohn, 1987; Jowell, 1998; Kuechler, 1998; Lynn, Japec, & Lyberg, 2006; Rokkan, 1969). The special difficulties often emphasized include challenges to "equivalence," multiple language and meaning difficulties, conceptual and indicator issues, obtaining good sample frames, practical problems in data collection, as well as the sheer expense and effort involved. A number of authors in the present volume also point to the organizational demands as well as challenges faced in dealing with the varying levels of expertise and the different *modi operandi*, standards, and perceptions likely to be encountered in different locations.

1.5 HOW COMPARABILITY MAY DRIVE DESIGN

The comparative goals of a study may call for special design, process, and tool requirements not needed in other research. Examples are such unique requirements as decentering, or *ex ante* input harmonization (cf. Ehling, 2003). But deliberately designed comparative surveys may also simply bring to the foreground concerns and procedures that are not a prime focus of attention in assumed single-population studies (communication channels, shared understanding of meaning, complex organizational issues, researcher expertise, training, and documentation).

What Lynn and colleagues (2006) conceptualize as a layer can usefully be seen as a central motivation for design and procedures followed, a research and output objective at the hub of the survey life cycle that shapes decisions about any number of the components and procedures of a survey from its organizational structure, funding, working language(s), researcher training, and quality frameworks to instrument design, sample design, data collection modes, data processing, analysis, documentation, and data dissemination. Figure 1.1 is a simplified representation of this notion of comparability *driving* design decisions. For improved legibility, we display only four major components in the circle quadrants, instead of all the life-cycle stages actually involved. The comment boxes outside also indicate only a very few examples of the many trade-offs and other decisions to be made for each component in a comparative-by-design survey.

1.6 RECENT CHANGES IN PRACTICE, PRINCIPLES, AND PERSPEC-TIVES

The practices followed and tools employed in the design, implementation, and analysis of general (noncomparative) survey-based research have evolved rapidly

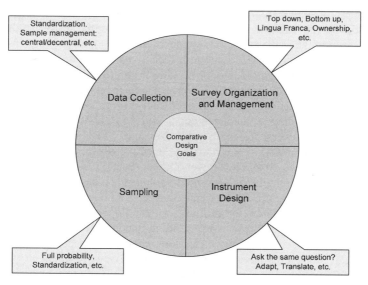

Figure 1.1. "Comparative-by-Design" Surveys

in recent decades. Albeit with some delay, these developments in general survey research methodology are carrying over into comparative survey research. A field of "survey methodology" has emerged, with standardized definitions and (dynamic) benchmarks of good and best practice (cf. Groves et al., 2009, pp. 1–37). Techniques and strategies emerging in the field have altered the way survey research is conceptualized, undertaken, and (now) taught at Masters and PhD level, quietly putting the lie to Scheuch's (1989) claim (for comparative surveys) that "in terms of methodology *in abstracto* and on issues of research technology, most of all that needed to be said has already been published" (p. 147).

The size and complexity of cross-national and cross-cultural survey research have themselves changed noticeably in the last 15–20 years, as have perspectives on practices and expectations for quality. Large-scale comparative research has become a basic source of information for governments, international organizations, and individual researchers. As those involved in research and analysis have amassed experience, the field has become increasingly self-reflective of procedures, products, and assumptions about good practice. In keeping with this, a number of recent publications discuss the implementation of specific projects across countries. These include Börsch-Supan, Jürges, and Lipps (2003) on the Survey on Health, Ageing and Retirement in Europe; Brancato (2006) on the European Statistical System; Jowell, Roberts, Fitzgerald, and Eva (2007b) on the European Social Survey (ESS); and Kessler and Üstün (2008) on the World Mental Health Initiative (WMH). Manuals and technical reports are available on the implementation of specific studies. Examples are Barth, Gonzalez, and

Neuschmidt (2004) on the Trends in International Mathematics and Science Study (TIMSS); Fisher, Gershuny, and Gauthier (2009) on the Multinational Time Use Study; Grosh and Muñoz (1996) on the Living Standards Measurement Study Survey; IDASA, CDD-Ghana, and IREEP (2007) on the Afrobarometer; ORC Macro (2005) on the Malaria Indicator Survey; and, on the PISA study, the Programme for International Student Assessment (2005).

Comparative methodological research in recent years has turned to questions of implementation, harmonization and, borrowing from cross-cultural psychology, examination of bias. Methodological innovations have come from within the comparative field; evidence-based improvements in cross-cultural pretesting, survey translation, sampling, contact protocols, and data harmonization are a few examples. Recent pretesting research addresses not just the need for *more* pretesting, but for pretesting tailored to meet cross-cultural needs (see contributions in Harkness, 2006; Hoffmeyer-Zlotnik & Harkness, 2005; Chapters 5 and 6, this volume). Sometimes the methodological issues have long been recognized—Verba (1969, pp. 80–99) and Scheuch (1993, 1968, pp. 110, 119) could hardly have been clearer, for example, on the importance of context—but only now is methodological research providing theoretical insights into how culture and context affect perception (see, for example, Chapters 10–12, this volume). Design procedures have come under some review; scholars working in Quality of Life research, for instance, have emphasized the need to orchestrate cross-cultural involvement in instrument design (Fox-Rushby & Parker, 1995; Skevington, 2002).

The increased attention paid to quality frameworks in official statistics comprising, among others, dimensions such as relevance, timeliness, accuracy, comparability, and coherence (Biemer & Lyberg, 2003; Chapter 13, this volume), combined with the "total survey error" (TSE) paradigm (Groves, 1989) in survey research, is clearly carrying over into comparative survey research, despite the challenges this involves (cf. Chapter 13, this volume). Obviously, the comparability dimension has a different meaning in a 3M context than in a national survey and could replace the TSE paradigm as the main planning criterion in such a context, as Figure 1.1 also suggests.

Jowell (1998) remarked on quality discrepancies between standards maintained in what he called "national" research and the practices and standards followed in cross-national research. Jowell's comments coincided with new initiatives in the International Social Survey Programme to monitor study quality and comparability (cf. Park & Jowell, 1997a) as well as the beginning of a series of publications on comparative survey methods (e.g., Harkness, 1998; Saris & Kaase, 1997) and the development of a European Science Foundation blueprint (ESF, 1999) for the European Social Survey (ESS). The ISSP and the ESS have incorporated study monitoring and methodological research in their programs; both of these ongoing surveys have also contributed to the emergence of a body of researchers whose work often concentrates on comparative survey methods.

Particular attention has been directed recently to compiling guidelines and evidence-based benchmarks, developing standardization schemes, and establishing specifications and tools for quality assurance and quality control in comparative survey research. The cross-cultural survey guidelines at http://www.ccsg.isr.

umich.edu/ are a prominent example. Numerous chapters in the volume treat such developments from the perspective of their given topics.

Initiatives to improve comparability and ensure quality are found in other disciplines too. The International Test Commission has, for example, compiled guidelines on instrument translation, adaptation, and test use (http://www. intestcom.org/itc_projects.htm/); and the International Society for Quality of Life Research (ISQoL) has a special interest group on translation and cultural adaptation of instruments (http://www.isoqol.org/). The European Commission has developed guidelines for health research (Tafforeau, Lopez Cobo, Tolonen, Scheidt-Nave, Tinto, 2005). The International Standards Organization (ISO) has developed the ISO Standard 20252 on Market, Opinion, and Social Research (ISO, 2006). One of the purposes with this global standard is to enhance comparability in international surveys. We already mentioned the cooperation between Eurostat and national agencies on the European Statistical System (http://epp.eurostat.ec.portal/ page/portal/about_eurostat/european_framework/ESS/).

New technologies are increasingly being applied to meet the challenges of conducting surveys in remote or inhospitable locations: laptops with extended batteries, "smart" hand-held phones and personal digital assistants (PDAs) that allow transmission of e-mail and data, phones with built-in global positioning systems (GPS), pinpointing an interviewer's location at all times, digital recorders the size of thumb drives, and geographic information systems (GIS) combined with aerial photography that facilitate sampling in remote regions, to name but a few. It is easy to envision a future when these technologies become affordable and can be used much more widely for quality monitoring in cross-national research.

Both the software tools and the related strategies for analysis have also changed radically for both testing and substantive applications. Statistical applications and models such as Item Response Theory (IRT) and Differential Item Functioning (DIF) have gained popularity as tests for bias, as have, in some instances, Multitrait Multimethod (MTMM) models. The increased availability of courses in instruction, also online, makes it easier for researchers to gain expertise in the new and increasingly sophisticated software and in analytical techniques.

Documentation strategies, tools, and expectations have greatly advanced. One needs only to compare the half-page study reports for the ISSP in the late 1980s with the web-based study monitoring report now required to recognize that a sea change in requirements and transparency is underway. Proprietary and open access databanks help improve consistency within surveys across versions and speed up instrument production, even if the benchmarks for question or translation quality remain to be addressed.

The improved access to data—which itself tends to be better documented than before—is also resulting in a generation of primary and secondary analysts who are better equipped, have plentiful data, and have very different needs and expectations about data quality, analysis, and documentation than researchers of even a decade ago.

Critical mass can make an important difference; the current volume serves as one example: In 2002, regular attendees at cross-cultural survey methods symposia held through the 1990s in ZUMA, Mannheim, Germany, decided to form an annual workshop on "comparative survey design and implementation." This is now the International Workshop on Comparative Survey Design and Implementation (CSDI; http://www.csdiworkshop.org/). CSDI's organizing committee was, in turn, responsible for organizing the 2008 international conference on Multinational, Multicultural and Multiregional Survey Methods referred to throughout this volume as "3MC" (http://www.3mc2008.de/) and were also the prime movers for this present volume. Moreover, work groups at CSDI were the primary contributors to the University of Michigan and University of Nebraska CSDI initiative on cross-cultural survey guidelines mentioned earlier (http://www.ccsg.isr.umich.edu/). Finally, although the survey landscape has changed radically in recent years (see Table 1.1), readers not familiar with vintage literature will find much of benefit there and an annotated bibliography is under construction at CDSI (http://www.csdiworkshop.org/).

Table 1.1 outlines some of the major developments that have changed or are changing how comparative survey research is conceptualized and undertaken. The abbreviations used in the table are provided at the end of the chapter.

Some of these changes are a natural consequence of developments in the general survey research field. As more modes become available, for example, comparative research avails itself of them as best possible (see Chapter 15, this volume). Other changes are a consequence of the growth in large-scale multipopulation surveys on high-stake research (health, education, policy planning data). The need to address the organizational and quality needs of such surveys has in part been accompanied by funding to allow for more than make-do, ad hoc solutions. Developments there can in turn serve as models for other projects. Finally, the increasing numbers of players in this large field of research and the now quite marked efforts to accumulate and share expertise within programs and across programs are contributing to the creation of a body of information and informed researchers.

1.7 CHALLENGES AND OUTLOOK

3M survey research remains challenging to fund, to organize and monitor, to design, to conduct, and to analyze adequately than research conducted in just one or even two countries. We can mention only a few examples of challenges related to ethical requirements by way of illustration. For example, countries vary widely in official permissions and requirements, as well as in informal rules and customs pertaining to data collection and data access. Heath, Fisher, and Smith (2005) note that North Korea and Myanmar officially prohibited survey research (at the time of reporting), while other countries severely restricted data collection on certain topics or allowed collection but restricted the publication of results (e.g., Iran).

Regulations pertaining to informed consent also vary greatly. American Institutional Review Boards (IRBs) stipulate conditions to be met to ensure respondent consent is both informed and documented. IRB specifications of this particular kind are unusual in parts of Europe, although as Singer (2008) indicates, European regulations on ethical practice can be rigorous. Some European survey practice standards recognize that refusals to participate must be respected, but definitions of what counts as a reluctant or refusing respondent differ

TABLE 1.1. Changes in Comparative Survey Research

Developments	Examples	Effects
Programs & Projects		
Size, number, ubiquity of 3M survey projects _See Chapters 25–31_	ESS; ISSP; WHO-WMH; SHARE; PISA; PIRLS; PIACC; TIMSS; Barometer survey families; WVS/EVS	Critical mass of (1) researchers engaged in ongoing programs, (2) methods research, (3) organizational experience results in capacity building
Frameworks		
Organizational structures _See Chapters 13, 15, 25–31_	Top-down, bottom-up models; centralized & decentralized models, specifications and organization	Shared ownership; public access; cumulative knowledge & capacity-building for all involved
Quality assurance & control _See Chapters 13, 16_	Quality movement (conferences); SHARE central monitoring system	Application of quality frameworks & total survey error paradigm; continuous quality improvement
Question Design & Translation _See Chapters 2–9_	Deliberate procedures & steps; teams; tailored pretesting; translation protocols based on theory and practice	Improved designs, testing, and versions; improved comparability
Sampling _See Chapter 14_	Proof for probability samples; notion of effective sample size; sampling expert panels	Cumulative knowledge about efficacy of sample designs & implementations; improved comparability
Data Collection _See Chapter 15_		
Kinds of information collected in addition to answers	Biomarkers; physical measurements; performances; collateral data to enrich analysis; metadata; paradata	Simpler, more affordable; also on general populations, contributing to prevalence information, policy planning, and so forth.
Where data are collected	Almost everywhere in the world	Theoretically founded selection of populations possible
How data are collected	Emerging, new & mixed modes; C-A modes; responsive designs	Rapid technological advances enhance options for isolated locations; integrated QA/QC; real-time monitoring & responsive designs possible
Documentation _See Chapters 16, 17_		
Kinds of data; documentation standards; related tools	C-A documentation; integral documentation of survey data, metadata, paradata	Required standards, user expectations and access, & survey practice all change. Documentation better and easier; new research & interventions facilitated

TABLE 1.1. *Continued*

Developments	Examples	Effects
Tools Simple & sophisticated tools for phases of survey life cycle and activities	Sample management systems; CARI; control charts; documentation tools like SMDS; DDI structure & codes; data wizards; translation tools; question databanks; data bank systems for version production; procedural guidelines	Diverse tools enhance implementation and QA/QC; shorter production time; enhanced version consistency and documentation
Analysis *See Chapters 20–24*	Forms of analysis to test & analyze; training options; soft- and hardware	Tailored analyses possible; more choice of procedures; software improved and more accessible, user know-how and options increased
Access to data *See CSDI website*	Online; download low cost or free; metadata & paradata included	Worldwide access to better documented & easier to use data & to better software; growth in scientific competition & review
Initiatives *See Section 1.6 & CSDI website*	CSDI, CCSG, DDI; ZUMA symposia, ESS, ECQOS conferences, ISO 20252; ITC Guidelines	Growth in methodological knowledge & critical mass, initiatives, & evidence-based methods research
User/research expertise	ESSi & EU PACO courses, ESF QMSS seminars; 3MC, ECQOS conferences; MA and PhD degrees	Improved research

across countries and this affects options for "refusal conversion." Multinational panels or longitudinal studies are a further special case. Challenges met here include variation across countries in stipulations regulating access to documents and to data that would enable a respondent to be tracked from wave to wave.

Although much remains to be done, expectations about quality are rightly increasing and these call for greater efforts at every level. At the same time, tools to help deal with the complexity of 3M surveys are evolving and experience in dealing with large-scale projects is growing. Some of the challenges once seen as basic—such as translation—can be much better handled today, provided projects decide to do so. Even the core issues underlying "translation"—those of meaning, meaning in context, and comparability—are being approached differently (see Chapters 2, 3, 10, and 11, this volume). In numerous areas, research from other disciplines is being brought to bear on surveys. Such cross-disciplinary work typically reveals a multitude of new considerations and may seem to complicate the picture. Much in the way the cognitive aspects of survey methods movement once seemed overwhelming to some, researchers may not relish embracing fields such as discourse analysis, linguistics, sociolinguistics, cultural theories, or content analysis into question design. Ultimately, however, the expectation is that these will help identify both problems and viable solutions.

At national and supranational levels of funding and research planning, recognition of the need for global research is strong. Recent initiatives at national and international level specifically target training and capacity-building courses, as well as research into how better to deal with cross-national and within-country cross-cultural research (see Table 1.1). Much "national" research is rightly seen as comparative at both subnational and supranational levels (cf. Smith, forthcoming). We are thus cautiously optimistic that it will become easier to find funding for methodological cross-cultural and 3M research than in the past, when funders at the national level were less aware of the national benefits and need for such research.

APPENDIX

Table 1.1 refers to surveys, initiatives, tools, organizations, and other entities often in abbreviated forms which could not be explained within the framework of the table. In the alphabetical list below we provide the full form of abbreviated terms and, as relevant, a few words of explanation or a website for further information.

Abbreviation	Full form	Additional information
C-A	Computer-assisted	As in CAPI (computer-assisted personal interview) and CATI (computer-assisted telephone interview)
CARI	Computer Audio-Recorded Interviewing	
CCSG	Cross Cultural Survey Guidelines	The guidelines posted at http://ccsg.isr.umich.edu/ are designed specifically for cross-cultural research, aiming to support and develop best practice for comparative surveys.
CSDI	The International Workshop on Comparative Survey Design and Implementation	An annual workshop on comparative survey methodology; http://www.csdiworkshop.org/
DDI	Data Documentation Initiative	An initiative to provide a generic structure and codes for documenting surveys; http://www.icpsr.com/DDI/
ECQOS	European Conference on Quality in Official Statistics (e.g., "q2006")	These conferences are held about every two years, the latest in Finland in 2010.
EFTA	European Free Trade Association	The association was set up in 1960; http://www.efta.int/.
ESS	European Social Survey	An EU-supported biennial survey which covers between 25 and 30 countries; http://www.europeansocialsurvey.org/
Eurostat	The European Statistical Agency	http://ec.europa.eu/eurostat/

Abbreviation	Full form	Additional information
EU PACO	The European Union Panel Comparability	Training workshops organized in connection with longitudinal cross-national panels; ftp://ftp.cordis.europa.eu/pub/tmr/docs/socecolong970252.pdf
IMF	International Monetary Fund	http://www.imf.org/external/index.htm
ISSP	International Social Survey Programme	An ongoing annual survey on five continents; http://www.issp.org/
ISQoL	International Society for Quality of Life Research	http://www.isoqol.org/
ISO	International Standards Organization	http://www.iso.org/iso/home.htm/
OECD	Organization for Economic Co-operation and Development	http://www.oecd.org/
PISA	Programme for International Student Assessment	This international triennial survey began in 2000; http://nces.ed.gov/surveys/pisa/
QA	quality assurance	
QC	quality control	
SHARE	Survey on Health, Ageing and Retirement in Europe	http://www.share-project.org/
TIMSS	Trends in International Mathematics and Science Study	http://timss.bc.edu/
LSMS	Living Standards Measurement Study	The survey was set up by the World Bank in 1980; http://www.worldbank.org/lsms/
MIS	Malaria Indicator Survey	http://www.searo.who.int/EN/Section10/Section21/Section1365_11100.htm/
MTUS	Multinational Time Use Study	This survey started in the early 1980s; http://www.timeuse.org/mtus/
ITC	International Test Commission	http://www.intestcom.org/
UNESCO	The United Nations Educational, Scientific and Cultural Organization	http://:www.unesco.org/
	UNESCO Statistics	http://www.uis.unesco.org/
WHO	World Health Organization	http://www.who.int/en/
WMH	World Mental Health Initiative Survey	A global survey begun in the late 1990s http://www.hcp.med.harvard.edu/wmh/
WFS	World Fertility Survey	An international survey in 41 countries in the late 1970s–early 1980s.
ZUMA	The Center for Survey Research and Methodology (Zentrum für Umfragen, Methoden, und Analysen), Mannheim, Germany.	Now part of GESIS-Leibniz Institute for the Social Sciences.
3MC	International Conference on Multinational, Multiregional, and Multicultural Contexts	http://www.3mc2008.de/
3M	Multilingual, multicultural, multi-national	A more economic way to refer to multilingual, multicultural, multi-national surveys.

2

Equivalence, Comparability, and Methodological Progress

Peter Ph. Mohler and Timothy P. Johnson

2.1 INTRODUCTION

This chapter discusses the development of comparative surveys from a methodological standpoint. Section 2.2 looks at the rise of cross-national survey research as this pertains to methods and methodological challenges. In Section 2.3 we present five methodological landmarks for comparative survey research and in Section 2.4 consider the terminology of comparability and propose a simplified terminology for cross-national and cross-cultural survey research. Our discussion focuses on standardized, closed-ended questionnaires in surveys following the most common model of a source questionnaire and translated versions for other languages and locations, (see Chapter 3, this volume).

2.2 SIXTY YEARS OF COMPARATIVE SURVEY METHODOLOGY

2.2.1 The Rise of Comparative Survey Research Methods

As with many other technological advances, World War II served as a catalyst for the development of survey research methods. It perhaps comes as no great surprise, therefore, that the earliest known attempts to conduct survey research beyond a single cultural context took place during that terrible conflict, when the U.S. government established the Morale Division within the United States Strategic Bombing Survey. The immediate purpose of this division was to understand the psychological effects of allied bombing on the morale of civilians in occupied countries (United States Strategic Bombing Survey, 1947a, 1947b), a mission that required modification of domestic survey research techniques for

---

administration in cultural and geographic environments unfamiliar to the researchers (Converse, 1987). Subsequent analyses employed these data to compare the effects of bombing on civilians in Germany, Great Britain, and Japan (Janis, 1951).

An important motive for cross-national research in the immediate post-war years was the need to study the effectiveness of international communications and Cold War propaganda (Lazarsfeld, 1952–53). During most of the Cold War era, for example, the U.S.-funded Radio Free Europe regularly conducted population surveys across Eastern and Western Europe, relying on "independent opinion research institutes in a number of West European countries" (Radio Free Europe-Radio Liberty, 1979). A sampling of the numerous survey reports generated by Radio Free Europe revealed no examples in which issues of measurement comparability across nations was discussed or otherwise considered (Radio Free Europe, 1968; Radio Free Europe-Radio Liberty, 1979; 1985).

Non-governmental funding also became more readily available to undertake expensive cross-national surveys in the post-war years. An early academic collaboration, supported by the Ford Foundation, was the Organization for Comparative Social Research (OCSR; Katz & Hyman, 1954), which was concerned with comparing the social and psychological effects of international (i.e., Cold War) tensions and threats on individuals within seven European nations. Another important early study was the 1948 Buchanan and Cantril (1953) survey funded by UNESCO that compared public perceptions of other nations across 9 countries. Privately financed cross-national work also became common. In 1948, *Time* magazine funded and published findings from a 10-nation survey of international public opinion conducted by Elmo Roper (Wallace & Woodward, 1948).

In general, this interest in and demand for cross-national survey research during the early post-war years grew at a far quicker pace than did the methodologies necessary to conduct comparative work. Frustrations were expressed regarding the lack of methodological experience with which to conquer this new frontier (Duijker & Rokkan, 1954; Wallace & Woodward, 1948), with Lowenthal (1952: vii) suggesting that researchers (including himself) were "fumbling and stumbling." These early pioneers also confessed to the difficulties and asymmetries that emerged from language difficulties among investigators representing various participating nations. Duijker and Rokkan (1954, p. 19), in discussing differential English language (the "working" language for OCSR) proficiencies of the researchers, noted that:

> Some of the team leaders were clearly handicapped by their language, they felt that they did not succeed in making their points as they should and they felt frustrated because they were not able to influence decisions the way they would have wanted.

The hazards of properly translating survey questionnaires was also quickly recognized (Buchanan & Cantril, 1953; Ervin & Bower, 1952–53; Stern, 1948), as were other methodological challenges to collecting "equivalent" measurements across nations.

Within the United States, recognition that the quality of data being collected from survey questionnaires might vary by the racial background of respondents was also becoming evident. One of the most pressing domestic problems in the United States during this period was the public health crisis triggered by the increasing rates of poliomyelitis among children (Oshinsky, 2005). Research focused on this crisis also recognized the potential problem posed by noncomparable survey data. One epidemiologic study investigating the incidence of this disease, upon finding considerable variation between black and white families, commented that "there is the possibility…that reports from the colored were not as complete or in other respects were not comparable with those from the white" (Collins, p. 345). Practical solutions to these measurement disparities, though, remained unexplored (or at least went unreported).

Despite these serious methodological challenges, cross-national surveys became relatively commonplace during the ensuing decades. By 1968, more than 1,600 academic research papers reporting findings from cross-national surveys and/or surveys conducted in developing nations had been published (Frey, Stephenson, & Smith, 1969). A landmark academic study, based on parallel data collection activities in five countries, was published early in that decade by Almond and Verba (1963) in *The Civic Culture: Political Attitudes and Democracy in Five Nations.* Although later criticized in design and implementation (see Verba, 1980) this study reflected what was then considered state-of-the-art survey methodologies for the conduct of cross-national research. In the published presentation of their study, Almond and Verba considered at length the "problem of equivalence" and the possibility of measurement artifacts, noting: "differences that are found in response patterns from nation to nation or group to group within nations have little meaning if they are artifacts of the interview situation" (Almond & Verba, 1963, p. 57). Almond and Verba's accomplishment has in many ways served as an exemplar for conducting cross-cultural survey research in the decades that have followed, as researchers from numerous academic disciplines and cultures have grown to recognize the methodological challenges of comparative survey research. Many of the current cross-national and cross-national initiatives reviewed in the last section of this volume can be seen as intellectual products of this pioneering work.

2.2.2 The Development of Programs of Comparative Survey Research

A series of surveys oriented toward replication began during the three decades of the 20th century with the establishment of the Eurobarometer and the International Social Survey Programme (ISSP). The Eurobarometer program celebrated its 35th anniversary in 2008 and thus is the longest running continuous cross-national survey program. Despite its academic roots and ongoing accomplishments, it is a governmental survey under the auspices of the European Union (EU) Commission. The ISSP is the longest running cross-national academic survey program. The aim of the ISSP was to establish such a long-term program on a lean budget. This collaborative program of research began with four countries but grew rapidly. Methodological questions quickly followed, partly related to attaining comparability in the common set of questions and the socio-demographic questions used as background variables in the studies. Other issues related to organizational aspects of a growing membership and diverse target populations. Originally devised as a self-completion survey, for example, literacy problems in new member countries soon resulted in a change in ISSP rules to permit face-to-face interviews. Developments within the ISSP, today the largest worldwide academic survey, have resulted in individual and (ISSP) group reflections on best practices in the conduct and analysis of multicultural, multilingual, and multiregional surveys (hereafter 3M).

As early as 1987, Kuechler proposed several principles of practice based on experiences with the first round of ISSP data collection involving the first four countries. He recommended that participating nations should have similar experience with and exposure to attitude surveys. He later commented that "a country must recognize freedom of speech as a basic human right, and all its citizens, not just those in specific strata, must routinely exercise this right without fear of prosecution or discrimination" and that "cultural norms must favor the expression of individualism and tolerance toward variations in opinion and behavior" (Kuechler, 1998, p. 194). The subsequent expansion of the ISSP to 45 member countries by 2009 almost certainly contradicts these recommendations.

Kuechler (1987) also recommended that the specific tasks associated with instrument development be shared across all participating countries and more generally that each nation be represented by at least one local researcher. For methodological reasons, he advocated collecting survey data at more than one time point—in order to be able to assess the potential effects of sampling fluctuations and nation-specific events. Kuechler suggested that cross-national survey data should first be analyzed separately at the national level in order to establish contingency patterns; these could then be subsequently compared using survey data pooled across nations (see, too, Scheuch, 1968).

Additional early lessons were identified by Jowell (1998, pp. 174–176), who expressed concern that the ISSP and other cross-national survey projects of that time failed to recognize "just how fundamentally survey norms (standards)… differ between countries" and provided multiple examples of how these differing norms influence measurement comparability. To confront these problems, he proposed 10 general "rules of thumb":

1. Do not interpret survey data relating to a country about which the analyst knows little or nothing.
2. When analyzing survey data, resist the temptation to compare too many countries at once.
3. Pay as much attention to the choice and compilation of aggregate-level contextual variables as to individual-level variables.
4. Be as open about the limitations of cross-national surveys as you are enthusiastic about their explanatory powers.
5. Stringent and well-policed ground rules from comparable survey methods should become much more common in comparative studies than they are now.

6. When analyzing survey data, suspend belief initially in any major inter-country differences observed.
7. Confine cross-national surveys to the smallest number of countries consistent with their aims.
8. Collective development work, experimentation, scale construction, and piloting should be undertaken in all participating nations.
9. Provide detailed methodological reports about each participating nation's procedures, methods, and success rates, highlighting rather than suppressing variations.
10. Routinely include methodological experiments in cross-national research.

By the late 1990s, the ISSP had established numerous work groups to address methodological issues of 3M survey research (see Chapter 27, this volume). A number of the recommendations made by Kuechler and Jowell find an echo in the European Science Foundation Blueprint (European Science Foundation, 1999) for the development of the European Social Survey (ESS), first fielded in 2002 (see Chapter 26, this volume). In both the ISSP and the ESS, instrument design is based on an "ask-the-same question" model (see Chapter 3, this volume). This model requires that the common questions asked everywhere result in data which permits comparison. In the next section we turn to consider major methodological landmarks related to comparability.

2.3 FIVE METHODOLOGICAL LANDMARKS IN COMPARATIVE SURVEY RESEARCH

Certain methodological challenges for 3M survey research have been evident from the outset—such as collecting data in different languages and organizing project cooperation in a lingua franca in different cultural and institutional settings. Others, as just indicated, have emerged and been elaborated as 3M and within-country cross-cultural research expanded. In this section, we briefly review five methodological landmarks in 3M research. These are:

1. The use of *indicators* as the basis for comparison
2. The recognition of *context* as a relevant determinant for comparison
3. The application of *translation theory* and theories of meaning to the adaptation/translation of survey instruments
4. The acknowledgment of *probability multipopulation sampling* as the statistical prerequisite for comparison
5. Advances in *statistical methods* allowing for complex modeling such as multilevel analysis or identification of measurement invariance

The first three of these are directly related to the indicators and questions used. The last three (translation, sampling, and statistics) are discussed at length elsewhere in this volume. Our remarks below are thus directed more to the first two topics.

As we see it, each of these landmarks is an essential factor in conducting 3M survey research. In theoretical terms, we consider them in essence already addressed. Kish (1994), for instance, established the theoretical basis for comparative sampling; Gabler, Häder, and Lahiri (1999) demonstrated the correctness of his theory. Harkness (2003) and elsewhere demonstrated the relevance of a theoretical basis and a quality assurance framework for conducting survey translations (see also Chapter 7, this volume). Scheuch (1968) was quite explicit about the probabilistic nature of indicators, emphasizing that the equivalence of indicators across populations must be assessed statistically. We return to this below. Verba (1969) and Scheuch (1968) both demonstrated the unavoidable and pervading relevance of context (micro and macro) for interpretation and analysis, making it clear that nominally "comparable" questions at the level of language might be understood quite differently in different locations. Several chapters in this volume elaborate on this insight.

While each of these topics may have been directly addressed in prior litera-ture, survey research is still challenged to address them in design and implementa-tion. It is also by no means certain that these "principles" are known or, if known, accepted everywhere. Thus it happens that translation is often done as cheaply as possible with naïve ideas about expected outcomes and procedures; quota samples and unclearly defined and documented samples are by no means uncommon in international research. Indicators are often confused with question wording; and questions are used in multiple contexts without prior investigation as to their perceived meaning and discourse functions in a given location and language.

In our discussion of indicators and of context in Sections 2.3.2 and 2.3.3, we refer frequently to the sociologist Erwin Scheuch and the political scientist Sidney Verba. This is not to say the ideas they expounded are to be found nowhere else. However, Scheuch and Verba both focus in their writing on comparative survey research rather than comparative social research. They are also both theorists who were also heavily involved as practitioners in comparative survey research. They thus came to know survey methods and understand questions of methodology. Their writings foreshadow much of what is now common knowledge in survey research but also reflect insights either not yet shared by all or simply ignored in practice.

2.3.1 The Role of Indicators

Scheuch (1968) has this to say about indicators and questions:

> By now social scientists should have become accustomed to looking at questions as indicators—indicators that have a *probabilistic relationship to a property one intends to measure*. Indicators are interchangeable in terms of their functions, which are to express the property we want to ascertain. Hence, the criterion for maintaining that questions are comparable is not whether they are identical or equivalent in their commonsense meaning, but whether they are *functionally equivalent for the purposes of analysis*. Scheuch (1968, p. 113, italics added).

We note here that Scheuch talks about questions as indicators but implicitly draws a distinction between the wording of questions and their function for the purpose of analysis. Another and often useful perspective is to distinguish between the two. Indicators need not be questions, they can be actions such as weight-taking. In addition, different formulations (questions) can aim to address the same indicator (see Chapter 3, this volume). In addition, different indicators might prove to be functionally equivalent in the sense described by Scheuch. Different indicators for environmental behavior (whether one tried to conserve one or the other natural resource) would normally be clothed in different questions. The important point Scheuch makes is that what matters is the indicator's functioning in a statistical model and not, as he puts, the "commonsense meaning" of a question. Thus, if the questions (or indicators) used are not *functionally equivalent for the purposes of analysis, we cannot justifiably compare the data collected.*

2.3.2 Functionally Equivalent Indicators

What are functionally equivalent indicators as intended by Scheuch? How do we know they are that? Functionally equivalent indicators are revealed in analysis, they cannot be judged on the basis of face value similarity. This notwithstanding, contextual (cultural) knowledge can provide some insight into proposing indicators that might function similarly. Scheuch, Verba, and others talk about differences in who, for example, might qualify for the descriptor "friend" in different cultures. Differences that existed would mean that the relationships counting as a "friend/freund/amigo/ami" cannot properly be compared across groups. Such examples abound. Scheuch (1968, p. 114) reports how a question from the Bogardus (1925) social distance scale needed to be adapted for Germany. The Bogardus question asked about the extent to which a respondent would accept certain (ethnic and religious) groups of people in given degrees of social distance. One of these degrees was "have as a neighbor in the same street." In order to convey the same degree of social distance for Germany, the response category offered was "have as a greeting acquaintance (Grußbekanntschaft; we might say 'nodding acquaintance')." This category proved to work well for German respondents in the social distance scale.

Functionally equivalent indicators as Scheuch intended them to be understood, therefore, should behave in a similar manner in statistical analysis. To identify such indicators is not trivial, as it involves at a minimum a sequence of instrument design, pretesting, redesign and so forth in each participating country. In addition, statistical associations could be spurious. Scheuch emphasizes that substantial theoretical predictions are necessary, accompanied, if possible, with further empirical evidence (Scheuch, 1968, p. 114; see also Van Belle, 2008, p. 58).

As reflected throughout this volume, many comparative surveys do not undertake intensive quantitative and qualitative testing of source questions or indicators for multiple new locations. Budget and time constraints are common motivations. Verba (1969) reminds us that intensive development of equivalent indicators can be a cumulative task. We will return to this suggestion shortly.

2.3.3 The Significance of Context

In the social distance scale discussed earlier, the neighbor question had to be changed for Germany because of differences in social patterns of interaction and in the social meaning of "neighbor." Even if respondents are asked the same questions everywhere, as Verba notes (1969, p. 69), "The problem remains of interpreting the meaning of these results. The problem derives from the fact that the questions have been asked within different social and cultural context." This is a fundamental point.

2.3.4 Principles of Practice for Comparative Survey Research

Verba's emphasis on the importance of considering context in analysis and Scheuch's insistence on ensuring that indicators are comparable can be combined to identify three important principles of practice for comparative research in general and comparative survey research in particular:

1. The theoretical design and design implementation of comparative surveys needs to accommodate the realities of multiple contexts.
2. Instrument design must focus on ensuring comparable indicators are used. Here we suggest it is useful to distinguish between indicators and questions and to require that both are functionally equivalent in Scheuch's sense, while they may be nominally different (see Chapter 3, this volume).
3. Theoretical assumptions must guide indicator and question development, and statistical assessment employed to evaluate their merit before questionnaires are finalized.

This interplay between theory and statistics translates in survey terms into instrument design, qualitative/quantitative pretesting, and instrument refinement. At the same time, as Verba (1969; footnote 28) mentions, "at times we conduct the research simply in order to learn about those structural and cultural features that we need to understand in order to design the research." Thornton and colleagues (Chapter 4, this volume) state that it was their preliminary research in different contexts that enabled them to develop comparative questions at a much later date. In both instances, an argument for cumulative research can be found; learning from past experience, successes, mistakes, and solutions inherited or invented. Or, as Verba put it: "it is only after it has been done that one knows what one ought to have done, and the obvious conclusion is that research programs must build their store of understandings of the meanings of the variables that they measure" (1969, footnote 28). From a more explicitly concerned perspective, this interest in cumulative research also preoccupied Scheuch, to whom we now return.

2.3.5 Reinventing the Wheel?

In 1989, Scheuch complained with some emphasis that comparative surveys kept re-inventing the wheel instead of learning from previous research and that knowledge gained in one or the other project did not seem to become cumulative or shared knowledge, of advantage to the scientific community: "in terms of methodology in abstract and on issues of research technology, most of all that needed to be said has already been published" (Scheuch, 1989, p. 147).

One of Scheuch's major concerns was how to make knowledge cumulative so as to advance the field and the quality of research undertaken, given the compartmentalization of information and know-how within disciplines and the time lags in transfer of knowledge across disciplines. Even conceding that Scheuch was overoptimistic about everything already being available, these are concerns with a very modern ring. Methodological progress has been made in the meantime, as evident in the chapters on analysis, sampling, and translation procedures in this volume. In addition, the landscape of data collection modes and the technological tools to administer, monitor, document, and adjust procedures have all changed quite radically in recent decades, bringing with them potential challenges unthought of even 20 or 30 years ago. At the same time, Scheuch's clear statements about the status of indicators and his and Verba's insistence on the importance of context have had less impact on comparative survey methods than might be desired.

As comparative research grew, the design model to gain most favor was one that advocated keeping as much the same as possible in terms of question wording, modes, and samples, rather than in terms of what Scheuch would consider functionally equivalent (probabilistic) indicators and questions As several chapters in this volume reflect, innovation in comparative survey methods is hampered by a lack of overarching theoretical underpinnings and a lack of research either based on such theories or designed to test them.

2.4 COMPARABILITY IN SURVEY RESEARCH

In the foregoing sections, we repeated the terminology of earlier researchers without much explication. An ongoing issue for comparative research is obviously that of essential "comparability," whether this is discussed in terms of equivalence, functional equivalence, similarity, or some other frame of reference. It is to this that we now turn.

Much of the literature concerned with cross-cultural survey methodology has focused on the challenge of developing "equivalent" measures. In this section, we propose a paradigm and terminology. At the same time, to explicate our model, we must return to the roots of comparative methodology.

2.4.1 Establishing Equivalence?

While models of error and total survey error were emerging in general (noncomparative) survey research, models of equivalence including functional equivalence, often developed in other disciplines, gained currency in comparative research. The terms "equivalence" and "functional equivalence" have certainly gained huge currency in survey research and in diverse related disciplines. In a review of definitions of "equivalence," Johnson (1998) identified more than 50 different uses with and without extensive definitions. In many cases a variety of terms are employed to refer to one and the same concept while one and the same term is elsewhere frequently used to refer to different concepts.

In addition, "functional equivalence" is usually not intended in the way Scheuch described; the notion of (probabilistic) functionally equivalent indicators has found little uptake in comparative survey practice. In some instances, the models have leaned heavily on discussions in cross-cultural psychology and educational testing of "bias and equivalence." These may then distinguish between metric equivalence; construct (or structural) equivalence; measurement unit equivalence, and scalar or full score equivalence, aligned against construct, method, and item bias (see van de Vijver & Leung, 1997, p. 8ff). The methods and concepts used in comparative psychology and testing to demonstrate comparability (or in their terminology, "equivalence") rely on an abundance of items. This differs greatly from the situation in social survey research where the number of items measuring one attitude, trait, behavior, etc. is almost always very limited. Often only one item is used. This means that comparability has to be approached quite differently.

In addition, both the behavioral and socio-economic background information required in social science surveys is typically not measured using scaling techniques. Events, behaviors, or respondent characteristics are often counted rather than measured: years of schooling, personal income, number of persons in a household, visits to a doctor, voting, occurrences of buying products, etc. Very often single items or a cascading sequence of filter items are used for such reporting. This single-item approach is a serious challenge for any statistical assessment of reliability, validity, or similarity. Special efforts have to be made to establish conceptual comparability and statistical similarity (Saris, 1997; Saris & Gallhofer, 2007b).

Thus a one-to-one transfer of methods and concepts used in comparative psychology and testing is not possible. This situation should be reflected in the terminology used. Moreover, discussions of equivalence inherently assume or at least imply that it is possible to find two "fully equivalent" or "identical" objects with the "same" meaning. Empirically we consider this unlikely. Although we acknowledge that many investigators continue to search for *equivalent* measures, we view discussions of equivalent concepts as largely philosophical. This view is consistent with Verba, Nie, and Kim (1978), who referred to *complete equivalence* as a hypothetical achievement that will never be attainable in practice. Earlier, in discussing the comparability of the survey data they collected across five nations as part of their *Civic Culture* study, Almond and Verba (1963, p. 60) observed that "equivalent questionnaires—that one is *sure* are equivalent—appear impossible to obtain" (italics original). Since the concept of equivalence seems empirically unattainable, we propose below to focus on the more realistic goals of conceptual comparability and measurement similarity.

2.4.2 Comparability and Similarity

Equivalence or *identity (identicality)*, are ideal concepts. They entail the notion of complete or full equivalence or identity. Frustrating as this might be for our conceptualizations, there are no completely equivalent entities in our empirical world. Instead, we propose to use *comparability* as our heuristic concept in any discussion of whether *concepts* are comparable or not. We further propose to use the term *similarity* for investigations of how alike measurement components—constructs, indicators, and items—may be across groups. In this way, we avoid using one word at both the conceptual and the measurement levels and can usefully distinguish between them when this is necessary. We conceptualize comparability as a property of a concept across a multitude of populations (cultures, nations, etc.). We conceptualize similarity as the degree of overlap measures have in their representation of a given social construct, whether these are measures used in two or more cultural contexts or in a single cultural context.

Our approach follows the strategy of measuring latent constructs representing theoretical concepts via (probabilistic) indicators (see Figure 2.1). It seeks to establish the extent of similarity in two or more measures taken as indicators of a latent construct. In this approach, it is usual to ask how similar two objects are in respect to a set of well-defined characteristics. It has much in common with comparisons of two wooden tables in terms of their length, width, or height in meters or yards, and with measuring the intensity of starlight using magnitude scales. No list of characteristics will be exhaustive. It will never represent the "true nature" or "gestalt" of a given table or star (Popper, 1972). It will always be a specific *take* or *perspective* on the objects under observation. In this way comparability entails the possibility of measuring the similarity (or dissimilarity) of well-defined characteristics of two or more objects under observation using scientific methods.

Thinking about theory and methodology for comparative surveys will always require consideration of the notion of the similarity of the metric in question. Positive definitions of similar features may result in infinitely long lists. An alternative approach is to give priority to negative (not similar) statements. It is easier to falsify a hypothesis than to confirm one (Popper, 1935); in like fashion, it is easier to demonstrate scientifically that an object is not square, orange, large, or complex than it is to verify its positive counterpart. In statistical analysis, therefore, the null hypothesis is by far the most typically used approach for this kind of scientific statement.

In line with a null-hypothesis approach, general similarity scales can be constructed that range from zero (0) to infinity (∞), where zero indicates "not similar" (null hypothesis) and infinity indicates "highly similar, but not identical." The zero-to-infinitesimal similarity scales have useful formal properties. Similarity is conceptualized as unidimensional, unidirectional, and unbounded. The zero-point, which is "not similar," is taken as the theoretical anchor or statistical null-hypothesis definition. The notion of infinitesimal similarity, (unbounded similarity) is also important, because it allows for continual increases in gradations of similarity. Measurements are thus conceived of as having a degree of cross-cultural "similarity," from zero upwards.

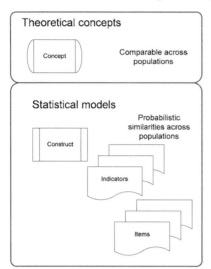

Figure 2.1. Comparable Concepts and
Similarity of Statistical Models

In analysis, it could also be useful to define a range for a similarity measure function, with 0–1 as the defined range. In this case zero (0) would be the origin and the similarity function would asymptotically converge toward one (1). Figure 2.2, in which the *x*-axis represents the number of characteristics/variables, and the *y*-axis represents similarity, illustrates such a function. High levels or similarity, for example, would enable us to compare Macintosh apples with Red Delicious apples, but not to compare Macintosh apples across cultures.

2.4.3 Applying the Similarity/Comparability Approach

If we define the set of characteristics in terms of which we wish to assess similarity, then the assessment can be applied to any measure. This offers an opportunity to consider more items from a possible universe of items than those produced on the basis of translation. It also means that the instrument design process aims to achieve the *best available comparative measurement at a given point in time in each culture or nation*. The measures chosen as "optimal" would then be those that provided this. This measurement could be based on multi-item scales if available or single items. A crucial step would be to identify and define thresholds of similarity that must be met to determine that a given degree of similarity is indeed present. If the latent constructs in the locations under observation, as well as the indicators and questions, prove to be sufficiently similar and no better items from the universe of items in a given culture are available (i.e., with demonstrably better measurement properties), translated items would be approved.

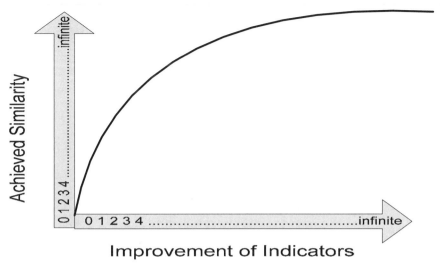

Figure 2.2. Example of Similarity Function

2.5 SUMMARY AND OUTLOOK

Considerable advancements have been made in comparative survey research procedures over the past 60 years. Five major insights were acknowledged in this chapter. Our chapter has focused on the importance of context and the relation between indicators and comparability in ways that complement discussions of context in other chapters in this volume (see, for example, Chapters 3, 7, 14, and 20).

For the purposes of comparative research, we see indicators as having probabilistic relationships with the constructs they measure. And, because cultural context can influence the meaning of constructs, we also see it as essential that comparative researchers proactively design their surveys to account for the varied contexts within which data will be collected. This is best done by using theory to guide indicator development and the design of instruments, so that functionally comparable indicators can be employed within each context, and the subsequent use of statistical assessment to identify optimal indicators of each.

Although many methodological challenges have been addressed and solutions continue to be refined, much work remains to be done, particularly at the measurement level of constructs, indicators, and questions. Conceptually, we believe that emphasizing the comparability of concepts and the similarity of survey measures is a more fruitful approach for comparative research. We encourage cross-cultural researchers to continue to explore new approaches to developing highly comparable concepts while at the same time taking advantage of up-to-date statistical strategies to test for measurement similarity.

PART II

QUESTIONNAIRE DEVELOPMENT

3

Designing Questionnaires for Multipopulation Research

Janet A. Harkness, Brad Edwards, Sue Ellen Hansen, Debra R. Miller, and Ana Villar

3.1 INTRODUCTION

This chapter presents an overview of current practice and options for designing survey questionnaires for implementation in multiple cultures and languages. We aim to help readers make informed choices about developing such questionnaires and provide pointers to assist in evaluating existing instruments. Toward this, we identify the potential and limitations of various options and identify areas in which research is needed. The chapter focuses on comparative design and thus has little to say about the "ordinary" business of questionnaire design, important as that is in any context. We remind readers only that irrespective of the design chosen for a comparative instrument, all the matters of "ordinary" questionnaire development must also be considered.

Harkness (1999) identified lack of documentation as an obstacle to methodological progress and research. While documentation of comparative studies is improving (see Chapter 16, this volume), publicly available material on instrument design provides only fragmented treatment of isolated specific details on the strategies, rationales, and protocols followed or recommended in designing comparative instruments. Although publications abound that discuss the validity and reliability of questionnaires used in comparative studies, very little is available on designing comparative instruments. It is therefore neither easy to learn about design options available for cross-cultural or cross-national survey research nor to find comprehensive and detailed information about procedures used in existing surveys.

Instrument design strategies differ quite strikingly across disciplines; approaches popular in one discipline may be quite unfamiliar in another. In part such differences can be understood historically, and in part they relate to the focus of a given discipline (e.g., on social structures, on traits at the individual level, or on skills and knowledge acquired). The goals of specific measurements also differ (testing ability, counting incidences, evaluating health, or exploring values and attitudes) as do the sample types and sample sizes commonly used in different fields (patient or student lists, convenience samples, or probability samples of the general public, etc.). Given our own expertise, more examples are presented here for questionnaires as developed in the social and behavioral sciences than in educational and psychological testing. We expect nonetheless that many points raised in this chapter will also be relevant for other fields.

In Section 3.1.1, we distinguish between comparative instruments and comparative research and in Section 3.1.2 consider areas of research relevant for design and available literature. In Section 3.2, we discuss the general challenges faced in comparative questionnaire design. Section 3.3 briefly discusses teams and management issues. Section 3.4 examines design fundamentals. Section 3.5 presents the key decisions faced and Section 3.6 introduces the principal models for cross-cultural, cross-lingual designs. In Section 3.7, we discuss a few particular aspects of design from a comparative perspective: answer scales, technical implementation, factual and socio-demographic questions, and vignettes. Section 3.8 considers pretesting procedures to assess questionnaire quality and suitability, with the emphasis on qualitative procedures. Section 3.9 closes the chapter with brief consideration of areas in which we feel research is needed and summary conclusions.

3.1.1 Comparative Instruments versus Comparative Research

Comparative instruments, as the term is used in this chapter, are instruments deliberately designed for use with multiple populations. Such instrument design must address a wide range of features, including conceptual coverage, selection of indicators, development of questions, level and choice of vocabulary used, adaptations of wording, format or mode, design of response categories, and technical features which are related in part to language issues and chosen mode(s) of application.

Many questionnaires used in comparative projects are not comparative by design in the sense just described (see Chapter 1, this volume). They may, for example, be designed with a single general population in mind but be fielded with other populations. The needs of these new populations might be nominally addressed through translation, inadequate as this might prove to be (see Chapter 7, this volume). In such cases, too, as implied earlier, cultural constraints on levels of language, on question and answer scale formats, on adaptations, and on viable modes all need to be addressed.

In any project, comparative in purpose or not, the most frequent approach used in developing questionnaires is to re-use questions which seem suitable that have already been used in other surveys. The next most popular strategy is to adapt questions which have been developed for other purposes to suit new needs or populations. The approach taken least is to write entirely new questions. The advantages and disadvantages of each of these strategies (adapt, adopt, create new questions) for comparative purposes is discussed in Harkness, van de Vijver and Johnson (2003).

Most surveys end up combining all three strategies to meet the requirements of a specific study. Irrespective of which strategies are chosen to arrive at question formulations, questions should be tested before use.

A number of the challenges encountered in multinational, multicultural, and multilingual (hereafter 3M) research result directly from implementing questions without testing whether they can provide comparable data across the populations sampled. Problems can arise at any of multiple levels. They may, for instance, stem from inadequate conceptual coverage, from inappropriate indicators, or from the formulations chosen for source items in a source-and-translation (see Section 3.6.1) design model. The 2001 Supplement to Mental Health (U.S. Department of Health and Human Services, 2001) documents the damaging effect inadequate measurement of different populations can have on large-scale data collections, even in only one country (see also Harkness, 2004).

Harkness, van de Vijver, and Johnson (2003) note that questions that have worked well in one or more studies gradually acquire a "pedigree of use." These may be selected for further studies without evaluation of whether they are truly suitable for new contexts. Since a pedigree encourages replication of existing wording, there may also be resistance on the part of design teams or clients to change the wording of such "pedigreed" questions even if this would make them more suitable for new contexts.

Even when researchers do attempt to develop a comparative instrument, the procedures followed may be inadequate from the start or may not be realized as originally (and adequately) planned. The results will then be suboptimal. Not only studies on a modest budget are subject to mishaps. For instance, LeTendre (2002) discusses planning and communication problems which prevented the interactive development intended for qualitative and quantitative components of the *Trends in International Mathematics and Science Study* (TIMSS) from being realized. Karg (2005) reports on problems in the implementation of the Organisation for Economic Co-operation and Development (OECD) *Programme for International Student Assessment* (PISA) study; Lyberg and Stukel (Chapter 13, this volume) report on problems in the design and realization of the *International Adult Literacy Study* (IALS).

When questions do not function as expected, measurement is compromised. In a World Health Organization (WHO) study conducted in Ethiopia in the mid-1980s, a question commonly used in Western contexts as a measurement of depression — *Is your appetite poor* — went awry in what, with hindsight, seems to have been a predictable manner. "Appetite," in contrast to "hunger," implies the availability of food. Given that food was in scarce supply at the time, Kortmann (1987) reports Ethiopian respondents interpreted the Amharic translation to be about the *availability* of food.

3.1.2 Literature on Comparative Questionnaire Design

As indicated, the literature dealing specifically with cross-cultural questionnaire design is quite small, despite both the burgeoning methodological literature on instrument design in noncomparative contexts and the ever-growing number of

3M research projects and accompanying substantive publications. Articles and book chapters on substantive research interests mention design aspects only briefly or in very general terms. Technical reports vary greatly in detail; they may provide specific information but generally give little space to methodological debate, design choices, or illustrative examples. Literature with a methodological focus on instrument design often advocates use of single population procedure(s) in a multi-population study or experiment. This is not surprising since it is not the kind of information normally recorded in detail in "documentation" or in substantive articles, nor, indeed, in technical reports. Where it would be found is in methodological handbooks. However, the field has long had only weakly developed methods on comparative questionnaire design.

At the same time, publications such as Georgas, Weiss, van de Vijver, and Saklofske (2003), Hambleton, Merenda, and Spielberger (2005), Harkness, van de Vijver, and Mohler (2003), Jowell, Roberts, Fitzgerald, and Eva (2007b), Karg (2005), Porter and Gamoran (2002), and Suzuki, Ponterotto, and Meller (2008) are indicative of a renewed interest in 3M research methods in general.

Comparative literature from the 1960s and 1970s, often reprinted later, places considerable emphasis on problems with conceptual coverage (e.g., Elder, 1976; Sechrest, Fay, & Zaidi, 1972; Smelser, 1973), functional equivalence and comparability (e.g., Berry, 1969; Verba, 1969, 1971; contributions in Warwick & Osherson, 1973), as well as the challenges raised by working across languages (e.g., Scheuch 1968,1989; Sechrest, Fay, & Zaidi, 1972; Verba, 1971; Werner & Campbell, 1970). Recommendations advanced, such as rules (of varying number) for writing questions to make them easier to translate, are repeatedly cited or presented in the literature without much critical review (e.g., Brislin, 1986; Brislin, Lonner, & Thorndike, 1973b; Smith, 2003; van de Vijver & Leung, 1997). Some of the topics raised in older literature, such as the drawbacks of targeting "equivalence" at question wording levels, remain highly relevant.

However, the conceptual and theoretical frameworks for survey question design and thus also for (pre)testing have changed considerably in recent decades. In addition, technological developments in Web, telephone, and other computer-assisted applications have revolutionized instrument design and testing in ways older literature could not envisage.

Current literature in cross-cultural psychology and in 3M educational and psychological testing continue to investigate challenges to conceptual coverage and to suggest how these might be addressed. They also discuss cultural adaptation of instruments to improve measurement (e.g., contributions in Georgas et al., 2003; in Hambleton, Merenda, & Spielberger, 2005; in Porter & Gamoran, 2002; and in Suzuki, Ponterotto, & Meller, 2008). From a social science perspective, Smith (2003, 2004b) provides examples of the many aspects of general question design to be considered in producing instruments in multiple languages and provides copious further references. Harkness, van de Vijver, and Johnson (2003) discuss advantages and disadvantages of major design options and offer a general framework for design decisions, a framework further developed here. An emerging literature on cultural patterns, cognition, and perception as these relate to survey research is reflected in Schwarz, Oyserman, and Peytcheva (Chapter 10, this volume) and in Uskul, Oyserman, and Schwarz (Chapter 11, this volume);

Yang, Chin, Harkness, and Villar (2008), and Yang, Harkness, Chin, and Villar (Chapter 12, this volume) consider possible connections between response styles and question design.

Time lags in knowledge transfer and in the reception of findings across disciplines are not unusual. Nonetheless, it is unfortunate that major recent advances in general survey methods research have had little effect on how comparative instruments are designed. There are, however, some notable exceptions, including the field of cognitive testing (see Section 3.8).

Research and theoretical frameworks in a variety of disciplines would be highly relevant for developing cross-cultural instruments but have to date received scant attention. Relevant fields include cognitive linguistics; comparative linguistics; sociolinguistics; semantics and pragmatics; intercultural communication and cultural competence, the cultural embedding of discourse norms and conventions; text and genre analysis; readability; and visual perception.

3.2 CHALLENGES FOR COMPARATIVE RESEARCHERS

Researchers setting out to design for comparative contexts face a number of special challenges; we identify six which seem especially important and discuss these in a thematically logical order.

3.2.1 Basic Question Design Principles

Question and questionnaire design in general survey research terms has developed from an "art" (Payne, 1951), through handcrafting and a set of techniques (Converse & Presser, 1986), toward a quality and survey error-oriented methodology (Groves et al., 2009; contributions in Biemer, Groves, Lyberg, Mathiowetz, & Sudman, 2004) and quantification of problems (Saris & Gallhofer, 2007b). Understanding question design challenges even in general terms is no small undertaking. Design recommendations have emerged for different kinds of questions and, increasingly, for different survey modes. Depending on the type of information sought, questions are sometimes classified into four categories (Bradburn, Sudman, & Wansink, 2004):

- Questions asking about behaviors or facts
- Questions asking about psychological states or attitudes
- Questions asking about knowledge and competencies
- Questions asking respondents to recollect autobiographical data

In each case, special strategies may be required to address such issues as social desirability, memory requirements, order effects, sensitivity of content, response styles, and the analysis planned.

In comparative research, comparability becomes a major design requirement. Any design chosen must ensure comparability across languages and contexts. As such, it should include a scheme for producing other language versions consistent with whichever principles of instrument design are followed, as well as any adaptation procedures envisaged.

As mentioned earlier, different disciplines prefer particular instrument formats and strategies. Psychological instruments may have long batteries of questions; opinion and attitudinal research often make do with one or two questions per construct. Educational tests must evaluate and calibrate comparable difficulty of questions across cultures and languages. This may involve calibration of the difficulty of knowledge items, but also include cognitive burden involving possibly sounds, alphabet letters, pictorial material, numerals, or, as in the PISA study, the reading and interpretation of textual passages. Ueno and Nakatani (2003, p. 216) discuss cultural sensitivities in the Picture Completion section of the Wechsler Intelligence Scale for Children (WISC-III), in which missing parts of people and animals (e.g., ear and foot) were replaced in the Japanese test with inanimate elements (vehicle tire and stocking). Educational test developers must also engage in extensive "alignment" procedures to ensure that questions are matched to curricula standards and curricula standards matched to questions (see contributions in Herman & Web, 2007). In the comparative context, alignment may be an arduous task (see contributions in Porter & Gamoran, 2002).

3.2.2 Knowing When to Supplement or Alter Design Procedures

Knowing the relevant literature in general survey methods does not sufficiently equip one to design successful comparative instruments. The second challenge is to be able to identify whether practices that have been useful in other contexts should be replaced or augmented. Examples sometimes mentioned in the literature are the "best" length of answer scales and whether forced choice questions and hypothetical questions are advisable in certain cultures.

3.2.3 Finding Guidance

The third and related challenge is that researchers find little guidance on how to identify strengths and weaknesses of procedures, or on how to select an optimal approach for given design needs. This holds for more general aspects of comparative questionnaire design, such as identifying and testing constructs, indicators, and items, as well as for such technical aspects as deciding on open or closed question formats, answer scales, or such visual arrangements and how to appropriately emphasize parts of a questionnaire for different languages and cultures. These decisions call for cultural knowledge. For instance, in order to decide whether respondents should be provided with the possibility to answer covertly (without the interviewer being told the response directly), researchers must understand sensitivity issues in each cultural context in which data will be collected.

At the same time, cultural knowledge will not always provide solutions. Cleland (1996) reports on demographers' early strategies to accommodate the needs of respondents in "less developed" countries asked to report on births, pregnancies, and deaths. Aware that birthdays did not have the same significance or paper trail in every culture and that recollection of such life cycle events can be complicated, demographers tried unsuccessfully to collect better statistical data on the basis of vernacular or personally oriented time lines instead of numerical period notation (e.g., "before weaning" versus "at age 3"). In the same article, after reporting on various strategies used with only mediocre success to elicit reliable data, Cleland ultimately suggests that good interviewers are more important than instrument design.

3.2.4 Establishing a Framework and Team

Fourth, in order to identify additional design needs, researchers need to create a design framework and infrastructure and recruit a design team. They must identify team members and create protocols of interaction and decision making for these members. Strategies will be needed to enable the team to go beyond the boundaries of their own perceptions, cultural patterns, and perceived wisdom. We return to this in Section 3.3.

3.2.5 Creating a Quality Assurance and Monitoring Framework

Fifth, the development of the questionnaire and its technical application(s) should be embedded in a quality assurance and monitoring framework. Lyberg and Stukel (Chapter 13, this volume) point to difficulties of establishing stringent quality assurance (QA) and quality control (QC) programs in cross-national studies; much work remains in this respect for multilingual instrument design.

The framework must extend to all instruments or any translated versions of a source instrument used in a project. It is not always possible to know in advance all the languages and cultures which may be involved in a study. Nonetheless, a design that includes consideration of the needs of a fair number of diverse cultures and languages stands a reasonable chance of being able to be adapted to accommodate further new needs that might emerge. At present, some studies employ multiple procedures without a clear rationale, while others skimp on such basic necessities as pretesting. Too often the testing undertaken focuses either on "translation" issues or on statistical analysis, both of which are concerned with the output end of development, rather than the developmental process itself.

3.2.6 Establishing Comparability

Lastly, and importantly, the need for comparable data across implementations in different languages and locations is a central concern in comparative instrument design. Some studies start from the premise that comparability can be best targeted by standardizing as much as possible. Standardizing is then usually understood as keeping formal aspects of design as much the same as possible. This may lead designers to promote replication of question wording, close translation, and repetition of other design features such as filter techniques, formatting, and diagrams. There is some support for such standardization. Smith (1995) points out how a changed diagram in a Dutch module of the International Social Survey Program (ISSP) affected respondents' answers, and Harkness (2003) illustrates how an adapted question in a German ISSP module radically affected respondents' answers. On the other hand, a number of authors (and disciplines) explicitly recognize that adaptation may be essential and that standardized replication in other languages may be a chimera (cf. Chapter 17, this volume, on socio-demographic variables; Hambleton, 2002; contributions in Hambleton, Merenda, & Spielberger, 2005, on educational and psychological test questions; Scheuch, 1968; and Verba, 1969, both on questions in the sociological and political science domains). Others again point to the negative consequences associated with inappropriate standardization (Harkness, 2008a; Lynn, Japec, & Lyberg, 2006).

3.3 INSTRUMENT DESIGN EXPERTS AND TEAMS

An array of expertise is called for to ensure a 3M design is viable and appropriate. In each phase of development cultural input may be necessary.

First important steps include (a) identifying the array of skills expected to be required; (b) deciding the language(s) and communication mediums to be used; (c) recruiting collaborators from the different populations involved to ensure both skills needed and local knowledge are available; and (d) establishing a community of cooperation and trust.

Cross-cultural input in instrument development should be procedurally determined, not left to chance, otherwise input and engagement may be suboptimal. Linn (2002), for example, comments on uneven input in the development of 3M assessment tests: "Contributions of potential items were also far from uniform across countries. The uneven distribution has been characteristic of all the international assessments" (p. 40). Participation can be stimulated by sharing comments and by specifically asking for comments on particular aspects. Specific quality assurance (QA) and quality monitoring (QM) strategies can thus be used to ensure that adequate input and exchange do indeed take place.

Teams supply the human capacity and range of expertise unable to be provided by one individual (cf. Mohler, 2006). At the same time, the degree of common ground (shared understanding) between members may be limited in the beginning. Project planning should therefore budget time for explanation and reiteration to ensure that relevant experience and knowledge can be shared among, say, those with a strong understanding of local potential and requirements, those with substantive research expertise, and those with experience in question design, technical implementation, or pretesting. To facilitate discussion and accelerate understanding, some degree of capacity building may also be necessary. Team membership will probably expand and contract at different phases; to avoid misunderstandings, such plans should be transparent to all involved. Especially when norms regarding communication forms and content differ, it is important to have agreement and transparency with regard to expectations and basic needs.

Organizational literature discusses relevant aspects of intercultural communication likely in 3M collaborations (see, for example, contributions in Gudykunst, 2005).

There are obvious advantages to having a common language (lingua franca) in which groups can discuss ideas and questions in real time. Using a lingua franca can have drawbacks, however. If questions are developed in a language foreign to most of the team, for example, conceptual and technical weaknesses can go unnoticed. There are strategies to counteract possible drawbacks, however. Draft versions of indicators or questions can be appraised in terms of cultures and languages beyond the lingua franca. "Advance translation"—a procedure by which draft translations are used to look afresh at source questions in terms of content, implications, and possible target language formulation—could be useful here (cf. Braun & Harkness, 2005; Harkness, 1995; Harkness & Schoua-Glusberg, 1998; and Harkness et al., 2007).

Team members whose first language happens to be the lingua franca may unintentionally dominate discussion. Moreover, without a strong understanding of linguistic and measurement features of a source instrument, a design group may not be well equipped to specify procedures and/or provide aids (such as annotations) for producing other language versions.

Importantly, too, the language in which studies and questions are developed also shapes the conceptual and cultural frame of reference. Irrespective of design chosen, collaborators will thus often need to acquire and maintain a duality of cultural awareness in developing and reviewing their design suggestions. This competence is something akin to skills acquired by cross-cultural mediators and intercultural communicators; it can be coached and developed over time. Procedures to promote and check understanding and to appraise cross-cultural viability should also be part of the developmental protocol.

3.4 INSTRUMENT DESIGN FUNDAMENTALS

Section 3.4 addresses comparative aspects of basic considerations for question-naire design; Section 3.4.1 presents the concept-construct-indicator-question chain; 3.4.2 looks at whether questions can be asked and answered; 3.4.3 discusses intended meaning and perceived meaning; and 3.4.4 considers mode and design issues in the comparative context.

3.4.1 From Concept to Questions

A distinction can be made between theoretical concepts (which cannot be measured), latent constructs (which can be only indirectly measured), manifest indicators (which can be measured) and the questions themselves. These are useful distinctions in considering comparability and possible adaptation at different levels of design (cf. Harkness, Mohler, & van de Vijver, 2003b). Figure 3.1 characterizes survey *questions* as measurement tools and language vehicles used by researchers to formulate enquiries about *indicators* chosen to measure specific *latent constructs*, so as to gain insight into theoretical *concepts*.

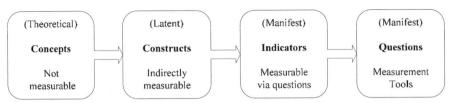

Figure 3.1. Concepts, Constructs, Indicators, and Questions

Once potential indicators for a given construct have been identified and assessed, it becomes clearer whether or not common indicators across populations are possible. The analyses intended also influence these decisions. Van Deth (1998b) discusses various options and constraints in relation to analysis and common or different indicators.

If the aim is to use common indicators for each population, then indicators must be found that provide adequate representation of a construct for all the populations under investigation. If only a few shared indicators are available to measure complex constructs, the questions presented may not provide an adequate representation of a given construct for certain populations. For example, in a religiously homogenous population, the latent construct of religiosity might be adequately captured by a few questions on key manifested aspects. In a population with several prominent religions of diverse character, a greater number of indicators or of questions regarding an indicator might be needed to adequately understand a variety of important aspects. Inadequate comparability of concepts or inadequate conceptual coverage could either disguise real differences or conceal real similarities.

In studies that plan to develop a source questionnaire and translate for other versions, language issues should be considered early enough to inform source question design (see Chapter 7, this volume).

3.4.2 Can Questions Be Asked and Answered?

It is always important to know whether questions can be asked and answered and whether they validly and reliably measure what they are intended to measure. We cannot assume questions that work well in one location will function well elsewhere. Questions may not be relevant or salient for a given population, a group may not have the information necessary to answer, or questions innocuous in some contexts may be threatening or taboo in others. Having a thorough understanding of local conditions and cultural contexts early in the design stage can help identify problems and simplify solutions.

For example, questions which are threatening or sensitive for one population might need to be presented differently, explained, or possibly avoided altogether. Questions about a respondent's children might be threatening or misinterpreted. Chinese respondents, for instance, might associate such questions with the one-child-per-family policy. Some populations may consider identifying children to strangers as risky, for reasons ranging from cultural beliefs associated with naming people to the prevalence of child trafficking. Cultural beliefs may complicate the collection of biospecimens: Hair, nail clippings, and blood could be perceived as giving others power over the person from whom the specimens stem.

3.4.3 Is the Perceived Meaning the Intended Meaning?

In developing questions, the aim is to ensure that respondents understand questions as they were intended to be understood. Research has illustrated well how misunderstandings arise, even when respondents and researchers, broadly speaking, share language usage and knowledge of the world (e.g., Belson, 1981). Misunderstandings can affect data quality: Statistics New Zealand (2002) states "Respondent misunderstanding could potentially be one of the largest sources of non-sampling error in our surveys" (para. 7).

Respondents' social reality and cultural framework shape their perceptions and survey responses in a variety of ways. In cross-cultural research we must especially expect that realities and cultural frameworks will differ from one population to another and that this may pose very much increased threats to a shared understanding across populations of what questions ask and what answer options mean.

Several contributions in this volume discuss facets of culture and communication: Schwarz, Oyserman, and Peytcheva (Chapter 10), as well as Uskul, Oyserman, and Schwarz (Chapter 11) discuss ways in which cultural grounding and established habits of attention influence perception and processing in American and Chinese populations. Yang et al. (2008) and Yang and colleagues (Chapter 12, this volume) discuss how response styles associated with certain cultural groups may introduce response bias. Braun (2003) interprets different responses to "the same questions" in western and eastern parts of Germany as reflections of the different social realities framing respondents' interpretations and response choices. It is thus becoming increasingly apparent that perceived meaning and culturally preferred patterns of response need to be carefully investigated.

3.4.4 Mode, Design, and Response

Modern survey studies may deliberately use a mix of modes or other tailored strategies to accommodate respondent preference or needs and thereby enhance response (see, for example, Dillman, Smyth, & Christian, 2008 on modes and design; on responsive study strategies, see Groves & Heeringa, 2006; Groves, Lepkowski, Van Hoewyk, & Schulz, 2009).

Pennell, Harkness, Levenstein, and Quaglia (Chapter 15, this volume) discuss modes in the context of cross-national data collection and the sparse research available. We simply note that the implementation originally envisaged may not always prove viable and that this can affect the realization of the design. Certainly, this might happen in any context but in comparative research, literacy, innumeracy, and language can pose special and frequent problems. Moreover, not all languages have a standard written form. If bilingual interviewers or interpreters are employed, this affects standardization of wording and stimulus (Harkness, Schoebi, et al., 2008; Harkness et al., 2009c). It could also mean that answers intended to be provided covertly (for example by using a code letter or a written reply rather than stating the reply verbally to the interviewer) are no longer covert. If such eventualities are considered in advance, alternative forms of response can be sought.

3.5 KEY DECISIONS ON INSTRUMENT DESIGN

In producing questions for multinational implementation, question design teams make three basic and interrelated decisions about commonality of questions, origin of questions; and how the project plans to ensure comparability and viability of instruments used in each location and language. The following discussion builds on the framework presented in Harkness, van de Vijver, and Johnson (2003).

3.5.1 Commonality of Questions

Many comparative studies in the social and behavioral sciences aim to ask the same questions of every population from which data is collected. The underlying model is thus one of common (shared) latent constructs, common indicators, and common questions (see Section 3.6.1). Alternatively a study might ask different questions of each population. Here the underlying model is that latent constructs are common, indicators might be common or different, and the questions used can differ in content (see Section 3.6.3). A mixed approach combines a set of common questions across locations/populations with other questions that are location- or population-specific (see Section 3.6.5).

3.5.2 Origin of Questions

A second key decision is whether researchers plan to adopt (replicate) existing questions, adapt (modify) existing questions, or develop new questions. In many instances, all three strategies may be used in one study. These decisions then determine which options are available for developing versions in other languages, what kind of adaptation, if any, is planned for any given version, and the timing and form of pretesting of items needed.

3.5.3 Degree and Timing of Cultural Input

In planning the design of an instrument—deciding the teams to be involved, the various design milestones and strategies to be used, and the testing and assessment procedures to be followed—researchers implicitly or explicitly decide on the form and degree of cross-cultural input in their study. Such input is usually seen as a means to help ensure comparability and viability of the instruments developed. Studies differ greatly on how they target and assess cross-cultural viability. Quality of life (QoL) researchers in particular have argued for including cross-cultural input at every stage of development and also outline elaborate schemes to do this (e.g., Bullinger, Schmidt, & Naber, 2007; Skevington, 2002; Skevington, Sartorius, Amir, & the WHOQOL group, 2004; and the WHOQOL Group, 1994; for psychological research see van Widenfelt, Treffers, de Beurs, Siebelink, & Koudijs, 2005).

QoL literature often distinguishes between what it calls sequential, parallel, and simultaneous approaches to question design. Differences can be found in the way these terms are used and explained in the QoL literature and in literature drawing on these (see, for example, Bullinger, Schmidt, & Naber, 2007; MOT, 1997; Skevington, 2002). We do not attempt to resolve these differences here. Generally speaking, the terms reflect something about the emphasis placed on cross-cultural input, the stages of development at which cultural considerations are addressed, and the strategies employed.

Sequential development approaches, as described in the QoL literature, place little emphasis on cross-cultural input during question development and basically only address multicultural considerations at the translation stage. The term assumes a source questionnaire is developed and other versions produced on the basis of translation.

Parallel development, as used in the QoL literature, targets cross-cultural input early in the conceptual and question development stages of a common source questionnaire. This can involve consultation with local experts, having a multi-cultural drafting group develop a set of questions then vetted by a larger cross-national group of researchers, or including items from all the participating countries in the pool of items for consideration.

Linn (2002) points to difficulties encountered in trying to realize common development of educational test items and an item pool drawn from all participating countries. The fact that the United States had greater resources to develop and test such items ultimately led to more items being supplied by U.S. educational agencies.

Simultaneous development, as the term is used in the QoL literature, targets the highest degree of cross-cultural involvement, with cultural input at every stage contributing to development and to assessment. Developmental strategies used include cognitive interviewing, focus groups, participatory community involvement, and expert group review. Although the literature mentions a wide range of procedures in describing simultaneous development, little insight is offered into the relative merit of the various techniques. Studies on a modest budget would need more parsimonious strategies.

3.6 MAJOR COMPARATIVE DESIGN MODELS

This section presents the most widely recognized approaches to comparative design: asking the same questions; asking different questions; and combining both approaches.

3.6.1 Asking the Same Questions: Source Questionnaire and Translations

By far the most commonly used method to develop survey questions for comparative research creates a source questionnaire in one language and then produces other language versions from this on the basis of translation. The success or failure of this ask-the-same-question (ASQ) approach is largely determined by the suitability of the source questions for all the cultures for which versions will be produced (cf. Harkness, van de Vijver, & Johnson, 2003).

One of the chief attractions of the ASQ approach is the potential power of analysis offered, namely full score comparability. Thus comparisons can be made across populations, question by question, or item battery by item battery. Source questionnaire and translation ASQ models also allow questions to be replicated; the source questionnaire can include questions from other studies and these may be "replicated" (within the constraints of translation) in different locations. ASQ models are relatively easy to implement: first the source is produced and then the translations. They also follow a traditionally endorsed (although ultimately questionable) "common-sense" model of enhancing comparability by keeping things the same.

One potential drawback in trying to develop shared questions for multiple populations is that the questions may become less specific in topic or formulation than would questions designed for a national study. This may result in inadequate coverage of the construct to be measured and in construct bias. One way to counteract conceptual undercoverage is to add country-specific questions, as described in Section 6.5.

A serious and related challenge to using source-and-translate ASQ is that the questions are required to "mean" the same thing and convey the same stimulus in different contexts, populations, and languages. As illustrated earlier, meaning is by no means determined by words alone. Development procedures for source questions must therefore ensure that the questions selected are understood similarly in the various languages and locations of the study. At the same time, the very limited pretesting practices currently followed in important 3M studies do not ensure this. In addition, Jowell (1998) reminds us that locations sharing the language of the source questionnaire also need to review the instrument for local suitability (e.g., the United States versus the United Kingdom).

The least elaborate form of a source-and-translate ASQ model consists of two distinct steps: (a) develop a source questionnaire, and (b) translate this to produce the other language versions needed. This procedure receives the most criticism from researchers advocating intense cross-cultural input during question development (e.g., Camfield, 2004; Skevington, 2002). However, a well-designed source-and-translate model can accommodate considerable cross-cultural input: drafting groups can be multicultural in composition; questions can be drawn from studies conducted with various populations and suggested by different cultural groups; cross-cultural discussion of adaptation can be incorporated; and pretesting of the draft source questionnaire can be multilingual and multicultural. An ASQ model that ensures adequate cross-cultural input at the conceptualization, drafting, and testing stages can offer a viable and affordable model for many projects.

In practice, however, the potential for multilateral input is badly underutilized. This may stem from practical issues such as time pressures but may also be encouraged by over-confidence in the suitability and immutability of pedigreed questions, a misjudgment of the (limited) potential of translation, and a lack of awareness of how context shapes interpretation and of the frequent need for adaptation.

3.6.2 Ask the Same Question by Decentering

A second strategy for keeping questions the same is decentering, which develops two questionnaires simultaneously for two populations and languages, an iterative process of translation-cum-paraphrase.

Our outline below describes only one basic form of how decentering can work. Even so it allows for two alternatives. One employs something like a Ping-Pong exchange: an item in language A leads to an item in language B and this leads to a further version in language A, and so forth. Alternatively, one item in language A can lead to sets of alternative versions in language B and also in language A. The steps identified below begin at the stage when questions are formulated.

1. First, a question is devised or chosen in language A.
2. The question is translated into language B so as to convey the essence of the measurement targeted in the language A question. The focus in translation is on conveying the conceptual essence of the questions, not on close translation.
3. Multiple paraphrases or further translations may be generated for the translated item in language B, providing a set of possible versions in language B. Whether one or several language B versions are produced is a matter of approach. Whatever is produced for language B is translated into language A.
4. As versions are produced and compared across the two languages, anything that causes problems is altered or removed. In this way, culturally or linguistically anchored obstacles are eliminated for each language.
5. If a Ping-Pong procedure is followed, modifications made in one language can immediately inform the next version produced in the other language. If a set of versions is produced in each language, developers can combine removal of perceived "obstacles" with matching up whichever version in language A best fits a version in language B.
6. At some stage, a version is reached in each language which is felt to match the version in the other language.

Decentering and similar procedures have been advocated as a way of avoiding cultural and linguistic bias and a means of enhancing comparability (Erkut, Alarcón, García Coll, Tropp, & Vázquez García, 1999; Potaka & Cochrane, 2004; Werner & Campbell, 1970). There are a number of drawbacks nonetheless (cf. Harkness, 2008a; Harkness, van de Vijver, & Johnson, 2003). Since decentering removes culturally specific material, specificity and saliency may be low. Consequently, decentered questions can in some respects be less appropriate for fielding in either context than questions developed by other means. Decentering is incompatible with replication because the wording of questions inevitably changes. It is also not suitable for simultaneous production of multiple translations. Apart from the practical difficulty of attempting this process across, say, 12 languages and cultures, construct coverage, indicator saliency, and pertinence of measurement would be at risk.

3.6.3 Asking Different Questions

Ask-different-questions (ADQ) approaches aim to provide comparable measurement of the targeted latent construct by asking questions that may differ in content across instruments used but are held to produce comparable data. The indicators chosen might be shared across populations, but different indicators might also be used if considered to serve the purpose of producing comparable data. ADQ approaches are sometimes described as "functional equivalence" strategies (see van Deth, 1998b; Przeworski & Teune, 1970; Triandis, 1976). We prefer the less ambiguous ADQ (cf. Harkness, 2003, 2008a).

One of the great appeals of asking different questions is that it obviates the need to translate.[2] A second attractive feature of ADQ models is that the country-specific questions used can relate directly to the issues, terminology, and perspectives salient for a given culture and language. Thirdly, conceptual coverage for a given location is then likely to be good. A fourth advantage is that the development of a questionnaire for a population can be undertaken as and when needed. Countries might therefore develop their instruments at the same time or, if joining an existing project at a later date, develop their own country-specific and country-relevant questions when these are required. At the same time, many researchers intent on across-country comparisons are rightly chary of the design challenges involved and the possible constraints on analysis.

A basic procedure for an ADQ model could be as follows: First, the design team decides on the concepts and latent constructs to be investigated and any other design specifications they might make, such as identifying common indicators or allowing location-specific indicators, or planning a mixture of both. Then, country- or population-specific questions are designed to collect locally relevant information for a given construct as informed by the chosen indicators. Finally, versions for different countries and languages are produced in a collective effort or developed by different teams at different times as the need arises.

3.6.4 Combination Approaches

Often enough a combination of ASQ and ADQ is used, in which a core of questions shared across countries is combined with country-specific questions that provide better local coverage of the concepts of interest. In various disciplines, the terms *etic* and *emic* are sometimes associated with such combined approaches.

---

[2] In actuality at the levels of documentation and data merging some translation may become essential.

Discussion of such emic and etic approaches to research is extensive (cf. Sinha, 2004). Sinha criticizes researchers who define emic concepts or questions as being culture-specific and etic constructs or questions as having universal or across-country relevance. At the same time, this is how the terms are often actually used (e.g., Gibbons & Stiles, 2004, p. 14). The terms emic and etic are also used somewhat differently in various fields (cf. Headland, Pike, & Harris, 1990; Serpell, 1990; Sinha, 2004). Such differences are very common; uses of terms develop over time, and not always in the direction of clarity.

Whenever researchers decide to ask the same question of different populations with the intention of collecting comparable data, they assume the question has etic status. If this is simply assumed, but not demonstrated beforehand, the literature speaks of an "imposed etic," reflecting the top-down approach (Triandis & Marin, 1983). If a study uses emic questions to gather local information on a phenomenon and the data permits identification of commonalities across cultures, the literature speaks of a "derived etic." More discussion can be found, for example, in Berry (1990), Peterson and Quintanilla (2003), and Sinha (2004).

3.7 SOME SPECIAL DESIGN ASPECTS

3.7.1 Response Options

Ongoing debate about answer category design in the general survey research context carry over into 3M design decisions. For comparative purposes, there is no strong body of research on these matters. Thus Smith (2003) reviews many of the pertinent issues but understandably cannot provide answers. These include the number of response options, forced choice, open-ended or closed-ended questions, rating versus ranking, fully versus partially labeled, and verbal versus numerical scales, visual depictions, visual heuristics, fuzzy measurements (e.g., vague quantifiers), Likert-type scales, and the statistical analysis permitted by different kinds of scale measurement (nominal, ordinal, interval, or ratio).

In multilingual research source language answer scales must be conveyed in other (target) languages. Harkness (2003) discusses how answer scale translation relates to design (see also Chapter 7, this volume). Again little literature exists on this topic. Strategies commonly used to produce target language versions are (a) translate as closely as possible; (b) adapt what cannot be translated, (c) use preferred local answer option formats and wording. In this last instance, researchers employ scales with which they are familiar which they consider comparable to source scales (Harkness, 2003; Harkness et al., 2007).

Szabo, Orley, and Saxena (1997) and Skevington and Tucker (1999) describe a further strategy. Source questionnaire terms to be translated were identified for the end points of answer scales. Participating countries then derived labels for intermittent points of the scale from local research that investigated the magnitude respondents associated with various verbal labels.

A German-American research project covering similar ground found differences in the degrees of intensity across populations for what at face value were "comparable" labels, including the end point labels (Harkness, Mohler,

Smith, & Davis, 1997; Harkness & Mohler, 1997; Mohler, Smith, & Harkness, 1998). The research was extended to Japan with similar findings (Smith, Mohler, Harkness, & Onodera, 2009).

Several discussions consider agreement scales used in the ISSP. Examining ISSP translations of *strongly agree/agree/neither agree nor disagree/disagree/ strongly disagree* scales in some 10 countries, Harkness and colleagues (2007) found that the face-value semantics of each scale point are often quite diverse. Villar (2006, 2008) found differences in response patterns within populations across modules, depending on how the second and fourth points on the scale were realized. Sapin, Joye, Leuenberger-Zanetta, Pollien, and Schoebi (2008) also reported differences in response depending on answer scale translation in data from the European Social Survey (ESS); Saris and Gallhofer (2002) report on multitrait-multimethod designs to test ESS questions in the context of modified answer scales in Dutch and English. South African ISSP researchers reported problems implementing the same ISSP agreement scale in some rural populations (Struwig & Roberts, 2006). The authors added Kunin faces to help respondents consider varying degrees of agreement or disagreement. Unfortunately, many ISSP questions for which faces were presumably used did not reference topics readily associated with smiles or frowns (cf. also Dasborough, Sinclair, Russell-Bennett & Toombs, 2008).

Pictorial or diagram images may indeed not be appropriate for all target populations. In the 2007 ISSP module on leisure time and sports, an optional question about body image used drawings of men and women in swimwear (http://www.issp.org/documents/issp2007.pdf/). The women were depicted in bikinis. The notion of bathing and the appropriateness of bikinis seem at least questionable for surveys of the general population in more than 40 countries on five continents. Bolton and Tang (2002) report greater success in rural Rwanda and Uganda with visual depictions and also with ADQ components. Locally developed lists of challenges were created for each community, resulting in some items being shared across some locations and others not. Pictorial aids helped participants select an answer scale option for degrees of task burden. The visual aids depict a person in local dress carrying a bundle on their back. As the bundle increases in size, the person bows further under its weight to indicate the increased burden involved.

Culture affects how respondents perceive and select answer options (see Chapter 11, this volume). On the level of language and culture, bilingual respondents have been found to answer differently when interviewed in their native language compared to answering in English (e.g., Gibbons, Zellner, & Rudek, 1999; Harzing, 2006). Yang and colleagues (Chapter 12, this volume) document discussions of answer scales in the response style literature, but also make clear that solutions for comparative answer scale design are not to be found there.

For 3M projects in particular, the challenges of designing answer scales and deciding answer options are many: Are specific formats suitable across target populations? Are 2, 4, 5, or more answer points suitable for all respondent groups? How will answer option adaptation affect comparability? Any approach chosen should be tested in a variety of languages and settings, with different populations and questions. Research in this area is badly lacking.

3.7.2 Technical Design Realization

The *technical realization* of a survey instrument is concerned with format, layout, numbering, and filter strategies, with possible guidance for interviewers and programmers, documentation requirements, design development tools, and a variety of other features. Such metadata are an essential part of instrument design.

The issues to be resolved relate to characteristics of a given instrument and to whatever language versions and mode(s) of administration are planned. Language and organizational features involved include typographical conventions, visual norms, and differences in space needed to present text or capture responses, as well as broader conventions of communication including colors, symbols, and sequencing of information. These considerations need to be addressed for each language and cultural system.

Technical design decisions include the locus of control—whether the survey is self-administered or interviewer-administered—and whether it is computer-assisted, paper-and-pencil, Web-based, or uses some other format, such as audio media. Decisions about data output, coding, and documentation also need to be addressed. Three main technical considerations must be addressed: (a) instrument usability (for interviewer or respondent navigation and response-capturing); (b) ease of programming and testing; and (c) ease of coding, data outputting, analyzing, and documenting survey data. Technical aspects of design often overlap with those related to question content and formulation, as is the case with answer scales. Mode decisions shape wording and organization of information.

The *Cross-Cultural Survey Guidelines* (CCSG, 2008) module on technical design proposes five guidelines in realizing technical aspects for a 3M design. We can only briefly explain the relevance of each guideline here; the website provides a full rationale and procedural steps.

Guideline 1: *Ensure that instrument design is appropriate to the method of administration and the target population*: Literacy issues and technological access limit the modes possible for certain populations; populations differ in familiarity with survey interviews and their purposes. For example, some populations may need more or fewer navigational guides than others.

Guideline 2: *Develop complete instrument design specifications for the survey instrument, indicating culture-specific guidelines as necessary*: Specifications outline all the components of the questionnaire and its contents, guide formatting or programming, ensure design consistency across instruments, provide guidance on technical adaptation, and facilitate post-production data processing. Formatting specifications also address language-specific character sets and differences in alphabetical sorting including diacritics (*é*), ligatures (*œ, æ,* and *ß*), and character combinations.

Guideline 3: *Develop interface design guidelines for computer-assisted and Web survey applications:* Interface design affects the respondent-computer or interviewer-computer interaction, influences user performance, and can affect data quality. In addition to maximizing usability, design should also be consistent across survey implementations at appropriate levels. For example, design specifications across language versions should follow the appropriate processing directionality for a given population (left to right, right to left, top to bottom, etc.)

Guideline 4: *Establish procedures for quality assurance of the survey instrument that ensure consistency of design, adapting evaluation methods to specific cultures as necessary:* All aspects of design can affect data quality positively or negatively. For instance, choices of colors should be validated by experts for particular cultures. This may involve harmonization to a set of "culture-neutral" colors across instruments, or adaptation of colors as necessary.

Guideline 5: *Provide complete documentation of guidelines for development of source and target language or culture-specific instruments:* Documentation should provide data users with necessary transparency.

3.7.3 Factual and Socio-Demographic Questions

Questions containing reference to facts of the world (an event, practice or institution) or asking about personal facts (educational or occupational status) may need to be adapted in ASQ approaches. Granda, Wolf, and Hadorn (Chapter 17, this volume) discuss designing socio-demographic questions in detail. Questions asking about medication used, types of cigarettes smoked, or varieties of cooking oil used, should reference brands or types relevant for a specific population. It is important that information requested can be provided by all populations. Willis and colleagues (2008) provide examples related to information available to smokers in different locations about light and heavy forms of tobacco.

3.7.4 Vignettes and Comparative Research

Vignettes, as used in survey research, are cameo descriptions of hypothetical situations or individuals. Respondents are asked to make evaluations about these hypothetical entities. In the context of comparative research, King and Wand (2007) and King, Murray, Salomon, and Tandon (2004) present vignettes as anchoring devices to adjust self-assessment responses (see also Chapter 4, this volume). Vignettes have also been used as pretesting strategies (cf. Gerber, Wellens, & Keeley, 1996; Goerman & Clifton, 2009; Martin, 2004, 2006).

In multicultural research, the question obviously arises whether vignettes should be population-specific formulations (ADQ) or developed as ASQ and translated. Martin (2006) emphasizes the need to ground the content and wording of vignettes in the respondent's social context and language. This would seem to question whether translation is a suitable procedure for producing appropriate vignette versions.

For example, the names of individuals featured in a vignette could pose comparability and perception problems. Associations for names differ across populations so keeping names identical would alter effects (imagine "William" in a Vietnamese vignette). Localizing names may be an alternative if cognates exist: *Willem, Wilhelm, Guillaume,* and so forth. In vignettes used in stigma research (White, 2008), the names of hypothetical individuals were "translated." However, it is by no means certain that the connotations associated with names are stable across countries. Even if we assume a prototypical "John" of Great Britain, he need not be comparable to a "Hans" of Germany or a "Juan" of Spain. Web lists of popular names (e.g., http://www.babynamefacts.com/) also reflect differing current popularity, not to mention the status or ethnicity associated with various names.

3.8 PRETESTING DESIGN SUITABILITY

3.8.1 Developments in Pretesting

Pretesting in general is underutilized in comparative research projects. This section aims to cover developments in pretesting of relevance for 3M studies but must often take developments in the general survey research context as its starting point, given the paucity of comparative research on pretesting.

The following definition of pretesting from the pretesting module in the *Cross-Cultural Survey Guidelines* (CCSG, 2008) could apply to a single cultural context or to multiple languages and cultures: "a set of activities designed to evaluate a survey instrument's capacity to collect the desired data, the capabilities of the selected mode of data collection, and/or the overall adequacy of the field procedures" (p.1). Presser, Rothgeb, et al. (2004) note that question design and statistical modeling "should work in tandem for survey research to progress" (p. 12). One form of testing is insufficient; both qualitative and quantitative analyses of instruments are needed in developmental phases and indeed multiple methods should be utilized.

First and foremost, many well-established 3M source questionnaires are not extensively pretested in the locations for which they are to be used. This stands in stark contrast to the oft cited recommendation intended for noncomparative research: "If you do not have the resources to pilot-test your questionnaire, don't do the study" (Sudman & Bradburn, 1982, p. 283).

Second, we note that progress in pretesting in the last decade has been uneven, both generally, and more recently in application to multicultural surveys. For example, cognitive interviewing has greatly increased in popularity and is now widely accepted as an important pretesting tool. It is a valuable tool for identifying problems related to question wording and for investigating the process of responding. However, it is much less useful for gaining insight into problems with metadata, interviewers, or factors not observable in a laboratory setting. Its methods are also not easily scalable for evaluating survey instruments or procedures across a number of subgroups.

In particular U.S. researchers have begun to review cognitive procedures in multicultural applications, even if the focus remains primarily on pretesting translated versions of existing instruments. Thus in 2002, the U.S. Census Bureau adopted guidelines for translating instruments and also quickly made pretesting of Spanish questionnaires mandatory (de la Puente, Pan, & Rose, 2003). Examples include Blair and Piccinino (2005b), Dean et al. (2007), Fitzgerald and Miller (2009), Goerman (2006a), Hunter and Landreth (2006), Miller (2004), Miller et al. (2005b), Miller et al. (2007), Willis (2004), Willis (2009), Willis et al. (2008), and Willis and Zahnd (2007). At the same time, the sobering comments by Presser, Rothgeb, et al. (2004) still hold true:

Although there is now general agreement about the value of cognitive interviewing, no consensus has emerged about best practices (…) due to the paucity of methodological research (…) [and] a lack of attention to the theoretical foundation for applying cognitive interviews to survey pretesting (p. 113).

Fourth, an array of additional strategies is now available for assessing instrument quality. Various kinds of paradata are potentially powerful sources of information regarding respondent and interviewer behavior and interaction (cf. Couper & Lyberg, 2005; and Chapter 18, this volume). In addition, at the other end of the spectrum, statistical analyses such as item response theory, differential item functioning, multitrait-multimethod models, and latent class analysis, have been used to explore how different items or versions of an item function for respondents and across respondent groups (see contributions in Part VII, this volume, and contributions in Presser, Rothgeb, et al., 2004).

3.8.2 Opportunities for Comparative Pretesting

As recognition of the usefulness of cognitive pretesting has grown, other pretesting strategies have regrettably received less attention in either general or 3M contexts. They tend to be underutilized and have also received little methodological refinement.

Split-ballot experiments are one example. These can provide definitive evidence to support decisions about many issues in question design (Fowler, 2004; Groves et al., 2004). These are currently not much used; at national level, the U.S. National Survey of Drug Use in Households (NSDUH) is one notable positive exception (cf. Groves et al., 2004). At the same time, split-ballot studies among bilinguals sometimes advocated for multilingual surveys must be treated with caution. Bilinguals' response processes and responses may differ from those of monolinguals for whom translations are mainly intended (on bilinguals see contributions in Kroll & de Groot, 2005).

Many forms of experiment are underutilized and have unsurprisingly received little methodological consideration for multilingual application. Experiments may be expensive in some modes and locations, but can be relatively inexpensive to mount in Web surveys and in locations where interviewing is comparatively inexpensive. Low-budget experiments have been used in national contexts to investigate incentives and response rates in in-person surveys as well as assessment of human-computer interface in Web and computer-assisted telephone interviewing (CATI) instruments (Christian et al., 2008; Edwards et al., 2008). Recent experiments on survey translation and survey interpreting (Harkness, Villar, Kephart, Schoua-Glusberg, & Behr 2009a; Harkness et al., 2009b) also reflect the considerable potential for low-budget comparative methods research.

Despite the high interest in new and multiple modes, testing that examines mode issues remains where it was a decade or more ago (cf. de Leeuw, 2008; Dillman, 2004). Literature on mode effects in comparative contexts is very sparse (see Chapters 15, 26, and 27, this volume) and only modest progress has been made on pretesting techniques in noncomparative contexts, despite the increased attention modes receive in general survey design.

Many technological advances have enabled researchers to investigate new aspects of the response process in survey research. Eye-gaze tracking and response delay tracking, for example, can provide important insights into user-interface issues relevant for instrument and application design, but these are rarely used in pretesting. Given the multiple differences in visual processing for different languages and differences in cultural perception, this would seem an important area for multicultural research.

Survey data dissemination has also developed considerably in the past decade, but dissemination approaches are rarely the subject of pretest activities. Pretesting typically focuses on the beginning and middle of the survey process; rarely are end-user needs pretested. However, focus groups and user debriefings could enable a better fit between survey products and user needs recommended in Section 3.7.2. Testing must also address technical aspects of design. We also expect differences across cultures in expectations and possible interface needs. As Section 3.7.2 indicates, pretesting plans must address items in their context. Metadata such as variable names, item instructions, item types, populations to be asked each item (i.e., skip patterns), question and answer source, and adaptation from source need to be carried with the item, and imply a complex item context important for analysts and methodologists. These aspects will be quite complex in 3M surveys.

Finally, pretesting methods such as usability testing, prototyping, scenario testing, regression testing, and database testing are rarely mentioned in the current pretesting literature in any context, despite being appropriate for various phases of instrument development. Prototyping would seem particularly useful for cross-cultural instrument development. Testing the performance of one or more small model boats in a laboratory setting can inform design for a full-size boat at much reduced cost. Similarly, prototyping permits quick and relatively inexpensive evaluation of an instrument or instrument section, at a stage before a large investment has been made in full-scale production.

3.8.3 Toward a Theory of Pretesting

There is currently no basic "science" of pretesting methods for surveys and few methodologists offer guidance on how a wide range of pretests should be designed or implemented to ensure all necessary features of an instrument are adequately assessed. This is a real drawback from comparative research, particularly since automatic application of methods used in noncomparative research has proved to be almost as ill-advised as has the prevalent neglect of pretesting.

Lyberg and Stukel (Chapter 13, this volume) summarize several broad quality concepts appropriate for pretesting, including fitness for use. In addition, literature on developing and testing computing systems drawn from project management concepts of risk assessment and cost-benefit analysis offer guidance for testing automated survey systems such as computer-assisted telephone interviewing (CATI), computer-assisted personal interviews (CAPI); audio computer-assisted self interview (ACASI), and computer-assisted Web interviews (CAWI). Such computer-assisted applications are becoming increasingly affordable and are by no mean restricted to surveys in affluent regions of the world.

In conclusion, we suggest that survey pretesting has proceeded on an uneven path over the past 80 years. As survey methods adapt to changing cultures and technologies, a more comprehensive and theoretical approach to pretesting in general is required. Problems encountered in cross-national and cross-cultural surveys may well hasten this progress.

3.9 DESIGN OUTLOOK

Survey research methodologists and others working on survey methods in various disciplines have discovered a great deal about questionnaire design and design realization in numerous forms and modes. At the same time, a general overarching theory for instrument design is lacking and much remains to be discovered.

We have valuable but piecemeal insights into the response process and how and why respondents perceive what they perceive and answer what they answer. Various chapters in this volume add further pieces to the jigsaw puzzle. Nonetheless, we have only begun to consider some aspects of interview "context" in theory-based terms. Few would now comfortably talk about *the* meaning of a question. It is increasingly understood that meanings may be multiple, that meaning is co-constructed in a context, and that whatever meaning is in a given instance, it is *not* a unique property of words or questions. At the same time, it is by no means clear what the consequences of these insights will be for question design. Perceived meaning is perhaps a much more pressing matter for comparative research, since it may differ systematically across groups, but it is obviously relevant for all questionnaire design.

In similar fashion, while there is a growing understanding of the relevance of context and culture for any design, we are only beginning to consider in a systematic fashion what the cultural frameworks of different populations of respondents may signify for design and analysis. Much therefore remains to be investigated before a comprehensive theory of design can be accomplished.

Design procedures used in comparative research often lag behind best practice in noncomparative research. Their needs, however, are more complex. As amply indicated here, comparative design is design for populations with different cultural orientations, world perceptions, knowledge, and experience, usually involving multiple languages. Thus issues of comparability are at the heart of cross-cultural and cross-national design efforts in ways not required for most other research.

In many studies, researchers aim to achieve comparability by using the same questions. We, like others, have pointed to drawbacks to this approach. At the same time, if properly developed and tested, ASQ models can offer affordable and robust strategies for many areas of investigation. Importantly, however, the research community is not yet able to identify the best (or minimally required) procedures for any such development. As projects are often under considerable resource constraints, an optimal but parsimonious model is both practically and scientifically essential. To complicate matters further, best practice procedures will differ from discipline to discipline, since the various types of instruments required also call for appropriate formats and contents and permit and need different assessment procedures.

Building a robust and adequate theoretical framework for multipopulation questionnaire design will take time and research. In some areas, nonetheless, the prospects of quite rapid progress look good. Insight could be gained quickly into the efficacy of various developmental and assessment procedures, for example. In this volume, chapters by Pan et al. (Chapter 6), Goerman and Casper (Chapter 5), Willis et al. (Chapter 8), and Dept, Ferrari, and Wäyrynen (Chapter 9) reflect new evaluative investigations into instrument and translation materials, while Thornton et al. (Chapter 4) describe their journey toward workable comparative strategies on a modest budget.

A strong methodological literature has also yet to develop. At the same time, research addressing methodological aspects of comparative instrument design is growing, as documented in recent survey conference and workshop papers, in the emergence of meetings and interest groups focusing on comparative survey methods, and in journal articles and other publications. The International Workshop on Comparative Survey Design and Implementation (CSDI) is one fast access point for developments in comparative methods research (http://www. csdiworkshop.org/). The CSDI website also provides links to disseminate information about activities. Such initiatives afford opportunities to share knowledge, and to build and sustain critical mass. The CCSG Guidelines website (http://www.ccsg.isr.umich.edu/) mentioned earlier is a unique source of hands-on information for researchers. Guidelines on design, translation, testing, and pretesting of different formats and different survey purposes are indeed emerging apace in diverse fields, documenting the interest in doing things as well as possible and, importantly, in spreading knowledge (see Chapter 1, this volume). The Q-Bank database hosted at the U.S. National Center for Health Statistics (http://wwwn.cdc.gov/QBANK/Home.aspx/) is a noteworthy national attempt to provide access to cumulative insights gathered in cognitive pretesting of questions used in U.S. federal research (Miller, 2005). It is now able to deal with questions in languages other than English.

This returns us to the topic of documentation, providing details of design, of testing, of the research intentions and the analysis motivating design, as well as version production. Instrument designers need information on specifications of typical (or special) design considerations, guidance on where to find existing information on technical conventions and their importance, and on answer scale realizations for various formats, and on up-to-date contextual information about survey infrastructures and realization potential for locations around the globe. An important activity for the immediate future will be to gather and share such information which will be of immense help in planning and executing instrument design, not only for those new to the field.

4

Creating Questions and Protocols for an International Study of Ideas About Development and Family Life[2]

Arland Thornton, Alexandra Achen, Jennifer S. Barber,
Georgina Binstock, Wade M. Garrison, Dirgha J. Ghimire,
Ronald Inglehart, Rukmalie Jayakody, Yang Jiang, Julie De Jong,
Katherine King, Ron J. Lesthaeghe, Sohair Mehanna,
Colter Mitchell, Mansoor Moaddel, Mary Beth Ofstedal,
Norbert Schwarz, Guangzhou Wang, Yu Xie, Li-Shou Yang,
Linda C. Young-Demarco, and Kathryn Yount

4.1 INTRODUCTION

This chapter describes our journey to create and test procedures and instruments for use in international comparative research. It describes how we began our work with no existing measures, even in one country, of our theoretical concepts and worked to construct and test a battery of measures for use in several diverse countries. Other researchers sometime embark on comparative research at a similar starting point, and we hope that this account of our journey may prove useful to them.

Our research was motivated by a desire to understand people's knowledge and perceptions of social and economic development. Our goal was to create and test questionnaires and protocols for measuring ideas and beliefs about development that would be appropriate for administration in a variety of countries. In Section 4.2, we briefly explain the developmental model and its basic propositions about social change. In Section 4.3 we describe our organizational approach and initial steps in designing projects in several countries. Section 4.4 explains how we used the experience and knowledge accumulated from our work in individual countries

---

[1] Survey Methods in Multinational, Multiregional, and Multicultural Contexts, edited by Harkness et al.
Copyright © 2010 John Wiley & Sons, Inc.
[2] Authors are listed alphabetically after Arland Thornton, the project coordinator.

to prepare questionnaires and protocols for use in deliberately comparative projects. In Section 4.5 we discuss specific problems we encountered, along with lessons learned. Section 4.6 provides preliminary evidence of the degree to which we were successful in measuring aspects of developmental thinking. Finally, Section 4.7 discusses the implications of our experience for other researchers who may design international data collections.

4.2 DEVELOPMENTAL CONCEPTS AND THEORIES

We began our work with the understanding that a key element of developmental thinking is the developmental paradigm, which is a model of social change that posits that all societies progress through the same universal stages of development (Burrow, 1981; Harris, 1968; Stocking, 1968, 1987; Nisbet, 1969; Smith, 1973; Sanderson, 1990; Mandelbaum, 1971; Thornton, 2001, 2005). The developmental paradigm indicates that at any one point in time there is a hierarchy of countries on a developmental ladder; this amounts to a ranked scale of nations. We do not present these developmental concepts as true and good here, but as ideas that can be important for people in everyday life whether or not they are true or good.

Advocates of this paradigm suggest that the most developed or modern societies are in northwest Europe and the northwest European diasporas, while other societies are seen as less developed, developing, or traditional. It is relevant to note that the United Nations currently ranks countries on a numerical Human Development Index (HDI). One of our research goals was to learn whether ordinary people know about and endorse this model of a developmental ladder on which societies are ranked.

Many western scholars using this developmental model associated certain social and economic characteristics of northwest Europe with *modernity* and *development*. These characteristics included an industrial economic organization of life, urban living, and high levels of education, as well as technological skill, high levels of consumption, and low levels of mortality (Millar, 1779; Mill, 1848; Guizot, 1890; Tylor, 1871). On the other hand, social and economic patterns predominant elsewhere, including features such as agricultural economic production, rural living, lack of education, low technological skills, low consumption levels, and high mortality were defined as *traditional* or *less developed*. A second goal of our research was to establish to what extent, if at all, ordinary people in diverse cultures used the concepts of *developed* and *less developed* to describe societal attributes in the same way as scholars using the developmental model.

It has long been accepted that family patterns within and beyond northwest Europe display considerable heterogeneity but also that certain characteristics of family life are particularly marked in northwest Europe. These attributes include individualism, the organization of many activities external to the family, limited respect for the elderly, nuclear households, an older age at marriage, affection as a component in the mate selection process, a higher status of women, and low and controlled fertility. In Western research, these family patterns and attributes have become associated with modernity or development. Family attributes common in other parts of the world have been characterized in this model as traditional or less developed. Among the family attributes seen as traditional are the organization of activities around the family, extensive family solidarity, great respect for the elderly, large and complex households, young age at marriage, arranged marriage, a low status of women, and high and uncontrolled fertility. The frequent use of these definitions of traditional and modern families in research and public policy prompted us to investigate whether ordinary people around the world know and use similar definitions.

At the same time, we recognized that in some cases models for family and social life long held outside the West may overlap with the Western models described above as modern. Indeed, intellectual and religious elites in some settings outside the West have sought to claim as indigenous certain desirable family attributes that have been labeled by Westerners as modern by locating the guiding principles for these family attributes in historical religious texts (e.g., Hoodfar, 2008; Yount & Rashad, 2008).

Western scholars have created theories about causal connections between the northwest European family system and the northwest European social and economic systems (Thornton, 2001, 2005). Most have viewed this hypothesized causation as the effect of socioeconomic development on family change, but some have hypothesized an effect of family change on development. In our research, therefore, we aimed to investigate the extent to which individuals in everyday life believe that modern societal attributes causally affect modern familial attributes and/or that modern family attributes causally influence modern societal characteristics.

Developmental ideas provide a framework not only for how development happens, but also for evaluating the value of certain societal and familial traits. The attributes of family and society specified as modern also are perceived to be desirable in this framework; it also assumes that the good things of development are attainable through discipline and hard work. Therefore, our study sets out to evaluate the extent to which people view certain aspects of modern family and societal life as better as or worse than traditional dimensions of family and society, as well as the extent to which people view the good family and societal things as attainable.

4.3 PROJECT ORGANIZATION AND COUNTRY-SPECIFIC WORK

The organization and implementation strategies international research projects adopt depend upon many factors, including the source and flow of funds, the experience and knowledge of the research team, and the availability of collaborators with the necessary interests and research infrastructure. In the beginning we had only a very small team of researchers and very limited resources. In addition, the goal of our research was to measure knowledge of and adherence to a complex and sophisticated set of beliefs and worldviews. This is potentially quite a different undertaking from asking respondents to answer questions about behavior, attitudes, or past experience.

Several design decisions grew out of these considerations. In order to obtain wide and diverse perspectives on the use of developmental thinking among ordinary people, we decided to adopt multiple methodological approaches. We decided to begin our work by utilizing the methodological expertise already available in our research team. Starting small by necessity and beginning in individual locations with individual designs, we planned eventually to accrue the expertise and knowledge necessary for a more ambitious comparative project across countries. We began by creating measurement procedures, protocols, and questions suitable for use in one country, without any explicit concern for international comparability, but expected to be able to use what we learned and created in one country as a foundation that could be modified and elaborated in various other locations and languages. Given our considerable constraints, this incremental approach seemed the most viable. At later points in our research, we were able to include greater consideration of comparative measurement in our design and implementation strategies.

This country-specific strategy was implemented in two ways. The first was the design of individual country studies devoted to the measurement of developmental ideas and beliefs. We conducted such studies in Nepal, Argentina, Egypt, and the United States. Our second strategy was to add small sets of our developmental questions into studies conducted for other purposes. We followed this approach in Vietnam, Taiwan, Iraq, Iran, Egypt, Saudi Arabia, Albania, and China. In both approaches our goal was to collect country-specific data without trying to make the data comparable across countries. We now describe the various individual-country studies using these two approaches, beginning with our first project in Nepal.

4.3.1 Nepal

A small team of sociologists with expertise in exploratory interviewing, focus groups, and survey research initiated a small mixed-methods study in Nepal. Fieldwork began in 2003 with informal exploratory discussions with residents in Chitwan Valley. These interviews provided insight into how ordinary Nepalese think about development and the factors associated with it. Informed by these insights, we next conducted semi-structured interviews and focus group discussions about developmental thinking and family life. Finally we moved to face-to-face survey interviews with Nepali adults. For more information, see Thornton, Ghimire, and Mitchell (2005).

4.3.2 Argentina

Informed by our research in Nepal, a mixed-methods project was launched in Argentina in 2003–2004. The Argentinean team conducted focus groups among high school students in the city of Buenos Aires and in rural schools in northern Santa Fe Province. Before each focus group began, participants were asked to complete a self-administered paper-and-pencil survey questionnaire. The questionnaire also was administered to other students in the same Buenos Aires high schools and in additional rural high schools in northern Santa Fe. For more information, see Binstock and Thornton (2007).

4.3.3 Egypt

In 2006, a similarly designed study was conducted with adults in Cairo, Egypt. This time, however, the questionnaire preceding the focus groups included an explicit sequence of open-ended freelisting questions about modern/tradition family and development, which were used to explore qualitatively the elements that constituted these cultural domains locally.

4.3.4. United States

In 2006, a two-phase project was conducted in the United States, beginning with a set of cognitive (probe) interviews in Washtenaw County, Michigan. In addition, a set of development questions was fielded as a supplement to the University of Michigan Survey of Consumers—a national telephone survey of adults. In the telephone survey we were able to include experiments to evaluate the effects of question wording and question ordering.

4.3.5 Vietnam, Taiwan, Iraq, Iran, Egypt, Saudi Arabia, Albania, and China

Several opportunities presented themselves in the form of adding small sets of our developmental questions to studies conducted for other purposes. In this way we were able to add a small battery of questions about development and family life to the 2003 Vietnam Red River Delta Family Survey. A new strategy was adopted here; we randomly probed a sub-sample of respondents, asking open-ended questions about what they thought modernization was, and what the positive and negative aspects of modernization were. In 2004, we collected data from Taiwanese college students by adding a small module to a self-completion questionnaire on democratic ideas and values. This Taiwanese data collection was embedded in a panel study that resulted in us having repeated measures of the same questions from the same respondents over time. In 2004 and 2006 in Iraq, and in 2005 in Iran, we added a small battery of our development and family questions to World Values Surveys being conducted in these countries. In 2005, developmental questions were incorporated into a national United Nations Children's Fund Survey on maternal and child health in Albania. Also in 2005, surveys of young adults were conducted in six large cities, three in Egypt and three in Saudi Arabia; our questions were added to questionnaires focusing on politics, religion, and gender. Finally, in 2006 we added a modest number of items about development, family, and inequality to a Family Policy Survey conducted in several Chinese provinces.

4.4 DESIGNING COMPARATIVE INTERNATIONAL PROTOCOLS AND QUESTIONS

As we accumulated information and insights from the country-specific projects mentioned above, we turned to consider producing questions of deliberately comparative design that would be asked in different countries and languages. The idea was to design questions useful for researchers working at the national level as well as those involved in cross-national research. We would also create a questionnaire that would be asked in surveys in five countries.

An essential requirement to accomplish this goal was a team of experts in relevant disciplines who brought to the project wide cultural diversity and expertise. Our project drew together scholars in the fields of anthropology, demography, political science, psychology, and sociology. The team members also brought expertise from a range of epistemological approaches, including ethnography and survey research. Team members had knowledge and experience in countries as diverse as Argentina, Belgium, China, Egypt, Iran, Nepal, Saudi Arabia, the United States, and Vietnam. The design team included people of different nationalities with different first languages. Several members were both knowledgeable of the research concepts and goals of the project and expert in two or more languages and cultures.

In designing the questions for the comparative project, we took into account the conceptual and implementation needs in diverse countries. Instead of designing questions for one specific country and then adapting those questions to other countries, we formulated questions that would work for many countries. In this work we focused specifically on the situations of Argentina, China, Egypt, Iran, Nepal, and the United States, but also drew on knowledge the team had of other cultures and populations.

Our task of designing a questionnaire appropriate for many diverse places was facilitated both by having team members from many places and by the fact that we had previously conducted country-specific projects in many different settings. We made both question writing and translation a team project, with translation being an integral part of the questionnaire design process. The questions were formulated in English, which served as the team's lingua franca. Translation of the questions from English into another language sometimes revealed problems that could only be resolved by making changes in the English version to facilitate comparability. Since more than two languages were involved, this required multiple and coordinated iterations across the various languages. Pretesting of questions in one setting could also find problems in question wording that required changes of the questions both in that particular setting as well as in other places.

We designed our study to be administered in a five-country "pilot study." We divided the approximately 60-minute questionnaire into two parts: a part consisting of about two-thirds of the questionnaire that was common across all countries to provide the comparative needs of the project; and a part consisting of about one-third of the questionnaire to provide relevant country-specific data for each of the countries. Construction of the country-specific part of the questionnaire was delegated to the people in charge of the project in a specific country.

Between 2007 and 2008, the survey instrument was fielded in Argentina, China, Egypt, Iran, and the United States. Severe budget limitations and methodological constraints in various settings resulted in different sampling and interviewing strategies in the five countries. Except for the United States, the surveys were conducted face-to-face. The study in Argentina was conducted with a sample of adults living in urban settings throughout the country; the Chinese data collection was conducted with adults living in Gansu Province; the Egyptian data were from samples of adult women and their husbands in one district in Qaliubia Governorate and one in Fayoum Governorate; the survey in Iran was of adult women in the city of Yazd. The 2007 U.S. data collection consisted of two separate 15-minute supplements appended to the Survey of Consumers, a nationally representative monthly telephone survey of American adults.

Differences in samples and interviewing modes mean that strict comparability across settings is thus not possible with these data.[3] Nonetheless, somewhat rough comparisons of distributions across settings can still be obtained. Although these data can be used to infer to their respective sampling universes, we consider them to be pilot studies in the context of our comparative international focus.

4.5 PROBLEMS ENCOUNTERED AND LESSONS LEARNED

We now turn to a discussion of some of the problems encountered and the lessons learned in our country-specific and comparative international work.

4.5.1 Conceptual Coverage of the Concept of Development

Our studies indicate that the concept of development has been disseminated broadly around the world, is widely understood by ordinary people, and is frequently used in everyday discussions. In all the countries studied, one or more phrases could be found that are very similar to the concept of "developed" or "modern" in English.[4] We also found considerable overlap in the meaning of development in many different countries; it is, for example, generally understood that development is strongly related to socioeconomic factors. Thus in open-ended questions about the meaning of modernization, most respondents in Vietnam defined modernization in economic terms, citing advantages such as having enough to eat and a good standard of living. Structured interviews and focus groups in Egypt similarly revealed that development there meant such things as education, science and technology, a sound economy, job opportunities, and high-quality and accessible medical services.

---

[3] Although each of the five data collections interviewed adults, they used different age cut-offs for the adult population.
[4] We list here phrases used in some of the languages to denote the English concept of "developed." Nepali: "bikas," Spanish: "desarrollado," Arabic: "tanmiya" and "takadum"; and in Vietnamese, "hiện đại hóa" (referring to both modernization and development).

Our research also indicates that the definition of development used in the places we have studied is very similar to the one used by the United Nations (UN) in its Human Development Index (HDI), which is a composite of income, education, literacy, and health. We ascertained this by asking survey respondents to rate several countries on their levels of development and found, in general, that survey respondents rate countries very similarly to the UN HDI. This suggests not only that people in many places around the world have a common concept of development but that their conceptualization of it can be successfully studied in surveys.

4.5.2 Variability Across Geographical Locations

In asking respondents to rate countries on their levels of development, it became clear that respondents in different parts of the world know different countries. In our early country-specific work, we had therefore tailored the countries asked about to fit those that would reasonably be expected to be known in a given location.

We modified this strategy in our comparative project and standardized the countries rated in each of the surveys. We chose countries to be rated that we thought would be relatively well known around the world and that also represented a range of scores on the UN HDI. The selection was then pretested; among those dropped after pretesting because they were not well known in some places were Sweden, Somalia, and Zimbabwe. The final list included Japan, Nigeria, India, the United States, China, Central African Republic, France, Brazil, and Pakistan. The Central African Republic was not well known in many places, but for those who said that they did not know the Central African Republic, we instructed interviewers to tell them that it was a country in the middle of Africa, and this explanation provided most respondents with sufficient information to rate the Central African Republic.

In addition to wanting respondents in our comparative surveys to rate the nine standard countries, we wanted them to rate their own country in order to ascertain their view of their own country's standing in the developmental hierarchy. For our surveys in China and the United States, this was accomplished automatically, as each of these countries was included in the standard list. To accomplish this in Argentina, Egypt, and Iran, it was necessary to add the respondent's own country at the end of the list.

4.5.3 Concepts, Indicators, and Questions

Cultural Context and Question Meaning. Our work in Argentina taught us about the importance of cultural context and perceived question meaning for questions asking about the causes and effects of development. A number of questions that had worked well in Nepal about how development in Nepal might change family life and how particular family changes in Nepal might affect the Nepalese standard of living were originally adapted for Argentina by simply changing the country reference to Argentina. However, the responses provided in Argentina suggested that the questions were perceived differently in Argentina than in Nepal because respondents were being asked to respond about changes in two different countries with different attributes—in one case, Argentina, and in the other case, Nepal. By making the frame of reference in one country Nepal and in the other country Argentina, we had made the questions incomparable in the two settings, because development or family change might be perceived as having one effect in Nepal and another in Argentina.

We therefore adopted a vignette approach which allowed us to standardize the frame of reference for all respondents in all countries. Vignettes have been widely and successfully used in the social and health sciences. For example, Sadana, Mathers, Lopez, Murray, and Iburg (2002) gave respondents in multiple countries descriptions of people with identical health states and asked them to rate the level of health among the described people—thereby permitting the researchers to evaluate differences in rating modes across countries. In another use of vignettes, King, Murray, Salomon, and Tandon (2004) and King and Wand (2006) asked respondents to rate a series of fixed vignettes to learn about the ways in which different respondents used the response categories. Then, the researchers used this information about differential response patterns to adjust the scores that respondents gave concerning their own situations and experience. In another use of vignettes, Correll, Benard, and Paik (2007) evaluated the effects of race, gender, and parental status on people's evaluations of job applicants. They did so by giving respondents descriptions of job applicants who had identical attributes and experiences except for race, gender, and parental status which were experimentally manipulated to see their effects on respondent evaluations.

In adopting the vignette approach, we asked respondents to consider a hypothetical country that had the attributes of being low-income, agricultural, and having high mortality. Respondents were asked to give their views on (a) how development would change family life in that country and (b) how certain family changes would influence development in that same hypothetical country. This allowed us to specify the baseline of the country that would be undergoing change and permitted the standardization of the situation in the surveys in different countries. However, it also made the questions hypothetical and less concrete and shifted the focus from the respondent's own country to another situation. This may be undesirable if perceived consequences of change in the respondent's country is the focus of the researcher's interest.

Measuring Attainability. As noted earlier, our theoretical model suggested the need to ascertain whether people in different countries thought that certain socioeconomic and family achievements were, in fact, attainable. However, the concept of *attainability* proved to be a challenge to implement in the cross-national project. *Fatalism*, the idea that things are outside of people's control, is related to attainability. We decided therefore to use culturally specific questions about fatalistic ideas for some countries using location-specific language and idioms. As a result, we also moved the section on attainability versus fatalism out of the shared core section to the optional section.

Conceptualizing Religion and Religiosity. It was important to include questions on religion and religiosity in our research, since these institutions have sometimes provided alternatives to developmental models. However, it again proved to be difficult to implement appropriate questions cross-culturally. This difficulty became apparent, for example, as we dealt with the complexity of religion in China and Vietnam. Historically, there was not one definite term to describe the phenomenon labeled in the West as "religion." Religion in China is composed of a combination of three religious traditions: ancestor worship, Buddhism, and Taoism (Teiser, 1995). Each of these religious traditions has its own saints, or gods, including deceased family members, Buddha(s), and locally worshipped god(s) (Tan, 1983). Chinese religion is also more of a "diffused" religion, where religious practices are mixed with nonreligious elements (Yang, 1962; Kazuo, 1995; Tan, 1983). The introduction of socialism into China further complicated the conceptual framework, since religion was afterwards defined as the opposite of science and contradictory to socialism (Linshu, 1994). Our resolution to this issue was to make religion and religiosity a country-specific topic rather than part of the core questionnaire across all countries.

4.5.4 Knowledge of Respondents

Many of the questions we wanted to ask respondents in all phases of the project focused on worldviews and belief systems rather than the direct experiences and circumstances of the respondents. As such, they may violate the basic principle that respondents should only be asked questions that they have the direct knowledge to answer. For example, we wanted to ask questions about development levels in places respondents had never visited. We also wanted to ask respondents to compare family attributes between developed and not developed countries, although their own experiences were probably limited to a small range of countries.

Informal discussions, structured interviews, and focus groups in several countries indicated that most people in everyday life can readily express opinions about development and other countries, even if they lack first-hand knowledge about them. Most of the people we have interviewed did indeed have ideas about developmental topics and about other countries and could discuss these things in interview settings, even when their knowledge was gained from second-hand channels, such as the media, schools, and word of mouth.

The issue of respondent knowledge arose in a very concrete way when we decided to adapt a question that we had used in Nepal to use in the United States. In 2003, we asked Nepalese respondents to compare Nepal and the United States on several dimensions. Although most Nepalese had never been to the United States, they had enough knowledge or impressions of the United States to provide comparisons that were, in general, quite consistent with the known differences between the two places. In designing questions for the 2006 U.S. data collection, we wanted Americans to compare Nepal and the United States as the Nepalese had previously done. However, because Nepal is a small rather than large and powerful country, it seemed unlikely that Americans would have the same level of knowledge about Nepal as the Nepalese had about America. This motivated our research team to supply more information to American respondents to assist them in making the comparisons. We experimented by telling American respondents that Nepal was a mountainous country located in Asia between China and India. However, this strategy encouraged respondents to base their answers on the descriptions we provided rather than draw on any impressions they themselves had of Nepal. Consequently, instead of this effort helping Americans to think about Nepal as a place, it led many of them to think about Asia, China, India, or mountainous countries in general. This realization led us to ask Americans to compare the United States with India—a large and relatively well-known country with many attributes similar to those in Nepal. This strategy seemed to work well.

4.5.5 Design of Response Categories

In Nepalese pretests, questions using multiple response category formats in which the respondent was asked to choose one proved difficult to implement. Multiple probes were often required to procure the information the questions were designed to elicit, lengthened the interview, and increased respondent burden. Questions using dichotomous answer options proved easier to administer and reduced response time and effort required. This approach also made it easier to create response categories that are consistent across languages (Smith, 2003). Also, dichotomous response categories have been found to provide greater reliability than items with multiple categories (Alwin, 2007). The pros and cons of dichotomous versus multiple response categories in international surveys have been raised by others, including Smith (2003).

In our 2006 U.S. survey, we asked respondents to choose whether certain family and societal attributes were more common in developed or less developed places. While the Nepalese research had suggested a forced choice between these two categories was the preferred format, for the U.S. survey we decided to include a split-ballot experiment to assess the effect of using dichotomous and trichotomous response categories. One half of the sample was offered a middle category "about the same" and the other half was presented with a dichotomous forced choice.

Analyses of these data indicate that more respondents chose the middle category "about the same" when it was explicitly offered than volunteered it as their preferred choice when it was not explicitly offered This result is expected, since respondents in surveys in general avail themselves of the response options that are offered. At the same time, most respondents in each of the U.S. experimental samples chose one of the two opposing answer options. In addition, the different response formats in each split did not affect the *ratio* between the number of respondents choosing one of the polar categories and the number choosing the other polar category. Both the dichotomous and trichotomous versions also produced approximately the same amount of missing data. These results suggest that the dichotomous approach is acceptable in the United States and the combined findings from Nepal and the United States were taken to provide support for using dichotomous response categories in our data collections in the comparative project.

4.5.6 Dealing with Limited Variance

Our research was motivated by the desire to understand the distribution of beliefs and values concerning development and to analyze the correlations of these things with other dimensions of life. Accomplishing both goals simultaneously has proven difficult whenever elements we wish to evaluate are endorsed either by an overwhelming percentage of respondents or by only a very small percentage. For example, 91% of Nepalese said that increased use of contraceptives would make Nepal richer. In China, only 10% of respondents said that women's status will decline as countries undergo economic development. Our choice to use dichotomous answer categories may have contributed to this skewness. For the current phases of our work, the descriptive goal of estimating levels of beliefs and values has priority. However, as we move toward estimating how various factors influence developmental beliefs and values and how these in turn influence behavior, questions will need to be refined in order to ensure sufficient within-country variance.

4.6 PRELIMINARY EVIDENCE FROM COUNTRY-SPECIFIC STUDIES

How successful have the strategies described here been in measuring developmental ideas? The five-country comparative pilot studies have only recently been completed and are only partially analyzed. Thus in addressing this question, our comments are limited to data from the earlier country-specific studies, focusing in particular on data from Argentina, Egypt, Iran, Iraq, Nepal, Saudi Arabia, and the United States. However, in these instances too, analysis is not yet complete and we can make only basic conclusions about data quality. As mentioned earlier, the analyses in some cases are based on small and purposive samples and the research needs to be replicated with larger and representative samples.

We have utilized several criteria to evaluate the success of our data collections. One criterion focuses on people's ability to use and apply developmental concepts in their responses. A lack of understanding and knowledge of developmental thinking would be revealed in respondents becoming frustrated, terminating the survey early, refusing to answer questions, saying that they do not know how to answer the questions, and providing answers that do not appear to be related to the questions. We also consider comments provided by interviewers and respondents on the interviews and comments provided by focus group moderators. The amount and type of item nonresponse, including break-offs, in any given survey are examined and we check for response patterns suggestive of faulty interviewing or response style behavior such as acquiescence.

We could reasonably expect well-designed questions tailored to specific settings to be understood and answered adequately. There is strong evidence across all the country-specific studies that questions about developmental ideas were generally well understood. Despite vast cultural, economic, and geographic differences, most respondents in each of the countries were comfortable with the survey, seemed to understand the questions, and answered straightforwardly. The surveys often contained questions about knowledge and beliefs, with many questions being about relatively complex and abstract ideas. Considering that the surveys ranged from around 17 minutes (United States) to 70 minutes (Nepal) in length, it is striking that very few respondents terminated the survey early and that the few terminations encountered were for reasons unrelated to the developmental questions. For example, the three break-offs in Nepal were because the respondents could not communicate in Nepali, the national language. We can conclude that, despite their complexity, the questions on developmental ideas did not lead respondents to terminate the interview.

The relatively low level of nonresponse suggests that respondents also felt able to answer the questions. By way of illustration, item nonresponse for the questions asking respondents to relate family and societal attributes in various places rarely exceeded 3–5% of the responses, and often remained at or below 1%. The data from the self-administered questionnaires with high school students in Argentina exemplifies the lowest levels of missing data, with most of the questions having only 0.5–1.5% missing data. Iran and Iraq, however, had examples of higher levels of missing data for these questions with 8–10%. The vast majority of missing data in our studies are the result of respondents reporting that they "don't know" the answer to a question. However, surveys that probed for a response after a "don't know" answer were often able to elicit a substantive response and had substantially lower levels of missing data than surveys that did not probe after a "don't know" response.

Scales that asked respondents to rate countries on an 11-point scale on a particular characteristic such as development, income, gender equality, education, freedom, or morality had slightly higher levels of missing data. This was primarily due to the fact that respondents were often asked to rate 5–10 different countries on specific characteristics and they sometimes said they did not know a country well enough to rate it on a given characteristic. Reassuringly, respondents tended to offer "don't know" less often when rating large countries and countries well known in their region of the world.

Both cultural patterns and the use or nonuse of probes contributed to the level of missing data. As reported above, countries in which interviewers did not probe after "don't know" responses had substantially higher overall levels of "don't know" responses. Data from the 84 Egyptian respondents who participated in the Cairo area focus groups illustrate some difficulties with questions asking respondents to rate countries from 1–10 on "economic development," "education," "income/wealth," and "gender equity." The percentage of nonresponse ranged from 1% (questions on Japan and the United States) to 29% (questions on Zimbabwe). Relatively high percentages of "don't know" responses also occurred for Nigeria, Sweden, and Brazil. We did not probe "don't know" responses for these questions in this Cairo study. Other Middle Eastern country surveys (Iran, Iraq, and Saudi Arabia) also usually had higher levels of "don't know" responses than other countries in the study, such as United States, Nepal, and Argentina, even before probes. This might indicate a greater reluctance on the part of respondents in these countries to answer such questions if they do not think that they have enough information.

Probing is a good way to reduce nonresponse for such country-rating questions. The standard probe that we created for use in the surveys was "Even if you don't know exactly, about where would you put [country X]?" One concern with using this probing procedure is that the quality of the responses solicited through probing could be of lower quality than the quality of nonprobed data. Nevertheless, we found that within countries, the distribution of ratings given after a probe was relatively similar to the ratings offered without a probe. For example, Nepalese had particular difficulty rating the education and development of Somalia. Ten percent failed to provide an initial answer for Somalia, but after the probe, only 4% did not provide a response. Notably, the respondents who provided a score after the probe rated Somalia only slightly lower than those who provided a response without a probe. For the other 9 countries asked about in the Nepal data collection, only 0.9–2.2% of respondents failed to provide an estimate of development level after a follow-up probe to "don't know" responses. Distributions of the pre- and post-probe responses were also very similar in the U.S. survey.

Despite difficulties some respondents had rating some countries on development, they still, on average, rated the countries very similarly to the ratings provided by the United Nations Human Development Index (UNDP, 2003). It is notable that this seems to hold for all of the countries investigated in this project.

The quality of the responses was estimated in some countries using methods to detect response patterns likely to reflect methodological artefacts rather than substantive information. For example, in Nepal we were able to ask some questions in the opposite direction of the rest of the questions in a section to see if respondents were using the same responses to oppositely worded questions. However, most of the Nepalese answered the reverse-coded questions consistently, providing evidence of response style bias for only a small number of the respondents (Thornton, Ghimire, & Mitchell, 2005). This further suggests that the questions asked were understandable and that the answer categories provided were meaningful to the vast majority of respondents. It also suggests that acquiescence or satisficing are not the reason for the low degree of nonresponse for the various questions.

The focus groups and in-depth interviews conducted in various countries also provide evidence for data quality. Moderators, interviewers, and respondents reported high engagement and interest in the topic. Discussions in Egypt, the United States, Argentina, Nepal, and other places suggested that respondents who did not have a formal name for developmental ideas were nonetheless able to respond to questions on the topic and rarely had trouble conveying their beliefs concerning such ideas.

At the same time respondents did find some questions challenging and in fact different questions proved challenging for respondents in each country. One example is particularly informative. In our data collections in the Middle East and some other places we discovered that respondents made important distinctions between traits that they considered Western and traits that they considered modern. This distinction proved to be important because respondents' evaluations of specific traits as good or bad appeared to be based partially on whether they considered the traits to be Western or "modern." It is also possible that a respondent's intrinsic positive or negative evaluation of a trait could influence whether or not he/she considers that trait to be Western. We are currently working to understand the ways in which people from different cultures attribute various traits.

4.7 CONCLUSIONS

This chapter describes our efforts to measure the ideas and beliefs that ordinary people around the world have about social and economic development. As described earlier, socioeconomic development involves a sophisticated set of concepts, beliefs, and values that include theories about many aspects of family and societal change, along with value judgments concerning the relative merits of various dimensions of family and social life. When we began this project, we were unaware of any existing measures of these models and concepts for even one country; thus formulating and evaluating questions for an international comparative research program was a major task. Beginning with a small research team and a modest budget, we worked in one country and then another, allowing our findings and lessons learned each time to inform our next research phase. The research strategies adopted in different locations were in part determined by the expertise available locally, but we also worked incrementally using insights gained from one project to inform the next.

With a number of studies completed and interest growing in the enterprise, we turned to the task of creating a pool of questions that could be used to produce a deliberately comparative questionnaire. In doing so, we drew on the team's accumulated wisdom from experiences in a substantial number of diverse settings. We also used the multinational, multilingual, and interdisciplinary group of collaborators that had come together with rich substantive and methodological knowledge of multiple settings around the world.

This multifaceted and incremental approach to questionnaire design was necessary to accomplish our goals, and it shows many signs of having been successful. We believe that it would have been exceptionally difficult—perhaps even impossible—to have skipped the country-specific work and to have moved directly to the preparation of comparative questions for many cultures around the world on a topic for which little previous research was available. By beginning with the design of country-level projects, we learned much about the requirements for different settings. This also allowed us to establish the international network of colleagues who then formed part of the collaborative design team for the later comparative questionnaire. In looking back, we believe that our own initial lack of country-specific experience measuring developmental concepts would not have boded well for an immediate comparative design. In addition, without the network of colleagues now existing across many countries, it would not have been possible to implement or learn from the research in the ways we did.

We recommend anyone wishing to engage in a similarly complex research agenda to identify mechanisms to explore and understand research needs at numerous local levels before engaging in developing a comparative instrument. Our experience has also shown how invaluable a team of colleagues in each national location is both for research at the local level and as experts to inform and support a comparatively designed instrument.

The iterative process of design we were able to use (question design, translation, question modification) was also very useful. Nonetheless, creating comparable measures across very different societies with divergent languages is an exceptionally challenging undertaking. While many things can be measured comparably, seeking perfect equivalence is problematic (see also Chapters 2 and 3, this volume).

In addition, the practicality of project implementation and time schedules can interfere with standards of perfection. Problems are sometimes discovered in the last phases of implementation of a project in a particular country. This may require modest modifications for that country that cannot be iterated back to the other countries. This problem can, of course, be minimized by generous budgets and lengthy preparation times.

The difficulties of making questions comparable across all countries in a project, of course, increase dramatically as the number and diversity of countries increase. Although we began the comparative part of our project with the ambitious goal of designing questions for any place in the world, our attention was focused most intently on the six countries of Argentina, China, Egypt, Iran, Nepal, and the United States. Our process of working iteratively across question design, translation, and question modification worked reasonably well for the limited and specified set of cultures and languages in our research, but in projects with more diversity and more countries such an iterative approach might well be impractical and a less satisfactory approach of proceeding sequentially from one country to the next might be required (see Chapter 3, this volume). Our research was also happily not burdened by pressures to replicate existing questions from other studies, which would also have been an obstacle to utilizing an iterative approach to creating questions.

Our plans include ongoing evaluation of the data already collected. We expect that this analysis will shed important light on the areas where we have succeeded and failed in producing cross-culturally comparative measures and data. This analysis also is likely to give us guidance for improvements for future data collections. Although considerable room for improvement doubtless exists, we believe that we are currently well positioned to design larger studies to document the distribution of developmental ideas, beliefs, and values around the world, and to analyze their determinants as well as their consequences.

ACKNOWLEDGMENTS

This project has benefited from the support of many individuals and organizations. We wish to acknowledge grants from the Population Studies Center of the University of Michigan, the National Institute of Child Health and Human Development, the National Institute of Aging through the Michigan Center for the Demography of Aging, the Survey Methodology Program of the University of Michigan, The Alfred P. Sloan Foundation through the Center for Myth and Ritual in American Family Life, Emory University, the Sloan Foundation, the China Population and Development Research Center, the United States Institute of Peace, and Eastern Michigan University. We thank Judy Baughn, Jana Bruce, Brittany Chulis, Rich Curtin, Rebecca McBee, Jim Lepkowski, and the members of the 2006 Survey Practicum at the University of Michigan. Errors of omission and commission rest with the authors.

The field work described in this chapter required the time and energy of numerous people and organizations in the various countries. We especially acknowledge the work of those directing the data collection in each of the countries: Albania, Arjan Gjonca; Argentina, Georgina Binstock; China, Guangzhou Wang and Yu Xie; Egypt, Kathryn Yount, Sohair Mehanna, and Mansoor Moaddel; Iran, Jalal Abbasi-Shavazi, Abbas Askari-Nodoushan, and Mansoor Moaddel; Iraq and Saudi Arabia, Mansoor Moaddel; Nepal, Dirgha Ghimire; Taiwan, Li-Shou Yang; the United States, Arland Thornton, Linda Young-DeMarco, and Jim Lepkowski; and Vietnam, Rukmalie Jayakody.

5

Managing the Cognitive Pretesting of Multilingual Survey Instruments: A Case Study of Pretesting of the U.S. Census Bureau Bilingual Spanish/English Questionnaire

Patricia L. Goerman and Rachel A. Caspar

5.1 INTRODUCTION

With increases in globalization and migration, survey research organizations around the world are facing a growing need to collect data from respondents of different cultural and linguistic backgrounds. Historically, due to the high cost of collecting the data, small and linguistically isolated groups have been excluded from surveys altogether. Excluding these subgroups from national surveys in the United States has become less acceptable for two main reasons. There is a marked increase in interest in topics relevant to these isolated groups (such as their access to health care, household composition). In addition, there is concern about declining response rates and the associated risk of nonresponse bias.

As the number of surveys conducted in multiple languages within the United States has increased, survey organizations have begun to recognize the need to pretest different language versions of survey instruments; many have also struggled to identify best practices for this type of pretesting, particularly in the absence of large amounts of literature directly addressing this topic. Multilingual pretesting can be significantly more challenging than monolingual pretesting at nearly every step of the process. To begin with, pretesting techniques such as cognitive interviewing do not always transfer directly and work appropriately across cultural groups without modification (Goerman, 2006b; Pan, Craig, & Scollon, 2005). In addition, there are also a number of logistical issues involved in the management of this type of research. Innovative and sometimes unique decisions must be made regarding matters such as the composition of the research team, the creation of

---

interview protocols, interviewer training, respondent recruitment, and the procedures followed to summarize and analyze data and report results.

Using a bilingual pretesting project conducted in the United States as a case study, this chapter discusses important aspects to be considered in pretesting a survey translation and/or a multilingual survey instrument. Through this discussion, we point to lessons learned and areas in need of further research. While not all multilingual research projects will face identical challenges, this chapter provides a list of issues and possible solutions that can be adapted for other multicultural and multilingual research projects.

5.1.1 Background on the Bilingual Questionnaire Pretesting Project

Within the United States, the ever-expanding immigration and the dispersion of immigrants across the country have highlighted the importance of providing survey instruments in multiple languages. Results from the 2000 Census revealed that 47 million people in the United States spoke a language other than English at home. Spanish was the most commonly spoken non-English language, with 28 million speakers. A large number of Spanish speakers resided in "linguistically isolated" households in which nobody over the age of 14 spoke English "very well" (U.S. Bureau of the Census, 2006; Shin & Bruno, 2003).

In about 2003, the Census Bureau began to develop a bilingual questionnaire for use in the 2010 Census. The questionnaire was created using a "swimlane" design, in which each page contains two side-by-side columns, one in English and one in Spanish (see Figure 5.1). Displaying both the source and translated versions of an instrument side by side in the same document is a somewhat unusual design. In fact, in early research, the Census Bureau actually considered four different layout options before choosing the "swimlane" design. The options considered were (1) the "Swimlane" option, (2) an "Embedded booklet" option, which contained a Spanish translation of each question directly beneath the English version of the question, (3) a "Separate forms" option in which respondents would receive separate English and Spanish forms, and (4) a "Back-to-back" option in which the questionnaire contained English on one side of each page and Spanish on the other side. Based on the results of focus group and cognitive interviewing research (Caspar, 2003), the Census Bureau ultimately chose to use the "swimlane" format to develop the questionnaire.

Caspar (2003) focused solely on choosing the *format* of the bilingual questionnaire. The bilingual questionnaire pretesting project (BQPP) research discussed here was conducted in order to test the question wording and comprehension.

The BQPP reported here was conducted jointly by a team of U.S. Census Bureau and RTI International researchers. Using test versions of the bilingual questionnaire, the team conducted two iterative rounds of cognitive testing. The initial goal was to test and possibly improve the wording of the Spanish-language version. The source language wording (in this case, English) had been tested and finalized prior to beginning work on the Spanish translation, and further testing of the English text was not a priority (see Goerman & Caspar, 2007 for further discussion).

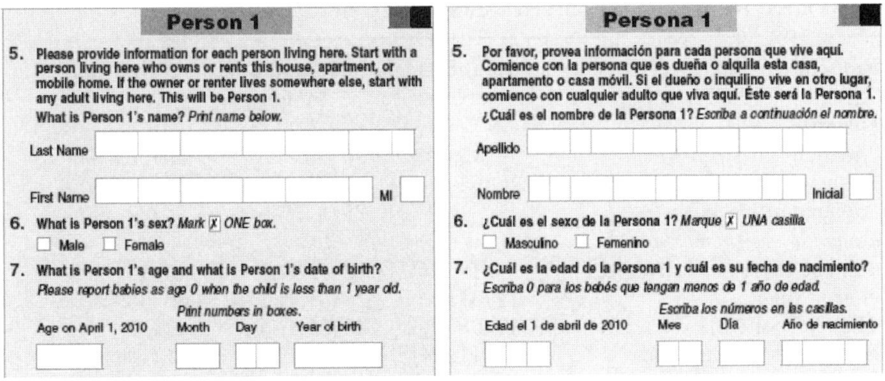

Person 1

5. Please provide information for each person living here. Start with a person living here who owns or rents this house, apartment, or mobile home. If the owner or renter lives somewhere else, start with any adult living here. This will be Person 1.
What is Person 1's name? *Print name below.*

Last Name

First Name MI

6. What is Person 1's sex? *Mark* ☒ *ONE box.*
☐ Male ☐ Female

7. What is Person 1's age and what is Person 1's date of birth?
Please report babies as age 0 when the child is less than 1 year old.
Print numbers in boxes.
Age on April 1, 2010 Month Day Year of birth

Persona 1

5. Por favor, provea información para cada persona que vive aquí. Comience con la persona que es dueña o alquila esta casa, apartamento o casa móvil. Si el dueño o inquilino vive en otro lugar, comience con cualquier adulto que viva aquí. Éste será la Persona 1.
¿Cuál es el nombre de la Persona 1? *Escriba a continuación el nombre.*

Apellido

Nombre Inicial

6. ¿Cuál es el sexo de la Persona 1? *Marque* ☒ *UNA casilla.*
☐ Masculino ☐ Femenino

7. ¿Cuál es la edad de la Persona 1 y cuál es su fecha de nacimiento?
Escriba 0 para los bebés que tengan menos de 1 año de edad.
Escriba los números en las casillas.
Edad el 1 de abril de 2010 Mes Día Año de nacimiento

Figure 5.1. Snapshot of Segment of U.S. Census Bureau "Swimlane" Questionnaire

In the first round of testing, cognitive interviews were conducted with 44 Spanish-speaking respondents who spoke little or no English. Respondents were recruited in approximately equal numbers from four different "national origin" categories: Mexico, Central and South America (Colombia, El Salvador, Guatemala, Honduras, Nicaragua, Peru, Venezuela, and Argentina), and an aggregate category comprised of Cuba, Puerto Rico, the Dominican Republic, and Spain. Respondents were recruited to reflect a variety of age and educational levels. They also varied by gender and by the amount of time they had lived in the United States.

During analysis of the first round of testing, we found that if we only looked at the Spanish wording in the form, we had no information about possible equivalency of interpretation. In addition, we lacked information about whether all types of respondents were able to successfully navigate the form by following the correct skip patterns. As a result it was sometimes difficult to make recommendations for improvements to the form as a whole (see Goerman & Caspar, 2007, for more details on shortcomings of this method).

During the second round of testing, the form was tested in its entirety, including both the original English wording and the Spanish translation. A total of 66 interviews were conducted in the second round, with approximately equal numbers of three types of respondents: (1) monolingual Spanish speakers; (2) bilingual Spanish-dominant respondents; and (3) monolingual English-speaking respondents. Our Spanish-speaking and bilingual respondents were demographically similar to those recruited for Round 1. About half of the 25 English speakers were of Hispanic origin and the other half were white and black. Table 5.1 provides select demographic characteristics of the respondents included in each round of testing.

In the second round of testing, we focused on Spanish terms that had been changed on the basis of findings from the first round of testing. In addition, we were able to examine equivalency between the two language versions and to make recommendations to improve both the original English text as well as the Spanish translation.

TABLE 5.1. Demographic Characteristics of Respondents in the BQPP

Respondent Characteristics	Round 1 (n=44)	Round 2 (n=66)	Row Total (n=110)
Linguistic Skills			
Monolingual Spanish	20 (45.5%)	20 (30.3%)	40 (36.4%)
Bilingual Spanish Dominant	24 (54.5%)	21 (31.8%)	45 (40.9%)
Monolingual English	0	25 (37.9%)	25 (22.7%)
Educational Attainment			
College or Advanced Degree	8 (18.2%)	8 (12.1%)	16 (14.5%)
Some College	14 (31.8%)	26 (39.4%)	40 (36.4%)
High School/GED	17 (38.6%)	26 (39.4%)	43 (39.1%)
Some Formal Education	5 (11.4%)	6 (9.1%)	11 (10.0%)
Age			
18 – 30	9 (20.5%)	20 (30.3%)	29 (26.4%
31 – 45	17 (38.6%)	31 (46.9%)	48 (43.6%)
46 – 65	13 (29.5%)	10 (15.2%)	23 (20.9%)
65 or older	5 (11.4%)	5 (7.6%)	10 (9.1%)
Region of Origin			
(Spanish speakers only)	(n=44)	(n=41)	(n=85)
Mexico	12 (27.3%)	17 (41.5%)	29 (34.1%)
Central America	14 (31.8%)	8 (19.5%)	22 (25.9%)
South America	10 (22.7%)	9 (21.9%)	19 (22.4%)
Cuba, PR, Domin. Rep., Spain	8 (18.2%)	7 (17.1%)	15 (17.6%)

In the remainder of this chapter we discuss our experiences with this project in contrast to monolingual pretesting projects in order to illustrate some of the issues that require consideration in a multilingual pretesting project.

5.2 THE RESEARCH TEAM

The formation of the research team is the starting point for every pretesting project, whether monolingual or multilingual. Special consideration must be given to forming a team whose skills address all aspects of the research task. With cognitive testing, the team typically includes researchers with knowledge of the substantive topic(s) covered by the survey and the specific goals of the survey. In addition, there is a need for researchers with experience in interview protocol development, respondent recruitment, conducting cognitive interviews, analysis of findings, and formulation and reporting of results and recommendations. It can be difficult to find a single researcher who can fill all of these roles, particularly when the project involves multiple languages. Forming the team for a multilingual cognitive testing project brings the added complexity of identifying staff with the required expertise not only in the areas listed above but also in both the source and target languages. Ideally all members of the team would be fully fluent in all languages included in the testing. However, this is an unrealistic requirement, particularly when multiple languages are involved.

In the United States, it is not uncommon for a monolingual researcher to serve as the lead researcher. This person provides the team with substantive expertise and/or expertise in study design and the development of cognitive interview protocols. Additional assistance from researchers in the target language(s) is then needed for protocol translation, respondent recruitment, the conduct of interviews, analysis of results, and the formulation of recommendations.

In our specific case, the BQPP team was comprised of both bilingual (English and Spanish) and monolingual (English) staff. The sponsors of the project (and ultimate users of the pretesting results) were mostly monolingual English speakers. Reporting had to take this fact into account. In the beginning, the inclusion of monolingual team members seemed to introduce additional work for the team, particularly for the first round of testing which was only conducted in Spanish. All materials (consent forms, the interview protocol, etc.) were initially prepared in English, reviewed by the team, revised, and then ultimately translated into Spanish. Had all team members been fluent in Spanish, the need for the English materials might have been unnecessary. However, in our case many of the project sponsors were monolingual English speakers and they too wanted to review some of the interview materials at this point. As a result, the English documents served an important and necessary function (see, however, Section 5.4).

As the number of languages involved in a given project grows, it is increasingly unlikely that the entire research team will be fluent in all the necessary languages. The total number of researchers needed for a given project will also vary based on whether individual researchers have expertise in multiple areas. It is nonetheless likely that a common language (a lingua franca) will be used by all members of the team to communicate with each other.

5.3 COGNITIVE INTERVIEWER CHARACTERISTICS

Interviewer effects are well-documented in the literature on standardized survey interviewing (see Cleary, Mechanic, & Weiss, 1981; Fowler & Mangione, 1986; Schnell & Kreuter, 2005; Tucker, 1983). However, the role of the interviewer in shaping a respondent's answers in a cognitive interview setting has not been similarly researched. This is likely due to the nonrandom, small sample sizes typically used for cognitive testing. It may also be due to the less structured nature of the cognitive interview and the fact that the focus of the interview is typically less on what the respondent's answer to a specific survey question is and more on issues of comprehension, terminology, recall, and item sensitivity. However, given that the dialogue between an interviewer and a respondent may be more in depth during a cognitive interview than in a standardized survey interview, it is especially important that the interviewer behave in a neutral, nonthreatening, and professional manner. The ideal demographic characteristics and behaviors of cognitive interviewers in relation to specific types of respondents is an issue in need of research. It seems likely that the interaction of traits such as gender, age, race, cultural and linguistic background, and the style of interaction on the part of both the interviewer and the respondent may have an even greater effect in the context of an in-depth, less structured exchange such as a cognitive interview than in a standardized field interview.

In all cognitive interviewing, the interviewer must be fully fluent in the respondent's preferred language in order to be able to converse easily and, perhaps more importantly, to be able to understand what may be subtle nuances in how the respondent comprehends a survey question. Native speakers or those with extensive formal study of the language that includes cultural immersion are likely to have a strong grasp of the language. Each interviewer must also be well trained in how to conduct a cognitive interview. It is often difficult to find researchers who have expertise in both survey methodology and the languages in question for a given study.

One possibility for assigning cognitive interviewers to respondents is to have them only conduct interviews with respondents who match their cultural and linguistic backgrounds. The benefit of this approach is that the opportunity for the interviewer to fully understand the respondent's comments is maximized. In addition, the respondent may open up more fully to an interviewer with the same cultural background as his/her own. At the same time, respondents sometimes feel more comfortable talking about sensitive issues with a person who is more of an outsider to his or her culture.

When testing a survey translation, it is ideal to have cognitive interviewers who are fluent in both the source and the target languages. Interviewers who understand only the target language would be unable to review the source text and might have difficulty fully understanding the intent of a specific phrase or question. They may then be at a disadvantage in determining whether a respondent's difficulties are due to a poor quality translation or to a fundamental misunderstanding of the concept of interest. If only a monolingual interviewer were available, he/she would need to work closely with bilingual team members. Given these drawbacks, we recommend that a best practice for conducting multilingual studies is to ensure all interviewers are fluent in the language of the source instrument as well as at least one additional language in which testing will be conducted.

For the BQPP, we explicitly chose interviewers who were capable of conducting cognitive interviews in both English and Spanish. Two interviewers were native Spanish speakers who had lived in the United States for many years. One interviewer was a native English speaker who learned Spanish in an academic setting and then spent time living in Spanish-speaking countries. The fourth interviewer was a native speaker of a language other than English or Spanish who had gained knowledge of both Spanish and English through academic study and cultural immersion. All four interviewers had a strong background in cognitive interviewing.

Based on debriefings with the interviewers in our project, we found that each interviewer felt she was able to effectively conduct the cognitive interviews in both Spanish and English. It was interesting to note, however, that the non-native Spanish speakers reported that being an "outsider" to the respondent's culture was particularly helpful in certain respects. Cognitive interview probes may sound very strange to respondents. For example, when a respondent hears a question such as

"What does the term 'unmarried partner' mean to you in this question?", he or she may feel that the interviewer is being disingenuous. If respondents believe that the meaning of a term is obvious, they could find this type of question insulting. If, however, the interviewer is not a member of the respondent's culture, respondents may attribute strangeness to this factor and find questions of this nature less insulting. This effect may also vary by language or cultural group. The non-native English-speaking interviewers, for example, did not report this phenomenon when interviewing English-speaking respondents in our project. This may be because the English-speaking respondents were more familiar with interviewing techniques in general and were more accustomed to taking surveys and to the types of questions presented in a survey interview setting. This might have made them less likely to find cognitive interview probes strange.

On the whole, we found it very effective to have cognitive interviewers who were fully fluent in both the source and target languages in our study; however, it should be kept in mind that this was a two-language project. It may not always be possible to find bilingual researchers in the context of multinational studies that involve large numbers of languages. In that case, one solution might be to have a small number of bilingual researchers who can report the findings back for their monolingual team members. If no bilingual researchers are available, it may also be necessary to work with monolingual interviewers through interpreters. More research is needed in terms of how to best handle this issue in the context of multilingual studies involving large numbers of languages.

5.4 INTERVIEW PROTOCOL DEVELOPMENT

One of the most important decisions that needs to be made in a cognitive interview study is which type of protocol to employ: either a completely scripted protocol, a combination protocol containing scripted probes (probes decided and formulated prior to the interview) and allowances for emergent probes (probes that become necessary during the course of the interview), or a completely open-ended protocol where interviewers create their own probes in real time. This becomes an even more important decision in the context of a multilingual study where it can be difficult to find team members who are both expert cognitive interviewers or survey methodologists and have the necessary linguistic knowledge.

A second and equally important issue in a multilingual pretesting project is that the interview protocol needs to be written in more than one language. Researchers must therefore decide whether to develop the different language versions of the protocol simultaneously so that they can inform each other or whether to develop a protocol in one language first and then translate it into the other(s). As a part of this process, the researchers must also consider how closely the different language versions of the protocol will mirror each other in terms of wording and content. In a two-language project simultaneous development may be quite feasible, especially if bilingual researchers are available throughout the development process (see, for example, Potaka & Cochrane, 2004). In a multi-language project, different developmental strategies and comparability goals will be needed if, for example, 15 protocols need to be developed simultaneously.

In the first round of testing for the BQPP, the interviewers used only a Spanish-language protocol to test the questionnaire with Spanish speakers. In the second round of testing, they used a protocol in Spanish and also one in English. For both rounds of testing, a protocol was first developed in English and then translated into Spanish. We chose to work in English first, because our project leads were native English speakers and because the English-speaking project sponsors wished to have input into the process. Budget and timing constraints led us to develop and circulate the English protocol for comments prior to having it translated. Our goal was to avoid having to translate multiple drafts of the protocol.

While there was not sufficient time or budget for simultaneous development of the two language versions of the protocol, several strategies helped us to ensure the protocols in Spanish were culturally appropriate. Because the set of problematic questions might differ by language, we first identified terms and questions in each language that we anticipated could be problematic. We then created one master list of issues to probe in both languages and included all of these in our first draft of the English protocol. Second, in drafting the English protocol, we kept in mind the need for an appropriate Spanish version and aimed for ease of translation. A bilingual project manager also reviewed drafts of the English protocol specifically keeping ease of translation into Spanish in mind.

A third strategy adopted was to aim for natural-sounding Spanish rather than insisting on close or literal translations, which might sound unnatural. We instructed our two translators to avoid having the Spanish sound like a translation. After the translators came to agreement on the translation, the Spanish-speaking researcher also reviewed it. In essence, an informal committee approach to the translation was followed (see U.S. Bureau of the Census, 2004, on the team or committee approach to translation).

Finally, and importantly, the English version of the protocol was not finalized until after it had been translated into Spanish. As a result, if during translation we felt something could be better expressed in a different way in Spanish, we were free to go back and make changes to the English version.

We also identified strengths and weaknesses to our approach. First, the development process was relatively quick in that we did not spend excessive amounts of time developing each language version. Far more time was spent reviewing and revising various drafts of the English version than of the Spanish version. However, the protocol might have been more appropriate and/or natural for Spanish-language interviews if it had been developed in Spanish rather than derived through translation. Making sure that the protocol sounds natural and covers relevant issues in all languages is of the utmost importance. Because of this, we recommend simultaneous development of different language versions of a protocol whenever feasible (on simultaneous development, see Potaka & Cochrane, 2004).

The other decision that must be made with regard to protocol development is the extent to which the protocol will be scripted as opposed to allowing for emergent probes and a more unstructured interview. In the field of cognitive interviewing there is a lack of consensus regarding the extent to which a cognitive interview should be scripted and planned in advance (CSDI, 2007). Many researchers point to the importance of giving interviewers freedom to explore unexpected issues that arise through the course of an interview. In addition, an overly scripted interview may cause the interaction to seem unnatural and/or repetitive (Goerman, 2006b). More scripted cognitive interview protocols are typically used when interviewers have less experience. At the same time, researchers must exercise caution to avoid the use of unnatural-sounding, scripted probes that are translated too literally or read as worded without any interviewer discretion.

The risk in using a completely unscripted protocol with experienced interviewers is that individual interviewers may go off in different directions, making it impossible to compare findings on a particular term, issue, or concept across cases. Since cognitive interviewing usually involves a relatively small number of interviews, it is ideal to be able to compare findings on a particular issue across all cases.

Protocol development was handled differently for each round of testing in the BQPP. In the first round of testing we developed a protocol that included the survey questions along with suggested probes to examine respondents' understanding of particular terms and questions. We planned to allow for emergent probes and also not to require the interviewers to read each probe exactly as worded with each respondent. This worked well during the course of the interviews themselves and interviewers picked up on different issues across the cases as they worked. However, in analysis we found that when we wanted to look at a particular term or question across the 44 cases, there were points on which not all interviewers had gathered the same type of information.

For our second round of cognitive testing, therefore, the protocol was similar but greater effort was made to use the same probes across cases. We again did not require the probe wording to be read exactly as scripted, but we did emphasize that we wanted the same issues to be probed across cases. We also allowed interviewers to add probes and pursue issues that came up in individual interviews. We found that this second approach made data analysis much easier.

Regarding the extent to which a protocol should be structured in a multilingual research project and the place of emergent probes, our findings suggest that it is best to have experienced interviewers in each language, because they are best equipped to conduct emergent probing. We recommend providing sample scripted probes and clear instructions for interviewers to probe on particular terms and issues that are of concern, regardless of how they actually word their probes or go about getting that information. When experienced interviewers are not available, we recommend including as much assistance as possible within the protocol (in the form of sample or scripted probes) and spending an adequate amount of time on interviewer training. The amount of time needed for training will vary according to the experience level of the interviewers in a given project.

5.5 COGNITIVE INTERVIEWER TRAINING FOR A MULTILINGUAL PROJECT

As with any cognitive interviewing project, it is important to train all interviewers in a multilingual project carefully. The training should consist of four components. First of all, novice interviewers should receive general training on cognitive interviewing methods. Secondly, all interviewers should receive project-specific training in which the protocol for the project is reviewed and details regarding correct completion of study materials (consent forms, incentive payments, methods of recording the interview, etc.) are discussed. A third segment of the training should address any issues of linguistic or cultural sensitivity that should be considered to ensure that respondents feel comfortable with the interview and that meaningful data are collected. A part of this cultural sensitivity training should also address how best to encourage specific cultural groups to respond (see Chapter 6, this volume). In the event that the lead researchers are not familiar with all cultures and languages to be included in the study, it is valuable to have cognitive interviewers spend time brainstorming and attempting to anticipate issues that might come up and possible solutions during the training. Finally, some time should be set aside for the fourth component: practice interviews in each of the languages of interest in the study. Such mock interviews, conducted with the trainer or other interviewers taking on the role of the respondent, are an excellent way to provide interviewers with hands-on experience with the protocol before they begin their interviewing assignments. It can also be useful to have new interviewers observe interviews conducted by an experienced interviewer prior to going into the field themselves. When using novice cognitive interviewers, we recommend a general training course of two to three days to be able to cover all of the basic techniques and to provide some hands-on practice.

In the case of the BQPP, all the interviewers involved had previous experience with cognitive interviewing; general cognitive interview training was therefore unnecessary. We did, however, provide each interviewer with a list of general, neutral probes that could be used to elicit additional details from respondents. Interviewers were instructed to review the list and become familiar with the probes so that they could be used in conjunction with the more specific probes included in the protocol.

The project-specific training for the BQPP involved having all the interviewers meet as a group with the protocol developers to review the goals for each round of testing. We found that the most useful protocol training method for our project was to go through each item in the protocol and discuss the goals of each question and suggested probe with the interviewers. We also found it best to let the interviewers know that while the exact probe wordings did not have to be followed, the intent of each probe did need to be reflected in the interview in order to have comparable findings about terms and questions across cases. We also instructed the interviewers to feel free to follow up on new leads or information that arose during their interviews and to inform the team about them so that other interviewers could probe on those issues as well.

All interviewer training in the BQPP was conducted in English to accommodate the monolingual team member who led the training. The interviewers kept the Spanish version of the protocol with them so that as each item was discussed in English they could also refer to the translation and raise any questions regarding the Spanish wording if needed. Because some potential Spanish language issues may have been overlooked this way, we recommend conducting interviewer training in the language in which the interview will be administered whenever possible.

With regard to issues of cultural sensitivity, one key area discussed was the possibility that undocumented immigrants might be especially concerned about how the information they provided during the interview would be used and with whom it would be shared. Interviewers were prepared for this possibility and were able to reassure and gain the participation of most of the small number of reluctant respondents they encountered.

Since all of our interviewers were experienced at conducting cognitive interviews in Spanish, we did not spend time talking about cultural issues related to probe wording or about conducting the interviews in Spanish. Had we started with inexperienced cognitive interviewers, we would have gone through the wording of the different probes in Spanish and talked to interviewers about ways in which we had experienced respondents to have difficulty with particular probing techniques. In addition we would have offered strategies to resolve those difficulties. If the lead researchers do not have experience with a particular language or culture, we would advise talking to interviewers about problems that interviewers anticipate and allowing them time during the training to come up with strategies that they might employ if they encounter difficulties.

In sum, we recommend three days of training for interviewers new to cognitive interviewing to learn general cognitive interviewing techniques, including time for practice through simulated interviews. Project-specific training can usually be completed in about a day, depending on the length of the interview protocol. An additional session of 1–2 hours should be devoted to cultural or linguistic issues specific to a given culture, and at least half a day should be devoted to practice interviews including feedback on how the interviewers are doing.

5.6 RESPONDENT RECRUITMENT

Cognitive interviewing is generally conducted with small numbers of respondents and the intent is not to create a representative sample. Nevertheless, respondent recruitment is an extremely important component of a cognitive interviewing project. Since only a small number of respondents are involved, it is especially important to verify that each individual accurately represents the population of interest. One challenge for recruitment is identifying efficient ways of screening for eligible participants without revealing the criteria that define eligibility. Since cognitive interview respondents are often paid to participate, masking the eligibility criteria is important to ensure that individuals are not selected who are interested in receiving the monetary reward but do not actually meet the criteria. This can be a complicated undertaking, especially if recruitment is by word-of-mouth and one participant may pass information on to others regarding how to be selected.

With multilingual projects it is necessary to have a recruiting team that includes linguistic and cultural competence for each language involved, so that someone is able to interact appropriately with each prospective respondent. In addition, when one of the primary goals of a project is to assess difficulties respondents have with a translated questionnaire, it becomes especially important that the respondents recruited are monolinguals, the people most likely to be interviewed with the translated instrument in the field. It is also essential to test a translation with monolingual respondents, since bilingual respondents are often able to compensate for a poor translation by bringing their knowledge of the source language to bear on their understanding of the translated questions. If respondents are bilingual the researchers may have a more difficult time identifying problematic translations through the cognitive interview results.

A particular challenge for the BQPP was developing a method for identifying eligible monolingual English speakers, monolingual Spanish speakers, and bilingual, Spanish-dominant speakers. We chose to interview these three types of respondents because these were the types of people who would receive the bilingual questionnaire as a part of the 2010 Census. In our bilingual category we chose to seek out Spanish-dominant bilinguals because the Spanish version of the form had previously received less testing and we wanted to focus more attention on the Spanish wording.

With interviewing taking place in four sites across the United States, the project team relied on staff familiar with each site to determine where eligible participants could be found. Nonetheless, it was still important to have an effective means of verifying the appropriate language group classification for each participant. During a telephone screening, potential respondents were asked whether they spoke English as well as Spanish and if so in which language they were most comfortable communicating. Once language proficiency was determined, additional questions were asked to classify the participant with regard to dimensions such as educational attainment, country of origin, gender, age, tenure in the United States, and household size in order to maximize the diversity of the group of cognitive interviewing respondents.

Our approach worked reasonably well for identifying monolingual English speakers. It worked less well with Spanish-speaking respondents. On several occasions, a participant who self-identified during screening as a monolingual Spanish speaker was found to speak reasonably good English upon arrival at the interview. These people would have been more appropriately assigned to the bilingual Spanish-dominant group. In a few cases participants clearly reported less proficiency in English because they were unsure what the interview would require and did not want to put themselves in the position of having to communicate in English if the task was too complex. A very small number of participants appeared to have "gamed" the screening process in order to qualify for the study. The overall result of these misclassifications was that more bilingual respondents and fewer monolingual Spanish speakers participated in the study than originally planned.

On the whole, we recommend that careful attention be paid to identifying the types of respondents best suited to test a particular survey instrument. When testing a translation, it is extremely important to interview monolingual respondents so that they will not be able to bring knowledge of the source language to bear on their interpretation of the translation, possibly masking problems with the translation. To this end, staff assigned to recruit participants must be carefully trained to assess language proficiency.

5.7 ANALYZING AND REPORTING OF RESULTS

In a cognitive interview project, it is common practice to write a summary of each individual interview. To analyze the results, researchers often use these summaries to make comparisons across cases. When a project involves data collected in more than one language, a number of decisions need to be made. Will the interview summaries all be written in a common language or will they be written in the languages in which the interviews were conducted? Having interviews written and analyzed in the interview language(s) provides a complete record of the terminology used by interviewers and respondents. In addition, not all terms or concepts will be directly translatable across cultures or languages, so it is best to examine each issue in the language in which it originates whenever possible. But findings may then need to be considered and reported in the project lingua franca, too, and this involves decisions that need to be made. When summaries need to be in a common language, will they be written first in the language in which they were conducted and then translated, or will they be written directly into the lingua franca? Many of the decisions will depend on the language skills of the research team and the intended audience(s) for the results.

For the BQPP, the interview summaries ultimately needed to be written in English because English-speaking researchers and project sponsors needed to be able to read them. In the first round of testing, summaries of the Spanish interviews were written directly in English rather than written first in Spanish and then translated into English, due to time and resource limitations. Each interviewer listened to her interview tapes in Spanish and summarized the interviews in English. This proved to be a reasonably efficient method, but some problems were encountered along the way, which we describe below.

The first summaries that were produced contained very little Spanish, and this reduced their utility because the words actually used by the respondents were missing. During the cognitive interviews, respondents were asked how they would express a concept in their own words, particularly in situations where they did not understand a term on the questionnaire. When these discussions were translated into English in the summaries, the alternative terms in Spanish recommended by respondents were lost.

In the second round of testing, we again needed all summaries to be in English. However this time the interviewers making the summaries were instructed to include key terms, phrases, and respondent quotes in Spanish. In order to permit non-Spanish speakers to read and analyze the summaries, they were also asked to provide an English translation for any Spanish text that they included. Interviewers were instructed to include all key Spanish terms respondents used that were different from those used in the questionnaire, and to include any quotes, stories, or explanations that might be helpful. This second method worked well and had the added benefit that interviewers did not need to review the audiotapes again during data analysis or report writing.

Once the interview summaries were written, analysis across cases could begin. Because there were four interviewers, we needed to develop a way to synthesize our results. We also needed to accommodate the fact that some results were in English and some were in Spanish. A coding scheme was developed to allow interviewers to compare results across cases. Each interview was coded for things such as the demographic characteristics of the respondent, whether he/she had interpreted a given term or question as intended, what type of misinterpretation had occurred (if any), along with any new terms the respondent had offered as alternative wording. Tracking these new terms, in both English and Spanish, enabled us to examine easily the frequency of particular problems across languages and other demographic characteristics of respondents. Especially in situations that require the completion of a large number of cognitive interviews, we recommend the development of some type of coding scheme.

The summary of results from our testing included in the final report contained many recommendations for alternative wording in Spanish. The term "foster child," which proved problematic in our cognitive testing in Spanish, provides a good illustration of the challenges and decision-making processes necessary in this type of project. In the United States, foster care is a system in which the local government places a child under age 18 in a household to receive parental-like care. The foster parents receive financial support from the state to help with the expenses of caring for the child. Foster children may live in a household for a brief time or for several years. In some cases the children are eventually returned to their biological parents or previous guardians and in other instances they are ultimately adopted.

The Spanish translation for "foster child" initially used in Census Bureau materials was "hijo de crianza." Testing revealed that almost none of our Spanish-speaking respondents associated this term with the U.S. government Foster Care Program. In fact, in Round 2, when we had 41 Spanish-speaking respondents from various countries, only three of them understood the term "hijo de crianza" to refer to a child placed with a family through an official government program. Puerto Rico does have a Foster Care type of program, but the three Puerto Rican respondents in Round 2 all interpreted the term "hijo de crianza" in an unintended manner. Two of the three Spanish speakers who interpreted the term as intended were from Mexico and the third was from Chile. Two of them said that they had heard people talking about the program, on television or through acquaintances who had participated in the program. One of them was bilingual and offered the English term "foster" herself.

The term "hijo de crianza" in fact has a salient meaning in Spanish, namely "child by upbringing or care," as opposed to a child related to someone by birth. However, this usage of the term is unrelated to any government-sponsored program. On the whole, we found that respondents who were from other countries and were unfamiliar with U.S. government programs had no idea that "hijo de crianza" was meant to refer to an official Foster Care Program. Many respondents related it to the situation of caring for the child of a relative, such as a sibling or cousin. The care situations they discussed were informal, to help out temporarily when a relative was experiencing hard times, or in the case of immigrants, often to supervise or care for the child of a relative who was still in the home country.

In reporting these results to the monolingual English-speaking project sponsors, we explained the meaning of this term and the contexts in which it is normally used in Spanish. We recommended that the translation be changed to refer explicitly to the government program in question: "Hijo de crianza del programa Foster del gobierno" (child by upbringing through the government Foster program).[2]

Ideally, interview summaries and reports should be written in the language in which the interviews were conducted. However, it is often necessary to provide all results in one common language in order to ensure that all relevant parties, such as monolingual project sponsors, have access to the reported information. If all project sponsors do not share a common language, it may be necessary to translate reports into one or more languages. In this type of case, it is also advisable to include terms and phrases in the other languages tested. Terms that have proven to be problematic in each translated version should be presented verbatim and then explained in the main report language. The over-arching goal is to report results in a way that will allow all relevant parties to participate in the decision-making process with regard to the implementation of results and recommendations.

As the BQPP data were collected in both Spanish and English but most of the project sponsors were monolingual English speakers, all results were reported in English. On the whole, in reporting results on Spanish terms, we found it necessary to include (a) the Spanish term, (b) a literal translation of its semantic meaning, along with its usage or cultural significance, (c) an explanation of how and why respondents were interpreting it in a certain way, (d) a recommendation for alternative wording, (e) an English translation of that new wording, and finally (f) an explanation of why we thought that the new wording would work better. Such information enables project sponsors, survey methodologists, and linguists to work together in deciding which recommendations to implement after pretesting research is completed.

In the case of multinational projects with sponsors from different countries, it will be necessary to think about the target audience(s) for research reports and the accompanying decisions that are required. It may be that reports need to contain multiple languages or entire reports might need to be translated or adapted into different languages. If the main components listed above are maintained in the different language versions, project sponsors should be able to work together to make decisions based on research results and recommendations.

5.8 AREAS FOR FUTURE RESEARCH

This chapter has highlighted a number of areas where further research is desirable. First of all, a great deal of literature exists on interviewer effects; however, very little has been undertaken specifically related to interviewer effects in the cognitive interview setting. This issue is of particular interest in a multilingual cognitive testing project where the same researchers may conduct interviews in more than one language and with respondents with varying degrees of acculturation. Some important research questions are: What are the ideal demographic and social characteristics of an interviewer in relation to specific types of respondents? What different kinds of techniques are successful when an

---

[2] Ultimately, space constraints led the project sponsor to decide to include the original translation, with the English term in parenthesis: "Hijo de crianza (Foster)."

interviewer belongs to the respondent's cultural group and what techniques are useful when the respondent and interviewer are from different groups? Are different social status issues at play between people from the same or different cultural groups? How might any of this affect interviewer/respondent interaction?

A second issue is the extent to which cognitive testing protocols should be scripted for use in multicultural and/or multilingual pretesting projects. Empirical research should be conducted that identifies and evaluates the types of results obtained from fully scripted protocols versus more open-ended protocols (cf. Edgar & Downey, 2008). This is of particular interest when a cognitive interview protocol has been created in one language and translated into another. It may be difficult to create culturally appropriate probes through literal translation. Similarly, open-ended protocols may allow for a more culturally appropriate cognitive interview that would uncover more issues specific to a given culture. However, Pan and colleagues (Chapter 6, this volume) find that open-ended protocols do not work well for every cultural group.

A final area for future research relates to the comparison of cognitive interview findings and data across languages in terms of strategies, tools and comparability challenges. The more languages involved in testing and reporting, the more pressing the need for careful procedures and budgeting of resources. Additional research could uncover new ways in which to analyze, compare, and report on multilingual data. This will be particularly relevant to large multinational studies in which data is being collected across languages and cultures (see, for example, Miller et al., 2008).

6

Cognitive Interviewing in Non-English Languages: A Cross-Cultural Perspective

Yuling Pan, Ashley Landreth, Hyunjoo Park,
Marjorie Hinsdale-Shouse, and Alisú Schoua-Glusberg

6.1 INTRODUCTION

One of the pretesting methods used in survey research is cognitive interviewing. This is an in-depth interview procedure in which researchers "study the manner in which targeted audiences understand, mentally process, and respond to the materials we present—with a special emphasis on breakdowns in this process" (Willis, 2005, p. 3). Cognitive interviewing typically uses probing questions to tap into respondents' thinking processes and their response formulation processes in response to survey questions.

The usual procedure in U.S. survey research to conduct cognitive interviews in languages other than English is to develop cognitive interview probes in English and then have these translated into the target languages. In doing so, translators are requested to follow the English original wording and structure, within the constraints of the target language. This practice has met with some criticism. Previous studies (e.g., Pan, 2004, 2008; Pan et al., 2005; Goerman, 2006b) have documented some difficulties encountered in administering cognitive interview probes translated from English into Chinese and Spanish, and have observed some puzzling phenomena. Recent studies found that if Chinese-speaking and Spanish-speaking respondents were asked to paraphrase questions, they tended to repeat questions verbatim, almost as if they did not understand the concept of paraphrase (Coronado & Earle, 2002; Pan, 2004).

To date there has been little research that (a) systematically examines how cognitive interview probes perform across language groups and (b) explores how effective translated probes are for generating informative data for cross-cultural research. This chapter aims to contribute to the knowledge gap in this research.

---

[1] Survey Methods in Multinational, Multiregional, and Multicultural Contexts, edited by Harkness et al.
Copyright © 2010 John Wiley & Sons, Inc.

The study reported here focused on seven cognitive interviewing techniques in four non-English languages (Chinese, Korean, Russian, and Spanish). Taking English data as a baseline for comparison, we examine respondents' linguistic behavior in cognitive interviews, investigating their use of certain words, expressions, and the syntactic structures favored in their responses. The specific research goals of this study were as follows: (1) to investigate how cognitive interviewing techniques that are routinely used in the U.S. English context perform in languages other than English; (2) to examine how effective translated probes are in gathering data from non-English-speaking groups; (3) to investigate cultural and linguistic barriers for interviews conducted according to (noncomparative) U.S. standards for non-English cognitive respondents; and (4) to identify potential strategies to address any barriers found.

6.2 COMMUNICATION NORMS FOR COGNITIVE INTERVIEWING

The Cognitive Aspects of Survey Methodology movement, sometimes called CASM (cf. Jabine, Straf, Tanur, & Tourangeau, 1984) introduced cognitive interviewing techniques to survey research. The fundamentals of these procedures are as follows: Depending on the specific issues under investigation, survey researchers prepare an interview protocol containing scripted probe questions. The probe questions are used to ask respondents to provide detailed information on how they understand survey questions, recall information to provide an answer, decide on the relevance of the answer, and formulate answers (cf. DeMaio & Rothgeb, 1996). The aim of the interview is to identify problems in the questionnaire on the basis of information respondents consciously or unconsciously provide (their verbal answers, their body language, and such details as hesitation). It is therefore critical for the outcome of a cognitive interview that the respondent is capable of articulating his/her thoughts, opinions, and feelings during the interview.

Some typical examples of U.S. probing questions are: *What do you think they meant by this question/statement?* (an example of a meaning-oriented probe); *Please tell me in your own words what this question is asking* (an example of a paraphrasing probe); and *How did you arrive at that answer?* (an example of a process-oriented probe).

Each of these probing questions is respondent-centered ("you"). They each request the respondent to participate in self-reporting and expect the respondent to articulate his/her thinking processes and to provide feedback on the issue. Syntactically, such questions are direct and straightforward. They are a key component of cognitive interviewing. Given that cognitive interviewing is a speech event that uses language to achieve a purpose, understanding the impact of a speech community's preferred communication norms in such a speech event is critical. In order to examine how cognitive interviewing techniques work in U.S. survey research, we need to examine some underlying assumptions for language use in the cultural context of American English.

Hall's concepts (Hall, 1959, 1976) of high and low context communication is helpful here for us to explore differences in communication styles among the five language groups studied. Hall used the term high-context cultures to refer to those he identified as relying heavily on contextual elements, background information, and interpersonal relationships for communication, and correspondingly relying less on the words or texts themselves. In low-context cultures, communication focuses more on the factual and informational aspects of an exchange. Meaning is determined then more by the factual content of the message and less on when, how, and by whom it is expressed.

Although Hall's concept of high- and low-context communication has its limitations, such as not exploring contextual or situational differences, it has proved influential and useful for cross-cultural comparisons of communication styles. As Shaules (2007) notes: "it acted as a criterion by which communication styles or cultural groups could be compared" (p. 28). For example, in terms of these distinctions, Japanese communication has been described as high context in tendency and Anglo-American communication patterns as tending more toward low context (Hall & Hall, 1987). In Japanese communication, therefore, a preference for an indirect style is reported, while a more direct and explicit style is ascribed to U.S. discourse (Hall & Hall, 1987). In comparing linguistic directness among cultures based on Hall's ideas of high- and low-context communication, Storti (1999) finds that cultures can be placed on a continuum of directness and indirectness with regard to expressing opinions, as indicated in Figure 6.1.

Storti (1999) grouped Asian cultures toward the indirectness end of the continuum and placed the "American culture" toward the directness end, while Russian and Spanish cultures were located somewhere in between in terms of directness in expressing opinions.

Cultural factors, including cultural value systems and social circumstances of personal experience, have been recognized as strong influences on survey quality and survey participation. However, much of the discussion of cultural factors is confined to the immediate interaction surrounding question-and-answer exchanges in a survey questionnaire. For example, Johnson et al. (1997) investigated how cul-

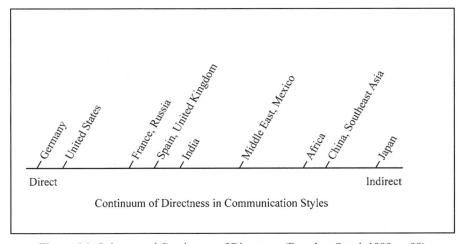

Figure 6.1. Cultures and Continuum of Directness (Based on Storti, 1999, p. 99)

tural norms, values, and experiences influence the processing of the four tasks of the response process, which are: question comprehension, retrieval of relevant information, use of that information to make required judgments, and selecting and reporting an answer (Tourangeau et al., 2000). Other studies specifically look at how cultural context affects question interpretation (e.g., Chapters 10 and 11, this volume; Braun, 2003; Schwarz, 2003a).

There is, however, little literature on how cultural differences in communication styles affect respondent behavior in either a survey field interview or a cognitive interview setting. In particular we lack empirical data on the possible tension between the Anglo-Saxon model of communication which underlies survey interviewing and the communication styles of other cultures. Although survey interviews are not everyday communication in American English, the design of survey interviews is based upon the preferred norms of communication in American English, which aims for clarity-brevity-sincerity or C-B-S style in professional communication (cf. Lanham, 1974; Scollon & Scollon, 2001). The research reported here uses empirical data on cultural differences to develop an adaptive approach to cognitive interviewing.

Hall's concepts of high-context and low-context communication prompted us to ask whether cognitive interviewing techniques, as developed in English in America, can be successfully transferred to other languages through a "direct translation" process. (As used here, direct translation refers to a word-for-word translation approach.) We also postulated that differences in communication styles could hinder data collection. For example, if the questions we ask our respondents to answer violate their communicative norms, can they and will they answer our questions as we intend? Will their responses to such questions give us the pretesting data we need to assess the adequacy of the questionnaires?

6.3 DATA FOR THE STUDY

Our research is based on a multilingual cognitive testing project undertaken at the U.S. Census Bureau to pretest translations of the advance letters and informational brochures for the American Community Survey (ACS). The translated ACS materials tested include: (a) an Introduction Letter that introduces the survey and includes important informed consent information; (b) a Thank You letter for respondents who completed an interview; (c) a short informational brochure; and (d) a more detailed brochure with Frequently Asked Questions about the survey. These documents were developed in English and translated into Chinese, Korean, Russian, and Spanish in order to better address the increased diversity of the U.S. population and the accompanying need to conduct surveys in languages other than English.

The cognitive testing project was designed to assess the adequacy and appropriateness of the translated versions of the ACS survey documents and, as relevant, to identify changes needed. A total of 113 respondents were selected for the project, comprised of 24 monolingual speakers each for Chinese, Korean, and Spanish groups, 25 monolingual speakers of Russian, and 16 monolingual English speakers. Table 6.1 summarizes respondent characteristics by each language group.

TABLE 6.1. Respondent Characteristics by Language Group

Characteristic	English	Chinese	Korean	Spanish	Russian
Educational attainment					
Less than high school graduate	2	16	4	18	1
High school Graduate, less than college graduate	10	4	9	6	12
College graduate	4	4	11	0	12
Year of Entry					
Since 2000	N/A	8	7	13	9
1990–1999	N/A	12	8	4	15
1980–1989	N/A	4	5	6	0
Before 1980	N/A	0	4	1	1
Gender					
Male	6	9	9	12	11
Female	10	15	15	12	14
Age					
18–24	0	1	1	6	0
25–34	3	0	3	5	1
35–44	5	5	5	8	4
45–54	5	5	5	2	3
55–64	3	8	4	2	3
65+	0	5	6	1	14
Total number of respondents	16	24	24	24	25[a]

[a] The Russian group conducted one additional interview above the goal of 24 total.

The respondents' demographic distribution mirrored those derived from the most recent ACS data at the time. Demographic characteristics included educational attainment, place of birth, year of entry to the United States, gender, age, country of origin for Spanish speakers, and also dialect for Chinese speakers. All respondents were monolingual speakers of the target languages. Technically, this assigns them to "linguistically isolated subpopulations" within the United States (Bureau of the U.S. Census, 2007a, 2007b, 2007c). Bilinguals were excluded from this study since the materials were primarily intended for such monolingual speakers. The study required specific demographic targets for each language group, and thus the demographics for each language groups were unique to that group For example, the Chinese- and Spanish-speaking respondents tended to have less than a high school education, while the Korean- and Russian-speaking respondents tended to have completed a college education. The Spanish-speaking respondents were predominantly younger and more recent immigrants (arrived in the United States in 2000 or later); the other language groups were older, most also having arrived in the United States prior to 1999. These differences match characteristics of actual respondents in the ACS who requested translated materials, since we aimed to recruit respondents who matched the demographics of people that, based on the American Community Survey data, could be expected to need translated materials. As a result, the groups recruited could provide good information for within-culture understanding but could not permit good comparisons across the groups. We recognize this may be a limitation of our study.

Bilingual researchers in the four language teams conducted the cognitive interviews. Each language team consisted of three bilingual researchers: one survey methodologist leading the team, and two language experts. Language team members were selected based on their language proficiency (native speaker's proficiency), cultural knowledge (education and work experience in the target culture), and translation experience. Based on these qualifications, the three bilingual researchers worked as a team to translate the interview protocol, analyze cognitive testing results, and recommend alternative wordings for the translation in each target language.

The cognitive interview protocol was developed in English and then translated into the four target languages, following the Modified Committee Approach[2] (Schoua-Glusberg, 1992). Thus, we were able to administer the interview protocol across all 113 cognitive interviews in five languages. A set of questions probing on the same concepts and messages was asked to speakers of five languages. In translation, the teams phrased the probes in ways their prior experience (both as native speakers and as cognitive interviewers) suggested was most appropriate for each specific language. Administering a protocol that probed on the same concepts and message gave us a rare opportunity to compare and contrast respondents' linguistic behaviors across language groups. Since the aim of the current study was to examine how cognitive interviewing techniques work in non-English languages, our analysis focused on the Spanish, Chinese, Korean, and Russian interviews.

6.4 ANALYTICAL FRAMEWORK AND METHODOLOGY

The study presented here examined cognitive interviewing techniques in four languages from a cross-cultural perspective. As explained below, we adopted an ethnography of communication framework for our discussion of respondents' linguistic behaviors and applied techniques developed in field of discourse analysis to analyze our data.

---

[2]A team of three language experts worked independently for each language, each translator translating one third of the protocol. After translations, team members met to review each translated item. Each translator contributed to the discussion with the aim of improving and refining the first translation, making sure that it reflected the intent of the English original probing questions and flowed well in the target language. The intent of the English probing questions was specified by research specifications set out in the cognitive testing research plan (see Pan et al., 2006).

6.4.1 An Ethnography of Communication Framework

Ethnography of communication (Saville-Troike, 1989) research draws on the anthropological field of ethnography in analyzing discourse. Within this field, a communicative event is defined as a bounded entity with the purpose of communicating a message. As understood in the ethnography of communication, a communicative event consists of the following salient components:

- Scene (setting, topic, purpose, genre)
- Key (tone or mood)
- Participants (who they are, roles they take)
- Message form (speaking, writing, other media)
- Message content (what is communicated)
- Act sequence (order of communicative acts)
- Rules for interaction and norms of interpretation (common knowledge, cultural knowledge, shared understandings)

Given that it uses language to obtain information from respondents, cognitive interviewing can usefully be viewed as a communicative event. Cognitive interviewing communicative events involve a particular social setting (research lab, respondent's home, or some other place), participants (interviewers and respondents), a message form (mode of interview), message content (subject matter of the interview), act sequence (question-answer), and a shared understanding of roles and interaction norms in a cognitive interview.

Each culture has its own set of norms and conventions that influence individual behaviors. When respondents participate in a survey interaction, they bring to the event their cultural knowledge, perceptions, and behavior norms about how to act. Communication can go awry if participants do not share the same perceptions, or if they have different expectations regarding individual components of the interview and their functions.

An ethnography of communication framework helps us to explore these characteristics. In the discussion that follows we focus on three components of this framework: sequence, participants, and rules of interaction. Each is an essential part of a cognitive interview act.

6.4.2 Discourse Analysis Techniques

We used analytical techniques developed in the field of discourse analysis to examine the data collected. Although there are different views of discourse analysis, it is basically concerned with language at the level of text and language in use (Paltridge, 2006). Discourse analysis (DA) considers how people manage interactions with each other, how people use language to achieve communicative goals, and how people communicate within and among groups.

DA has been successfully used as an analytical tool for studying verbal interchange of ideas in various social settings. Many of these have centered on "interviews" understood as question and answering exchanges, including courtroom interrogations, medical counseling sessions, educational environments, and gate-keeping situations. A variety of issues relevant for the success or failure of inter-action and communication have been identified, such as mismatch of communica-tion styles in gate-keeping interviews (Gumperz, 1982), use and abuse of language evidence in the courtroom (Shuy, 1993), and gender differences in communication preferences in medical interviews (e.g., Wake, 2006; Seale et al., 2008).

Of particular interest for our purposes is the use of DA techniques to describe the characteristics of interactions in interview settings and to reveal covert social or cultural issues, such as power struggles, social inequality, or miscommunication due to differences in cultural values and beliefs. Examining courtroom discourse in Australia, Eades (2006), for example, identified "gratuitous concurrence" as a feature of the Australian Aboriginal English communication style. "Gratuitous concurrence" is characterized by a speaker's tendency to say "yes" in answer to questions, regardless of whether or not he/she actually agrees with the proposition of a question. In survey terms, this comes very close to acquiescence response behavior (cf. Krosnick et al., 1996b). Eades (2006) reports that this behavior in the courtroom context resulted in inappropriate court judgments.

We used discourse analysis techniques, such as examining the relevance and informativeness of responses to probing questions, to identify linguistic character-istics of respondents' contributions to the interview in the four language groups. Our goal was to examine how cultural knowledge and aspects of interaction might influence their answers to survey questions and their behavior in a cognitive interview setting. As stated, we also wish to evaluate the effectiveness of specific interview probes for each language group.

6.4.3 Methods and Procedures

Our procedures consisted of two main components. One was an analysis of inter-view transcripts and the second was debriefing with cognitive interviewers in the target languages to obtain their input on cultural appropriateness and effectiveness of cognitive interviewing techniques.

In the first phase, all the cognitive interviews were transcribed using broad tran-scription conventions (Ochs, 1979). In analyzing the interview transcripts, three specific strategies were followed to improve comparability of the data. These strate-gies were then employed to analyze the transcripts across the four language groups.

First, seven types of specified cognitive interviewing tasks were analyzed for all interviews to look for patterns in respondents' responses, then four interviews were randomly selected and analyzed in detail. This analysis is time-consuming and labor-intensive work. An additional four interviews were then randomly selected and analyzed to see if the same features occurred as already found in the first four interviews. By sampling and reviewing these further interviews, we were able to gain more evidence to better warrant generalizations on the basis of our initial findings.

In the second phase of analysis, we investigated cultural norms of communication that govern an interview setting or question-answer sequence. We designed debriefing questions to obtain feedback from the language experts in our four language teams, each of whom were native speakers of the respective target languages. Interviewers were debriefed concerning their experience in conducting the interviews and their impressions of respondents' reactions to being interviewed. The debriefing sessions focused on topics that would elicit their input on cultural views regarding norms of speaking, norms of being polite, social distance between strangers, preferred style for communication, and power structure in interpersonal as well as institutional discourse. Qualitative methods such as these are extremely useful for identifying phenomena that have yet to be fully explored; they allow exploration into aspects and patterns that are less understood, and they help lay the groundwork for future quantitative studies.

6.5 SUMMARY OF FINDINGS

This section reports the main findings from the discourse analysis of interview transcripts and debriefing carried out with the interviewers about how speakers of the four target languages behaved in the cognitive interviews. The findings include information on the effectiveness of probing questions and linguistic behaviors of respondents across the four language groups. "Linguistic behaviors" here refers to how respondents answered a particular probing question, how they provided personal opinions, and how they defended their position in the interview process.

6.5.1 Question-and-Answer Sequence: Responses to Probing Questions Across Languages

From the perspective of the ethnography of communication, if participants of a communication event have a mutual understanding of what is expected in each turn of the exchange, communication will go smoothly. When participants do not share the same expectations, the communicative event may not have the expected outcome. To analyze the question-answer sequence in our cognitive interviews from this perspective, we focused on the effectiveness of the probing questions in generating the information sought and what responses the four language groups gave to these probing questions.

Effectiveness refers to whether a probing question can generate the right information to satisfy the research objectives of cognitive interviewing. We first examined seven types of probing questions across the language groups. Table 6.2 shows the example probes used in the interview protocol and the overall effectiveness by language group.

Based on the specific interviews analyzed, we rated how effective the types of probes were. Each type of probe was categorized as being "effective" (the probe generated the right information and the right amount of information), "somewhat effective" (the probe generated some information, but not sufficient detail, or partial information), or as "not effective" (the probe did not produce any relevant information).

Among the seven specific probes analyzed, only two types of probes were categorized consistently across the four language groups: the comparison probe and the paraphrasing probe. The comparison probe was both effective and easy for

TABLE 6.2. Effectiveness of Types of Probes by Language Group

Type of Probe	Example Probe from Interview Protocols	Effectiveness in Language Groups			
		Chinese	Korean	Russian	Spanish
Paraphrase probe	What do you think they are saying in this paragraph?	○	○	○	○
Interpretation of terms	What does the term "confidential" mean to you in this sentence?	●	●	●	○
Evaluative probe	Do the words, terms, and ideas used in this letter sound right or appropriate in your language or to your culture? Are there any that do not seem right or appropriate?	•	•	○	○
Preference probe	Was there anything you liked about the letter?	○	•	•	○
Sensitivity probe	Was there anything that caused you concern?	•	•	●	●
Comparison probe	Do you think there is any real difference in the message among all three versions? If so, what do you think the difference is?	●	●	●	●
Hypothetical probe	Let's pretend you were selected to participate in this survey and an interviewer handed you these materials. After you read them, do you think you would agree to participate in the survey? What things did you think about while making your decision just now?	○	○	●	●

[●] Effective ; [○] Somewhat effective; [•] Not effective

the participants to answer. However, despite the overall effectiveness, there were still subgroups among some of the languages that had difficulty articulating the reason for selecting the option they chose. This meant that the interviewer had to probe further to determine the justification for the selection. For the paraphrasing probe, respondents were asked to explain a short paragraph (two to three sentences) in their own words. Across language groups, respondents tended to repeat the original message or to comment on only one aspect of the message, rather than demonstrate that they fully understood the meaning of the paragraph.

As evidenced in Table 6.2, five other types of probes (interpretation of terms, evaluative, preference, sensitivity, and hypothetical probes) were effective for some language groups but not for all. With six out of seven probes being evaluated with similar effectiveness ratings, the Chinese and Korean respondents tended to react in a similar manner to the different types of probes. Likewise, with five out of seven probes having the same effectiveness, the Russian and Spanish respondents tended to demonstrate a similar pattern in response to the categories of probes. In contrast, the number of effective types of probes in common between the Chinese or Korean respondents, compared with the Russian or Spanish respondents, was only two or three types of probes, including the probes that worked equally well for all groups. Three probes aimed to elicit opinions and reactions from respondents on issues under discussion (evaluative, sensitivity, and hypothetical) all worked better for the Russian and Spanish groups than for the Chinese and Korean groups. The Chinese and Korean groups showed exactly the same pattern on these three types of probes; these probes were not a very effective means to elicit comments from the Chinese and Korean respondents.

Hall's distinction of high- and low-context communication may help explain the findings summarized in Table 6.2. The Chinese and Korean groups tended more toward high-context and indirect communication, and the Russian and Spanish groups tended more toward low-context and direct communication. As a result, probes that aim to elicit direct responses and opinion-oriented comments may not work well for Chinese and Korean groups, but could function relatively well for Russian and Spanish groups. The following sections look at the linguistic features of this communication style in more detail.

6.5.2 Common Findings for Chinese and Korean Groups

Discourse analysis of linguistic features of Chinese and Korean respondents in our data set shows that the Chinese and Korean respondents in our study shared some common behaviors when presented with a set of direct and focused questions. These behaviors can be summarized as (1) limited response, (2) avoidance of personal view, (3) ambiguous and vague answers, and (4) answering without confidence. Each of the examples presented in the following begins after the respondent has read the ACS introductory letter.

Limited Response. A striking feature of the Chinese and Korean respondents' linguistic behavior was that their responses to probing questions were very limited in length and scope, often consisting of one word, or a few words or one phrase. The respondents did not elaborate on their responses, nor did they provide detailed or specific reasons to explain their points of view. The following example from the Chinese interview data illustrates this point.

Example 1. Chinese interview[3]
ID#14, female; age: 18–24; college graduate; year of entry to the US: after 2000
Dialect preference: Mandarin

1. INT: OK, let me take this (letter) back. What was this letter about?
2. R: American population survey, right?
3. INT: OK, I'll give the letter back to you. Was there anything in this letter that you liked?
4. R: Do you mean its content or …?
5. INT: Anything, be it the content or the form. Was there anything that was appealing? Was there anything that left you a deep impression, or do you think it's meaningful, or well written? You can comment on anything.
6. R: Maybe this paragraph, this second paragraph. It's about the survey content, help something. And the last point, (one) can go online.
7. INT: If you were selected to participate in the American Community Survey, what parts in this letter made you want to participate in this survey?
8. R: It will help our community, it will improve our life environment, right?

This example demonstrates some characteristics of a high-context communication style. In it, the respondent used a short phrase "American population survey" to answer the first probing question, "What was this letter about?" (Turn 1). She also added a question word "right" to qualify her answer. In Turn 4, when being asked what she liked about the letter, the respondent asked for more clarification. The interviewer provided a series of probes to meet the respondent's request for clarification (Turn 5). In Turn 6, the respondent indicated parts in the letter that contained information that she liked, briefly suggesting what she liked about each. For example, referring to the second paragraph, she said "it's about the survey content, help something." She did not give a detailed explanation as to why she liked this paragraph and didn't specify what she thought the survey content was about. She used a word corresponding to "something," indicating vagueness in her response. She also indicates that she liked the last point (of the letter)—the fact that one can go online to check more information about the results of the ACS. But she does not directly state this; it can only be inferred from her contribution in context.

In contrast, let us now consider an example from an English interview.

Example 2. English interview
ID#1, male, age: 35–44, some college, black.

1. INT: Okay, alright. Was there anything that you liked about the letter?
2. R: It seemed pretty simple, I mean, basically telling me what's going to happen. Basically telling you something as far as trying to get some data to decide where these new schools, hospitals, as I said, emergency things are needed. But it seemed basically pretty simple. (Short pause) I don't

---

[3] Due to space constraints, examples in languages other than English are presented in English without including the original transcriptions.

know... I mean, I know they say you are required by the law to respond to the survey, but I don't know if this down here is Title 13 Section 9 of what they are showing up here...

3. INT: [Unintelligible]

4. R: As far as this title, or this code, I don't see it unless this is it right here

5. INT: Oh, okay, okay, okay. You don't see the (...) of what the title says. Is that it?

6. R: Let me see, let me see... Okay, so this is telling you what that Title is as far as them doing the Census Bureau, but I don't... (pause) I guess this is the Title here. The section, the code. Section 141, 193, 221. I wouldn't know what it is; you know what I am saying?

7. INT: Okay.

8. R: I guess the only thing I would be wondering is...What codes are those? But other than that, it seems pretty simple.

Three features differ markedly here from the Chinese example just considered. First, the English-speaking respondent was concrete and detailed in reporting his thought processes. He used the first-person pronoun "I" and verbs that indicate thought processes, such as "I mean," "I guess," "I don't know," "what I'm saying," and "I would be wondering." These linguistic terms report some of his (perceived) mental processing. The first-person pronoun indicates his position as an individual and identifies that he was stating *his* opinions and reporting *his* position in regard to the letter that he had reviewed. The verbs he used describe his mental state as to how much he understood of the letter.

Second, this respondent not only answered the probing, but also volunteered some details to back up his reply. For example, in Turn 4, he gave a direct answer to what he liked about the letter—the letter seemed pretty simple. He went on to describe what he meant by that ("basically telling you something as far as trying to get some data to decide where these new schools, hospitals, as I said, emergency things are needed"). He then talked about the part of the letter he had trouble with, the section discussing legal requirements and the citation of laws.

Third, he provided some evaluative comments that showed his reaction to and interpretation of the materials under review. For example, in Turn 2, he said that the letter "seemed pretty simple" and he repeated that statement twice in the same turn ("it seemed basically pretty simple") and in Turn 8 ("But other than that, it seems pretty simple"). The discourse markers in his response such as "I mean," and "you know what I'm saying," indicate his perception of the interview as an informal event, in which he could express his views freely.

These two examples are illustrative of how our Chinese and English-speaking respondents answer questions in an interview. Data analysis shows that the English-speaking respondents tended to focus on the task at hand and to talk at length about the issues under discussion. In considerable contrast, the Chinese-speaking respondents did not provide elaborate responses to the questions asked.

Avoidance of Personal View (Community-Based Argumentation). Chinese- and Korean-speaking respondents tended to avoid providing personal views or clearly stating their positions relating to issues under discussion. The Chinese group often reported their opinions in the voice of "vague other"—someone not specified by the respondent. The Korean group tended to provide group views rather than personal views in a manner similar to that of the Chinese respondents. When asked about their opinions or comments concerning the materials under review, both populations often shifted away from the use of the singular first person pronoun (I) to the plural form (we). In discourse analytical terms, they used a community-based argumentation style to present their views by making statements such as: "We think…," "We Chinese residents have….," or "Chinese/Korean people do not say this in this way." They also tended to repeat the words or phrases provided in the survey materials that they had just read or from the cognitive interviewer's questions.

The following example is from the Korean interview data.

Example 3. Korean interview[4]
[The interviewer asked a scripted probe to determine what, if anything caused concern. The Korean translation of the word "confidentiality" was the same as the Korean word for "secrecy." The discourse follows:]

1. INT: Was there any part of the brochure that caused you to become concerned?
2. R: This "secrecy" or this sort of thing should be left out. In other words, even though *American people* may think this ("secrecy"/confidentiality) is important, *Korean people*, may not read this carefully and if (they are) skimming through it quickly, it (the statement about secrecy/confidentiality) would arouse negative feelings.
3. INT: Okay. Do you think there is too much talk about "secrecy"/confidentiality?
4. R: There is nothing like this in Korean society in general.

This scripted probe was written to elicit respondents' personal reactions to the materials under review by emphasizing "you" in the question. The respondent's answers, however, are not framed as a personal view at all. Instead, he used generic noun phrases, "American people," "Korean people," and "Korean society" to contrast between two perspectives without involving reference to his own view.

Ambiguous and Vague Answers. Chinese-speaking respondents tended not to provide an overt or focused answer to a question. Their immediate responses to a question may have no direct relationship to the question. For example, one probing question asked for the paraphrase of a statement on the survey data uses ("The American Community Survey produces critical up-to-date information that is used to meet the needs of communities across the United States. What does this statement mean to you here?"). When asked what the statement meant, one

---

[4] Due to space constraints, examples in languages other than English are presented in English without including the original transcriptions.

respondent spoke of the highway noise in her neighborhood and the need to build a wall along the highway, an illustration of a generally noticeable tendency to provide nonrelevant information in the place of the requested direct response.

The term "pragmatic ambiguity" is used here to refer to a phenomenon in which the respondent provided a response that can be interpreted in many ways. Such responses did not contain the information or comments that the question was trying to elicit. Vagueness and ambiguity are typical of responses from the Chinese group. The following illustrates this well:

Example 4. Chinese interview[5]
ID#10, female, Age: 45–54, Less than high school graduate, Year of entry to US: 1990–1999, Dialect preference: Shanghai

1. INT: Were there any parts in the brochure that were particularly good, that made you want to participate?
2. R: (It) doesn't matter.
3. INT: What do you think of the design and the layout of the brochure? For example, do you think the cover page looks good? How about the inside cover, and the back cover?
4. R: (It) doesn't matter.
5. INT: So, if a census representative gave you these materials, after you read them, would you agree to participate?
6. R: (It) doesn't matter to me, so long as it doesn't involve [inaudible] ....

The respondent's response to the interviewer's three questions is the same short phrase "(it) doesn't matter." This phrase serves two pragmatic functions. First, it formally satisfies the interviewer's request for an answer, and, second, it enables the respondent to avoid providing a direct answer to the question. The phrase can be interpreted in many ways; it thus provides a response, but not an answer.

Huddleston and Pullum (2005) discuss the distinction between responses and answers (see also Sun, 2008). A response is an utterance produced "as a result of being asked some question." It is defined by its position in the question-answer sequence, but not defined by its semantic content. An answer, in comparison, provides the specific information in light of the question asked" (p. 162). In other words, responses "respond" to questions by filling the answering slot in the question-answer adjacency pair,[6] but they may well not actually answer what the questions are asking.

---

[5] Due to space constraints, examples in languages other than English are presented in English without including the original transcriptions.

[6] An adjacency pair (cf. Schegloff, 1972) is a unit of conversation consisting of one turn each by two speakers. The turns are functionally related to each other in that the first turn requires a certain type or range of types of second turn. Question-and-answer sequences are typical examples of adjacency pairs.

Responses instead of answers are obstacles to the main purpose of cognitive interviews which is to collect answers containing substantially relevant information for analysis.

Answering Without Confidence. A tendency could be noticed in the Korean group to respond to the questions with questions or make statements that implied a lack of confidence when answering the questions. While the interviewers uniformly provided explicit statements that there were no right or wrong answers, responses from Korean respondents were suggestive of a lack of confidence. This was especially the case for older, female, and lower-educated respondents. They often said "I don't know" or "I don't understand this." Even those relatively well educated (high school graduate or more) tended to give their responses by saying "It seems/ sounds like..." or "I'm not sure about it but I guess...," showing lack of confidence in their comments. Some of the Korean-speaking respondents tried to elicit confirmation whether their answers were correct after each question. Others did not directly ask for confirmation at each question, but asked the interviewer at various points during the interview whether they were answering correctly in general. This could show Korean cultures which highly value modesty, or could be "satisficing"[7] tendency for the Korean respondents in cognitive interviews in this study.

Analysis of the transcripts revealed that the Chinese and Korean groups shared a number of linguistic features in their response patterns. When compared to the data for the Russian and Spanish groups, it seems that the probing questions based on translations from English constituted greater communicative challenges for the Chinese and Korean groups. For these two language groups, the probing questions did not perform as well.

Summary of Debriefing Sessions with Chinese and Korean Cognitive Interviewers.
In summarizing debriefing discussions with the Chinese and Korean cognitive interviewers from an ethnography of communication perspective, several striking features emerged. First is the perception of the cognitive interviewing as a communicative event. Most of our respondents in these language groups completely lacked survey or interview experience. All the Chinese- and Korean-speaking respondents reported they had not participated in a survey interview or ever filled in a survey questionnaire in their home country. In the United States, they tended to be excluded from ongoing surveys or interviews because they could not be interviewed in languages they knew.

This lack of survey or interview experience meant the Chinese and Korean respondents were not familiar with the exchange expectations and "rules" of such

---

[7] Satisficing refers to a tendency of respondents to engage in the answering activity with sufficient attention to provide a nominally plausible answer to the given question but not to expend any further effort (which might be needed to fully process and respond optimally). See Tourangeau et al. (2000) and Krosnick et al. (1996b) for detailed discussion.

communicative events. For them, from their perception of the interaction, other frames of interaction may have been salient. By frame of interaction, we mean the frame of reference and behavior based on cultural knowledge and expectations about people, objects, events, settings, and ways to interact that influence language use in terms of comprehension and language production (Tannen & Wallat, 1993). Differences in frames of interaction pose a challenge to cognitive interviewing because respondents seemed to be quite confused by the event of the interview and the documents tested in the cognitive interview. They tended to use their usual indirect communication style in providing responses. The consequences for cognitive survey researchers are responses or data that may deviate from what the study is designed to investigate.

The second issue is the perception of the participant relationship in a communicative event. During the debriefing discussion, the Chinese cognitive interviewers commented on how Chinese tended to keep their personal opinions to the inner circle, not to reveal much about their personal view to the outside circle. According to Li (2003), the 40-year-long political oppression in contemporary China under the Communist rule led to the development of two distinct psychological spheres and two separate zones of expression in the minds of Chinese people. One is the public sphere for talking in public and one is the private sphere for talking with one's friends. When people are in their public persona, they do not reveal their own opinions, they only talk "standard political talk." They reveal more of their true "self" through the private persona. In the context of a survey interview, therefore, it is essential to determine whether people are responding to survey questions using "standard political talk" or according to their own views. Li (2003) also suggests it is very often difficult to make that determination.

This helps to explain why the Chinese-speaking respondents might come across as very passive or evasive in the cognitive interviews. Many of our Chinese-speaking respondents expressed surprise that the U.S. government wanted to listen to citizens' voices (a very different practice from the Chinese government). Some of them felt happy after the interview but were uncertain during it. They commented on this and often compared their cognitive interview experience with their experiences in China or their expectation of the Chinese government. They said that things like a government agency interviewing common citizens for feedback would never happen in China.

The third issue is the dimension of interpersonal relationships. Cross-cultural studies (e.g., Shaules, 2007) maintain that East Asian cultures tend to place a high value on maintaining harmony in social relationships. The concept of "self" is defined as fundamentally related to others and the self is identified by human relationships and social roles (Schwarz, 2003b). This is reflected and communicated in the frequent use of "we" rather "I" among the Chinese and Korean respondents.

In addition, we found that the Korean respondents tended to place a high value on politeness; one of the standard ways to show politeness is to be modest and acquiesce to others' opinions. This is particularly true for older generations. During the interviews with the low-educated older women, we often observed respondents tended to agree with the interviewer irrespective of the content of the question (acquiescence bias). One of the main challenges in such interviews was to elicit answers from the respondents who almost always responded "yes" or "don't know" to questions.

6.5.3 Common Findings for Russian and Spanish Groups

Both the Russian and the Spanish groups in our data set shared more features in their response patterns with the English group than the Chinese and Korean groups did when compared to the English. This agrees with the ratings of effectiveness of probing questions across language groups presented in Table 6.2 and follows the theory of cultures and continuum of directness presented in Figure 6.1. In other words, for language groups that are closer to English in communication styles (i.e., Russian and Spanish), the translated questions and probes performed better in terms of effectiveness. There were, however, two marked differences in response patterns for both the Russian and the Spanish groups, as discussed below.

Providing an Opinion Even If Uncertain: Russian Group. In contrast to Korean and Chinese participants, the Russian group seemed confident, readily providing their personal views, usually at some length. At the same time, they often provided vague or unrelated answers or answered in ways that clearly indicated that they had misunderstood either the materials or the probes. Analysis revealed a frequent tendency to answer with the same words as used in the materials, without providing the requested response rephrased "in their own words." This general tendency could be interpreted as a strategy to avoid admitting that they did not understand a given statement.

Example 5. Russian Interview[8]
ID#5, Female, Age: 65+, High School Graduate, Year of Entry to US: 1990–1999

1. INT: Look, please, at the sentence: "You are required by U.S. law to respond to this survey (Title 13, United States Code, such and such Sections)." What do you think is the meaning of this statement. Again, in your own words, please.
2. R: I think that they must clarify what I need to pay attention to, what remarks I must make, what I'm satisfied and dissatisfied with. And, in general, I think these clauses must clarify what I must do and what is required of me.

The probing question requested the Russian respondent's own paraphrase in order to assess whether she had understood the message. The respondent gave a relatively lengthy response to the question but focused on what "they" (the U.S. government) should do or what they should clarify (Turn 2), not how she understood the statement. It is possible that she did not understand the statement. At the same, she did not state that she had trouble understanding it.

---

[8] Due to space constraints, examples in languages other than English are presented in English without including the original transcriptions.

Providing an Opinion Even If Uncertain: Spanish Group. Of all the groups, the Spanish-speaking cognitive interview respondents behaved most similarly to English-speaking respondents. They usually elaborated on responses, focused on the questions, and tried to provide answers to the specific questions asked. Although at times they moved from "I" to "we" in giving their answers, they only did so when they wanted to speak for the Latino community as a whole, not to deflect focus from themselves as individuals.

However, one question reveals a clear-cut difference between them and the English language respondents. The probe on the ACS introductory letter that asked "What is this letter about?" elicited responses from Spanish respondents that overwhelmingly focused on mentioning the uses of the ACS data that were mentioned in the letter. This is, in fact, the question in which the Spanish respondents differed most from the English respondents in this project. The English-speaking respondents recognized the intention of the question was to elicit their opinions on how the letter was formulated. As a result, they commented on what they liked or disliked about the letter. For example, when asked the question, "Can you tell me what the letter is generally about?" one Spanish-speaking respondent did not respond to the question and focused instead on the ACS uses of the data. She responded that "It talks about … they will do the Census, they will take random samples to ask or find out about community needs, whether it is because with this information they want to know if there is a need for more schools, medical aspects, transportation. Basically to see the needs of the community." This tendency to focus on the uses of the ACS data was found among Spanish-speaking respondents irrespective of level of education.

Summary of Debriefing Discussions with Russian and Spanish Interviewers. Debriefing discussions with the Russian interviewers revealed that Russian respondents did not behave deferentially toward the interviewers and were not shy or quiet about expressing their views and opinions. The Russian respondents were openly critical about what they felt was good or bad in the documents and were not reluctant to offer personal opinions. Interestingly, they seemed to expect their suggestions could directly lead to changes in the materials. According to the interviewers, in most situations, social position or wealth does not influence expression of opinion in the former Soviet Union. This might differ in the authority contexts they felt; people would be less open with superiors at work and with government officials. In other contexts, people would only shy away from being candid when doing so could hurt their own position.

Another unique feature of the Russian respondents for this study is that they tended to have been in the United States longer than respondents in the other groups and to be older than other groups. The impact of this difference on their acculturation is not known. At the same time these were monolingual Russian speakers and this would hinder acculturation, since they could not speak or understand English and they tended to associate with other speakers of Russian. Because they had come to the United States later in life, they were not, or were only briefly part, of the U.S. workforce.

Most Russian immigrants we interviewed had not lived in their country of origin in the last two decades. This population was generally comfortable expressing personal opinions or viewpoints in the interviews. However, as our team of Russian language experts reported that in the former Soviet Union, people usually disclosed true opinions around trusted people, very close friends, or family, and never to government officials or their representatives.

Different nationalities in Latin America exhibit different styles with regard to expressing their views and opinions freely. As a result, for what we call here the Spanish group, it is difficult to characterize the perceived participant relationship and rules of interaction. In the debriefing discussion, the Spanish interviewers indicated that depending on their nation of origin and the circumstances in their home country when they left, respondents could be more or less reluctant to give opinions about politics or economics openly.

The Spanish interviewers also felt that if respondents perceived social distance with an interviewer (higher status or more power for the interviewer), this could have an impact on open communication, as the respondent could feel intimidated. However, the interviewers did not feel this was the case in the cognitive interviews they conducted. In addition, interviewers reported that age will make a difference in the perceived participant relationship and interaction. They reported that the young and educated are more open with each other than they are with the older generations. Among lower education immigrants, a wife often asks her husband before responding, even in younger couples, and particularly among immigrants from rural areas. The interviewers also felt that the younger generations may be less intimidated about giving opinions, especially after being in the United States some time.

6.6 DISCUSSION

This study shows that when the same set of cognitive interviewing techniques or questions was administered to four language groups, there were remarkable differences in the ways members of these language groups provided responses. Through discourse analysis and debriefing sessions, we identified some patterns of linguistic behaviors for respondents of these language groups. For example, the Chinese and Korean groups tended to provide limited responses and their answers were not focused on topic, while the Russian group showed a tendency to give "confident" answers which nonetheless might or might not be pertinent. These different response patterns reflect preferred styles of communication and cultural norms of interaction in a respondent's culture. Respondents' linguistic behaviors are a reflection of how history, social practices, and cultural norms are crystallized in language use.

Effective cognitive interviewing depends, to a large extent, on the respondent's cooperation and active participation. It is quite possible that the respondents from the language groups investigated here all intended to be cooperative, but the patterns in their answers and responses are very different from the expected norms of English cognitive interview behavior. Such behavior may be perceived as either uncooperative or evasive within the context of western cultural norms and expectations.

Our results also raise questions about the data quality collected through cognitive interviewing and the comparability of data across language groups. Differential linguistic behaviors across language groups make it a challenge to draw comparisons on the dimension of how respondents perceive questions and materials presented to them in a cognitive interview. We are not sure if probing questions serve their intended function, including if they effectively elicit information. It is therefore important to have a good understanding of the linguistic behaviors non-English-speaking respondents in an interview setting and the sociocultural paradigms that govern these behaviors. Based on this understanding, we can find ways to design interview questions that can be effective and appropriate in eliciting the desired information.

Findings from this research provide first insights into which cultural and linguistic issues are worthy of attention prior to and during cognitive interviewing, and what techniques may be applied to address these issues. For cultural barriers of communication style, we need to consider culturally appropriate ways to engage non-English-speaking respondents in an interview, including practice sessions, proper introduction and grounding of the task, explicit explanations and instructions, adjusting interviewer's communication style based on the target population, and providing feedback in an appropriate way to encourage respondents.

For linguistic issues, we need to pay closer attention to the types of questions asked in an interview and carefully craft questions to offset perceived potential problems. The current research documents and sheds some light on the nature of (problematic) responses generated by cognitive interviewing. Further research is needed to investigate response behavior more and to test various approaches to solving these issues with alternative types of probing. For now, however, we offer these a priori suggestions for developing alternative probes:

- Provide follow-up probes to reduce limited response or unrelated answers.
- Craft probes that focus on the issues under discussion. Remove the focus on the respondent (e.g., What part of the letter left you a deep impression? What part of the letter does not sound natural to Chinese people?).
- Move away from the focus on "you" as an individual when interviewing Asian language groups; instead, focus on "you" as a group (e.g., Korean people, Chinese people like you, or everybody) to make the interview less intimating.
- Develop more specific questions asking certain concepts/terms rather than general questions and prepare follow-up probes for anticipated ambiguous answers.
- Narrow the scope and focus or start with specifics to reduce vagueness or evasiveness.
- Provide a set of alternative probes on one task so that if one probe fails, the interviewer could try the next one.
- Avoid "yes/no" questions to reduce the tendency of short and brief answers. Instead, craft questions that can induce an answer rich in details.
- For paraphrasing probes, focus on the message such as "what's the intention of this statement?" instead of the direct questions such as "what

does this statement mean to you?" This will reduce the respondents' tendency to borrow words or text from materials under review and could be more likely to elicit their own thinking.

- Avoid asking for likes/dislikes on (say) government documents. Instead, ask what is good/positive and bad/negative in a letter or brochure.

6.7 IMPLICATIONS AND FUTURE STUDY

The examination of cross-cultural and cross-language cognitive interviews through the lenses of discourse analysis and the ethnography of communication focuses on detailed analysis of issues being studied. Other pretesting methods in multilingual studies, such as behavior coding, can be informed by our findings about communicative styles and cultural norms. For instance, in our transcript analysis we looked at the extent to which respondents across different cultures challenge interviewers or their questions. We identified patterns of differences across language groups. This has strong implications for behavior coding because many behavior-coding studies assume that respondents behave in the same way when a question is problematic, regardless of the language of the interview.

Our study is limited in scope and in magnitude; we have only touched upon one type of questionnaire pretesting method. In order to validate findings from this study or to apply the same approach to other language groups, we need to consider combining this approach with other methods. Once the patterns of the effect of communication styles and cultural norms on cognitive interviewing are identified and explained through the discourse analysis process, a coding scheme for quantitative analysis can be developed to code the linguistic characteristics described by discourse analysis. Quantitative analysis, then, can provide evidence on the magnitude of issues we reported in this chapter.

Additionally, because our study sample was a purposive sample, it did not allow us to investigate the impact of social factors such as education and acculturation level. These are doubtless important. Future study is needed to look into the relationship between the respondents' social and demographic characteristics and the quality of their responses in structured survey interviews.

The ultimate goal of research along these lines is to exploit empirical evidence to improve cognitive interviewing in cross-cultural studies. One essential research question is whether we need to argue for a more universal design of cognitive interview protocols or for culture-specific tailoring probes and techniques. Our study is limited in the data available and it is too early to suggest any conclusive approach. But we believe this is area that deserves attention and needs further study. We close with our first recommendations for multicultural cognitive interviewing studies.

1. Conduct an expert review of the survey questions and materials to be pre-tested to identify issues, and pinpoint potential problems, which includes cultural experts' input on the appropriateness of how to ask questions in a target language.

2. Design a cross-cultural interview introduction that takes into consideration sociolinguistic conventions of question-answer sequences as a speech event. Make necessary adjustments on how to introduce the topic, what to include in the introduction, and how to explain the purpose of the interview.
3. Develop a universal interview protocol as a general guide. Then develop probing questions in the target language instead of translating from "source probes."
4. Use tailored probes and techniques for each language to address culture-specific issues, provided these serve the intended purpose and achieve the objectives of the cognitive interview.

PART III

TRANSLATION, ADAPTATION, AND

ASSESSMENT

7

Translation, Adaptation, and Design

Janet A. Harkness, Ana Villar, and Brad Edwards

7.1. INTRODUCTION

The opening chapter in this volume discusses the extent to which deficits in practice, awareness, and theoretical underpinnings can be major obstacles to making progress in developing suitable methods and benchmarks for comparative survey research. Such deficits have certainly created obstacles to improving translation practice and translation outputs in survey research.

This is not because researchers have considered translation unimportant. In decades of writing about cross-national and multinational research, language and translation issues are almost inevitably presented as serious challenges. In addition, a quite large body of literature on the validity and reliability of individual translated instruments references instances and sometimes procedures of survey translation. These publications occasionally provide general recommendations on translation strategies or on translation assessment procedures. More commonly, they cite other authors as justification for the procedures followed to produce and test the translated instruments whose validity and reliability they examine (cf. Harkness, Villar, Kephart et al., 2009a, 2009b). A prime example of such citation practice is found in articles that cite publications by Richard Brislin (1970, 1976, 1986) to justify using back-translation procedures.

A successful survey translation is expected to do all of the following: keep the content of the questions semantically similar; within the bounds of the target language, keep the question format similar; retain measurement properties, including the range of response options offered; and maintain the same stimulus. Such matters as burden and form of disclosure are also meant to be kept constant. The question design stage determines whether most of these have any chance of being realized in translation (see Chapter 3, this volume). Even with an appropriate source design, however, this is a fairly tall order for translation. For example, survey translation generally aims to render the semantic content of a question in

---

one language in another language, and this goal may stand in conflict with some other expectations for translation, such as maintaining the stimulus.

An informed understanding of what translation involves is the basis for setting translation goals and specifications. Any translation quality framework and any assessment of translations in terms of this framework beg the question of a theory of translation and expectations derived from this theory for a given translation (cf. Hönig, 1997).

Nonetheless, survey translation is often undertaken and discussed without a strong understanding of either the principles of translation or of current theories of meaning. As a result, there continues to be a disjoint between theories, practice, and benchmarks acknowledged in the admittedly diverse translation sciences and the various approaches taken to translations and to assessment of translation in survey research. "Established" translation practice in survey research, in the sense of what is commonly done, is, we argue, by no means good practice.

One of the goals of this chapter is to make clear what may happen in survey translations. Another is to describe current best practice in survey translation in a manner that explains the motivations and strengths of certain strategies against the drawbacks of others. Further goals are to point to areas in which we believe improvements can be made and where changed tactics and policies are required. Although the chapter presents theory-based as well as practical perspectives, we aim to be as nontechnical as possible. For all aspects of this chapter and some considered only briefly or not at all (such as translator training or language harmonization), readers may also wish to consult the translation modules in the Cross-Cultural Survey Guidelines at http://ccsg.isr.umich.edu/. Harkness was the lead author of these.

7.1.1. Implicit Assumptions About Translation and Languages

All translation guidelines, expectations, assessments, and translation outputs are based on explicit or implicit assumptions about the goals and potential of translation. In survey translation, both translation and assessment are commonly shaped by implicit assumptions about language, translation, and quality. An example can illustrate this quickly. In a survey manual laudable in that it identifies the translation and adaptation tasks for a project and what preparation is involved, we find translation assessment "guidelines" such as the following:

- "Translated text should have the same register (language level, degree of formality) as the source text;
- Translated text should neither clarify nor omit text from the source text, nor add additional information;
- Translated text should contain equivalent qualifiers and modifiers, in the order appropriate for the target language" (AIR, 2002, p. 16).

Each of these statements assumes quite complex things about language which we will not elaborate here. They also reflect implicit assumptions about the character of translated instruments; we explore only a few of these in brief.

The first statement assumes, for example, that register can be matched across source and target language and, moreover, that register should be matched. Populations with lower literacy levels might, however, require simpler vocabulary than in the source questionnaire. The second statement assumes it is possible not to omit source "text" and not to add additional "information," and that this is indeed the course to follow. Again, leaving the complex implications about language and translation aside, we merely note that it seems feasible that one population might need an explanation that another would not. The third statement assumes that "equivalent" is a meaningful term to use with reference to quantifiers and modifiers and, in addition, that there will be quantifiers and modifiers in target languages that meet the intended criteria in relation to source language components. Researchers in both translatology and in survey-based research disagree on the suitability and scope of "equivalence" vocabulary and frameworks. Kenny's (2009) brief review of uses in translation studies makes the diversity of meanings associated with the term clear. Herdman, Fox-Rushby, and Badía (1998) and Johnson (1998) demonstrate the diversity of uses of the term in respect to survey research. We opt not to use it here, preferring in different contexts the terms "appropriate" and "adequate" with reference to translations.

7.2. SURVEY TRANSLATION: FORMS AND TERMINOLOGY

Correctly or not—and we emphasize that discussion of this needs to intensify in the social and behavioral sciences—survey translations are largely expected to stay close to the source text, as the "guidelines" just discussed reflect. This section therefore aims to illustrate different forms of "staying close" to a source text and to identify terms sometimes used for these forms in survey research. Hopefully, this will provide more insight into options and results than currently available.

A number of these approaches (such as word-for-word translation as used here) are usually unsuitable for questionnaire translations; researchers sometimes mention them in the sense of disclaimers—to indicate what they are *not* aiming to do. At the same time, we note that most survey publications do not clarify how terms used are to be understood and do not actually demonstrate the form of target language version attempted or produced. The distinctions below are developed specifically in relation to survey research. Many of these terms have, in addition, multiple uses in the translation sciences which we do not cover here.

7.2.1. Translation Distinctions and Terminology

Source and Target. Throughout the chapter we use the term "source language" to refer to the language translated *out of* and "target language" for the language translated *into*, hence also related references to "source questionnaire" and "target questionnaire." Apart from such frequently used terms, a common vocabulary has still to be found for talking about different kinds of translation or translation problems in survey research, as is, indeed, a framework for defining survey translation quality. Below we assign a given form of translation to a term that is often used for it. The discussion reflects, however, that some of the terms are sometimes used indiscriminately for any of several translation approaches.

Transparent or Covert Translations. Various terms are used in different theoretical frameworks to indicate, in simple terms, that a translation "comes across" like an original text in the target language and does not signal that it is a translation. The usual aim in survey translation is that translations do not reflect that they are translations. In the translation sciences, this is a more complex discussion than we need go into here (see, for example, House, 1977/1981 and 1997 on "covert" and "overt" translation, and Venuti, 1995, on "foreignization" and "domesticization").

Word-for-Word Translation. This is translation that operates at the level of words[2] and the most salient "meanings" or senses associated with these words. The translation closely follows the sentence structure and semantic content of the source text, not the target text.

Example 1. *Geben Sie dem Befragten die Liste 15: Give you (to) the respondent the list 15.*

The English version does not have normal English sentence structure (syntax), nor is "list" the appropriate English term to render what is intended. "List" is, instead, the etymologically cognate term to what is used in the German (*Liste*). A suitable corresponding English term would be something like "showcard." The determiner obligatory in German (*die*), translated here with "the," would normally not be needed in English. Word-for-word translations can also be acceptable translations for given contexts provided an absolute match of elements is possible across the languages and conveys the intended meaning. This is usually not possible across large stretches of text.

Example 2. *Sein Name ist Paul: His name is Paul.*

Word-for-word translation can be useful to reveal how the grammatical and semantic features of a source text are organized. Phillips (1996) made word-for-word English translations of songs by famous German composers alongside a further English translation that observed English style, vocabulary, and syntax needs. The three texts could thus help non-German singers know which words corresponded to which (or did not correspond to anything), so they could phrase their vocal interpretation better.

Close or "Faithful" Translation. If we distinguish between word-for-word and close translation (as we propose to do here), then in close translation the translator

---

[2]We point out that what constitutes a word in one language might involve several words in another.

tries to remain close to the semantic import, the vocabulary, and the structure of the source text but also to meet target language requirements regarding vocabulary, idiom, and sentence structure.

Example 3. *Geben Sie dem Befragten die Liste 15: Give the respondent showcard 15.*

The English translation uses appropriate English words and English syntax but stays closer *structurally* to the German than the alternative and also acceptable translations in (4i–ii) and it stays closer to the German vocabulary ("give": *geben*) than 4iii.

Example 4i.   *Give showcard 15 to the respondent.*
    4ii.   *Present showcard/flashcard/card 15.*
    4iii. *Hand the respondent showcard 15.*

It is possible for a semantically and structurally close translation as defined here to be fully appropriate and idiomatic in terms of target language purposes. However, since the focus of close translation is more to convey the source text in another language and less to meet the needs of a target language respondent, close translation might not address *pragmatic* needs of a new target population. Omission or addition of material, even if pragmatically justified, for example, would not conform to the demands of a close translation.

Too Close Translation. Too close translation is a term we propose for translation that disregards normal usage in the target language, usually inadvertently (poor translation), or possibly in order to stay "close" to the source text for some reason. In this way it may verge on word-for-word translation and not be appropriate for questions to be fielded. Example (5) below relates to vocabulary (see example 1). However, sentence structure can equally be "too close" to the source text, then sounding odd or stilted for the target language, at best.

Example 5. *Geben Sie dem Befragten die Liste 15: Give the respondent list 15.*

The English translation stays close to the German in sentence organization, as in the acceptable translation of example 3, but breaks with normal English idiom by using "list." For interviewers working in English, therefore, the translation is suboptimal. Karg (2005) presents striking examples of too close translations from the Programme for International Student Assessment (PISA) educational test translations.

Idiomatic Translation. Idiomatic translation conforms to the familiar expression, usage, and form of the target language/culture. It does *not* need to contain phrases themselves considered to be idioms.

A close translation and even a word-for-word translation may or may not be idiomatic, depending on how much various elements of the source and target languages match up directly. Every word matches in the following:

Example 6. *Ich bin sehr zufrieden: I am very satisfied.*

We remind readers that matching and idiomatic source and target questions do not necessarily achieve comparable measurement.

Literal Translation. This term is frequently used for either close/faithful translation but also for word-for-word translation as this is described above. In the translation sciences it is sometimes used for straightforward translation with a focus on information (as in technical translation) but also in various other ways (cf. Roberts, 2002, p. 435).

Direct Translation. The translation researcher Gutt (1991) defines direct translation in terms of an intention to convey the communicative choices of the source text in the target language. Direct translation in this sense therefore reflects the characteristics and "flavor" of the source text with available target language means. McKay and colleagues (1996) use "direct" to describe a single source to target translation (cf. Harkness & Schoua-Glusberg, 1998). Others seem to be using it for close or faithful as described above or, alternatively, for word-for-word translation.

Conceptual Translation. It is not easy to know what survey researchers mean when they use this term. It may be being used in contrast to translations that operate at the level of words. It is sometimes juxtaposed to "linguistic translation," which again might be understood as either translation at the level of words or as a reference to translation approaches that focus on grammatical and lexical aspects of a language and less, for example, on pragmatics and meaning in context (cf. Catford, 1965; Saldanha, 2009; Snell Hornby, 2006, p.151).

Adaptation Versus Translation. The aim of adaptation in survey instruments is to tailor questions better to the needs of a given audience but still retain the stimulus or measurement properties of the source. Adapted questions might be called for in a one-language context (e.g., adapting questions for children). At the same time, the need to produce a new language version often brings with it the need to tailor (adapt) for a new population and context. Two examples in English, the first as source, the second as adapted, demonstrate the principle:

Example 7. *Do you have difficulty walking several blocks?* (U.S. source question): *Do you have difficulty walking 100 yards?* (U.K. adaptation)

British towns are not organized in blocks and adapting the U.S. indication of distance makes the intended distance (if "100 yards" is indeed this) clearer to British respondents.

Translation, Adaptation, and Design. Translation as intended here is interlingual, that is, across languages; it enables an instrument to be used with new linguistic groups. Adaptation may be interlingual or intralingual, that is, also concerned with changes required to tailor materials *within* a language. Adaptation may change instrument design on one level to maintain it on another.

The relationship between design, adaptation, and translation can be close. A population that requires translation may well require changes to accommodate new social realities, cultural norms, or respondent needs (e.g., level of vocabulary). In some instances, the dividing line between changes undertaken to accommodate cultural or social realities and those required for reasons related to language cannot be neatly drawn. Answer scales provide a variety of examples. For example, if by translating a scale as closely as possible, a bipolar scale becomes unipolar, translation has altered the design. All changes need to be tested before use.

7.3. CURRENT THINKING, PRACTICE, AND RESEARCH

In some respects, thinking and practice regarding survey translation have changed noticeably over the last decade. There is an increased acceptance that survey translation practice needs serious concrete attention and that there are specific ways to improve translation outputs. In both older and recent literature on survey translation, there is some agreement that planning for translation should be part of the study design. Whenever possible, it should also be an integral part of questionnaire design rather than something separate and subsequent to it (Erkut, Alarcon, Garcia-Coll, Tropp, & Vázquez-García, 1999; Potaka & Cochrane, 2004; Werner & Campbell, 1970). These are welcome developments. Emerging research considered in Section 7.4 suggests that numerous further insights and related changes in perspective and practice can be expected.

7.3.1. Current Thinking and Theory

Despite this progress, there is still not a strong understanding among survey researchers of the theory and practice of doing translations, nor of the special needs in this respect of survey translations. Equally, there is still not a strong understanding of the needs of survey research among those trained in the translation sciences.

As indicated earlier, the vocabulary used to discuss survey translation is often under-defined and in part contradictory. Moreover, standard starting points in the translation sciences such as a definition of the translation goal (purpose or function), genre, medium (audio-visual, paper, aural), and the intended audience are rarely explicit *starting points* in survey research translation.

Instead, discussion often focuses on translation challenges at the level of words. "There's no word for …" debates are typical of this, as we see it, misguided focus on words and their assumed meanings. For example, authors have described the depression item *Have you felt blue or down recently* as hard to translate because there is no word for "blue" (as sometimes stated about Welsh) or because "blue" is not used to convey the sense of "somewhat depressed" in a given language and in fact may be used to convey something quite different (such as "drunk" in German). This is discussed at the level of words and is unlikely to take us far in survey translation. Instead, we suggest, the focus should be on ascertaining intended meaning and intended measurement and trying to convey those (cf. Chapter 3, this volume).

The role of context in shaping the meaning respondents perceive also requires more informed discussion with regard to design, translation, and adaptation. Certainly, a number of publications take note of the importance of context in how questions are interpreted, that is, how context affects the way respondents perceive information presented and assign readings or "meanings" (e.g., Sudman, Bradburn, & Schwarz, 1996; Schober & Conrad, 1997; Suchman & Jordan, 1991; Tourangeau, Rips, & Rasinski, 2000). At the same time, survey research has not yet developed a comprehensive theoretical framework that fully accommodates context. As a result, there are no established systematic procedures to take context into account in developing questions for comparative research or in translating them for multiple linguistic groups.

Evaluative research is sparse on the translation procedures frequently used in survey research. In addition, survey research understanding of translation issues lags well behind developments in the translation and interpreting sciences, even accepting that translatology is a young and evolving field (cf. the historical over-view of developments and trends in Snell Hornby, 2006, and accounts of different national schools and traditions in Baker & Saldanha, 2009). Thus, although the use of oral translations (bilingual interviewers or interpreters) complicates the survey interview process and may compromise data quality, little research has appeared on either the practice or consequences of using these forms of translation (cf. Harkness, Schoebi, et al., 2008; Harkness, Villar, Kruse, et al., 2009a, 2009b). Similarly, although the term "back translation" seems inevitably to occur in any discussion of survey translation (as documented in Harkness, Villar, Kephart, et al., 2009b), very little basic or applied research looks at back translation as a method of assessment or compares the effectiveness of this assessment procedure to other forms of translation assessment (but see Brislin, 1970; Cantor et al., 2005; Forsyth et al., 2007; Harkness, 1996; Harkness, Villar, Kruse, et al., 2009a, 2009b; Kim & Lim, 1999. See also Section 7.3.3 on back translation).

7.3.2. Current Practice

This and the following sections in 7.3 present questions that practitioners may have about those involved in a translation effort and their individual contribution, about how efforts can be organized, and about various options on the form of translation. More detailed discussion of many of the points covered below can be found at the Cross-Cultural Survey Guidelines website (http://ccsg.isr.umich.edu/).

Who Should Translate?

Machines? Machine translation (MT) is a re-energized subfield of computational linguistics able to capitalize on sophisticated technology and huge language databases to reduce the involvement of humans in translated text production. Interest in using machine translation is related to costs, speed, and ease, and to providing access for all. MT is a complex field with numerous issues we cannot discuss here. Its potential lies currently in highly standardized and constrained fields of discourse where word-based matches can be identified and, it is assumed, cultural and dynamic aspects of meaning are reduced. This is obviously not the case for survey research, where we still lack even basic consensus on what needs to be conveyed in other language versions.

Nonetheless, we do expect software tools such as translation memory and databanks to play useful roles in survey translations in coming years.

Do It Yourself? Projects commonly utilize people without training in translation but who (presumably) have language competence in both the source and target languages. Language competence is certainly a prerequisite for translation, but not a sufficient condition for working as a translator. Nor is previous experience in translation—"having done it before"—a guarantee of quality or competence.

Professional Translators? Trained translators, on the other hand, will not necessarily know how to tackle survey translations. One part of the problem is that while survey researchers may not understand enough about the potential of translation to ask for and explain the right things, translators may not understand enough about surveys to deliver the right things. Inappropriate presuppositions translators bring to the task about what is required may go unrecognized and uncorrected.

Teams? Well-organized team translation is an ordered and multistage process with built-in quality checks and integrated refinement. With different team members engaged at different stages, a team can provide the various kinds of expertise needed to arrive at a good translation. Committee translation is another term used for team translation (cf. Schoua-Glusberg, 1992), but this sense should then be distinguished from procedures in which a group of people get together to translate as a committee.

Don't Translate—Use a Lingua Franca? Some surveys use a language that all respondents of a basically multilingual survey population can follow (a lingua franca). This strategy has been used in cross-national and within-country research. Hofstede (2001, p. 43) reports on a 26-country study from 1967 using 4 languages (English, Spanish, Japanese, and Portuguese) to interview IBM employees. In countries with highly diverse linguistic groups, surveys may be fielded in the language(s) most target populations can understand and whatever other languages can be budgeted for. Little research has specifically addressed the effects on survey estimates of using a lingua franca. However, research on how bilinguals respond to surveys when interviewed in their several languages finds that people respond differently depending on the language they answer in (e.g., Erwin Tripp, 1964; Harzing, 2006; Richard & Toffoli, 2008). In therapeutic counseling, similar findings have been reported Lijtmaer, 1999; Ramos-Sanchez, 2007).

People using more than one language may use or prefer one of the languages for certain contexts or topics. They might, for example, use one language at work and another at home, or one with parents and another with their children. If in such contexts the survey topic is not usually spoken about in the language of the interview, various effects are likely, including language switching (Lijtmaer, 1999; Ramos-Sanchez, 2007).

7.3.3. Options for Target and Source Instruments

Written Translations. Good practice in survey research calls for written translations; they can be reviewed, refined, pretested, documented, distributed, and used again. The quality assurance and control measures available for oral translations (see below) are very much weaker than for written translations (Harkness, Schoebi, et al., 2008).

Double Draft Target Translations. Best practice also advocates having more than one draft translation. In team efforts, these foster a rich discussion in the review to produce a final version (Harkness 2008b). However, double drafts cost more than one draft. To reduce costs, the source text can be divided (split) between translators as described next.

Splitting up the Source Text: Splitting a questionnaire between translators can save time and effort, particularly if a questionnaire is long (cf. Harkness & Schoua-Glusberg, 1998; Schoua-Glusberg, 1992). This procedure assumes a team translation review and finalization. Split procedures are less recommended for novice teams (see Harkness, 2008b). If splitting is used, the source text is divided up between translators more or less in the alternating fashion used to deal cards in card games. By thus sharing material from one section between translators, possible translator bias is reduced and translator input is maximized evenly across the questionnaire. Each translator translates his/her own section in preparation for a review session.

Care is always needed to ensure that material or terms which re-occur across a questionnaire are translated consistently provided they refer to the same entity or notion. Split questionnaires may require particular care and checking in this regard (Harkness & Schoua-Glusberg, 1998; see Cross-Cultural Survey Guidelines website http://ccsg.isr.umich.edu/).

Oral Translation: Oral translation takes two forms in survey research: an interviewer translates as he/she conducts the interview or a bilingual interpreter acts as a intermediary between the interviewer speaking language A and the respondent speaking language B. Harkness, Schoebi, et al. (2008) and Harkness, Villar, Kruse, et al. (2009d) indicate the considerable risks and drawbacks of using oral instead of written translations. When oral translation is unavoidable, it should therefore be accompanied by extensive preparation and training. This in turn involves considerable expertise, time, and budget allocations. The more usual motivation for oral translations is, however, to reduce, time, effort, and/or costs.

Double Source Text. Sometimes two "source" instruments are made available for translating countries, as is the case in Eurobarometer surveys and in the Organisation for Economic Co-operation and Development (OECD) Programme for International Student Assessment (PISA).

The sparse literature on double source questionnaires leads to many questions and few conclusions. Various relationships could exist between the two "source" documents, as could differing specifications toward "comparability." Decentering (see Chapter 3, this volume) does not seem to be the basis of development. Fetzer

(2000) states that Eurobarometer source questionnaires are in English and the second "source" French version produced through translation. Eurobarometer "Flash" surveys (http://ec.europa.eu/public_opinion/flash/FL162en.pdf/) present a "bilingual questionnaire" in their technical reports without further explication. Also referring to Eurobarometer surveys, Duch (1994) speaks merely of developing "equivalent questionnaires" in English and French. Our own examination of Eurobarometer questionnaires suggests that occasional oddities in English and French "source" versions might be explained if some questions are developed in English and others in French.

Citing Reif and Melich (1991), Fetzer (2000, p. 36) also points to differences between the two Eurobarometer source versions—and further versions—in questions asking about acceptance of immigrants in one's "neighborhood." In English (for the United Kingdom), reference is made to "neighborhood," in French to "quartier" (not a match for "neighborhood"), and in German to "Nachbarschaft (a cognate term with the English but substantively different). Fetzer suggests that the conceptual differences are likely to affect respondents' responses.

Countries participating in PISA were requested to work from two "source" texts in a particular fashion, choosing one as a source text and the other as a check. Grisay (2003) speaks positively of these. Karg (2005), however, criticizes the logic and theoretical basis of PISA double source documents, the resulting quality and appropriateness of the two source versions, as well as the procedures followed by various countries to translate and the quality of resulting translations.

Iterative Procedures. A number of surveys in different disciplines report using "iterative" and multistage translation procedures, most commonly in connection with back translation (see below). A fairly basic iterative procedure in this sense would be translation into a target language, translation back to the source language, comparison of the source text and the (source language) back translated text, and adjustment. How adjustment needs are ascertained, what questionnaires are involved, what changes are made, how often the procedures are repeated and also why, is often unclear (Harkness, Villar, Kephart, et al., 2009a, 2009b). Depending on a number of factors, including whether the source is open to modification, changes made might be to the source or the target texts. If the source is also modified, this resembles decentering procedures for design described in Chapter 3 (this volume).

The Translation, Review, Adjudication, Pretesting, and Documentation (TRAPD) model described in Section 7.3.4 is iterative in a different sense. Here each phase of refinement of the target questionnaire may lead to modifications which are then reviewed, approved, and documented before the next stage is attempted.

Back Translation. Countless projects engage in procedures they describe as "back translation" (cf. review in Harkness, Villar, Kephart, et al., 2009a, 2009b, and discussion in Section 7.4). In back translation (BT), the target text is translated back into the source language and differences between the two source language versions are taken as possible evidence of problems in the target language text. Harkness (e.g., 1996 and 2003) argues that direct appraisal of the target translation is both theoretically sounder and practically more valuable than BT.

Harkness and Schoua-Glusberg (1998) noted that pressures on researchers to apply BT could inflate its reputation as a "standard." At the same time, criticism of the adequacy of BT has grown over the last decade or so (Hambleton, 1993; van Widenfeldt, Treffers, de Beurs, Siebelink, & Koudijs 2005; Maxwell, 1996; Harkness, Villar, Kephart, et al., 2009a, 2009b).

7.3.4. Current Best Practice: Team Translations

Strategies currently most favored for survey translations recommend a team approach, with different players collaborating in ways that serve to maximize their mix of expertise. Variations of team approaches are described in guidelines such as those for the European Social Survey (e.g., Harkness, 2008b), for the U.S. Bureau of Census (Pan & de la Puente, 2005), and on the Cross-Cultural Survey Guidelines website (http://ccsg.isr.umich.edu/). In the TRAPD team translation model (Translation, Review, Adjudication, Pretesting, and Documentation; Figure 7.1) translators provide the draft materials for the first discussion and review with an expanded team. Pretesting is an integral part of the translation development. Documentation of each step is used as a quality assurance and monitoring tool. In this way, notes on problems faced and their resolution are kept at each stage and consulted at the next step to inform decisions (Harkness, 2003, 2008b).

Team approaches usually exploit a direct and iterative exchange between various team members. For example, substantive reviewers or those responsible for signing off on a translation may return to translators with queries even at late stages of translation completion. At the same time, it would be counterproductive

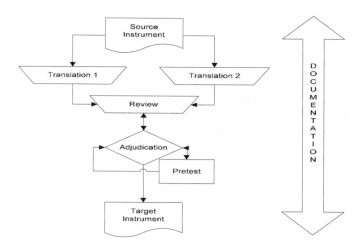

Figure 7.1. The Translation, Review, Adjudication, Pretesting, and Documentation Model (TRAPD)

to engage every team member in every decision, and this is certainly not the intent in a team approach. Proper planning can ensure that input from relevant players is available at various stages. All instruments should be pretested before use, once the translations have reached an advanced enough stage to warrant pretesting.

Team Composition, Briefing, and Training. Good procedures cannot remedy lack of competence but built-in quality assurance and monitoring steps are essentialcomponents and are useful in helping train less experienced members in a team. Within a quality assurance and monitoring framework, the suitability of anyone involved should be checked before they are engaged in the effort. Everyone should be briefed and trained as necessary on their required roles and on the procedures and schedule envisaged. In addition, output and performance (e.g., of translators and reviewers) should be checked at early stages in order to address any problems found (see the Cross-Cultural Survey Guidelines website http://ccsg.isr.umich.edu/ for more detail).

Translation Production and Review. Translators can produce draft translations and reviewers prepare for review meetings wherever they have access to the project materials and tools. Given the increasing potential to meet in virtual workspaces, review discussions need not always call for face-to-face meetings between those involved in the translation effort. However, if players have little experience of review procedures, managing review feedback and discussion can be more challenging in virtual meetings than in face-to-face meetings. In all instances, review meetings must be carefully prepared and moderated. Extended meetings only by telephone are problematic. For instance, participants and moderators only have oral/aural exchanges to guide their discussion.

Budgeting for Translations. Good translations do not ensure a good survey but bad translations do guarantee a bad survey. Despite the essential role translations play, projects sometimes speak of translations as costly and look for ways to reduce expenditure on translation and assessment in an often unprofessional manner. In fact, translation costs usually make up only a very small part of total survey costs. If this is not the case, project planning and implementation should be reviewed (see, too, guidelines at /http://ccsg.isr.umich.edu/).

Most projects cannot afford multiple assessment procedures but must nonetheless find robust ways to assess quality. Careful allocation of funds dictates that surveys use effective ways to produce, assess, and refine translation quality. Modest budgets are a strong reason to invest at the outset in robust translation and assessment procedures, guided by an appropriate theoretical framework. A parsimonious choice of procedures, however, assumes an understanding of options. Costs will inflate if multiple procedures are used indiscriminately.

7.4. RECENT AND EMERGING RESEARCH

This section concentrates on recent research on survey translation which has provided new insights. It is not intended as a review of survey translation literature. A number of publications reflect pronounced efforts to develop translation procedures and assessment methods that improve survey translation quality. Some treatments are holistic, dealing with survey translation efforts as a whole; others focus on particular features or strategies, such as role of pretesting in developing translated instruments.

7.4.1. Research on Translation Production

Harkness (2008b, 2003) and Harkness, Pennell, and Schoua-Glusberg (2004) have advocated translation procedures based on quality assessment and control (QA and QC) frameworks that use documentation at each step as part of the development and review process. The TRAPD procedure introduced earlier makes review and pretesting integral to the translation production process. Willis and colleagues (Chapter 8, this volume) present challenges discovered at each stage in projects based on TRAPD-like models.

In basic research on the team production process, Harkness and Behr (2008a, 2008b) made transcripts of audio-visual tapes of team translation review discussions, providing the first data of this kind we are aware of. The research examines strategies teams use to resolve translation challenges and the dynamics of such interaction, providing insight into the translation process and informing team member training. The video-taped reviews of German modules for the International Social Survey Programme (ISSP) followed TRAPD specifications: documentation of each stage to inform the next, two separate draft translations, team review, adjudication in review, and after review, pretesting, and further refinement. Behr's dissertation (forthcoming) is a case study using these data to identify various kinds of problems encountered, resolution strategies used, and the basis taken for judgments and decisions. This kind of process-oriented, evidence-based research promises to provide useful practical and theoretical insights into team translation that can inform training, benchmark development, and practice.

Survey translations are not always produced in written form; bilingual interviewers or interpreters are sometimes used. There is not yet much research on these real-time translation situations. Harkness, Schoebi, et al. (2008) investigate translations produced by bilingual interviewers during telephone interviews. When bilingual interviewers are used as translators, the interviewer silently reads a source language questionnaire and conducts the interview in the target language, translating as he/she proceeds through the interview. Such oral translations in surveys have usually been seen as a way to save time and money. Harkness and colleagues (2008) compared participants' performance and the interaction in orally translated interviews with those of interviews conducted with written scripts. Their findings suggest that oral translation threatens data quality within a language and threatens comparability across languages.

In a related project, Harkness, Villar, Kruse, and colleagues (2009c; 2009d) analyze transcripts from interpreted telephone interviews. Interpreted interviews are prototypically conducted between three individuals, with a monolingual interviewer speaking language A, a monolingual respondent speaking language B, and a bilingual interpreter mediating between these two in each language. The findings in this project, which used questions from the Harkness, Schoebi, et al., (2008) research, mirrored those from the earlier study. The process and outputs considerably complicated the interview procedure in terms of successful interviewing and compromised data quality and comparability.

We thus recommend caution in using either bilingual interviewers or interpreters in survey research. However, sometimes oral translation is the only option (with unwritten languages, for example) and further research along these lines is needed to better understand and meet the inherent challenges of interpretation and oral translation in surveys.

Finally, we note that translation has also been advocated as a refinement tool for question design (e.g., Harkness, 1995, 1996; Harkness & Schoua-Glusberg, 1998; Harkness et al., 2008; Braun & Harkness, 2005). In essence, this is motivated by the fact that the deep processing called for in translation helps developers notice features they might otherwise overlook. Here, too, however, more research is required to improve strategies to elicit the relevant information.

7.4.2. Research on Translation Assessment

A small number of research projects have looked at current translation assessment techniques and/or the assessment outputs (that is, the suggestions, queries, etc., resulting from the assessment step). Interested in what various assessment steps might reveal, Willis and colleagues (Chapter 8, this volume) discuss the kind of mistakes discovered at different stages of translation review and pretesting in projects more or less based on a TRAPD model. Chapters 5 and 6 (this volume) discuss insights gained from cognitive pretesting into problems related to translation. They implicitly underscore the relationship between design and translation, with relation to questions (Chapter 5, this volume) and to supplementary materials (Chapter 6, this volume). As we see it, they also prompt questions about the form pretesting strategies might need to take with different target populations.

Within the fields of educational assessment, there is considerable discussion of instrument design, cross-cultural and cross-curriculum viability, and assessment of translation quality (cf. references in Chapter 3, this volume). Dept, Ferrari, and Wäyrynen (Chapter 9, this volume) describe centrally controlled review procedures undertaken in international education projects by a company specializing in such reviews. In contrast to TRAPD, the review input was provided by external reviewers, independent of the people involved in initial translation production.

In a number of projects, Harkness and colleagues have investigated how reviewers review; what benchmarks they use in review; how they perceive the translations they review; and their preferred assessment strategies (Harkness, 1996, 2005; Harkness & Behr, 2008b). Harkness, Villar, Kephart et al., 2009a, 2009b) compare the output and usefulness of external reviews of survey translations against project-internal reviews. They also compare qualitative reviews of translations by different kinds of survey experts with reviews based on back translation.

7.4.3. Research on Translation and Answer Scale Design

Answer scales do not lend themselves well to translation (cf. Harkness, 2003; Harkness et al., 2004; Harkness, Chin, Yang, et al., 2007). Translations are sometimes not viable (e.g., *neither agree nor disagree* in Swahili and Hebrew) or may automatically result in altered designs (Harkness, 2003; Harkness et al., 2008). Nonetheless, research on alternatives for developing answer scales and on strategies used to translate answer scales is quite limited. Research on the intensity respondents associate with answer scale labels in German and English (e.g., Harkness, Mohler, Smith, & Davis, 1997), later extended to Japanese (Smith et al., 2009), indicates that nominally "matched" expressions such as "very good" (E)/sehr gut" (D)/"très bon" (F) are assigned different degrees of intensity of "good" by different populations. We note that this research concentrated on individual response category labels and that application of the findings to actual scales and survey questions remains to be conducted.

Szabo, Orley, and Saxena (1997) also investigated the degrees of intensity populations associated with response category labels, leading the World Health Organization Quality of Life (WHOQOL) project to adopt a novel strategy for developing response categories across languages. End labels were fixed for the source language scales and these were translated for other language versions. Any intervening response category labels were decided in a country-specific (language appropriate) manner. In this way, the project hoped to reduce difficulties of matching up multiple labels across languages. This approach assumes that endpoint labels could be successfully translated so as to identify comparable cut-off points across populations and that the intervening labels used would also be interpreted comparably across languages. Villar's (2009) research on translated and adapted answer scales raises some doubts about this, as does findings on endpoint translation (Harkness, 2003; Harkness et al., 2005).

Obviously, much remains to be investigated regarding answer scale version production, also in connection with the next topic, response category choice.

7.4.4. Research on Translation and Response Category Choice

Why populations might tend to associate different degrees of intensity with terms that appear to be "matched expressions" is under debate. Factors identified as potentially affecting response behavior include features of language, including differing options for negation (Harkness, 2003; Harkness et al., 2008); response style tendencies (see Chapter 12, this volume); cultural orientation and associated habitual perception (see Chapters 10 and 11, this volume); the design of answer scales (e.g., Schwarz, Hippler, Deutsch et al., 1985; Schwarz, Knäuper, Hippler et al., 1991; Chapter 11, this volume; Villar, 2009). For instance, Villar's (2009) examination of data and translations from a number of ISSP countries finds that within- and across-country response preferences differ significantly depending on whether the *agree/disagree* source answer scale was translated or adapted. Different levels of macro-, meso-, and micro-context have also been considered relevant (Braun, 2003; Braun & Harkness, 2005; Chapter 10, this volume).

A growing body of work considers how culture may mediate response choices. Haberstroh, Oyserman, & Schwarz (2002), for example, draw on cultural differences in how people perceive and apply conversational norms to survey questioning and answering to understand respondents' interpretation of questions and their resultant responses. Schwarz, Oyserman. and Peytcheva and Uskul, Oyserman and Schwarz (Chapters 10 and 11, this volume) demonstrate how culture may affect respondent perception and response choices in individualist, collectivist, and "honor"-oriented cultural contexts. In their discussion of response style research (Chapter 12, this volume), Yang and colleagues suggest that the habitual propensity a respondent may have toward a given response style is shaped by both individual characteristics/traits and his/her experience. Cultural context and exposure to cultural norms are considered to be important contributing factors to this "experience."

Finally, the effect of individual respondent motivation and burden on response behavior is discussed under the rubric of "satisficing," a combination of "satisfy" and "suffice" (e.g., Krosnick, 1991; Krosnick, Narayan, & Smith, 1996a). It is also possible that motivation to respond could differ systematically across populations (cf. the fear factor encouraging "undecided" responses referred to in Luz Guerrero & Mangahas, 2004).

7.4.5. Research on Adaptation

The recently re-energized field of adaptation studies focuses in the first instance on changes required to develop material for other mediums, in particular adaptations of literature and drama to film (cf. Leitch, 2008). In survey terms, this is modestly comparable to thinking about changes required when the mode or form of application of a one-language survey changes or when an instrument for adults is adapted for use with children. Bastin (2009) provides a brief overview of adaptation in translation studies disciplines, also identifying some forms of adaptation and strategies which could be useful in survey research.

Adaptation in the survey context is often concerned with deliberate changes to source material in order to meet new needs of various kinds. In the first instance, it involves modification of question content, format, order, or instructions to meet the needs of a new population, location, mode, or any combination of these.

Smith (2007a), for example, discusses adaptations required to update English questions in the U.S. General Social Survey (GSS). Maxwell (1996) considers changes required in educational testing. Contributions in Hambleton, Merenda, and Spielberger (2005) address adaptation and translation in educational testing. Contributions in Georgas, Weiss, van de Vijver, and Saklofske (2003) also document changes made in the *WISC-III* intelligence test instrument. However, changes are not always the result of deliberate decisions to modify. As discussed earlier, translation may force a design change.

Research on the theory and practice of survey adaptation is nonetheless meager. In general, detailed documentation on survey adaptations (or inadvertent changes) is also sparse and not easily located. The U.S. GSS codebooks, which provide detailed notes for users on changes made in questions from one GSS survey to the next, are a notable exception. At the same time, we note that documentation of changes is different from explanation or assessment of changes; those trying to learn about adaptation strategies or the adequacy of adaptations would ideally need both.

Table 7.1 presents features which frequently require to be adapted to fit a given location and population, mostly independent of (interlingual) language change (cf. Harkness, 2008a; Harkness, van de Vivjer, & Johnson, 2003).

The Cross-Cultural Survey Guidelines website (http://ccsg.isr.umich.edu/) provides a work in progress toward a typology of instrument adaptation strategies and rationales; see, too, Maxwell (1996) and Bastin (2009).

Access to any changes made in comparative surveys between source and target or different versions of a questionnaire may be complicated by the fact that questionnaires exist in different languages, not all of which researchers and other users may understand. The likelihood of simply spotting changes across question-naires is thus reduced. The European Social Survey (ESS), a comparative survey often praised for its documentation, provides an example. Centrally held records are available for each round on any changes (including unintended deviations) in target instruments that participating countries provide. Countries, however, have been known not to notice or to omit to report differences between the source and their target version(s). In such instances, no record is available for users.

7.4.6. Research Related to Translation Documentation and Tools

Documentation to inform survey life-cycle QA and QM must be available in a timely fashion (see Chapters 13 and 16, this volume; Harkness, 1999; Mohler & Uher, 2003). In the iterative TRAPD model (see Figure 7.1), documentation is envisaged as a tool to guide decisions throughout the translation process, as well as to provide documentation for users. However, maintaining documentation can be burdensome (cf. Mohler & Uher, 2003; Mohler, Pennell, & Hubbard, 2008). The documentation provided by country teams on the development of ESS translations was at times meager. This may be because those involved were not familiar with how and what to document, but it is also likely that the effort involved in manual documentation played a role. Little is also available publicly on why the answer scales that Villar (forthcoming) examines in different countries and languages were adapted in some modules and translated in others, often for one and the same organization.

"Localization translation" aims to translate and adapt source software text to ensure that translations fit new locations and languages (cf. Esselink, 2000; Schäler, 2009). The shared translation tools, project management schedules, and automatic documentation components used in localization very much structure the

TABLE 7.1. Examples of Some Adaptation Types and Topics

Topic	Details	Concrete Examples
Measurements	Distance; temperature; weight; length/height	100 yards; 20°C; 2lb. (imperial); 6ft
Components and order of calendar dates, names, initials, addresses	Order of day/month/year; order of first and last names and initials of these; components of addresses and their order	13/1/2011; Harkness, Janet; HJ (Hungarian initials order); street number, apartment/house number, zip code, place name, regional division, country
Institutions	Schools; institutionalized services and practices; government bodies	Grammar school; VAT; social security; welfare benefits; Parliament
Product & food terms, brand names	Cigarettes, cooking oil, vegetables and fruits	Light-tar cigarettes; names of products
Sexual terminology	Sexual orientations, sex practices, parts of the body	
Emotions and conditions	Instances and intensities of feelings or states	Anxiety; "brain fog"
Socio-economic questions	Education; income; type of housing	(See Chapter 17, this volume)
Answer scales	Endpoints; midpoints; other points	(See Chapter 3, this volume and this chapter)
Visual images	Saliency, focus, processing habits; semiotics	Perceived foreground & background; direction of reading/ processing; socially acceptable depictions.
Direct and indirect disclosure formats	Habitual indirect disclosure; questions sensitive only in some populations	Names of children
Cultural and social conceptions and conventions	An extremely wide range of topics could be involved	Hygiene practices; taboos, health care

entire translation effort for all those involved, also automatically embedding this development in a QA/QC framework. While the nature of localization translation work differs from that required for survey translations, such management, project assignment, and documentation subtools help ensure timely and consistent production of translations. Technological aids reduce the burden on those recording information and make it easier for others to access existing records and decisions on outputs. Tools can also make production faster, which can also provide an important competitive edge in a cost-conscious market.

Translators in a growing number of fields increasingly use a variety of tools which aid them in producing and checking translation outputs. Diverse tools are available free on websites that translators use. A number of worldwide survey organizations, in part in collaboration with translation software providers, use translation technology to speed up and systematize large-scale multiple language translation projects (cf. Gallup, http://www.sdl.com/en/customers/case-studies/gallup.asp/; de Voogd, 2008). We expect the use of software tools to grow in survey research. There is, for example, a growing awareness that tools that enhance production and incorporate QA and QM documentation features will simplify process and product documentation for multiple users (cf. Harkness, 1999; Harkness, Dinkelmann, Pennell & Mohler, 2007; and Chapters 13 and 16, this volume).

In a European Social Survey report to the European Union, Harkness, Dinkelmann, et al. (2007) identify the components ideally called for in a survey translation tool which would automatically provide documentation of different developmental stages and thus support production and end documentation. Their review of project management and translation tools found that none of those then available was ideal for survey translation. That report cautions against submitting unvetted materials to translation memory files. Translation memory (TM) is a common feature in translation software packages. It can be understood as a database that stores units (words, phrases, or more) that have been previously translated alongside the corresponding source text components as "translation units." Given the modest research on answer scales, for example, a TM database of answer scale labels would currently only be a collection of unvetted translations.

Tools can take many forms; even very modest tools help enhance quality. The procedures described in Chapter 9 (this volume), for instance, could essentially be maintained on Excel spread sheets (cf. Dept, Ferrari, & Wäyrynen, 2008). The tables originally used in ESS and ISSP translation work to align source and target text were only simple Word tables. A number of translation websites that translators use to advertise their services also offer software tools, tips, and discussion boards free of charge.

7.5 COMMON MYTHS OR MISCONCEPTIONS

At various points in the chapter, we have implicitly or explicitly referred misconceptions about survey translation. We briefly review these here:

- That written survey translations are expensive

We noted that translation costs should be small in comparison to other cost factors of a survey. Good translations support the survey goals. Good translations can also be considered as candidates for re-use. With appropriate documentation they could also be stored in a translation memory (TM) database.

- That oral translation procedures can be used with impunity

The small body of research on oral translation by interviewers and using interpreters indicates how burdensome the undertaking can be for all those involved. Better strategies need to be developed for when oral translation is unavoidable. Written translations are the much preferred option.

- That back translation (BT) should be a preferred assessment tool

Recent research (Harkness, Villar, Kephart, Schoua-Glusberg et al., 2009b; cf. Harkness 1996; Chapter 8, this volume) indicates that other forms of translation review and assessment are more efficient than back translation. The effort, time, and costs involved in alternative approaches are likely to be the same. Numerous projects, for example, use BT as only one of several procedures, possibly reflecting that more have been considered necessary. Multiple procedures obviously increase outlay.

- That retaining semantic content of questions in translated versions will retain measurement properties

Meaning is more a property of people than of words. Multiple levels of context contribute to how we interpret what the words people produce are intended to mean. We therefore cannot assume that two questions with ostensibly the same semantic content automatically "ask the same thing" and prompt comparable responses.

To these we add a further five, again providing brief explanations:

- That good survey questions will be robust in comparative contexts

Precisely the qualities that make a question effective and salient in a given context for a given population may prevent it from traveling well. It may exactly fit the conceptual, linguistic, and pragmatic context for which it was designed but fail to do this elsewhere. Mobility questions asking about "difficulty playing golf" might work well in contexts in which golf is popular, affordable, and accessible to both genders. The meaning of "having difficulty" changes, however, in contexts when the sport is unusual and expensive.

- That anyone can take on survey translation

Questionnaires are, at their best, deceptively simple. Designers often aim for straightforward language, fairly basic vocabulary, and short sentences. This might suggest that anyone with a reasonable grasp of source and target languages could produce good translations. That is not the case. In fact, experience shows that even trained translators have trouble producing appropriate survey translations. Question-naires are a complex text type. It is often challenging to produce workable trans-lations that carry out the required acts of measurement in the guise of questions and answers. Translations must also take into account that questions are intended to be offered (said) once and to require only a normal degree of textual processing.

In various places in this chapter, we also recommend the use of translation tools; translation checking software, language corpora, and so forth. Translators are trained in how to translate and also in how to make the best possible use of the growing array of tools of their trade.

- That translators or survey researchers can decide survey adaptations

In some fields, adapting texts to new mediums or cultures is the business of translators versed in the constraints and requirements of the mediums. In survey research, the strategies and forms of adaptation are not well defined and there is no established protocol for evaluating the merit of selected adaptations. Translators are often not conversant with survey instruments as measurement tools. As things currently stand, we suggest team input should decide adaptations and that adaptations should be tested with target populations.

- That there is one "right" or "best" translation

First and foremost, the purpose of a translation should determine how one and the same source text is translated. Differences in purpose can motivate numerous differences in target versions. Second, pretesting reports consistently demonstrate that questions can be understood to mean different things. *Do you have difficulty reading a newspaper?* is intended to be a question assessing vision. A person with a low literacy level might indicate they had difficulty reading newspapers because of that. Questions can be read in multiple ways and can also often be translated in multiple ways. Arguably any translation that manages *effectively* to do what it is intended (collect the required information) and does not do unwanted things (e.g., increase burden) is an adequate translation. It is possible that more than one translation can do this.

- That statistical analysis will suffice to address translation problems

Elsewhere we emphasize that translations must be subjected to both qualitative and quantitative assessment (Harkness et al., 2004). The obvious route to take is to make translations as good as one can using design and translation strategies and qualitative testing and then test whether statistical analysis verifies qualitative assessments or not or reveals new aspects not found in qualitative appraisal. Contributions in Hambleton, Merenda, and Spielberger (2005) discuss various aspects of implementing and interpreting statistical evaluations of educational instruments across languages; see too Chapters 20–24 in this volume.

- That documentation can be postponed until "after the event"

Translation is a strongly decision-based undertaking (cf. Wilss, 1998). Good practice dictates that survey translation products are reviewed before use. If documentation is available, questions raised about translation and adaptation decisions can be easily and reliably answered. Given the multiple and many-faceted decisions taken, trying to reconstruct them after the event will be error-prone. In team efforts, each phase of development builds on the decisions of the foregoing (see also Chapter 16, this volume, on documentation).

7.6. RESEARCH IN OTHER DISCIPLINES

First and foremost, survey translation efforts need to be based on a strong understanding of translation as a process with an array of useful procedures to transfer meaning from one language to another. Practitioners need to become more informed about the nature of translation as variously understood in the translation sciences and about the numerous options available to render a text in another language for a given purpose. These are essential steps if survey research is to move closer to defining in a useful manner the goals of survey translation as well as the goals of instrument designs which envisage translation. Solid translation sciences review articles on topics of relevance for survey—such as what the unit of translation might be or the notion of translatability and what that involves—are available in up-to-date handbooks and encyclopedias (e.g., the admittedly expensive Baker & Saldanha, 2009).

At the same time survey translations need to accomplish comparable measurement. Both researchers and translators need to understand better how and why aspects of a new context and language might affect measurement. Disciplines such as ethnography, participatory research, sociolinguistics and corpora research have insights of value to offer. And outputs need to be tested appropriately. Context at the level of the instrument includes co-text—the text surrounding the text under consideration—and the carry-over effects of previous question content and ordering (cf. Harkness, 2003). It extends to a wide variety of features we could subsume under "the measurement event" (cf. Yang et al., Chapter 12, this volume), including physical details of instrument and the interview. Theories of survey response already mentioned cover further relevant aspects; but these need to be considered in the dichotomous situation of target and source instruments and populations. Thus survey research must also do some groundwork; we need to begin to explore systematically the potential effects of context in terms of source and target instruments and populations.

The broad field of pragmatics provides a variety of approaches and theoretical frameworks to analyze human interaction. Some of these have been used in basic research on how respondents view and behave in survey interactions (Gricean maxims, for example) and in studies of interview interaction (such as the principles and procedures of conversation analysis). Other aspects of linguistic pragmatics, such as speech act theory, discourse pragmatics, and the negotiation of meaning in intercultural and intercultural contexts, have so far received less attention. In addition, it is by no means clear that received wisdom in one area of pragmatics for one context can be generalized across contexts (see, for example, Klungervik Greenall, 2009, on Gricean implicature in translation). Moreover, fields encompassed by "pragmatics" are diverse, dynamic, and in debate with themselves (cf. Rajagopalan, 2009). Nonetheless, survey research is some way away from preparing to bring these diverse streams of knowledge and debate to bear in any integrated fashion on the design of questions and their translation.

7.7 OUTLOOK

In terms of viable survey translation *procedures*, there has been important progress recently. If we can succeed in also defining what it is that translations are meant to convey and, importantly, if source questions were also designed or adapted so as to accommodate this, a great deal more will have been accomplished. Before this can happen, however, disciplines using survey research and translated questionnaires will need to become cognizant of their options and of the consequences of choice in terms of perceived question meaning, response categories, and measurement error.

This is a more challenging undertaking than demonstrating that certain approaches to translation and translation assessment are more useful to survey research than are others. Full demonstrations of even these are not yet available, since we are only beginning to collect evidence on the efficacy of one procedure over another, as illustrated by contributions in this volume.

Translation studies offer us a number of benchmarks and strategies as well as theoretical perspectives from which to start. Indeed their current debates share much with topics raised in this chapter; how to incorporate cultural aspects into translation; how to include findings from other disciplines relevant for translation; debating the different possible goals of translation and the implications of these; and, in terms of translator output and careers, a focus on special, somewhat technical translation. Survey translations can certainly in some respects be considered technical, although not in any sense that would bring them within the realm of MT. They are also especially complex in ways we are, encouragingly, beginning to identify and address—not just at the level of recruiting better translators, but also in terms of the frameworks adopted for question translation and its verification. As we see it, the next real challenge is one which holds transformatory promise: to link a theory and practice of question design to the specifications made for translation and adaptation.

8

Evaluation of a Multistep Survey Translation Process

*Gordon B. Willis, Martha Stapleton Kudela, Kerry Levin,
Alicia Norberg, Debra S. Stark, Barbara H. Forsyth,
Pat Dean Brick, David Berrigan, Frances E. Thompson,
Deirdre Lawrence, and Anne M. Hartman*

8.1 QUESTIONNAIRE TRANSLATION OVERVIEW

The survey world is currently witnessing a surge of interest in multilingual research. In the United States, this development is propelled by an increasingly diverse population, and U.S. surveys that track the well-being of the population are now expanding from one, or a few languages, to an increasingly larger number. For example, in 2001 the California Health Interview Survey was conducted in six languages: English, Spanish, Chinese (Mandarin and Cantonese), Khmer, Vietnamese, and Korean. In Europe, a similar interest is motivated by the growth of the European Union and its movement toward unification, standardization, and harmonization of surveys (Alaminos, 2004; Harkness, 2003; Hoffmeyer-Zlotnik, 2008). Studies that base version production on translation assume that by doing so, the same question can be asked in the different language versions produced.

In terms of standardization and harmonization procedures, this translation process can be thought of as *"ex ante* input" harmonization, in which a common measure (adapted into multiple languages) is intended to fulfill data objectives across groups, cultures, or countries (see Chapter 17, this volume). This is contrasted with output harmonization, in which different measures are developed across groups, and reconciliation is accomplished after data collection. In order for the former "one size fits all" approach to achieve success, the source questions to be translated must meet multiple requirements (see Chapter 3, this volume). As articulated by the U.S. Bureau of the Census (2004), the resulting translation should be reliable, complete, accurate, and culturally appropriate, and also

---

[1] Survey Methods in Multinational, Multiregional, and Multicultural Contexts, edited by Harkness et al.
Copyright © 2010 John Wiley & Sons, Inc.

demonstrate measurement comparability across versions (also see Braun & Harkness, 2005; Johnson, 1998; Miller, Willis, Eason, Moses, & Canfield, 2005a; Nápoles-Springer, Santoyo-Olsson, O'Brien, & Stewart, 2006; Pan & de la Puente, 2005; Saris, 2004; Schmidt & Bullinger, 2003; Singelis, Yamada, Barrio, et al., 2006; Yu, Lee, & Woo, 2004).

Despite the seeming logic and simplicity of this approach, there are a number of obstacles to achieving it, as numerous authors have pointed to pitfalls that arise in producing questions based on translation of a source version. Some focus on phrases that in translation fail to convey the idea or concept intended in the source version. For example, Nápoles-Springer, Santoyo-Olsson, O'Brien, and Stewart (2006) found that the phrase "medical tests and procedures" as translated into Spanish was particularly problematic for Latinos in the United States because it failed to bring to mind the kind of events intended. They modified both the source (original) and tar-get (translated) questions to include examples of procedures (blood test, X-ray, and cancer screening tests) so that the intended construct would be better understood.

Recent approaches to *how* to enact a survey translation largely endeavor to attain linguistic, semantic, and conceptual accuracy, while also avoiding an overly literal, word-for-word rendition that may fail to communicate the measurement goals of the question similarly across versions. Team translation procedures in particular offer a novel approach to both the translation effort itself and to the protocol used to evaluate the translation. These procedures include the addition of both rule-based guides to questionnaire appraisal (Dean, Caspar, McAvenchey, Reed, & Quiroz, 2005) and empirical methods of pretesting translated versions of a source questionnaire (Blair & Piccinino, 2005a; Carlson, 2000; U.S. Bureau of the Census, 2004; Harkness, 2008b; Harkness, van de Vijver, & Mohler, 2003; Harkness & Schoua-Glusberg, 1998; McKay, Breslow, Sangster, et al., 1996; Willis, Lawrence, Hartman et al., 2008).

Specifically, cross-disciplinary research has assisted in the development of multistep, team-based translation methodologies. Harkness (2002, 2003, 2007) describes the framework for a five-step process that involves multiple levels of review and reconciliation: *Translate, Review, Adjudicate, Pretest, and Document (TRAPD)*. The TRAPD framework encompasses a multifaceted view, emphasizing both linguistic and sociocultural elements. Similarly, the five-step process advanced by the Census Bureau (2004) employs a team approach: *Prepare, Translate, Pretest, Revise, and Document (PTPRD)*. Both the project guidelines provided for the European Social Survey and the Census Bureau guidelines address specific operational issues at a relatively high level of detail, aiding researchers to organize and execute a methodologically sound translation project.

However, there is currently a paucity of literature that presents actual data on the contribution of the TRAPD or PTRPD process to the overall quality of the translated survey instrument. This chapter describes an initial process evaluation of the Westat–National Cancer Institute (NCI) adaptation of the TRAPD methodology, presenting the results of five studies in which the translation efforts were based on the TRAPD framework. An evaluation of the documentation and outputs available from three specific procedural steps in the TRAPD process was made, and we report on the types of questionnaire design issues that were identified, and potentially remediated. A mixed-method (quantitative and qualitative) approach was used to develop conclusions concerning the effectiveness of the methodology in the translation process.

8.2 THE WESTAT–NCI TRAPD TRANSLATION PROCESS

The current TRAPD application generally entails a five-step process that affords opportunities at each step for evaluating and revising translated materials (see Forsyth, Kudela, Lawrence, Levin, & Willis, 2007). The revisions that derive from a given step and any relevant documentation are used as inputs for the following step (see Table 8.1). Although "D" appears last in the acronym, documentation is an ongoing component of each step.

The applications of TRAPD described in this chapter were executed as follows:

Step 1. Translation. In general, TRAPD incorporates team-based organization, translation production, and documentation. In the surveys considered here, a variety of specific procedures were in fact used, largely depending on the languages involved. The English-language source questionnaires were translated by either an in-house team of translators, by a translation vendor, or by a consultant (one individual or a team working in concert). Instructions to the translators varied by project. We do not view this variance in translation procedure as a serious limitation to the research given that our focus was not on the exact activities used within that step, but rather on the effectiveness of a multistep process that included features beyond simply translating the questionnaire. It is also important to note that for these projects in this study, translators generally conducted the translation process without producing supporting documentation describing problems within the source questions. Hence, later developmental steps relied solely on a critique of the translation by both the reviewers and the adjudicators (although each of those steps produced its own documentation).

Steps 2 and 3. Review and Adjudication. At each of these steps, bilingual staff reviewed the initial translations, in conjunction with the source (English) version, and suggested modifications based on their judgment and expertise. Adjudicators had the additional task of selecting between the version suggested by the original translator(s) or the reviewer, or deciding in favor of yet another version. Reviewers and adjudicators came to decisions on the basis of their (a) fluency in the source and target languages, (b) familiarity with cultural norms, (c) knowledge of survey measurement goals, and (d) experience in questionnaire design and survey methods. Having a team that fulfilled all elements of this skill set was seen as essential to arrive at decisions based on relevant survey considerations that are often beyond the expertise of the initial translator(s). Reviewers and adjudicators were required to document the anticipated defects in the translated items and their rationale for specific suggested revisions. In some cases, reviewers/adjudicators worked with a template that aligned the source and target text, thereby facilitating simultaneous note-taking. In other projects, they recorded their comments and suggestions in a format of their choice. In those cases, the documentation was not available in a readily accessible and comparable form. Given that the reviewer and adjudicator in each instance mainly focused on the quality of the translation, we

TABLE 8.1. Description of the Westat–NCI TRAPD Survey Translation Process

1. Translation	Development of a survey translation using a team approach.
2. Review*	Expert review of the translation(s) to identify problems and additional translation options.
3. Adjudication*	The adjudicator makes decisions to reconcile options from the preceding two steps.
4. Pretesting: cognitive interviewing*	Intensive interviews of language-appropriate respondents to identify difficulties in understanding and answering the questions and to identify translation issues that impede comparability.
Behavior coding of survey pretest	Field test of the survey translation and use of observational methods to identify potential problems with translated versions.
5. Documentation	At each step, compilation of qualitative and quantitative data that may be reviewed and further coded for purposes of report writing, quality control assessment, and evaluation research.

* Indicates the step was evaluated in the current investigation.

expected that their focus would principally be on flaws in the target (as opposed to source) questionnaire; that is, on the extent to which the target version effectively represents the original version. In the adjudication process, the initial translated version and the version produced through the review process were reconciled on the basis of the adjudicator's assessment of the most workable approach (with management team concurrence) and the translations were modified accordingly. For the current studies, qualitative data concerning identified problems with the items evaluated at both Review and Adjudication steps enabled us to undertake systematic analysis of the products of each of these steps (examples are provided in the Results section below).

Step 4. Pretesting. Table 8.1 illustrates a comprehensive pretesting approach in which cognitive interviewing (described in depth in Willis, 2005) and behavior coding (Fowler & Cannell, 1996) are applied in turn. Because behavior coding was used in only one of the studies described here, the current chapter is limited to evaluating the outcomes of cognitive interviewing, as it constituted the Pretesting step. Goerman (2006a) provides a description of the application of cognitive interviewing to translated questionnaires; the procedures she describes were generally followed for this research (see also Carrasco, 2003).

Step 5. Documentation. Finally, documentation of the potential problems that were identified was available from the Review, Adjudication, and Cognitive Testing steps. These three steps were therefore the focus of the current evaluation.

8.3 APPLYING THE TRAPD PROCESS IN FIVE STUDIES

Variants of the five-step process in Table 8.1 to develop questionnaire translations in five studies were applied. The studies involved different target languages (across all studies: Spanish, Cantonese and Mandarin Chinese, Korean, and Vietnamese) and varied survey content (tobacco use, diet, acculturation to U.S. society, and physical activity). Data from the Review, Adjudication, and Cognitive Interviewing (Pretesting) steps were analyzed to assess the problems identified at each of these steps, for each study. First, each study will be described briefly:

Study 1. The Tobacco Use Supplement to the Current Population Survey (TUS-CPS). The Tobacco Use Supplement (TUS) is administered both face-to-face and over the telephone. The English-language TUS was translated into the Asian languages by translation vendors and into Spanish by in-house Westat translation staff. These translators worked independently to produce four target language questionnaires (Spanish, Chinese, Vietnamese, and Korean). Review of these translations was conducted by survey language consultants (SLCs) with proficiency in the respective target languages. The Review step was designed to be in-depth and intensive, with detailed suggestions for revisions as the outcome of this process. The SLCs were given a template to assist them in structuring and documenting their comments. Adjudication was conducted by (tobacco-use) subject-matter experts having both translation expertise and a strong survey methods background. Cognitive interviews were then conducted with 41 respondents (see Willis, Lawrence, Hartman, et al., 2008).

Study 2. The National Health Interview Survey (NHIS), Cancer Control Module. The NHIS Cancer Control Module collects data on a wide array of cancer risk factors (Pleis & Lethbridge-Cejku, 2006) and was translated from English into Spanish. Professional translators from the Library of Congress produced an initial Spanish translation, using previously translated questions as a guide. The Review step was then conducted by a multiagency bilingual review team with content, language, and methodological expertise. Adjudication was conducted by an independent bilingual staff member at NCI who had extensive knowledge of the measurement objectives of the NHIS. This person also made the final decisions about revisions prior to cognitive testing. The current study focused on a subset of dietary questions within the NHIS for which bilingual survey methodologists conducted two rounds of cognitive interviews (nine respondents per round); the instruments were revised between the two rounds of cognitive testing and again after the second round.

Study 3. California Health Interview Survey (CHIS). The CHIS is a population-based telephone general health survey. Although it is translated into several languages, we focus on the translation process for a subset of items on diet and nutrition for the Chinese version. A professional translation vendor produced one Chinese-language questionnaire (intended for both Mandarin and Cantonese speakers) that was delivered to Westat for review. Review was conducted by external bilingual (Mandarin-Cantonese) experts hired and trained by Westat project staff. Adjudication was then conducted by an in-house expert who was fluent in both dialects and who had experience translating surveys from English into Chinese. In her role as the adjudicator, she made the final decisions about changes to the instrument. Subsequently, two rounds of cognitive testing were conducted (nine respondents in Round 1 and four respondents in Round 2) by bilingual staff with varying degrees of survey research experience. Respondents were of varying ages, income levels, countries of origin, and education levels. The questionnaire was revised between the two rounds of cognitive testing and after the second round.

Studies 4 and 5: Acculturation and Physical Activity. Westat simultaneously tested two independent sets of questions for the NCI: one centered on acculturation of various Hispanic groups to U.S. society and the other on questions on physical activity in everyday life. Although these two sets of questions were translated, reviewed, adjudicated, and pretested together, we treat them as separate studies for the purposes of this chapter, given the very different natures of the two topics and potential for fundamentally disparate varieties of problems to emerge. Translation was initially done by a single native Spanish speaker of Mexican origin who had prior survey translation experience. Review was carried out by two independent bilingual specialists with survey research experience and Latin-American Spanish language backgrounds. Adjudication was conducted by an independent expert who had prior experience as an adjudicator and a strong background in survey methods; this person made the final decisions regarding all revisions prior to the Pretesting step. A total of 18 cognitive interviews were conducted with Latinos by three bilingual survey methodologists, one of whom also had extensive translation experience. An additional nine interviews were conducted by both monolingual and bilingual interviewers. The respondents varied by mono- versus bilingualism and length of domicile in the United States. One respondent group consisted of 18 individuals who were interviewed in Spanish; and the other was composed of nine who were interviewed in English and who did not speak Spanish.

8.4 METHOD

For this evaluation study, we reviewed the documentation relating to the Review, Adjudication, and Pretesting steps for each project, to characterize the types of findings that emerged at each step. We documented the outcomes of each step for each question and for each study, using a general coding system that has proved useful in the past in providing a summarization of findings from cross-cultural, multilingual investigations. The *Translation, Cultural Adaptation, Generic Problem (TGC)* system (Willis & Zahnd, 2006; Willis, et al., 2008) is based on the finding that problems identified generally fall into three relatively distinct categories, defined as follows:

1. *T:* Problems that are linguistic in nature and due to defects in the *Translation* process (e.g., a term that is mistranslated or the presence of grammatical errors in the translated or target version).
2. *C:* Problems that transcend language in the structural sense, but represent problems of *Cultural* adaptation in which questions do not function appropriately in one or more groups due to specific features of that group (e.g., problems of worldview or structural differences between societies that create logical defects in the item).
3. *G: Generic* problems of question design that are found or judged to affect multiple cultural groups and are not culturally specific (e.g., general difficulty in recall of long-ago events); these are, therefore, problems that are contained in the source version that carry through to target translations.

8.4.1 Coding Procedures

Our study documentation provided detailed qualitative data that we coded according to the TCG coding system. We were able to use as raw material the written justification for recommendations (i.e., the anticipated or observed problems with each item) at the Review, Adjudication, and Cognitive Testing steps. To evaluate the unique contribution of the Review and Adjudication steps, only new problems that were identified at each step were coded. That is, if the adjudicator simply agreed with the reviewer that the original translation contained problems, this was not considered a problem first identified by the adjudicator (as it had already been identified during review). Similarly, for cognitive interviewing, coding was conducted for all problems that were first identified as a result of that Pretesting step. In summary, coding was conducted in a way that endeavored to identify the unique contributions of *each* evaluation/pretesting step, independent of the other steps, so that we could determine the efficacy of a given point in the question evaluation process.

Table 8.2 summarizes the analytic approach and contains information about the processes and products associated with each step. The heart of the analytic approach in Table 8.2 involves the identification of problems at each step (that is, the RAP steps of the total TRAPD process) by systematically coding all problems first uncovered at each step. Note that we have in all cases collapsed across items; rather than retaining information about particular survey questions, we chose to quantify at the code level to reveal the distribution of these codes across step and study.

As depicted in analysis step 2 of Table 8.2, coders identified problems whenever an item description indicated the potential for misinterpretation or erroneous answers. Coders were four Westat survey methodologists who had previous coding experience. Coder training consisted of group review of a training document to learn the coding system, and then distributing the documents from the five studies for independent coding. A set of coding rules was applied such that (a) a questionnaire item could have more than one problem and could, therefore, receive more than one code; and (b) a recurring problem that was essentially a single issue was coded only once. For example, an erroneous translation of a term that was repeated throughout the questionnaire was coded as such once, rather than every time it occurred. Once coding was completed, the codes were entered into a spreadsheet where they were aggregated and cross-tabulated.

Final assignment of problems to a T, C, or G code was done through consensus rather than by completely independent coding, largely because coding of this type was found to be difficult and required considerable discussion of nuance, This was a level of judgment that we felt was best adjudicated as a group process whenever coders had doubts about code assignment. In particular, it is sometimes difficult to judge whether a problem is culturally specific as opposed to generic. If the questionnaire had been pretested in both the source (English) and target languages, and there was empirical evidence indicating that a problem affected all groups, that problem was then classified as generic rather than culturally specific. However, there were other cases in which source-language testing had been conducted much earlier or not at all, and we lacked sufficient documentation concerning the English-language outcomes. In these cases, a decision as to whether

TABLE 8.2. Overview of the Three-Step Analytic Approach

Analysis Step	Process	Purpose
1. Review all documentation and detect problems	Identify problems based on documented evidence in each document at review, adjudication, and pretesting steps.	To identify problem items in preparation for coding of problem types.
2. Code each identified problem by problem type	Apply TCG coding system to describe types of problems found at each step.	To characterize precise issues found in each study at each step.
3. Tabulate problem codes and make comparisons across steps and across studies	Compare frequency and type of problems across steps and across study, collapsing over tested questions.	To explore the similarities and differences of results, across steps and studies.

a problem was generic or culturally specific was left to the judgment of the investigators based on a qualitative analysis of the nature of the problem.

As an example of this critical point, the testing of the TUS-CPS tobacco questions revealed that a question on how long one waits before smoking the first cigarette of the day often produced responses that fell outside the codeable range intended by the researchers. Rather than reporting a precise time such as "20 minutes," a respondent might answer "as soon as I get out of bed." This finding was replicated across each tested group—an empirical indication that it was generic in nature. On the other hand, for the Study on Acculturation and Physical Activity, interviews were conducted with Hispanics only. As such, the decision about whether problems identified were culturally specific versus generic in nature required a judgment about whether the finding represented a phenomenon that was unique to Hispanics or a more general violation of questionnaire design practice. For a question on physical activity that included items on "vigorous activity," "light or moderate" activities, "leisure" activities, and "physical activities specifically designed to strengthen your muscles," cognitive testing revealed that respondents had problems consistently recalling information and estimating activity frequency and activity duration. Given that the types of activities commonly reported (e.g., walking) were not specific to Hispanics, and because these results are consistent with those from prior cognitive testing of physical activity questions by different researchers (e.g., Willis, 2005), we concluded these to be generic in nature.

8.5 RESULTS

8.5.1 Problems Identified at Each Step: Review, Adjudication, and Pretesting

The first issue addressed in the analysis concerned the frequency of coded problems in translated items that were identified at each of the three evaluation steps (collapsed across individual codes). Table 8.3 summarizes the distribution of these problems for each study, broken down by step. Overall, each of the steps was generally successful in identifying problems, although the relative frequencies of these problems varied across studies. In all studies, either the Review or Adjudication step identified the highest frequency of problems, and the Pretesting step was either second or third, quantitatively (for example, in the NHIS, the respective percentages of problems identified at Review, Adjudication, and (cognitive) Pretesting were fairly even, at 38.6%, 25.3%, and 36.1%, respectively). Given that the cognitive interviews were done last in the series (as a "catch-all"), it may not be surprising that they generally produced a smaller number of problems.

8.5.2 Problems Identified by the TCG Coding System

Next, we ascertained the distribution of TCG codes for each of the five studies, to determine the degree to which these problem categories were common, as opposed to specific to the study (Table 8.4). Translation was by far the most frequently applied code in all five studies, and the relative frequencies of cultural and generic problems varied across the studies (e.g., in the CHIS, we identified a single generic problem, but nine cultural-adaptation problems; whereas, in the Physical Activity Study, the pattern was opposite).

8.5.3 Problem Distribution Across Steps

Finally, to examine the extent to which the evaluated steps detected similar types of coded problems, Table 8.5 reveals the distribution of the TCG codes across the

TABLE 8.3. Distribution of Coded Problems as a Function of Study and Step

Step	TUS		NHIS		CHIS	
	n	%	n	%	n	%
Review	188	39.0	32	38.6	33	55.0
Adjudication	228	47.3	21	25.3	9	15.0
Pretesting	66	13.7	30	36.1	18	30.0
Total	482	100.0	83	100.0	60	100.0

Step	Acculturation		Physical Activity	
	n	%	n	%
Review	13	32.5	22	53.7
Adjudication	17	42.5	14	34.1
Pretesting	10	25.0	5	12.2
Total	40	100.0	41	100.0

TABLE 8.4. Frequency and Percentage of TCG Problem Type, by Study

TCG Problem Code	TUS		NHIS		CHIS	
	n	%	n	%	n	%
Translation	444	92.1	58	69.9	50	83.3
Culture-Related	5	1.0	5	6.0	9	15.0
Generic Design	33	6.9	20	24.1	1	1.7
Total	482	100.0	83	100.0	60	100.0

TCG Problem Code	Acculturation		Physical Activity	
	n	%	n	%
Translation	26	65.0	30	73.2
Culture-Related	6	15.0	0	0.0
Generic Design	8	20.0	11	26.8
Total	40	100.0	41	100.0

TABLE 8.5. Distribution of TCG Problem Codes Identified at Each Evaluation Step

TCG Problem Code	Review Step		Adjudication Step		Cognitive Interview Step	
	n	%	n	%	n	%
Translation	282	97.9	280	96.9	46	35.6
Culture-Related	1	0.4	2	0.7	22	17.1
Generic Design	5	1.7	7	2.4	61	47.3
Total	288	100.0	289	100.0	129	100.0

Review, Adjudication, and Cognitive Interviewing steps, collapsed over study. The Review and Adjudication steps were found to have focused almost exclusively on translation problems. This is not surprising as these steps are nominally designed to locate issues of language translation. In particular, reviewers and adjudicators are primarily concerned with the degree to which the translation is a reliable linguistic representation of the source instrument. These evaluators may also be sensitive to cultural issues that make translation difficult, but they are generally not led to consider generic problems in the source questionnaire. Pretesting was much more balanced with respect to the categories of problems identified; overall, it located more generic questionnaire problems than either translation or cultural problems, yet the latter two categories were still fairly frequently represented. In all, almost 85% (83 of 98) of generic and cultural problems were identified through cognitive testing. As such, cognitive interviewing appears to provide the widest "net" of any of the evaluated steps in terms of variety of problems identified. Interestingly, although cognitive testing might not be expected to reveal translation issues because earlier steps had already presumably identified such problems, and because the focus of cognitive testing is not to identify translation problems, translation problems were, nonetheless, identified at a nontrivial level.

8.6 QUALITATIVE NATURE OF PROBLEMS DETECTED

In this section, we provide key examples of problems that were identified at each step.

8.6.1 Problems Identified During the Review Step

1. *Translation problem.* For the Chinese versions of the CHIS instrument, reviewers suggested replacing an incorrect translation (the equivalent of "Did you eat other vegetables?") with the correct form (in effect: "How many times did you eat other vegetables?"); and replaced formal characters for foods with more common characters.
2. *Generic (source instrument) design problem.* For the acculturation-related question "How do you feel about being [Hispanic]? Would you say you feel very good, good, somewhat good, or not very good?" The review documented a typographical error in the last English-language response category, which was originally stated as *no very good.*
3. *Culture-related problem.* For the CHIS item: "During the past month, how many times did you eat other vegetables? Do not include rice or foods you already told me about," translators had used examples of foods specific to the Chinese culture as examples of *other vegetables* (daikon/turnips, leafy vegetables, and lotus root) instead of the examples contained in the original English version (tomatoes, carrots, and broccoli). However, the Review documentation suggested that inclusion of *leafy vegetables* as a cultural adaptation might produce a further problem because *leafy vegetable/salad* was already asked about elsewhere in the questionnaire, and could result in confusion or in a food being counted twice. As such, an attempt to adjust for cultural differences, though well-intended, resulted in a further problem for a particular (Chinese) group.

Overall, the fact that the Review was successful in identifying basic translation errors, a mistake in the original English version, and a culturally oriented difficulty, illustrates the value of this step.

8.6.2 Problems Identified During the Adjudication Step

1. *Translation problem.* To better express in Spanish the intended notion of the English term *describes* in the question "Which of the following best de-

scribes the people in your neighborhood? Would you say all of them are Hispanic, most of them are Hispanic, half of them are Hispanic, few of them are Hispanic, or none of them are Hispanic?" translators noted that a clarifying noun needed to be added after the phrase, "the following." The first translation added the word *declaraciones* (i.e., "which of the following declarations…"). This was replaced after Adjudication with *frases* (phrases) because the adjudicator felt that *declaraciones* was too formal (i.e., it would be more appropriate for a "Declaration of Independence").

2. *Generic design problem.* One problem concerned the distinction between the words *and* versus *or*: For the introduction to the question series, "The next questions are about physical activities like exercise, sports, *and* physically active hobbies that you may do in your LEISURE time," the Adjudicator proposed changing the English version *and* to *or*, so as to indicate that the question could be about any of the examples (rather than all). A corresponding change was made to the Spanish version [replacing *y* (*and*) with *o* (*or*): ("Las siguientes…como ejercicio, deportes, o aficiones físicamente activas")].

3. *Culture-related problem.* The following question presented a problem related to how commonly a given term might be misunderstood by a particular group: "How often did you drink 100% FRUIT JUICE or 100% fruit juice blends, such as orange, mango, apple, and grape juices? Do NOT count fruit drinks. Read if necessary: INCLUDE only 100% pure juices. Do NOT include fruit drinks with added sugar, like cranberry cocktail, Hi-C, lemonade, Kool-aid, Gatorade, Tampico, and Sunny Delight." The adjudicator noted that *arandano* is the correct translation for *cranberry*, so there was, in strict terms, no translation problem. However, the fruit is not likely to be well known in Latin America. The adjudicator therefore suggested adding the English word *cranberry* as well, since it is likely that it would be better recognized by Hispanics residing within the United States.

We note that the translation problem concerning *declaraciones* above also does not represent an actual error, but rather a more subtle issue of nuance of meaning. Hence, the adjudicator's role sometimes involved revisiting issues that involve judgment and choice, instead of identifying outright errors. Interestingly, for the physical activity item for which *and* was replaced by *or*, the adjudicator's contribution was to identify a subtle error in the original English version, demonstrating how a comparison across multiple language versions can result in a modification to any of these. Presumably, a bilingual expert noticing a potential issue in a Spanish version is led to the source version to check the original objective and may conclude that the problem is not one of mistranslation but of unsuitability of the original (a "garbage-in, garbage-out" phenomenon). Ironically, translators may be faithful in their use of terms across a questionnaire, without noticing that this is ultimately counterproductive if the source version contains an error. Similarly, in deciding that *cranberry* should be inserted into the Spanish text alongside *arandano,* the Adjudicator, as one would hope, made judgments other than simply whether the translation was technically correct. Here, the value added by the Adjudicator is the recognition that a correct but little-known term may prevent respondents from recognizing what is meant.

8.6.3 Problems Identified During the Pretesting Step

1. *Translation problem.* Within the Tobacco Use Supplement, the Chinese translation of one item was found to be clearly incorrect, resulting in a completely different question. The English version read, "Have you ever switched from a stronger cigarette to a lighter cigarette for at least 6 months?" However, the Chinese translation was "Has it been more than half a year since you have switched from regular to light cigarettes?" reflecting both a serious conceptual error (switching for six months versus six months since switching) and a more subtle inconsistency (six months versus half a year).

2. *Generic source design problem.* For "Which is the MAIN reason you switched from a stronger to a lighter cigarette—as a way to try to quit smoking, or in order to smoke a less harmful cigarette?" the cognitive testing report indicated that some respondents reported that both less harm to their health *and* to help quit smoking were equally important reasons for switching from stronger to lighter cigarettes. However, the original (and, therefore, the translated) response options did not include "both" as a choice.

3. *Culture-related problem.* The item "Have you EVER SWITCHED from a stronger cigarette to a lighter cigarette for at least 6 months?" appeared to pose a problem for respondents who started smoking in their native country using a Korean brand of cigarettes which does not list tar and nicotine amounts on the package and then later switched to an American brand which is labeled.

Notably, and consistent with other reports (Willis et al., 2008), cognitive interviewing identified outright translation errors that should, in theory, have been identified in previous steps. The fact that the question concerning switching for 6 months was completely mistranslated had slipped by both the reviewers and adjudicators. The reasons for this are unclear because there is no way to tell from documentation at these steps *why* a problem would have been overlooked (even if the documentation includes specific notes on this item, these notes may not indicate the reasons that the problem was missed). On the other hand, it is fully understandable that generic problems are first identified at this point, as such identifications of basic errors is one of the fundamental functions of cognitive interviewing. Similarly, the culturally specific problem that respondents are unable to report on the strength of Asian-made cigarettes may not be recognized even by culturally knowledgeable translators, reviewers, or adjudicators (especially if they are non-smokers), and is precisely the type of problem that is best identified through an investigative, empirical method such as the cognitive interview.

8.7 DISCUSSION

Based on the patterns of results we observed, we make several conclusions and recommendations for translation, in the context of appropriate caveats and suggestions for further research that will clarify unresolved issues. First, based on the finding that multiple steps each identified nontrivial levels of problems, we believe it is useful to rely on a multistep approach. Overall, as suggested by both the qualitative and quantitative results, a variety of problems were identified, and each fundamental type of the problem may be identified at any step. This observation reinforces the notion that *a multistep method provides a series of filters that function in tandem to ferret out both obvious and subtle defects in the questions, relating to both the source and the target versions.*

Given that each of the steps in the TRAPD process serves a useful function, it is difficult to envision a comprehensive procedure that eliminates any particular step. It may be possible to combine the steps devoted to expert-based review of the instrument—e.g., Translation, Review, and Adjudication might all be done by the same team, as the use of multiple team members may promote some degree of self-criticism and consideration of a range of options. However, it still seems that a preferred alternative would be to have separate groups or individuals carry out these steps, as it is likely that "more eyes are better." Or, based on the finding that cognitive testing satisfied multiple objectives, one might consider translating and then submitting the questionnaire to a cognitive testing process that is designed to incorporate Review and Adjudication as well as Pretesting. For example, cognitive interviewing staff could evaluate the translated questionnaire as part of the process of preparing it for cognitive testing. However, this may be asking too much of cognitive interviewers, whose expertise may be too specific to fulfill all of these functions. Or, a combined approach might well prolong the process so that little cost or time savings result.

Our second basic conclusion is that, despite the prior efforts of the developers, generic problems of question design that exist within the source questionnaire are common and difficult to eradicate. Certainly the process of effective translation may not resolve these problems, as they are simply carried through from the source version to each translation. However, the mechanisms used to evaluate the translations—and especially cognitive interviewing—appear to be useful in identifying these issues.

Significantly, problems of cultural adaptation—where between-group differences in worldview or social behavior transcend language translation and produce noncomparability of survey items—appeared to be less frequent than either pure translation problems or general design issues. This could be because all five studies took place within the United States and so exhibit a greater degree of shared common ground than is the case in cross-national studies. Of course, one can reasonably argue that to the extent that culture-related problems do exist, they pose particularly severe threats to data quality, as they very likely produce nonignorable bias in cross-cultural comparisons (unlike generic problems, which would be expected to simply introduce error equally to all contrasted groups). As such, it is reassuring that these problems are identified, again largely as a function of cognitive interviewing, which may be especially proficient at bringing a range of otherwise hidden problems to the surface.

Finally, and in summary, we note that the evaluation steps, within a TRAPD framework, differ in what they accomplish. Pretesting may be the least productive quantitatively, as it identified a smaller absolute number of problems in most cases than either Review or Adjudication [and this finding is consistent with other research that quantifies the results of pretesting (Presser & Blair, 1994; Willis, 2005)]. From a qualitative point of view, however, cognitive interviewing may be vital in focusing on what DeMaio and Rothgeb (1996) have called "silent misinterpretations" that are only elucidated through the process of verbal probing of actual test respondents.

8.8 CAVEATS

The conclusions above must be tempered by several limitations of the current study. First, the current focus involves process evaluation, as opposed to evaluation of verifiable outcome measures. Therefore, we possessed no absolute criterion measure of either item quality or response error by which to determine whether the problems that were recorded and then coded as part of each of the five studies were real ones or artifacts of our research procedures. However, the types of problems identified are very similar to those described in previous cross-cultural studies (e.g., Miller, et al., 2005a), and, therefore, appear to be fairly ubiquitous in such studies. Future studies would benefit from the inclusion of criterion data in the form of flaws that are purposely embedded in evaluated questionnaires; this procedure would provide a metric of the degree to which known (or at least, expected) problems are identified at each evaluation step.

Second, the basic categorization of the type of problems identified—translation, cultural, or generic—is very rough; we have not established it as a reliable coding system (although it is very similar to one developed independently by Fitzgerald, Widdop, Gray, & Collins, 2008). Rather, we have viewed this as an initial indicator that leads to viewing problems in translated instruments according to a broad conceptual model. However, to more systematically and rigorously evaluate such a system, it is vital to precisely define the coding parameters and definitions and to empirically assess inter-coder reliability.

A third limitation involves scope: despite the considerable amount of effort expended in aggregating and then coding results across five investigations, the current study is limited to one country and to a single testing organization and involved a single instantiation of the core TRAPD process. It is, therefore, unclear whether the results extend to the broader environment consisting of multiple locations (or countries) and across multiple questionnaire design and evaluation organizations that may implement key procedures in a different manner than we did. We were also constrained in scope in the sense that our evaluation of the TRAPD translation and evaluation model focused on only three of the key steps (Review, Adjudication, and Pretesting). In some ways this may be appropriate to the extent that the Translation step in particular is viewed as procedural rather than evaluative, and may not give rise to data of the type that are analyzable using the current approach. Perhaps this situation could be rectified to the extent that translators are explicitly instructed to document any problems that they see in the source questionnaire or problems they experience in conducting translation, as recommended in the European Social Survey guidelines (Harkness, 2002, 2007).

Another limitation that should be pointed out concerns the availability of resources, funding, and time necessary to accomplish each step in our evaluation process. Although we did not track the specific level of resources required for each step in any of our five studies, the resources available to adequately accomplish the Translation, Review, Adjudication, and Pretesting steps were, of course, limited for any study. We developed the necessary procedures and networks of contacts to identify experts, but that process required significant time. In particular, Survey Language Consultants (SLC) needed to identify additional experts who could provide the necessary level of independent review. Based upon our experiences, screening criteria for identifying these experts are critical, and the most important factor for accomplishing this type of work may be reliance upon an SLC (or other language coordinator) who has survey experience; that is, an individual who is experienced in survey-oriented translation and understands the requirements of designing a survey instrument.

The SLC needs to have access to the staff necessary to implement the various steps in the process. Proper supervision and training of this staff is crucial to the success of the translation and evaluation project. To complete the project in a timely fashion, all of the staffing considerations need to be considered prior to the start of the project. Overall, a network of language experts who are trained in survey methods is essential to the successful implementation of this approach.

In conclusion, a multistep model such as TRAPD appears to be effective in identifying potential problems in the translation process, across a range of studies. This approach may represent an effective broadening of expert review and pretesting procedures in a way that will lead to enhancements of data quality across multilingual, multicultural, and cross-national studies.

9

Developments in Translation Verification Procedures in Three Multilingual Assessments: A Plea for an Integrated Translation and Adaptation Monitoring Tool

Steve Dept, Andrea Ferrari, and Laura Wäyrynen

9.1 INTRODUCTION—BACKGROUND

In the past two decades, there has been a proliferation of international comparative studies that assess knowledge and skills in student and adult populations. The data collection instruments developed for international surveys usually consist of cognitive tests and background questionnaires: The outcomes of the cognitive tests are reported on competence scales, while data from the background questionnaires are reported as sets of variables or indices that may explain differences in performance. This chapter describes recent developments in verification procedures used to monitor the quality of translated/adapted versions of cognitive tests in three international surveys and advocates the use of a single monitoring tool to document the entire life cycle of each language version of a test item. The proposed model of such a monitoring tool could be generalized to different types of multilingual and multicultural data collection instruments.

The test instruments considered here may assess respondents' component skills, their competencies in science or mathematics, and their ability to read, retrieve information, make inferences, and analyze texts. They are used to generating indicators, rankings, and international reports that are regarded as highly sensitive at national and supranational levels. While health status and intelligence test instruments are sometimes referred to as high-stake instruments because the outcomes may decide respondents' access to amenities, life course opportunities, career and educational benefits, the tests discussed here are only high-stake at the political level, in that results may be used to inform education policies. They are also widely used by economists to measure the human capital in various regions, or by policy makers to target reforms. Thus every effort must be made to ensure that the data are reliable, valid, and comparable across countries. This hinges on four methodological areas of concern: reaching a consensus on the assessment framework, defining sampling procedures, standardizing test administration conditions, and ensuring that the translated/adapted instruments are equivalent across countries (Hambleton, 2002). Recent developments in the fourth methodological area are examined here.

In briefly describing the translation and adaptation methods and discussing the verification procedures used, the authors will identify practices in each phase that (a) have been used to detect problems in translating and adapting the instruments, (b) resulted in thorough documentation, and (c) provided good proxies for indicators of quality in translation and adaptation. This will lead to the final section, in which a model for an integrated monitoring tool is proposed: This is a file in which every stage of the life cycle of a translated/adapted item is recorded.

Procedures and examples from three international surveys are described in this chapter to illustrate how translation verification practices evolve toward an increasingly structured process. In all three surveys, (OECD PISA 2006; IEA TIMMS 2007; UNESCO LAMP 2006–2007) the production of national versions of test instruments relies extensively on translation of one or several international source versions, which is hence viewed as a high-stakes process. As a result, the central organizations responsible for producing the instruments have been willing to finance rigorous assessment of the translations to target stringent quality standards. The procedures presented here are the result of this investment and include the verification of the translated/adapted instruments and the documentation of the verification process. The three studies are multilingual, cross-cultural projects using complex methodologies to translate and adapt the tests and to document this process.

OECD/PISA (2006)—The Organization for Economic Cooperation and Development (OECD) Programme for International Student Assessment (PISA) measures reading literacy, scientific literacy, and mathematical literacy in 15-year-old students. Released sample items are available on http://www.pisa.oecd.org/dataoecd/30/17/39703267.pdf.

IEA/TIMMS (2007)—The International Association for the Evaluation of Educational Achievement (IEA) Trends in International Mathematics and Science Study (TIMMS) aims to measure trends in student achievement in mathematics and science by collecting data from students in fourth to eight grade at 4-year intervals. Released sample items are available on http://timss.bc.edu/TIMSS2007/frameworks.htm.

UNESCO/LAMP (Pilot test in 2006–2007)—The United Nations Educational, Scientific, and Cultural Organization (UNESCO) Literacy Assessment and Monitoring Programme (LAMP) aims to measure a spectrum of literacy levels in developing countries. The tested populations are adults and young adults (15 or older) not attending school.

The authors of this chapter were involved in the translation verification of PISA 2000 instruments from 1998 on. Across survey cycles they have gradually developed more sophisticated verification frameworks, training modules, and follow-up instruments. They are currently responsible for the verification of PISA 2009 survey instruments and are preparing the verification of instruments for PIAAC (Programme for International Assessment of Adult Competencies), TIMSS 2011, and PIRLS 2011 (Progress in International Reading Literacy Study). They have also coordinated the translation verification of a number of other multi-lingual surveys, and lessons drawn from those exercises have greatly impacted veri-fication practices as well as the preparation and documentation of the same. This chapter describes these developments and discusses possible uses of verification outcomes.

9.2 QUALITY ASSURANCE AND QUALITY CONTROL

When the PISA survey was launched in 1998, the *quality assurance* framework for translations (i.e., the methodology developed to ensure that the translation proce-dures would be capable of delivering equivalent tests in all languages) hinged on (i) the double translation procedure rather than the translation/back translation proce-dure and (ii) the use of two source languages rather than one. In the back translation procedure, which was widely used before PISA and TIMSS, the first translator was the only one to focus on the source version and the target version simultaneously. In a double translation design, equivalence between the source language(s) and the tar-get language must be achieved by three different players (two translators and a recon-ciler), who all work on both the source and the target version. The PISA Consortium produced a set of detailed translation and adaptation guidelines (Grisay, 2003).

The *quality control* framework (i.e., the methodology developed to ensure that the result of the double translation and reconciliation procedure was indeed satisfactory) was practically nonexistent. Since the translators and reconcilers of the participating countries were requested to follow detailed translation and adaptation guidelines, quality control had to consist in verifying whether the translated versions complied with those guidelines. An international team of independent professional translators with experience in teaching and/or in psychology, expert knowledge of their mother tongue, and proficiency in both French and English—the two source languages in PISA—was hired to assess the target versions and report on their findings. They received a long verification checklist of aspects that needed to be verified. This key document was based on the PISA translation and adaptation guidelines.

In essence, the quality assurance framework for translation and adaptation of the three multilingual surveys described here consists of a well-defined set of stand-ards, rules, and recommendations. The quality control framework consists of a set of measures designed to monitor whether the standards were met, whether the rules were observed, and to what extent recommendations were followed. The relevance of verification can, in part, be assessed through the quality of the final versions of the instruments as expressed through item communalities across language versions, and, in part, through systematic and standardized reporting of the verification work.

In the early days of such translation verification, the deliverables were (i) a proofread and annotated target version and (ii) a short, subjective, qualitative report. We describe here developments that led to our proposing a more standardized form for the second deliverable, one that is designed before the translation/adaptation process begins and which can also generate statistics on the categories of issues identified by verifiers. In this way it adds a quantitative dimension to the evaluation of the national versions. However, translation verification is only one set of quality control measures; it needs to be combined and interconnected with other methods used to assess cross-country and cross-language comparability.

9.3 THE TERMS "TRANSLATION" AND "ADAPTATION"

The words *translation* and *adaptation* are used jointly throughout this chapter because the term translation was deemed too restrictive to describe the process of culturally adjusting a test rather than just translating it literally (Joldersma, 2004). While professional translators can receive specific training to translate test material and do so successfully if they are made aware of a number of survey-specific issues (e.g., the relative length of the key and the distracters in multiple choice items), adaptations typically require the input of domain experts, psychometrists, and/or test developers. The difference between translations and adaptations is that the latter consist in deviations from the source version(s) that are deemed necessary to maintain psychometric equivalence.

An adaptation may consist in changing the picture of a stimulus, in changing the combination of July/summer to July/winter (or, alternatively, January/summer) for the Southern hemisphere, in changing a coeducational school context to a boys' or girls' school context for certain countries, and so forth. It may involve a change of wording, register, context, currency, measurement unit, or form of address.

One interesting example was found in the LAMP pilot test: A test designed to measure respondents' competencies in sentence processing included a set of short statements, which the respondents had to label as true or false. One of the sentences was "All plants need light to grow." The Tuareg from sub-Saharan Africa claimed that, since in the desert there is a lot of light all year and nothing grows at all, it was less obvious for Tuareg respondents to identify this statement as true than it was for somebody living in a place where sunlight is at times scarce. Since the test item was intended to assess sentence-processing skills rather than scientific literacy, the statement could be adapted to "All plants need water to grow." In the same test, the statement "A stranger is someone you know well" was intended to be labeled as false, but that would not work if the statement were merely *translated* into Hausa (West Africa), because in this language the word for *stranger* is identical to the word *guest*, and a guest *could* be someone you do know well. An "adaptation," that is, an intentional deviation from the source, was therefore needed. In this particular case, the test developers were made aware of the problem and proposed "A guest/stranger you meet for the first time is someone you know well."

Also, within one language, there may be several speech communities in which word usage may vary. Therefore, lexical or cultural adaptations may be needed even when the test language for a given country is the same as the source language. In the United States, luggage may be stowed in the trunk of a car, but in Great Britain it would be stowed in the boot of the car.

9.4 WHAT IS TRANSLATION VERIFICATION?

In the three multilingual surveys referred to in this chapter, every translated/adapted target version, or national version, was submitted for verification. The word *verification* as used in this chapter refers to a combination of checking the linguistic correctness of the target version and checking the "equivalence" of that target version against the source version. While the linguists who perform the verification were trained to be aware of specific aspects that may affect psychometric properties of test items, the authors are not convinced that it is actually possible to check the psychometric equivalence of test items merely by comparing a translated/adapted version to the corresponding source version. Thus "equivalence" refers to linguistic equivalence, including equivalence in quality and quantity of information contained in a stimulus or test item, as well as equivalence in register or legibility for a given target audience.

When test instruments are piloted, including cognitive laboratory testing, it is often the case that some items do not work the way the test developers expected. Analysis of field test results using, for instance, the Item Response Theory model (IRT) may reveal cases of Differential Item Functioning (DIF) for certain test questions in certain countries, and this may lead to identifying residual translation issues or ambiguities that verifiers may have overlooked. The case analysis has to be carried out *post hoc*. When verifiers check equivalence *ex ante*, they can check whether all the information contained in the source is also present in the target version, whether the register remains the same, whether the level of difficulty of the test items is likely to have been affected by linguistic or syntactical artefacts, whether hints for the correct responses are not added or removed, and so forth. The verifiers follow the verification checklist and try to assess as many aspects of equivalence as possible, but this process has limitations.

However, useful information was derived from each verification exercise allowing us to develop and refine the translation verification practices. The current procedures have a predominantly empirical foundation: For example, residual bugs identified in test items after translation verification have led to reformulation of the verification guidelines and checklists. Training of verifiers includes hands-on exercises in the form of target versions in which a number of errors were introduced by the trainers. In these exercises, verifiers need to identify translation/adaptation issues that may typically affect item difficulty, and they also need to report their findings in a standardized way.

9.4.1 Development of an Empirical Typology of Verification Interventions

The authors and their team carried out an in-depth analysis of 5,380 comments made by verifiers in the course of the PISA 2006 Field Trial verification. Based on these findings, a set of categories was defined to accommodate all the verifiers' interventions ("interventions" here covers both corrections and suggestions). The resulting classification consists of eight intervention categories, whereby every correction proposed by verifiers from over 40 countries could be assigned to one category. This empirical typology was drawn up to reduce variability in the way verifiers report their findings. It is now integrated in the tools used to document the verification and provides information on the quality of translated/adapted versions that can also be interpreted by persons not conversant in the target language. This framework is currently used in both PISA and LAMP verification. Table 9.1 presents the eight categories used and their definitions.

A different method was used in verification of TIMSS. Here verifiers had to use severity codes to qualify their interventions. The verifiers selected a severity code from four different codes: Code 1 for major deviation or error (likely to affect item functioning); Code 2 for minor deviation or error (e.g., purely linguistic error that does not affect content or equivalence); Code 3 for suggestions for improvement (a translation is adequate but verifier suggests alternative wording); and Code 4 to record an acceptable change (also known as "appropriate but undocumented adaptation"). Verifiers were also trained with hands-on exercises to work with typical examples of issues that would call for each code.

9.4.2 Relevance of Verification Feedback for Item Development

Using the typology developed for PISA, the number of interventions for each category and in total can be reported per national version, but also per test unit. (A test unit is a stimulus, one or several questions related to that stimulus, and possibly a set of coding rules). Reports per national version are certainly informative, but they must be interpreted with care for a number of reasons: A very strict verifier, for instance, may choose to comment on very minor issues and make many suggestions even in a well-translated version. Any resulting high number of verifier interventions would then not mean that there were actually serious flaws in the translation.

One of the most interesting aspects of this type of analysis may be the indicator of "translatability" or "adaptability" of the test material. Researchers and test developers may find it useful to identify items or parts of items that result in translation difficulties in several languages. Additional analyses can be carried out on verifier intervention statistics by unit, after sorting the national versions by language groups. Item developers and researchers could then identify what items led to a high number of verifier interventions in a given language group. Such cases could be further investigated at the country or verifier level to determine what the nature of the issue involved (translation issue, cultural issue, etc.).

Regrouping the verification outcomes in this way also makes it possible to identify items that elicited many verifier interventions in almost all language groups. In these instances, item developers would be prompted to re-examine the item's reliability or relevance.

The severity codes used in TIMSS to distinguish between major error, minor error, and suggestions for improvement have greater operational usefulness than do intervention categories, but provide much less information. We later propose that verification monitoring tools should combine both systems to obtain and centralize as much feedback as possible from each verification exercise.

TABLE 9.1. Categories and Definitions Used in PISA and LAMP Verification

Added information	Any information given by the target version not given by the source version. It can consist of one word or a group of words, such as an explanation of a preceding word.
Missing information	Any information given by the source version not given by the target version. Can consist of one missing word or a missing group of words
Layout/visual issues	All layout-related aspects: unit formatting including stimulus layout, item headings and styles, item labels, question numbering, **boldface**, <u>underlining</u>, *italics*, graphs and legibility of captions, tables, answering lines, and the relative length of multiple choice responses.
Grammar/syntax	Grammar: Grammar mistakes that may affect equivalence, e.g., wrong subject-verb agreement, wrong case (inflected languages), conjugation error (wrong mood, wrong verbal aspect).
	Syntax: Syntax-related deviations from the source, e.g., a long (source) sentence is split into two (target) sentences or two (source) sentences are merged into a single (target) one; different tense is used; passive voice/active voice. Or syntax-related awkwardness due to overly literal translation of the source version(s).
Consistency	Within-unit consistency: Literal matches and synonymous matches that occur in the target version must reflect the *pattern* in the source versions. If a given word is used several times in the source version of a unit, the (single) translation of that word should occur several times in the target version. Conversely, if a synonym is used, the target version should also use a synonym (and not repeat the word).
	Across-unit consistency: Recurring elements that occur in a number of units should always be translated the same way, measurement units should be written the same way, etc.
Register/wording	Register: Difference in level of terminology (scientific term >< familiar term), difference in level of language (elevated >< casual).
	Wording: Choice of vocabulary, choice of appropriate wording to convey the same information as in the source version.
Adaptation	All intentional deviations from the source version made for cultural reasons or to conform to local usage. An adaptation is needed when there is a risk that respondents would be disadvantaged or advantaged if a translated version without adaptation is used.
Mistranslation	Mistranslation refers to a wrong translation which seriously alters the meaning. Mistranslations should always be reported together with a back-translation that conveys the mistranslation involved. A vague or slightly inaccurate translation is not a mistranslation in this strict sense.

9.5 TRANSLATION AND ADAPTATION GUIDELINES

In both PISA and LAMP, quality assurance for translations was underpinned by the comprehensive translation and adaptation guidelines, whereas in TIMSS quality standards were defined but the guidelines were briefer and more general. In the tension between standardized and customized procedures—an important and ongoing debate within the framework of international surveys—the way TIMSS was clear rules and recommendations. Conversely, in LAMP, translators had less leeway to adapt test items, because very detailed item-per-item guidelines were provided.

Due to the sophistication of the procedures required, the PISA Translation and Adaptation Guidelines are presented to the national centers (NCs) of participating countries in the form of a comprehensive manual. This document contains general indications about the type of materials, security requirements, criteria for the selection of translators and reconcilers, file management, harmonization, and errata. A section titled "Recommendations about translation traps in test materials" provides a list of translation problems that are specific to assessment material together with advice on how to address them. It also includes real examples drawn from previous survey material. It provides the translators with instructions on layout and graphics (issues related to page setup, typographical cues, etc.); linguistic difficulty level (factors that may affect the difficulty level of the item: length of the sentences, use of common vs. scientific terminology, idioms, metaphorical expressions, etc.); and common psychometrical traps (such as relative length of key and distracters in multiple-choice items, literal matches, and order of information).

In IEA/TIMSS (2007), translators were provided with a rather brief document. In addition to general guidelines for translating test material (e.g., advice about maintaining the same difficulty level), a limited number of more specific guidelines for adapting the items to national context were included, such as adapting names of people, geographical names, date formats, and measuring units. The few examples given were nonetheless illustrative. One was a question in a mathematics test asking "How many sides does a hexagon have?" If in the target language the word "hexagon" can only be translated as "six-sided figure," there will be a problem with the corresponding national version of that item. To avoid having the question also contain the answer in that language, it is clear that an adaptation (or the suppression of that item) would be necessary. However, the absence of specific item-per-item instructions for IEA/TIMSS proved to be a weakness: Translators from different countries stumbled over the same obstacles.

Conversely, the UNESCO/LAMP translation and adaptation guidelines document describes authorized and ruled-out adaptations item-by-item, providing detailed information about when it is acceptable to adapt, for instance, the measurement unit, the name of a person or of a fictional organization, or graphic material. They also indicate required literal or synonymous matches. For example, in a text about a marathon swimmer, it is specifically stated that 28-mile laps can be adapted to 45-km laps, but also that what the swimmer ate during her marathon must occur in a sentence in the third paragraph of the first column of the stimulus. For a question asking what the swimmer *ate* during her swim, it is specified that the translator should avoid using the verb *ate* in the stimulus.

9.5.1 Relevance of Translation/Adaptation Guidelines for the Verification Process

The three different translation and adaptation guideline documents provide different kinds of support but, in essence, they consist of general instructions, for instance, about maintaining the same difficulty level in cognitive tests, as well as more specific instructions, such as on adapting names or currencies. Our experience leads us to advocate a more condensed general section and to focus the greatest possible effort in assisting translators to spot those issues that require special attention. This will not greatly increase a project's budget, but the benefits in terms of translation quality are immense. In terms of cross-language comparability, it is useful if the guidelines specify strictly what can be adapted and what not, and provide test developers' hints, wherever as possible, about the type of adaptation that would be acceptable. For example, synonymous or literal matches between the stimulus and the questions should be identified as well as whether it is crucial to reflect these matches in the translated/adapted version.

Verifiers perform a sentence-by-sentence comparison of the translated/adapted version against the source version and either eliminate or report discrepancies. In addition, they are required to report whether each item-per-item guideline was taken into account. Verifiers acknowledge they have checked whether the specific guideline was addressed in check boxes or drop-down menus in a predefined monitoring instrument that echoes each guideline. Requiring verifiers to refer to guidelines repeatedly like this greatly increases the likelihood that translation flaws or problems that may affect equivalence will be detected.

9.6 TRANSLATION AND ADAPTATION PROCESSES

To gain insight into the translation/adaptation processes involved, it is convenient to present them as a necessary interaction between two levels of test administration: the International Project Centre (IPC) and national centers (NCs). The IPC develops instruments, defines standards, provides translation and adaptation guidelines, and organizes quality control. The NCs of participating countries have to organize the data collections and, before that, must recruit translators locally to translate and adapt the source version of instruments into the languages in which they will be testing.

For PISA 2006, a paper on issues that might compromise the translatability of test instruments was prepared for the test developers. They also received some training in writing tests that avoid potential translation traps. This stage comes well before the translation/adaptation process and was found to contribute to the overall quality of the source version, insofar as a number of typical ambiguities were avoided from the outset.

The IPC may decide to take over part of the translation responsibility, for example, if several participating countries use the same "language of test." In PISA, the IPC developed a "parallel" French source version: Two independent translations from English into French were merged into a final French version.

This was submitted to bilingual domain experts and reviewed until it was held to have the same status as the original English source version (Grisay, 1999).

The French source version served as the "generic" version for French-speaking countries that used it after making limited country-specific adaptations in it. More importantly, it was used by many participating countries as the source version for one of their two translations, enhancing the benefits of the double translation and reconciliation design. For TIMSS, a "generic" Arabic version was developed by the IPC because 14 participating countries used Arabic as a testing language. The NCs of those countries were requested to introduce national adaptations into this "generic" Arabic version. For LAMP, the IPC provided French and Spanish versions of the data collection instruments to French- and Spanish-speaking countries, but these versions did not have the status of a "final" version.

The parallel French source version of PISA, the generic Arabic version of TIMSS, and the French or Spanish versions of LAMP were each prepared by the IPC prior to the translation and adaptation process in the participating countries. In all cases, this early translation process was instrumental in detecting translatability problems and residual errors in the international source version. This information was used to amend the source version to resolve those problems, to edit or complete the translation and adaptation guidelines, and to provide useful translator's notes.

9.6.1 Relevance of the Type of Translation/Adaptation Process for the Verification

Parallel development of a second (or more) source versions and/or "generic" versions is instrumental in detecting potential translation/adaptation problems at an early stage. Those issues can then be listed in an annex to the translation and adaptation guidelines, and the verifiers can regard these as areas of special focus. In addition to a sentence-by-sentence comparison of source and target versions, verifiers are then asked to check and report how each potentially problematic issue was addressed in the target version they verify. If in doubt, the verifier could also compare both source versions to have a more precise idea of the degree of freedom in translation that would be considered as acceptable.

One difficulty lies in the subjectivity of the perception of this degree of translation freedom. A verifier may suggest a correction to bring the target version closer to the English source even though the translated version is a faithful rendering of the French source version, and might, therefore, be considered acceptable. Another interesting debate would concern the choice of the second source language. For example, if there are many participating countries testing in Arabic, perhaps the idea of simultaneously developing the tests in Arabic and English may be worth considering, including team translations and extensive cross-checks by bilingual domain experts. The verification would thus be reduced to a minimum for Arabic-speaking countries: Verifiers would be asked to comment on conformity with translation/adaptation guidelines only for national adaptations made to the Arabic source version.

9.7 VERIFICATION PROCESS

As described earlier, before the verification procedure was introduced in its current form, the back translation design was widely used to assess the quality of translated versions of data collection instruments in multilingual studies. In clinical tests, quality control of translated questions may include up to eight consecutive translation/back translation rounds. The term translation verification was first used to describe back translation. Then, in IEA studies, the term was introduced to describe the operation of submitting translated versions of test instruments to an independent organization for extensive proofreading. No special training was envisaged for proofreaders/verifiers hired to check translations. The verification process described in this chapter is, in fact, an extension of that practice. The role of back translation in the procedures outlined here is to give the referee (see below) more insight in the issues spotted by the verifiers, not to assess translation quality.

The first major innovation was to organize a training seminar for all verifiers involved and to show them in what aspects the translation of data collection instruments might differ from the translations with which they were more familiar. For the first seminar, considerable input was requested from test developers, who contributed by describing in simple terms what the test items were supposed to measure. Question intents were also added for each item. Linguists and psychometrists combined their expertise to add translation and adaptation notes to the source versions, so that the verifiers would base their work on more concrete instructions: They were asked to systematically check that those notes were taken into account.

Whether the guidelines are minimal or detailed, experience has shown that training verifiers, whether face-to-face, in group seminars, or remotely (e.g., PowerPoint presentations and Skype) considerably increases the extent to which the guidelines are internalized. It is a challenge to train verifiers to focus on certain adaptation issues that reach beyond the commonly accepted scope of translation. However, when the training modules are illustrated with examples and when a hands-on exercise is proposed, verifiers can effectively be made aware of specific aspects of test adaptation.

Although the IPC has high expectations as regards translation verification, it usually leaves the responsibility of the final instrument to the NCs, that is, the persons or organizations in charge at the level of participating countries. Verification is thus a process whereby external consultants make suggestions to improve the quality of the translated/adapted instruments. The extent to which such suggestions are treated as binding recommendations or as indicative advice varies.

For all three projects described here, verification was required for every translated instrument: The NCs submitted their materials for verification to an international team of verifiers commissioned by the IPC to assist NCs in producing the best possible translated/adapted versions of the data collection instruments. An alternative for modestly funded projects can be to require external verification of only a sample of translated instruments. This could be followed up by requests for further verification, depending on the sample-based outcome. A large-scale European survey adopted this procedure recently: The IPC selected a subset of questionnaire items that were potentially problematic, and these were verified for linguistic quality and for linguistic equivalence against the international source version. The NCs were advised that if the verification revealed serious translation and adaptation problems, a second subset of items would need to be verified, thus generating additional costs and possible delays.

9.7.1 Development of a Monitoring Tool

After the first verification exercises, in which the NCs selected and trained translators and the IPC appointed the authors of this chapter to select and train verifiers, an overlap between instructions to translators and instructions to verifiers was identified: The item-per-item guidelines and the translation/adaptation notes proved crucial for both. An Excel worksheet was developed in which item-specific guidelines and notes were entered. An interesting feature of this translation/ adaptation/verification monitoring tool is that it was designed to accommodate the entire life cycle of translated versions of test items.

Gradually, more players were involved in preparing and in using this tool, more information was entered in it and more information could be extracted from it. Translators (or the NC staff supervising the translation) were asked to report how all translation notes and item-specific guidelines listed in the monitoring tool were addressed. Verifiers were asked to enter their verification feedback in the same tool and to comment on the information given by NCs. Additional columns were provided to accommodate discussions between the NCs and verifiers or between NCs and a translation referee appointed by the IPC.

In the first versions of this Excel monitoring tool, called the Test Adaptation Spreadsheet (TAS) in PISA and the Verification Follow-up Form (VFF) in LAMP, verifiers were requested to enter descriptive comments. The main principle was that they would report and describe the issue/deviation they had identified. The correction they suggested was entered directly in the instrument, usually in "track changes" mode. More recent versions continue to prompt verifiers for this type of feedback and drop-down menus now allow them to choose the category to which their correction belongs, making it possible to generate statistics about the verification process.

Typically, this instrument may include:

- One column to identify the unit or locate a passage or item within the unit.
- One column for the source version.
- One column for the adapted national version.
- One column in which the NC can explain the rationale for an adaptation.
- One column in which the IPC can enter item-per-item guidelines and recommendations on how to treat a specific issue.
- Two columns for the verifier: the "verifier intervention" column has scroll-down menus from which the verifier can select one out of eight verifier intervention categories. The "verifier comment" column is used to report the issue that required an intervention.
- In the latest version, an additional column is added so that the verifier can also enter a "severity code" as used in the TIMSS verification.

- One or two columns for the IPC and/or for the NC to comment on the verifier's feedback, e.g., by confirming that the issue is addressed or by explaining why the verifier's correction can be ignored or rejected.
- An optional column to list last-minute observations arising from a final check of the verified instruments.

This type of form was used in PISA 2006 and in the LAMP Pilot. Because a great deal of the information needed for the verification was centralized in the Excel monitoring tool, the verifiers could concentrate on the task and cycle back and forth between the source and target versions—comparing sentence by sentence—and the monitoring tool, in which they both looked up and entered information. Scroll-down menus proved to be a very useful innovation in terms of quality and standardization of the verification feedback, since if verifiers have taken the trouble to reflect on the category to which their correction belongs, they find it easier to formulate the descriptive comment clearly.

In PISA, verifier comments were reviewed by a translation referee appointed by the IPC, who labeled some of the verifiers' findings as "key corrections," meaning corrections that must be implemented. The referee's work was based on the verifiers' choice of category, on their descriptive comments, and, possibly, on back translations. As mentioned, back translations were not used to assess translation quality but to provide the referee with more insight into whatever the verifiers spotted. Typically, verifier comments had the following form:

[category: mistranslation] "donkey" was translated as "monkey."
[category: missing info] "old" was missing in "old donkey."
[category: grammar/syntax] wrong subject/verb agreement in third sentence.
[category: layout/visual issues] donkey was not underlined.
[category: register/wording] typo changes the meaning—*traje especial* (special suit) instead of *traje espacial* (space suit) makes it difficult to answer question 3 correctly.
[category: adaptation] "Pedro's girlfriend" adapted to "Pedro's uncle."

Verifiers are instructed to describe deviations as they see them, not to describe the operation needed to improve the material. For example, they should write "Old" was missing in "old donkey" rather than "Please add old."

The IPC translation referee's basic task was to point out potentially crucial corrections to NCs. On some occasions, a back-and-forth exchange took place between an NC and the referee, using the monitoring tool as support, until agreement was reached on more controversial corrections. Note also that the referee was not necessarily conversant in the target language. Therefore, the verifiers were requested to make clear, intelligible comments that, in combination with the verifier intervention category, would make it possible to assess the relevance of the correction.

In the TIMSS verification, this type of monitoring tool was not used. TIMSS verifiers were asked to enter both their corrections and their comments directly into pdf documents. IEA had equipped the verifiers with proprietary software that allowed them to simulate a revision mode in a pdf document: color coding was used to indicate sentences or words that were edited by the verifiers and every correction, major or minor, was accompanied by a "severity code" and a short comment.

In the TIMSS verification, comments typically resembled the following examples:

[Severity code 1–major error] "donkey" was translated as "monkey."
[Severity code 1–major error] "old" was missing in "old donkey."
[Severity code 2–minor error] wrong subject/verb agreement in third sentence.
[Severity code 1?–may be a major or a minor error] donkey was not underlined.
[Severity code 1–major error] typo changes meaning and makes it difficult to answer Q3 correctly.
[Severity code 4–acceptable change] "Pedro's girlfriend" adapted to "Pedro's uncle."
[Severity code 3–suggestion for improvement] literal translation/awkward wording.

The IEA severity code scheme is relatively simple and makes the feedback directly useful for both the NC and the IPC. However, when—as in TIMSS—the verifiers' comments are inserted directly in verified target files (of which there may be several dozen for each national version), the review and follow-up by the IPC is more difficult at the practical level than when—as in PISA—the corrections and accompanying comments are listed in a single monitoring instrument per national version. The verifier's work was reviewed by an IEA referee before the verified files were sent back to the NC. In TIMSS 2007, the final check performed at the IPC covered layout issues only.

Since both the verifier intervention categories and the severity codes yield useful information, it is suggested that these two frameworks be used jointly. If every verifier correction or suggestion comes with both an intervention category and a severity code, the statistics that can be generated may give good proxies for quantitative and qualitative indicators of item translation quality.

9.8 VERSION MANAGEMENT

Version management is one of the challenges that come with periodical surveys in which some items are used across survey cycles for trend evaluation purposes, or in which items from other surveys are integrated in the test for linking purposes. Test developers and NCs may change over time and each new player may have the urge to make improvements and last-minute changes to instruments. Unfortunately, such changes may jeopardize the link. Besides, these changes may be undocumented, which makes it very difficult to keep track of the different versions. In addition, when there are a large number of language versions and within each language version there are also many successive versions, it may prove difficult to retrieve final versions of a given instrument at a given point in time.

In the future, more tests are likely to be computer-based or web-based. In a web-based environment, it is possible to introduce an automated procedure to "lock," approved field test versions that do not need to undergo any changes, so that the same versions can be used again for the main data collection. This is currently implemented in the Electronic Reading Assessment (ERA) component in PISA 2009 and in the new OECD Programme for the International Assessment of Adult Competencies (PIAAC). With this feature, anyone wishing to implement edits for an item must formally request that the item be "unlocked," and the change must then be documented. For example, if a country spots a residual translation ambiguity in its final field test version of a given test item, it would request authorization to change this item for the main study. This type of quality assurance and control procedure can reduce workload (and costs) for verification at the main data collection phase and for the verification of trend items. For instance, verifiers would only need to check whether the proposed edits conform to the translation and adaptation guidelines and are correctly implemented, but they would not need to check previously verified versions for undocumented changes, since the previous versions would not have been accessible after the tests were administered.

In all three international surveys, version management seemed to present too many risks. When units need to be re-used (from field test to main test or from one survey cycle to the next in the case of "trend" or "link" items), it is important that correct final versions be easily retrieved. A case can certainly be made for centralized archiving or, as discussed in Section 9.9, for upgrading the monitoring tool that up to now has *accompanied* translated materials. This tool could become the main or perhaps even the sole support for the translated materials (doing away with target-version Word or pdf files) until it is time to break out the translated and validated "raw" texts into formatted and assembled test booklets and coding guides, subject to a final check.

9.9 CONCLUSION

Within the framework of sophisticated translation and adaptation procedures implemented in the specific context of large-scale assessments, the players involved (translators, verifiers, NCs, IPC, researchers) can be provided with a combination of clear and concise translation and adaptation guidelines and item-by-item indications of issues including:

- Compulsory adaptations; recommended adaptations; and adaptations that are ruled out
- Literal and/or synonymous matches to be replicated
- Patterns in multiple-choice responses to be replicated (insofar as possible)
- Emphasis elements (bolding, underlining, etc.) to be replicated
- Translation "tips" for difficult or ambiguous words (e.g., words that have several meanings in the source language)
- Question intent, as relevant
- Coding rules, as relevant

The players need to be generally aware of the specific challenges involved in test translation. In any case, the same item-by-item guidelines should be used as a reference throughout the process. The present authors feel that, if existing models of translation/adaptation/verification monitoring tools were further developed and fine-tuned, so as to reflect item-by-item instructions and accommodate spaces to document interventions for every stage and every player, this would considerably enhance standardization, quality, and documentation.

If such a monitoring tool is developed with care before the translation/ adaptation process begins but already takes into consideration the different stages of the process, and if its use becomes a requirement, the translator's/verifier's attention will systematically be drawn to certain aspects that require special treatment: An example would be a synonymous match that needs to be echoed, or a particular pattern in multiple-choice responses. The purpose is that such aspects will be processed several times in every language version. One can envisage preparing these monitoring instruments in such a way that dichotomous interventions (e.g., "done/not done") could be used more often, but that space is always available for open comments, too. This integrated instrument would facilitate documentation of discussion on translation and adaptation: translators, NCs, verifiers, and IPC can all enter their comments and queries in one and the monitoring instrument; the output becomes part of the history of a translated test unit or item.

The format of this tool could be based on the PISA TAS or the LAMP VFF, but would be very carefully prepared by the IPC, preferably in consultation with test developers. The intention is that it would be used by the translator(s), by the reconciler in the event of a double translation procedure, by national experts or editors, by verifiers and then by a referee and/or the test developers, by the NC staff responsible for the national version, and by the reviewers responsible for a final check. A variety of intervention categories adapted to the various roles would allow frequent use of scroll-down menus to document various stages. This would also be an invaluable documentation for researchers and, in combination with item statistics, would provide a good measure of the effectiveness and efficiency of linguistic quality control procedures.

With some research and investment in IT solutions, the integrated translation and adaptation monitoring tool could also become the main support for translated text, without target versions in Word or pdf format, simplifying version management and reducing the associated risks. This is also suggested in a report to the European Union (EU) on a blueprint for a translation tool (Harkness, Dinkelmann, Pennell, & Mohler, 2008). This would require finding workable solutions for exporting validated "raw" texts from the monitoring tool to formatted templates for the test instruments that would include graphics and other layout elements, and possibly recurring elements as well. Additionally, the solutions should include a mechanism to "lock" validated versions so as to avoid undocumented additional edits or last-minute improvements. The players would have to request an authorization to "unlock" the files if they wanted to edit them, and could be asked to document the modifications they wanted to make.

There is evidence that methods based on systematic documentation, discussion, agreement, and external quality control are successful in addressing many issues before a test instrument is administered. The documentation of translation, adaptation, and verification processes should be centralized and made available to researchers with a view to triggering new advances in test adaptation methodology and obtaining item batteries that have an extensively documented translation/adaptation history. The main objective, however, must remain to collect data that will be largely comparable across countries because translation flaws were minimised and construct equivalence ensured.

PART IV

CULTURE, COGNITION

AND RESPONSE

10

Cognition, Communication, and Culture: Implications for the Survey Response Process

Norbert Schwarz, Daphna Oyserman, and Emilia Peytcheva

10.1 INTRODUCTION

Since the early 1980s, psychologists and survey methodologists have made considerable progress in understanding the cognitive and communicative processes underlying survey responding (for reviews see Schwarz, 1999; Sirken et al., 1999; Sudman, Bradburn, & Schwarz, 1996; Tourangeau, Rips, & Rasinski, 2000). To date this research has paid limited attention to cultural differences. However, there is increasing evidence that there are cultural differences in how information is processed (for a review, see Oyserman & Lee, 2007, 2008a). In this chapter we provide a brief overview of the relevant research and explore its implications for survey response.

We focus on the contrast that has received the most attention in cultural psychology, namely the contrast between East Asian and Western (Western Europe and North American) societies. These societies have been described as differing in their chronic or dominant focus on collectivism (embeddedness of individuals within social frames, interdependence among in-group members) vs. individualism (separation of individuals from social frames, independence of the self from others). While there is some evidence that results from East Asian samples cannot always be generalized to other collective societies (see Chapter 11, this volume), to date most of the relevant research on culture's consequences has focused on this comparison. Even if generalization is somewhat limited, using East Asian collectivism and Western individualism as a focal comparison allows us to build on this solid basis of well-developed conceptual frameworks and experimental evidence. Although the experimental tasks used by cultural psychology researchers do not directly parallel the tasks or situations studied by survey researchers, this body of research is relevant in that it illuminates cultural differences in processes known to be involved in answering survey questions. We offer conjectures about the likely survey measurement implications of cultural psychology research and outline an agenda for future theory-driven research more directly tied to the needs of survey researchers. Needless to say, our focus on one set of cultural axes—individualism and collectivism—does not imply that variation along other cultural dimensions is irrelevant to survey measurement; it merely reflects that the cognitive consequences of other variations are not yet sufficiently understood to lend themselves to a fruitful discussion.

The chapter is organized as follows. We first review core features of Western (individualist) and East Asian (collectivist) cultures and summarizes key differences in basic cognitive and communicative processes. We then provide an overview of respondents' tasks (question comprehension, recall, judgment, response formatting, and editing) and address how individualism and collectivism may influence each of these. In discussing this body of research, we use the terms individualism and collectivism when discussing between-country comparisons, assuming that between-country differences are due in part to chronic differences in levels of individualism and collectivism. For clarity, when discussing the results of priming tasks and experiments which highlight the processes underlying such average cross-national differences, we describe the participants as using individual- and collective mindsets (see Oyserman, Sorensen, Reber, & Chen, 2009).

10.2 COLLECTIVISTIC AND INDIVIDUALISTIC CULTURES: BASIC DIFFERENCES

A solid body of experimental research has documented pervasive differences in basic psychological processes between East Asia and Western Europe and North America (for reviews see Fiske et al., 1998; Kitayma & Cohen, 2007; Nisbett, 2004; Oyserman, Coon, & Kemmelmeier, 2002a). In the social domain, Western cultures conceptualize the self as autonomous and relatively independent, characterized by unique internal attributes that are largely independent of the momentary social situation (Markus & Kitayama, 1991). Relationships with others are assumed to operate on an equity basis and to be to the mutual benefit of both. Even family relationships can be severed if they become too imbalanced, draining, or unfulfilling. In contrast, East Asian cultures conceptualize the self as a mutually interdependent piece of a larger whole that is constituted in relationship with others. Relationships with others are assumed to be largely fixed by important group memberships. Relationships are, in that sense, obligatory. Unlike the individualistic model in which relationships that are unfulfilling are severed, within a collectivistic model relationships are understood as necessary to group memberships. Engagements with others follow set relational rules. Relationships are maintained because they are obligatory not because they are pleasant (for a review, see Oyserman et al., 2002a).

Given these tacit metatheories, Westerners explain social behavior primarily in terms of individuals, their traits and characteristics, whereas East Asians are more likely to draw on the social field of which an individual and his or her behavior is a part, resulting in reliable differences in causal attribution, impression formation, and prediction (see Nisbett, 2004; Oyserman et al., 2002a, Oyserman et al., 2009, for reviews). This higher emphasis on the social field among East Asians is further reflected in between-group differences in both the structure of autobiographical memory (e.g., Han, Leichtman, & Wang, 1998) and in individuals' knowledge about their own and others' behavior (e.g., Ji, Schwarz, & Nisbett, 2000), as reviewed in Section 10.3.2. Moreover, differences in metatheories about the self foster differences in self-protective biases and self-presentational strategies (e.g., Lalwani, Shavitt, & Johnson, 2006). In a series of studies, Lalwani and colleagues (2006) demonstrate that while Americans and those higher in individualism use strategies that allow for positive self-presentation, those higher in collectivism are more likely to use strategies that allow for reduced chances of other's seeing the self in a negative light. In the following sections, we discuss each of these differences in more detail in the context of the survey tasks to which they are relevant.

From a cognitive perspective, different cultural orientations or mindsets require different cognitive procedures for their efficient execution (for a review, see Oyserman & Lee, 2007, 2008a; Oyserman & Sorensen, 2009; Oyserman et al., 2009). As outlined by Oyserman and her colleagues, an individual mindset is associated with procedures that facilitate focus on an isolated stimulus and its unique attributes, pulling the stimulus apart from the field. In contrast, the collective mindset is associated with procedures that facilitate the identification of relationships, emphasizing the embeddedness of a stimulus in its field.

The application of cognitive procedures that facilitate either the isolation of individual stimuli or the perception of their embeddedness in a context is not limited to social tasks and results in pervasive differences in perception, judgment, and memory in the social as well as nonsocial domain. While members of all cultures have command of the respective procedures, cultures differ in the chronic accessibility of these procedures and the likelihood of their spontaneous use. For example, East Asians show higher field dependency than Westerners on a variety of social and nonsocial tasks.

At the same time, chronic cultural differences in cognitive procedures can be overridden by contextual influences. When a collectivistic focus is temporarily induced among Westerners, their cognitive performance mirrors the spontaneous performance of Asians; conversely, when an individualistic focus is temporarily induced among Asians, their performance mirrors the spontaneous performance of Westerners (for a review see Oyserman & Lee, 2008b; Oyserman & Sorensen, 2009). Indeed, individual and collective mindset can be systematically produced through a number of priming procedures as well as by language used in context (for a review see Oyserman & Lee, 2007, 2008a; Oyserman & Sorensen, 2009). For example, Oyserman and Sorensen (2009) find that whereas Asian respondents are better at spontaneously recalling spatial relations among objects than American respondents, their recall is impaired when an individual mindset is temporarily induced. Conversely, American respondents' recall is improved when a collective mindset is temporarily induced (Oyserman et al., 2009).

Observations like these have two important implications. On the methodological side, they highlight the *causal* influence of differences in cultural orientation. Given that any two cultures differ in numerous respects, the mere naturalistic observation of a cross-national (cultural) difference does not allow us to identify the causal role of any particular characteristic, which requires experimental manipulations of the characteristic of interest. On the substantive side, these observations indicate that many key cultural differences in *cognitive procedures* do not require extensive socialization in the intellectual traditions of a culture; instead, they are better portrayed as efficient responses to culturally dominant tasks, consistent with theories of situated cognition (see Oyserman & Lee, 2007; Oyserman & Sorensen, 2009; Oyserman et al., 2009, for more detailed discussion). Between-society differences in how everyday tasks, including the communication tasks relevant to survey research, are pragmatically understood are likely to be reflected in between-society differences in responses. We discuss this further below.

In using the shorthand of individualism or collectivism to describe societies, we do not intend to imply that individualism is the opposite of collectivism. Rather collectivism and individualism are orthogonal in the sense that societies socialize participants for both but differ in the extent that each of these dimensions is chronically or habitually salient. Given our focus on East Asian and Western societies, it is useful to note that a meta-analysis of the available data (Oyserman et al., 2002a) documents consistent, large, and homogeneous differences between China and the United States on these dimensions. Relative to American participants, Chinese participants report high on collectivism and low on individualism across a variety of measures. Thus, comparisons between these two countries provide clear examples of countries with predominantly collectivist or individualist orientations.

10.3 CULTURE AND SURVEY RESPONSE

Next, we address how these cultural orientations affect the survey response process, following the sequence of respondents' tasks from question comprehension, recall, and judgment to response editing and self-presentation (Strack & Martin, 1987; Tourangeau, 1984). We review both cross-national and immigrant-population studies and studies comparing results when using native language and language of adopted country. As will become clear, results have implications both for cross-national research and for studies including immigrants who may be interviewed either in their native language or the language of their adopted country.

10.3.1 Making Sense of Questions: Pragmatic Inference Processes

As a first step, respondents need to understand the question to determine what information they are to provide. The survey literature on question comprehension has long focused on semantic issues, urging researchers to avoid unfamiliar terms and complex syntax. While this is good advice, it misses a crucial point: Language comprehension is not about words per se, but about speaker meaning (Clark & Schober, 1992). When asked, "What have you done today?" respondents understand the words, but they still need to determine which behaviors the researcher might be interested in before they can give a meaningful answer. To infer the intended or *pragmatic* meaning of the question, respondents make extensive use of contextual information, from the researcher's institutional affiliation and the topic of the survey to the content of preceding questions and the nature of the response alternatives (for a review see Schwarz, 1996). Reliance on contextual information is licensed by the tacit assumptions that underlie the conduct of conversations in daily life (Grice, 1975), where contributions are expected to be meaningfully related to the goal of the conversation, the content of preceding utterances, and the questioner's interest and background knowledge.

While the general use of contextual information in pragmatic inference is assumed to be universal, members of collective cultures are more sensitive to conversational context than are members of individualist cultures. The limited available evidence suggests that this results in cultural differences in response patterns when the relevance of the contextual information needs to be detected, but not when its relevance is obvious, as the examples reviewed below will illustrate. For survey researchers, these cultural differences in sensitivity to the pragmatic context imply that methods that merely ensure the adequate translation of the literal meaning of a question are insufficient and need to be complemented by methods that assess the pragmatic equivalence of questions (for guidelines see Harkness, van de Vijver, & Mohler, 2003).

Detecting Redundancy. One condition under which the relevance of contextual information needs to be detected is the presentation of partially redundant questions. Conversational norms (Grice, 1975) require speakers to provide information that is new to the recipient, rather than to reiterate information that the recipient already has. This gives rise to a specific pattern of question order effects. For example, Schwarz, Strack, and Mai (1991) asked participants to report their marital satisfaction and their general life satisfaction in different orders. When the life satisfaction question preceded the marital satisfaction question, the answers correlated $r = .32$, but this correlation increased to $r = .67$ when the question order was reversed. This pattern of correlations reflects that judgments are based on the information that is most accessible when the judgment is formed. To evaluate their general life satisfaction, respondents can draw on numerous aspects of their lives, including their marriage. When the general question is asked first, *some* respondents may spontaneously consider their marriage, whereas others may not, resulting in a modest correlation. In contrast, information about their marriage is on *all* respondents' minds when they answered the marital satisfaction question first, resulting in a markedly higher correlation. In a third condition, Schwarz and colleagues drew respondents' attention to the conversational norm of nonredundancy by placing both questions explicitly in the same conversational context. For these respondents, the questions were introduced with a lead-in that read, "We now have two questions about your life. The first pertains to your marital satisfaction and the second to your general life satisfaction." Under this condition, the correlation between the two questions dropped from $r = .67$ to $r = .18$. Apparently, these respondents interpreted the general life satisfaction question as if it read, "Aside from your marriage, which you already told us about, how satisfied are you with other aspects of your life?" and hence disregarded information about their marriage, information which they had already provided, to consider other aspects of their life. Confirming this interpretation, a condition that presented this reworded version of the general life satisfaction question yielded a nearly identical correlation of $r = .20$.

If collectivistic respondents are more sensitive to conversational context than individualistic respondents, they should be more likely to notice the potential redundancy of their answers even in the absence of a lead-in that draws their attention to it. Empirically, this is the case. Haberstroh, Oyserman, Schwarz, Kühnen, and Ji (2002) asked students in Heidelberg, Germany, and in Beijing, China, to report their academic satisfaction and their general life satisfaction, either in the academic-life or the life-academic order. In the German sample, the correlation *increased* from $r = .53$ in the life-academic order to $r = .78$ in the academic-life order, replicating the previously describe pattern (Schwarz et al., 1991). In contrast, the correlation *decreased* from $r = .50$ in the life-academic order to $r = .36$ in the academic-life order for Chinese respondents, indicating that they spontaneously recognized the redundancy problem and disregarded previously provided information. To isolate the causal role of social orientation, a subsequent experiment temporarily induced individualism or collectivism among German students (Haberstroh et al., 2002). When primed for individualism, the answers of German students correlated $r = .76$ in the academic-life order, paralleling the correlation of $r = .78$ previously observed in the German sample; but when primed for collectivism, this correlation dropped to $r = .34$, paralleling the correlation of $r = .36$ previously observed in China.

In combination, these findings highlight several important points. First, chronically or temporarily collectivistic individuals are more sensitive to the conversational context than chronically or temporarily individualistic individuals. Second, differences in sensitivity to the conversational context can give rise to differential question interpretations, which can result in differential question order effects. Third, the underlying difference in question interpretation reflects differences in the pragmatic inference process, not differences in the literal meaning of the question. Such pragmatic differences can emerge even when the literal meaning of a question is perfectly equated through backtranslation procedures, as was the case in these studies. Careful translation of the *literal* meaning does not safeguard against differential interpretations of the *pragmatic* meaning in context. All participants understood the questions but only chronically or temporarily collectivistic participants assumed that the second question included the implied text, "aside from what you have just told me before" and so attempted to disregard information that they had already provided in response to the earlier question.

These findings also highlight the pitfalls of taking answers in cross-cultural studies at face value. Had the questions only been presented in the academic-life order, we might conclude that academic satisfaction figures more prominently in the lives of German than of Chinese students, apparently confirming that individual achievement plays a more important role in individualistic than in collectivistic cultures. Yet no such difference was observed in the life-academic order and the parallel findings with temporarily collectivistic German students indicate that the obtained pattern merely reflects differential sensitivity to conversational context.

How Pervasive a Problem? Pragmatic inferences about the intended meaning of a question are at the heart of many context effects in survey measurement (see Schwarz, 1996, for a review). Are all of these effects more pronounced in interdependent than in independent cultures? On theoretical grounds, we do not think this is the case and the available data are compatible with this (optimistic) conjecture.

On theoretical grounds, pragmatic inference is likely to be universal. When facing an ambiguous question, *all* respondents need to draw on contextual information to make sense of it. All respondents turn to the available information to arrive at an interpretation. For example, they use presented response alternatives to infer which behavior or opinion they are to report on (Schuman & Presser, 1981) and they attend to the numeric values of rating scales to infer what verbal scale labels mean (Schwarz, Knäuper, Hippler, Noelle-Neumann, & Clark, 1991). Pragmatic inferences of this type make use of information that is an integral part of the question itself; this information is attended to by *all* respondents and no particular sensitivity is needed to recognize its relevance to the question with which it is presented.[2]

In other cases, the relevance of contextual information is less obvious and needs to be detected by the respondent. Observance of the conversational norm of nonredundancy, for example, requires that respondents recognize the redundancy problem in the first place and chronically or temporarily collectivistic respondents are more likely to do so. By the same token, we assume that collectivistic respondents are more likely to consider background information about the questioner that may bear on the likely common ground and epistemic interest. For example, collectivistic respondents may be more sensitive to the questioner's institutional affiliations (Norenzayan & Schwarz, 1999) and the overall topic of the survey (Smith et al., 2006). We therefore conjecture that cultural differences in pragmatic inference will emerge when the relevance of contextual information needs to be *detected*, but not when its relevance is relatively obvious.

10.3.2 Recall and Judgment

Once respondents determine which information they are supposed to provide, they need to recall it from memory. This takes somewhat different forms for behavioral questions and attitude questions.

Autobiographical Memory and Behavioral Reports

Content and Organization of Autobiographical Memory. Cultural differences in the construal of self are reflected in the content and organization of autobiographical memory. These differences can already be observed at an early age. For example, Han and colleagues (1998) asked four- and six-year-old American and Chinese children to report on daily events, such as the things they did at bedtime the night

---

[2] Note, however, that the same pragmatic inference at the question interpretation stage can nevertheless result in differential substantive answers. For example, all respondents may infer from negative numeric values of the rating scale that the corresponding verbal endpoint label has a particularly negative meaning—yet their willingness to rate close others in these terms may differ as a function of cultural values (see Chapter 11, this volume). The latter effect reflects cultural differences in socially appropriate responding, rather than cultural differences in question comprehension.

before or how they spent their last birthday. Three striking differences emerged: differences in target of focus on self versus others, differences in depth versus breadth of memory, and differences in focus on internal states versus context.

With regard to target of focus, while all children made more references to the self than to others, the proportion of self to other references was more than three times higher for American than for Chinese children. With regard to depth versus breadth of memory, while the Chinese children talked about many minute details of the specific event in a succinct fashion, the American children talked at length about a few isolated aspects of personal interest rather than the event as a whole. Finally, with regard to differences in focus on internal states, American children's narratives contained twice as many references to their internal states, their emotions, preferences, and desires than was the case for Chinese children.

These differences are paralleled when adult participants are used. Wang and Ross (2007) review relevant recall literature that suggests parallel cultural differences in autobiographical memory. Adults of European descent recall earlier and more detailed childhood memories than do adults of Asian descent. These differences fit what would be expected if, in childhood, individualists' memories are more likely to be self-focused, focused on internal states, and detailed (as suggested by the Han et al., 1998, research summarized above). Similarly, Wang and Ross (2007) find first, that when asked to recall childhood events, adults of European descent recall events that they date to about three-and-a-half years of age while adults of Asian descent recall events that date on average to the period between ages four and five. Second, when asked to write down as much as they could about their early years before age five, European Americans and English participants produced more memories within the five-minute time limit than did Chinese participants, suggesting that memories are more self-linked in the former than in the latter case. Findings of this type indicate that accessible content of autobiographical memories varies with the salient cultural frame (see also Weintraub, 1978).

Such by-country differences may reflect differential processing at the encoding and/or recall stage. On the one hand, chronic differences in levels of individualism and collectivism may influence what people attend to and how they organize information while an event unfolds, resulting in differences at the encoding stage. Furthermore, chronic differences in individualism and collectivism (or other aspects of culture) may influence both what people attempt to retrieve and how they organize retrieved information in narrative form at the recall and reporting stage. These possibilities are not mutually exclusive and the available data do not allow us to estimate their relative contributions. Several studies show, however, that the language of survey administration is sufficient to elicit differential autobiographical reports, presumably because language serves as a prime that brings associated cultural conceptions to mind.

For example, Ross and colleagues (2002) observed that Chinese students at Canadian universities reported more collectivistic memories when the questions were presented and answered in Chinese rather than English. Moreover, their reports of daily moods showed a preponderance of positive moods under English language conditions, but equal levels of positive and negative moods under Chinese language conditions, consistent with cultural norms. To study this effect with autobiographical memories cued with standardized primes, Marian and Kaushanskaya (2004) had participants pull slips of paper with words such as "balloon" on them. Participants were asked to describe a memory involving the word. When randomly assigned to use English rather than Russian, participants who were Russian immigrants to the United States describe memories that focus on the self significantly more often than when randomly assigned to use Russian. Effects are not due to whether the event occurred in the United States or Russia or to language proficiency (as tested by a linguist). Taken together, these studies suggest that language used in the survey may produce both temporary differences in retrieval and reconstruction as well as differences in self-presentation vis-à-vis an in-group (home language) or out-group (English language) member. Effects are also not limited to studies of groups in North America. Trafimow and colleagues (1997) found that bilingual Hong Kong students reported more private traits and fewer social roles when describing themselves in English than in Chinese, consistent with the associated cultural emphasis on individual vs. collective aspects of identity.

In each of these studies, responses in English were compared to those in another language rooted in a home culture presumed to be higher in collectivism. While, as noted above, the processes underlying the found differences in response await more detailed investigation, the available evidence suggests that social relations and roles figure more prominently in the memories of people in collective rather than individualistic cultures, whereas the reverse holds for individual characteristics and experiences. That parallel effects can be found by priming individualism and collectivism suggests that effects cannot simply be due to differences in what information is stored in memory. Instead, it is likely to be some combination of how information is stored and how it is cued for recall. It may be that culturally prominent characteristics are both represented in more detail and linked to a larger amount of other material than less prominent characteristics, making for differential recall unless less prominent characteristics are cued. Taken by itself, this suggests that auto-biographical recall may be facilitated by recall cues that take advantage of the observed cultural differences. It is currently unknown, however, whether higher cultural prominence of an attribute is associated with higher accuracy or with higher recall and reporting bias, rendering recommendations about the use of differential recall cues premature. We consider this a promising avenue for future research.

Finally, it is worth noting that autobiographical events are more likely to be recalled when the language of the interview matches the language spoken during the relevant life period (e.g., Marian & Neisser, 2000). This is consistent with the general principle that recall is facilitated when the context of recall matches the context of encoding (e.g., Tulving & Thompson, 1973). It suggests that surveys of immigrant populations may benefit from matching the language of survey administration to the language spoken during the life period (such as pre- vs. post-immigration) or in the life domain (e.g., home vs. work) of interest. It should also be noted that language can cue individualism or collectivism or something else, depending on the pragmatic meaning of language in context. Oyserman and Lee (2008a) suggest that when the language used appears natural in context, elicited content is congruent with language. However, when language choice is perceived as an influence attempt, elicited content contrasts with language. Thus while studies such as that by Ross and colleagues (2002) suggest that Chinese language cues collectivism-relevant responses and English language cues individualism-relevant responses, effects in the opposite direction have also been observed when the request to speak the non-native language reminded respondents of their country's colonial past (see Oyserman & Lee, 2007, for a review).

Recall and Estimation: Public versus Private Behaviors. As already noted, collectivistic cultures require a higher degree of attentiveness to others in the social context and this need for attentiveness is further compounded by an emphasis on "fitting in" and maintaining harmony in relationships (e.g., Triandis, 1995). To ensure that they "fit in," individuals need to monitor their own behavior as well as the behavior of others to avoid unwanted discrepancies. Note, however, that this need only applies to public behaviors, which are visible to others and hence need to be monitored. In contrast, private behaviors, which others cannot observe, neither require nor allow monitoring for fit. Accordingly, Asians may know more about their own public behaviors than Westerners, attenuating the need to rely on contextual cues when asked to provide behavioral reports. Empirically, this is the case, as Ji, Schwarz, and Nisbett (2000) observed in a study of behavioral frequency reports.

Numerous studies with Western samples demonstrated that respondents often rely on the numeric values of frequency scales to arrive at a frequency estimate (Schwarz, Hippler, Deutsch, & Strack, 1985). This results in higher frequency reports when the scale presents high rather than low frequency values (for a review see Schwarz, 1996). This effect is more pronounced when the behavior is poorly represented in memory because poor memory representation forces respondents to rely on an estimation strategy (Menon, Raghubir, & Schwarz, 1995). Taking advantage of this general observation, Ji and colleagues (2000) demonstrated a cross-cultural difference. After pretesting to choose behaviors of similar frequency in both countries, they demonstrated differences in reliance on scale information to estimate. Specifically, they asked students in China and the United States to report the frequency of various public and private behaviors along scales with high or low frequency values. Several findings are worth noting.

First, Chinese as well as American students reported higher frequencies along high frequency scales than along low frequency scales when their reports pertained to private, unobservable behaviors (such as the frequency of dreams or negative thoughts about others). Moreover, the size of the scale effect was almost identical in both countries. This indicates that respondents in both cultures relied on the same estimation strategy; it also supports our earlier contention that individualistic and collectivistic respondents are similarly sensitive to contextual information that clearly pertains to the task at hand (see Section 10.3.1). Second, American students were as influenced by the scale when they reported on public behaviors as when they reported on private behaviors. This is consistent with earlier findings and suggests that neither class of behaviors enjoys an advantage in memory for Westerners. Third, in stark contrast, Chinese students were unaffected by the response scale when they reported on public behaviors (like visiting the library or being late for class) and provided nearly identical frequency reports in an open response format and along high and low frequency scales. Much as the monitoring rationale would suggest, these behaviors were apparently well enough represented in memory to eliminate the need for context-based estimation strategies.

These cultural differences in response strategy resulted in reports that would invite opposite conclusions in a cross-cultural survey. When presented with an open response format, American and Chinese students reported similar frequencies of public behaviors, consistent with the selection criteria for the behaviors used in this study. But when presented with a frequency scale, American students reported either higher or lower behavioral frequencies than Chinese students, depending on whether the scale presented high or low numerical values. As a result, a researcher might conclude that Americans engage in the behavior just as often, less often, or more often than Chinese, solely depending on the response format of the question. No such cross-country differences were observed when the behavior was private and all respondents relied on contextual cues to arrive at an estimate.

Attitude Questions. When the question is an attitude question, researchers often hope that respondents recall and report a previously formed opinion. In most cases, however, respondents will not find an appropriate answer readily stored in memory and will need to form a judgment on the spot. In doing so, they do not retrieve all information that may be relevant to the topic, but truncate the search process once enough information has come to mind to form a judgment (Bodenhausen & Wyer, 1987). Accordingly, their judgment is based on the subset of potentially relevant information that is most accessible, which is often information brought to mind by preceding questions. How this information influences the judgment depends on whether it bears on an applicable norm or on features of the attitude object. We address both cases in turn.

Norm Activation and the Language of Survey Administration. In the late 1940s, Hyman and Sheatsley observed that Americans were more likely to endorse the right of a Soviet reporter to report freely about the United States when they had first been asked about the right of an American reporter to report freely about the Soviet Union. Presumably, this question sequence activated a norm of reciprocity or even-handedness and later studies consistently found that norm activation affects survey response (for a review see Schuman & Presser, 1981). While the norm of reciprocity is widely shared across cultures, cultures differ in which other specific norms they endorse and the degree of importance they assign to them. Accordingly, a given question may be differentially likely to evoke a norm in different cultures, giving rise to pronounced differences in context effects.

One often overlooked variable that can affect the accessibility of culturally shared norms and meaning systems is the language of survey administration. For example, in a study of Greek students attending an American school in Greece, answers to the same questions administered in English and in Greek showed good correspondence in domains where American and Greek norms converged, but poor correspondence in domains where the norms diverged (Triandis et al., 1965). Apparently, the questions were answered within the cultural frame evoked by the language of the questionnaire. On the other hand, respondents may affirm their own cultural identity through more culture-consistent answers when the interview in a foreign language is perceived as part of an ingroup-outgroup juxtaposition (e.g., Bond &

Yang, 1982). These issues are of considerable applied importance for surveys of immigrant populations, which are often conducted in more than one language. Systematic experimentation is required to understand the underlying dynamics.

Constructing the Attitude Object. While the activation of norms through preceding questions can have a profound impact on survey responses, most question order effects reflect that preceding questions bring information to mind that bears on the nature of the attitude object. How this information influences respondents' judgments depends on how the information is *used* in forming a mental representation of the attitude object and of a standard against which the attitude object is evaluated (for a more detailed discussion see Schwarz & Bless, 2007; Sudman et al., 1996, Chapter 5).

Information that is *included* in the temporary representation formed of the attitude object results in *assimilation effects*; in this case, the judgment is more positive when positive rather than negative information comes to mind. In Section 10.3.1, we discussed a question order experiment with marital satisfaction and life satisfaction (Schwarz et al., 1991) and noted differences in correlation as a function of question order. These differences are also reflected in mean satisfaction levels: Happily married respondents reported higher, and unhappily married respondents reported lower, mean life satisfaction when the preceding marital satisfaction question brought information about their happy or unhappy marriage to mind (Schwarz et al., 1991). Conversely, happily married respondents reported lower, and unhappily married respondents reported higher, mean life satisfaction when a joint lead-in induced them to disregard previously provided information about their marriage. This is referred to as a *subtraction-based* contrast effect (or a "part-whole" contrast effect in Schuman & Presser, 1981): Subtracting positive (negative) information from the representation of the attitude object results in less positive (negative) judgments. As seen in Section 10.3.1, interdependent respondents are more sensitive to conversational contexts that require subtraction and more likely to show part-whole contrast effects (Haberstroh et al., 2002).

In addition, respondents may not only exclude accessible information from the representation formed of the attitude object, but may also use this information in constructing a standard of comparison. If the information is more extreme than other information used in constructing a standard, it results in a more positive (or negative) standard, relative to which the target is evaluated less positively (or negatively, respectively). For example, thinking about a politician who was involved in a scandal, say Richard Nixon, decreases trust in politicians in general. In theoretical terms, the exemplar (Nixon) is included in the representation formed of the superordinate category (American politicians), resulting in an assimilation effect. If the trustworthiness question pertains to a specific other politician, however, say, Bill Clinton, the primed exemplar cannot be included in the representation formed of the attitude object—after all, Clinton is not Nixon. In this case, Nixon serves as a standard of comparison, relative to which Clinton is evaluated as more trustworthy than would otherwise be the case (Schwarz & Bless, 1992). Such *comparison-based* contrast effects generalize to all items to which the standard is applicable, whereas *subtraction-based* contrast effects are limited to judgments of the object from which information is subtracted.

Any of the numerous variables that influence the categorization of information in general (for a review see Smith, 1995) can also influence whether information is used in forming a representation of the attitude object, resulting in assimilation effects, or a representation of the standard, resulting in contrast effects (Schwarz & Bless, 2007). We may therefore expect that recently documented cultural differences in categorization influence the emergence of assimilation vs. contrast effects in judgment. In general, individualistic individuals (Westerners or Asians induced into a temporary individualistic orientation) form more narrow categories and excel at separating stimuli, whereas collectivistic individuals (Asians or Westerners induced into a temporary collective orientation) form broader categories and excel at connecting stimuli (for a review see Oyserman & Lee, 2007, 2008a; Oyserman et al., 2009). These observations suggest several hypotheses that may be fruitfully explored in future research. First, Asians' tendency to form broader and more inclusive categories suggests that they may include information in the representation of the attitude object that Westerners exclude from this representation. Second, given that the impact of a given piece of information decreases with the amount of other information considered, any given piece of information should ceteris paribus exert less influence on Asians than on Westerners. Accordingly, Asians should be more likely to show assimilation effects than Westerners, but the size of these assimilation effects should be smaller. Third, Westerners' tendency to form narrow categories and to parse information into distinct units may facilitate the construction of comparison standards that are distinct from the attitude object. Hence, comparison-based contrast effects should be more likely in Western than in Asian samples.

Moreover, Westerners categorize objects on the basis of class membership whereas Asians categorize information on the basis of functional relationships (see Nisbett, 2004, for a review). For example, when asked to sort a cow, a dog, grass, and a tree into groups that go together, Western sortings (cow & dog vs. grass & tree) reflect membership in the general class of animals vs. plants, whereas Asian sortings reflect relationships (cow & grass vs. dog & tree). This use of different categorization rules may result in different mental representations of attitude objects and corresponding downstream differences in attitude judgments.

In sum, how respondents use accessible information in constructing representations of attitude objects and standards is a key determinant of the direction and size of question order effects in attitude reports. Basic research into cultural differences in categorization suggests that the underlying processes are culture sensitive, giving rise to differential context effects. Data bearing on these conjectures are not yet available.

10.3.3 Response Formatting and Editing

Members of all cultures attempt to present themselves in a favorable light. However, acceptable strategies for doing so, and the specific content that is considered favorable, differ between cultures (Heine et al., 1999; Lalwani et al., 2006). Individualist cultures encourage a view of the self in unique and positive terms that gives rise to numerous self-enhancement biases in form of unrealistically positive self-views and a preference for information that bolsters those views (for a review see Baumeister, 1998). They further value honesty in interaction with strangers (Triandis, 1995) and the available evidence suggests that unrealistically positive self-views are held with sincerity, although embellished when communicated. In contrast, collectivist cultures emphasize the maintenance of harmonious relationships with others and are more concerned with fitting in and saving face, which discourages Western forms of self-enhancement as well as potentially controversial utterances. Moreover, limited "editing" of the truth is considered acceptable in the interest of maintaining harmony and saving face (Ho, 1976; Triandis, 1995). Accordingly, collectivism is associated with impression management measures, and individualism with self-enhancement measures, of socially desirable responding (Lalwani et al., 2006). Using the Eysenck Lie Scale (Eysenck & Eysenck, 1964) as an indicator of impression management behavior, van Hemert and colleagues (2002) observed a zero-order correlation of $r = -.68$ between 23 countries' mean individualism and mean Lie Scale scores.

The differential emphasis on maintaining harmony and avoiding controversy may also underlie the observation that Asian respondents are less likely than Westerners to use extreme values on rating scales (e.g., Chen, Lee, & Stevenson, 1995). Note, however, that this (usually small) difference in the use of rating scales may also reflect differences in scale anchoring. The previously discussed differences in cognitive process render it likely that Westerners focus on the unique features of the stimuli at hand, whereas Asians consider them in their broader context. If so, Asians would evaluate the stimuli relative to a more varied set, which would result in more moderate ratings of all but the most extreme stimuli. Any observed differences in ratings would reflect actual differences in perception in the latter case, but differences in response editing in the former case. Systematic experimentation is needed to determine the relative contribution of these processes, which are not mutually exclusive.

10.4 SUMMARY

As our discussion indicates, cultural differences in basic cognitive and communicative processes have the potential to affect respondents' performance at each step of the survey response process. Hence, any observed cross-country differences in the obtained answers may reflect true differences in attitudes and behaviors, differences in the response process, or an unknown mixture of both. While recent progress in cultural psychology and survey methods has set the stage for a fruitful investigation of these issues, the available research is often limited to global country comparisons. Because cultures differ along many dimensions, such comparisons provide little insight into the underlying processes and usually fail to isolate the causal contributions of specific variables. Experimental manipulations of the variables assumed to differ between cultures provide a more promising approach, and the observation of parallel effects in experiments and country comparisons offers some assurance that the relevant variables have been identified. We consider this a promising avenue for future CASM (cognitive aspects of survey methodology) research.

11

Cultural Emphasis on Honor, Modesty, or Self-Enhancement: Implications for the Survey-Response Process

Ayşe K. Uskul, Daphna Oyserman, and Norbert Schwarz

11.1 INTRODUCTION

We ask and answer questions every day. But beneath these seemingly straight-forward interchanges lie a series of cognitive and communicative processes, which when better understood allow for better understanding of how cultures and questions influence answers (for reviews see Schwarz, 1999; Sirken et al., 1999; Sudman, Bradburn, & Schwarz, 1996; Tourangeau, Rips, & Rasinski, 2000). In answering questions, people take into account what the question likely meant, bring to mind relevant information, and then edit this information to form a response (Strack & Martin, 1987; Tourangeau, 1984). Each of these steps may be influenced both by features of the questionnaire and research context as well as by the culture within which the research is taking place.

What the question likely means, its pragmatic meaning, influences both what comes to mind and the response-editing process. Advances in two fields, cultural (and cross-cultural) psychology and cognitive survey methodology, provide important insights into these processes. Unfortunately these fields have not converged so their insights have not been integrated. This integration is addressed in here and in the preceding companion chapter, Chapter 10. Much of the current cultural and cross-cultural literature focuses on the contrast between Western individualism and East Asian collectivism and the Schwarz and colleagues' chapter provides an insightful overview of this literature.

In the current chapter, we move beyond East Asian, Confucian-based collectivism, to address another form of collectivism, honor-based collectivism, a kind of collectivism prevalent in other parts of the world—including the Middle East, Mediterranean, and Latin American countries. Because relatively less

---

[1] Survey Methods in Multinational, Multiregional, and Multicultural Contexts, edited by Harkness et al.
Copyright © 2010 John Wiley & Sons, Inc.

empirical work has focused on honor-based collectivism, we emphasize this literature in the next section of this chapter, providing an overview comparing collective cultures of honor with collective cultures of modesty and individualistic cultures that could be termed cultures of self-enhancement. Much of this literature is ethnographic and even when quantitative research exists, it does not have in mind the needs of survey researchers. However, this literature does highlight issues that survey methodologists should start attending to. To begin to create a bridge between this literature and the concerns of survey methodologists, in the second section of this chapter we briefly summarize the communicative and cognitive processes involved, making predictions about how culture of honor should influence pragmatic meaning, judgment and recall, and response editing. Because direct evidence is limited, we highlight work of our own in this area.

11.2 HONOR, MODESTY, AND SELF-ENHANCEMENT: DISTINGUISH-ING CULTURE'S BASIC DIFFERENCES

Though societies differ in many ways, researchers have been interested in identifying a few key dimensions of culture that are associated with systematic differences from which general predictions can be made (see Oyserman, Kemmelmeier & Coon, 2002 for an integrative process model). To date the individualism-collectivism dimension has captured most popular appeal and concerns whether cultures emphasize individuals or groups across a variety of domains (e.g., Hofstede, 1980). Simply defined, *individualism* is the extent to which individuals are perceived as a basic unit of analysis while *collectivism* is the extent to which groups (and individual membership within groups) are perceived as a basic unit of analysis (see Oyserman & Sorensen, 2009, for a review). Individualism highlights separateness, each person is a unique and worthwhile individual. Collectivism highlights connectivity between and among persons; persons gain meaning and worth through connection.

While early research on collectivism was informed by its Mediterranean-based forms (see Triandis, 1989), the form of collectivism most often studied is Confucian-based. In this form of collectivism, focus is on harmony—modesty, fitting in, not sticking out, and not bragging (Markus & Kitayama, 1991). Schwarz, Oyserman, and Peytcheva (Chapter 10, this volume) summarize the literature comparing Western Europeans and North Americans with East Asians and the implications of these differences for survey response. This comparison is valuable and forms the bulk of the empirical cross-cultural literature.

However, understanding Confucian-based collectivism is not sufficient for survey researchers conducting studies elsewhere, including areas of emerging interest such as the Mediterranean region (including Spain, Greece, and Turkey), Latin America, the Middle East, and Africa. In these regions, an alternative form of collectivism, focused on honor, has been reported as we describe below. Within a culture of honor, the central collective dimension is maintaining a good reputation—both within the group and with regard to relationships with out-groups. Like Confucian-based cultures of modesty, cultures of honor are collective —groups and group membership matter and reputation is both gained and lost not only through one's own actions, but also through the actions of others with whom one is closely associated (typically kin but also other social groupings). Because cultures of honor are collective in focus, it is likely that at least some of the literature on cognitive consequences of collectivism is generalizable beyond East Asia. By examining differences between collective cultures of honor and collective cultures of modesty it will be possible to specify more specific predictions about how cultural dimensions or syndromes are likely to matter for survey researchers. In the following section, we focus on cultural differences in norms for self-presentation since these are likely to be influenced by whether cultures focus on maintaining harmony or maintaining a good reputation and to influence how questions are understood, what comes to mind, and how information is edited and communicated within a survey.

11.2.1 Individualism

Individualism prescribes a worldview in which individuals are encouraged to define themselves and others as unique and separate individuals with different goals, preferences, and attitudes (for reviews see Markus & Kitayama, 1991; Triandis, 1989). Individualism makes salient norms of self-confidence and self-enhancement (Heine, Lehman, Markus, & Kitayama, 1999; Heine, 2007; Kitayama, Duffy, & Uchida, 2007; Markus & Kitayama, 1991; Suzuki & Yamagishi, 2004; Yamaguchi, 1994). Individuals are assumed to be responsible for themselves and a key self-presentational goal is to positively present oneself (for reviews see Heine, 2007; Oyserman, Kemmelmeier & Coon, 2002).

Indeed, the mostly American literature on self-valuation demonstrates that Americans tend to have positive self-views (e.g., Baumeister, Tice, & Hutton, 1989) and to prefer information that maintains or enhances these positive self-views (e.g., Swann, Pelham, & Krull, 1989). This preference for positivity extends to family members. Westerners evaluate close family members more positively (Endo et al., 2000) and are less critical in evaluating their children's performance (Stevenson & Stigler, 1992) than East Asians. However, there is no reason to assume that this preference for positivity is not even more general. Because individuals, not groups, are salient, and relationships between individuals are based on joint interest, people in individualistic cultural settings are less likely to process information in terms of in- or out-group memberships; today's stranger could be tomorrow's friend (Oyserman, 1993; Oyserman, Kemmelmeier & Coon, 2002). This implies that there are no strong prescriptions for the evaluation of strangers (e.g., Bond & Smith, 1996; Iyengar, Lepper, & Ross, 1999; Yamagishi & Yamagishi, 1994).

11.2.2 Collectivism

Collectivism focuses attention on the importance of the social interface—groups, how one fits into them, one's position within the group, and the ways to maintain positive status as a group member. Recent reviews of the literature demonstrate a reliance on East Asian samples to study collectivism, although some data have also been collected with other samples, including Latino or Hispanic American and Mexican participants (see Oyserman, Coon, & Kemmelmeier, 2002, for a review). Theoretical perspectives on cultural differences in psychological processes are rooted in research using Chinese and Japanese samples (e.g., Markus & Kitayama, 1991; Nisbett, 2003), and there is little evidence that these can be generalized to other cultural contexts (perhaps with the exception of cognitive differences involving salience of contextual information, see Oyserman & Lee, 2008a, for a review). In the section below, we focus on differences in self-presentational norms between collectivism emerging from East Asian and from other contexts.

East-Asian Collectivism. Confucian-based collectivism makes salient connections, nestedness of individuals within relationships, self-effacement, and modesty as ways of fitting in (Heine et al., 1999; Heine, 2007; Kitayama et al., 2007; Markus & Kitayama, 1991; Suzuki and Yamagishi, 2004; Yamaguchi, 1994). Within Confucian-based collective societies, key self-presentational goals are to be modest, and not stick out (Heine, 2007; Markus & Kitayama, 1991; Triandis, 1989), and not offend others (Suzuki & Yamagishi, 2004). The difference in self-presentational goals between Western and East Asian contexts is important for survey researchers who might otherwise interpret modest responses among East Asian respondents as reflecting less positive self-evaluation. A series of studies using more implicit measures of positive self-evaluation underscore the importance of taking into account norms of self-presentation. In these studies, Japanese respondents were more modest than Americans in their explicit responses, but no differences were found when more implicit measures such as the Implicit Association Test (Kitayama & Uchida, 2003) or tests assessing preference for letters in one's own name and numbers corresponding to one's birthday (Kitayama & Karasawa, 1997) are used, suggesting that differences are in self-presentation rather than true differences in self-valuation. Just as self-ratings are likely to be influenced by modesty and norms concerning not offending others, these norms are also likely to influence positivity of rating in-group and close others given the large overlap between the self and in-group in Confucian-based collective societies. For example, when East Asian parents and teachers were asked to rate the performance of their children, their ratings were more negative than were those of American parents and teachers, in spite of the fact that the objective performance of East Asian children was better than that of American children (Stevenson & Stigler, 1992). These results suggest that survey responses about oneself as well as proxy responses about others to whom one is connected are likely to be filtered through a norm of modesty. The norm should be relevant whenever the question cues a connection to self or group membership—there would be no need for modesty in appraising others who are irrelevant to self or group membership.

African, Latin American, Mediterranean, and Middle Eastern Collectivism. While East Asian Confucian-based collectivism highlights the need for modesty in self-presentation, in other regions of the world, another form of collectivism has been studied: honor-based collectivism. Honor is a form of collectivism based on social image and social reputation (Abu-Lughod, 1999; Cohen et al., 1996; Gilmore, 1987; Peristiany, 1965; Rodriguez Mosquera, Manstead, & Fischer, 2000; Rodriguez Mosquera, Fischer, Manstead, & Zaalberg, 2008; Stewart, 1994). Honor-based collectivism does not highlight modesty but rather emphasizes the public nature of self-worth and the need to protect and maintain honor through positive presentation of oneself and in-group members. Honor is a social psychological construct in that having, maintaining, losing, and restoring honor involves others; honor requires that others respect the self and view the self as having positive moral standing, and only when this occurs can one feel self-pride (Nisbett & Cohen, 1996; Pitt-Rivers, 1965; Stewart, 1994).

Honor was originally studied by anthropologists in regions such as Spain, Greece, Cyprus, Egypt, and Algeria using ethnographic methods such as participant observation (e.g., see Peristiany, 1965). Across locations, these studies highlight honor as maintenance of good reputation—maintained through good family reputation, social interdependence, and maintenance of gender-specific codes of behavior (e.g., Abu-Lughod, 1999; Gilmore, 1987; Pitt-Rivers, 1965, 1977). Honor has also been studied extensively in Turkey, also primarily using qualitative methods (e.g., Kardam, 2005; Bagli & Sev'er, 2003). According to existing studies, honor is central to Turkish culture. A rich vocabulary to define and discuss honor is likely to be a reflection of the centrality of the concept in this culture (Sev'er & Yurdakul, 2001). In Turkish culture, one's honorable deeds are a valued possession; they reinforce close ties binding the individual, family, kin, and community (Ozgur & Sunar, 1982). Studies on the conception of honor in Turkey point to its strong relational form and reveal that honor belongs to individuals as well as family members (Kardam, 2005; Bagli & Sev'er, 2003) and that individuals strongly feel to defend their honor when attacked. Indeed, Turkey is one of several countries in which honor crimes persist (Kardam, 2005; Pervizat, 1998; Yirmibesoglu, 1997).

Moving beyond qualitative research on honor, social psychologists Cohen and Nisbett and their colleagues (Cohen, 1998; Cohen & Nisbett, 1994, 1997; Cohen, Nisbett, Bowdle, & Schwarz, 1996; Cohen, Vandello, & Rantilla, 1998; Nisbett, 1993) and Rodriguez Mosquera and her colleagues (e.g., Rodriguez Mosquera et al., 2000, 2002; Fischer et al., 1999) focused on the concept of honor using more quantitative methods. While Cohen and Nisbett focused on the United States, Rodriguez Mosquera and her colleagues focused on Spain. Taken together, this quantitative body of work on honor-based collectivism is important because it highlights manifestations of honor-based cultural norms in a variety of modern societies.

In particular, Cohen and Nisbett argue that honor norms are likely to develop anywhere where law enforcement is weak or absent, wealth is portable, and economic outcomes are both variable and uncertain (Cohen & Nisbett, 1994; Nisbett & Cohen, 1996). They focused on the United States, examining existence of a culture of honor in the southern and western United States. (Nisbett & Cohen, 1996). Honor in this social context is characterized by the willingness to use force or violence to protect one's social status and position. If that is the case, then laws and policies should allow for such forms of violence, adults should support it, and behavioral traces of honor responses should be observable in laboratory situations. Across a series of studies, the impact of honor was found across each of these domains. Action to protect honor is safe-guarded in the laws and social policies of the American South and West more so than in the American North and East (Cohen, 1998; Cohen & Nisbett, 1997). Survey data collected in telephone interviews with adults demonstrated that American Southerners and Westerners voiced greater support for honor-related violence (and not violence in general) than did American Northerners (Cohen & Nisbett, 1994).

This correlation between geographic location and honor-based values was further tested in a series of experiments with students from Southern and Northern states who were all attending the same mid-Western university. In these experiments, male students were randomly assigned to an insult or noninsult condition. Cohen and colleagues (1996, 1998) demonstrated that Southerners perceived insults in terms of threats to honor—they were both more likely to see insults as damaging their masculine reputation and more likely to engage in domineering and aggressive behavioral responses than Northerners. These results are likely to generalize to Latino or Hispanic cultures (e.g., Vandello & Cohen, 2003).

In a series of studies, Rodriguez Mosquera and colleagues have demonstrated that within Europe, the expected differences can also be shown. Thus, social conceptualizations of honor are more salient in Spain than in the Netherlands (e.g., Rodriguez Mosquera et al., 2000, 2002; Fischer et al., 1999). Spanish participants rate honor and honor-related values such as social recognition as more important than do Dutch participants (Fischer et al., 1999). When asked to describe honor, Spanish participants describe honor in relation to family and social inter-dependence; for Dutch participants, honor is not socially contingent (Rodriguez Mosquera et al., 2002). Spanish participants respond more intensely to standardized insult vignettes than Dutch participants when insults threatened family honor, and this between-country difference is mediated by individual differences in concern for family honor (Rodriguez Mosquera et al., 2002).

As shown by Rodriguez Mosquera and colleagues, honor in such societies includes both the individual and closely related others. In honor-based collectivistic societies, honor is shared with close others and those in the individuals' important social groups (Mojab & Abdu, 2004; Rodriguez Mosquera, Manstead, & Fischer, 2002). Honor is a form of collectivism in that one's own honor is implicated by the honor of close others; social respect can be lost through one's own failures as well as through the failures of close others or can be gained or enhanced through one's own successes as well as the successes of close others (Gregg, 2005, 2007; Stewart, 1994). Thus the extent to which one's personal worth is determined interpersonally is a distinct feature of honor cultures (Rodriguez Mosquera, Manstead, & Fischer, 2002).

In honor-based collective societies, reputation matters, and reputation is a social construct that includes the esteem to which one's group is held, not simply personal attainments. Thus, in honor-based societies, positive evaluation of one's in-group is quite critical (Abu-Lughod, 1999; Rodriguez Mosquera et al., 2008). Just as in other forms of collectivism, self- and social identities are highly connected (Markus & Kitayama, 1991; Triandis, 1989, 1995). This means that protection of social image is a core psychological concern in honor cultures. Social situations in which the personal or social self may be negatively evaluated are threatening and this threat needs to be responded to; not responding properly can lead to dishonor (e.g., Gilmore, 1987; Peristiany, 1965). Whereas among Confucian-based collectivism, the way to maintain positive relations is through a norm of modesty, for honor-based collectivism, the way to maintain positive relations is through a norm of positive representation of the self and in-group and negative representation of out-groups.

11.3. CULTURE AND SURVEY RESPONSE

Next, we address how these cultural orientations affect the survey response process. Whereas the survey response process can be divided into three broad sections: question comprehension, recall, and response editing (Strack & Martin, 1987; Tourangeau, 1984), in the current chapter we focus in particular on the first and last parts of this process.

11.3.1 Making Sense of Questions: Pragmatic Inference Processes

As a first step, respondents need to understand the question to determine what information they are to provide. Here, respondents need to figure out what the researcher likely intends to find out (Clark & Schober, 1992). This can be called the *pragmatic* meaning of the question, and it comes not simply from the words that are used but also from the context in which the question is presented (for a review see Schwarz, 1999). On the one hand, everyone uses context at least to some extent, on the other hand, given that collectivism highlights the importance of social context, it seems reasonable to predict that members of collective cultures might be chronically more sensitive to features of the social context. We detail the implications of this, focusing on one aspect of context, scale format.

In a sense, filling out a questionnaire can be thought of as a form of conversation, albeit a conversation in which only the researcher is asking questions and only the respondent is replying. Just as in any conversation, respondents rely on a number of tacit assumptions to make sense of their task and provide sensible answers given their understanding of the pragmatic meaning of questions in context (see Schwarz, 1994, 1999). Research conversations are one-sided in the sense that the researcher cannot be directly queried by the respondent, either because responses are elicited via a self-administered mechanism such as a questionnaire or because interviewers have been trained not to provide interpretations so as to standardize response. Therefore, respondents must draw pragmatic meaning from larger cultural context and the proximal contextual cues present in the research context. These contextual cues include what may at first glance appear to be "formal" features of questions, such as the numeric values used to represent points on the scale (Schwarz, 1999; Chapter 10, this volume).

Suppose participants are asked in a survey to report on their success in life using a rating scale anchored with "not at all successful" and "extremely successful." To provide a rating, they have to determine the intended meaning of the end labels. For example, does "not at all successful" refer to the absence of outstanding achievements or to the presence of serious failures? Given that survey contexts offer little opportunity to clarify the meaning of questions, to infer the intended meaning, participants may draw on the numeric values provided in the rating scale. Using German participants and survey-based experimental methods, Schwarz and his colleagues (1991) tested this possibility. They found that respondents did give systematically different assessments of how successful they have been in life when the numeric format of the rating scale is varied. On average, scores were lower and about a third of individual respondents used the lower half of the range in responding when the scale was from 0 to 10. In contrast, when the scale was from −5 to +5, many fewer respondents used the lower half of the range and when the scores were recoded to range from 0 to 10, the average score was higher. Why would this be?

When the rating scale ran from 0 to 10, respondents seemed to understand the question as being one about the extent of success, as a unipolar construct—one could have more or less success. When the rating scale ran from −5 to +5, respondents seemed to understand the question as being one about the extent of success or failure, a bipolar construct—one could have more or less success (positive numbers) as well as more or less failure (negative numbers). Thus the numeric values used to make up the rating scales seemed to have affected participants' interpretation of the intended meaning conveyed by the anchor labels. To further test this interpretation, Schwarz and colleagues (1991) asked another set of German respondents to draw inferences about a target person based on the target persons' description of academic success. In all cases, the target person's rating was in the third position on an 11-point scale. What differed was whether the scale was a 0 to 10 scale or a −5 to +5 scale. A random half of participants read about a target person who rated his prior success as a 2 on a 0 to 10 scale. The other random half of participants read about a target person who rated his prior success as a −3 on a −5 to +5 scale. Though all respondents viewed formally equivalent information (the third lowest response on an 11-point scale with the exact same verbal anchors of "not at all successful" and "extremely successful"), would the pragmatic meaning be the same? Not if the −5 to +5 scale implied that success is a bipolar construct and the 0 to 10 scale implied that success is a unipolar construct. If pragmatic inference differed then respondents should understand a −3 response on the −5 to +5 scale as reporting some failures and a 2 response on the 0 to 10 scale as reporting not much success. Indeed, in the former case, respondents predicted that the target had experienced more academic failure, specifically that he needed to repeat more exams because he had failed them, than in the latter case.

Taken together, these studies, as well as a larger body of research on context effects on pragmatic inference suggest that research participants take into account even seemingly formal features of questionnaires in making inferences about what the questioner likely means. Once inferences are drawn, however, respondents still have to decide how they will respond. While the research on cultural differences simply suggests that higher collectivism should increase sensitivity to context effects (see Chapter 10, this volume, for a review), as we have outlined in our section on culture's effects on self-presentational norms, there are likely to be effects of culture on this last phase of questionnaire response as well.

11.3.2 Recall, Response Formatting, and Editing

Once respondents have figured out what a question is likely about, but before providing a response, they need to recall relevant information and figure out how to fit their own response into the format of the question and to edit their response to fit norms of propriety. This is a universal process, just as the search for pragmatic meaning is universal. All things being equal, members of all cultures attempt to present themselves in a favorable light. However, as we have outlined in the section on cultural norms for self-presentation, acceptable strategies for doing so, and the specific content that is considered favorable, differ between cultures.

Specifically, while individualist cultures encourage a positive view of the self and others, they also further value honesty in interaction with strangers (Triandis, 1995). In contrast, Confucian-based collectivist cultures emphasize the maintenance of harmonious relationships with others and are more concerned with fitting in and saving face while honor-based collectivist cultures emphasize positive presentation of self and in-group. For both forms of collectivism, some "editing" of the truth is considered acceptable in the interest of appropriate norm fulfillment (Ho, 1976; Triandis, 1995). Because norms differ, this would imply a specific pattern of culture by target interaction. Whereas individualistic positivity norms would result in positive ratings regardless of the target and modesty norms of Confucian collectivism would result in dampened ratings of self and in-group, but not influence the evaluation of out-groups, honor-based collectivism positivity norms would result in heightened ratings for self and in-group and lower ratings for out-group members. Thus respondents from individualistic, modesty-based and honor-based collective societies would edit their responses differently depending on whether the target of judgment was the self, a close other, or not an in-group member. Both individualistic and honor-based societies should promote self-enhancement (of self and in-group members, particularly close others) compared with modesty-based societies. Members of modesty-based societies would notice the different implications of unipolar and bipolar scales, but given the cultural imperative to be modest, respondents from modesty-based societies should be less likely to attempt to correct for the negative implications of the bipolar scale when rating themselves or close in-group others. Instead, the bipolar scale may even highlight concerns about modesty, resulting in lower self and close family ratings. Conversely, members of individualism and honor-based societies should be loath to use the lower end of the bipolar scale when rating themselves or close family members. With regard to strangers, members of modesty-based societies would have no reason to rate them in a way that may imply failures in their lives; to the contrary, one's own modesty may be expressed in positive ratings of strangers. However, self-enhancing individualistic societies offer no strong prescriptions for the evaluation of strangers, whereas derogation of out-groups is more acceptable in collective, honor-based societies.

In a direct test of these hypotheses, Uskul, Oyserman, Schwarz, Lee, and Xu (2008) replicated and extended Schwarz et al.'s (1991) design in a pilot and two experimental studies. Whereas the goal of Schwarz and colleagues' (1991) initial work was to demonstrate the impact of pragmatic meaning, the goal of the research by Uskul and colleagues (2008) was to demonstrate the interaction between pragmatic meaning and cultural norms. Whereas the Schwarz et al. (1991) studies included only German participants and did not explicitly take a cultural perspective, Uskul and colleagues (2008) compared participants from societies marked by individualism (Americans), honor-based collectivism (Turks), and Confucian-based collectivism (Chinese). Because a culture-based framework would lead to different predictions depending on whether a respondent is asked to report on self, in-group, or nongroup relevant others, they also moved beyond Schwarz and colleagues' (1991) initial focus on self and own parents to also examine ratings of strangers of the same age as parents. To clarify that the dependent variable, success in life, was equally desirable across the three cultural groups, they asked college students in each country, how desirable being "successful in life" was to them, finding that life success was equally desirable— slightly higher than a five on a seven point scale—in each of the three cultures.

Results highlight the importance of using a culturally informed model. Culture-relevant effects were found for scales and pattern of responses in ways that suggest that effects are not due simply to differences in what unipolar and bipolar scales imply about the relative presence of positive attributes but also to differences in culturally appropriate use of the affordances provided by the scales to represent the self and close others. In both honor-based collectivistic and individualistic cultures, appropriate responses are positively enhancing of self and close others. In modesty-based collective cultures, modest descriptions of self and close others are appropriate responses. Results followed this pattern.

Specifically, Chinese respondents gave more modest ratings of their own success and that of their parents than either Turkish or American respondents, who were equally positive in their ratings of parents and self. With regard to the interaction of scale and question target, while Chinese respondents were modest in their assessment of self and parents independent of whether the scale was unipolar or bipolar, the assessments of Turks and Americans were higher when the scale was bipolar, just as were German participants in the original Schwarz and colleagues (1991) studies. Turks, Americans, and Germans all rated themselves and their parents as more successful on the bipolar scale than on the unipolar scale. Ratings of strangers of the same age as parents followed the expected pattern. Having been freed from modest self-presentational concerns, Chinese respondents showed the scale effect and rated strangers more positively when the scale was bipolar while Turkish respondents did not rate strangers more positively when using the bipolar scale as they did when they evaluated their parents. As expected, American respondents did not differentiate between in-group and out-group members and showed the scale effect in evaluating all three question targets.

In sum, for individualistic (American, German) and culture of honor (Turkish) groups, the implication of the negative numbers (presence of varying degrees of failure) was enough to shift responses about oneself or about parent's success up to the positive numbers (presence of varying degrees of success). Chinese participants also understood the scale in the same way, as can be seen by the fact that when there was no cultural modesty imperative (when providing a proxy report on out-group members), Chinese also gave more positive responses when using the bipolar scale.

11.4 CONCLUSIONS

Whereas cognitive survey research to date has either ignored culture altogether or focused on a contrast between Western individualism and East Asian collectivism (for a review, see Chapter 10, this volume), in the current chapter we have suggested that survey methodologists should also consider other forms of collectivism, particularly if their research participants are from Southern Europe and the Mediterranean, the Middle East or Africa. Our review of the culture literature highlights the influence of cultural norms on likely responses, even when the pragmatic meaning of questions does not differ. In particular, we focused on culture-based differences in both presentation style and distinctions between the self and close others on the one hand and distal or out-group others on the other hand. The literature on honor-based responses suggests that when cultures make salient an honor-based collectivism, respondents will focus on positive presentations of themselves and close others.

Our own research in this area, however preliminary, provides support for this prediction and suggests that honor-based and Confucian, modesty-based collectivism likely draw attention to different norms relevant to survey responding. While participants all try to put their best foot forward, this entails modest self- and close other deprecation for Confucian groups, but not for honor-based groups. Moreover, these same underlying processes will produce differing results for proxy reports about distal, nongroup relevant others. For Confucian groups, the modesty norm becomes irrelevant but for honor groups, positive statements about any others are unlikely to be viewed as irrelevant to honor, resulting in more negative proxy reports about distal others.

Culture of honor research has documented that honor-based responses are relevant to a broad array of societies, including southern Europe, the Mediterranean, Latin America, the Middle East, Africa, and the American West and South. While current knowledge cannot address whether pragmatic understanding of questions differs, it is clear that the editing process is likely to differ across honor, modesty, and positivity cultures. Future research targeting greater understanding of honor-based norms is highly relevant to the field of survey methods.

12

Response Styles and Culture

Yongwei Yang, Janet A. Harkness, Tzu-Yun Chin, and Ana Villar

12.1 INTRODUCTION

This chapter presents a critical discussion of the literature on patterns of response behavior that are often referred to as *response styles*. In particular it discusses connections between response styles and cultural situations presented in the literature of the last half century.

Response styles are commonly defined as consistent and stable tendencies in response behavior that are not explainable in terms of question content or what a given question aims to measure (cf. Bachman & O'Malley, 1984a, 1984b; Hui & Triandis, 1989; Watkins & Cheung, 1995; Cheung & Rensvold, 2000; Baumgartner & Steenkamp, 2001; Fischer, 2004). Such stable tendencies independent of question content are seen as biased reporting. Pronounced preference for some answer categories can, of course, be substantively motivated. If, for example, the selection of extreme response categories is restricted to questions which tap constructs or dimensions assumed to be theoretically related, the selection of extreme responses may reflect a respondent's true position regarding a construct of interest. If, on the other hand, extreme categories are selected across questions assumed not to correlate or if answers chosen contradict others, biased reporting is often considered as the explanation.

If a respondent consistently favors certain response options or scale positions in response scale, irrespective of question topic, this behavior is usually taken as evidence of either a response style as described or, alternatively, as evidence of response behaviors triggered by factors such as satisficing or social desirability. These other response behaviors can only be mentioned briefly here and in Section 12.1.2. For some general discussion, see Krosnick (1991; 1999) and Krosnick, Narayan, and Smith (1996a) on satisficing and DeMaio (1984), Tourangeau, Rips, and Rasinski (2000), Lensvelt-Mulders (2008), and Johnson and van de Vijver (2003) on social desirability; the last authors also consider the comparative perspective.

[1] Survey Methods in Multinational, Multiregional, and Multicultural Contexts, edited by Harkness et al.
Copyright © 2010 John Wiley & Sons, Inc.

12.1.1 Major Response Styles

Three response behaviors are most frequently discussed as "response styles": *acquiescence, extreme* responding and *middle category* responding.

Acquiescence: The acquiescent response style is characterized by a consistent tendency to select one side of an answer scale, usually the positive side. Acquiescing respondents will tend to agree with statements that are presented to them, regardless of their content.

One consequence of acquiescence is that respondents could well endorse contradictory statements, even if these come close together in the questionnaire. They might agree, for example, with two statements that indicate opposite loci of control, such as "What happens to me is my own doing" and "I have little influence over the things that happen to me" (Ross & Mirowski, 1984). Another consequence could be that respondents would prefer positive sides of answer scales, irrespective of question topic. If asked, for example, about their satisfaction with a variety of aspects of a visit to a hospital, acquiescent respondents would tend to choose the "satisfied" side of the scale rather than the "dissatisfied" answers. Acquiescence is also sometimes discussed in terms of *yea-saying* and *nay-saying,* the nay-saying term referring to a persistent tendency to disagree or disacquiesce (Couch & Keniston, 1960; Bachman & O'Malley, 1984b; Baumgartner & Steenkamp, 2001).

Extreme responding: Respondents who consistently choose endpoints of answer scales are often held to have an extreme response style (ERS). They choose an endpoint of a scale, representing an extreme or strong degree of endorsement of whatever response dimension is presented in either a positive or negative direction (e.g., agreement/disagreement). ERS respondents are typically understood to prefer endpoint positions representing high or low scores, inde-pendent of the entity being scored and how they in actuality might evaluate it.

Middle category responding: This is the term used to describe the consistent tendency on the part of a respondent or group to choose the middle point on an odd-numbered scale (e.g., 1–5) or the middle portion of an even-numbered scale (e.g., 1–6).

12.1.2 Response Styles and Other Responding Behaviors

Response behavior patterns held to be response styles can be distinguished from other response behavior patterns also usually regarded as bias, such as a preference for socially desirable answers or response behavior associated with "satisficing". Socially desirable responses are responses that conform with or endorse social norms or socially preferred behavior of various kinds (Tourangeau, Rips, & Rasinski, 2000; Lensvelt-Mulders, 2008; Cannell, Miller, & Oksenberg, 1981). Respondents responding in a socially desirable fashion might, for example, underreport unhealthy behaviors and over-report healthy behaviors. Under satisficing theory (Krosnick, 1991, 1999; Krosnick, Narayan, & Smith, 1996a), biased response behavior is seen as the result of respondents minimizing their total effort in completing the interview or survey while still complying at some minimal level with the question and answer request. Baumgartner and Steenkamp (2001) discuss related behavior as "noncontingent responding," which they describe as "the tendency to respond … carelessly, randomly or nonpurposefully" (p. 145). Other response behaviors associated in the literature with satisficing or social desirability include high item nonresponse without obvious substantive motivation; frequent selection of the same category across many consecutive questions; and documented under- or over-reporting of behaviors (Holbrook, Green, & Krosnick, 2003). These behaviors are usually held to be related more to individual measurement occasions and features of a particular measurement event, or a given respondent's engagement in the event, than to stable response selection preferences. At the same time, since low "civic commitment" or factors such as a general dislike of surveys might well coincide with noncontingent responding and satisficing, the behaviors associated with satisficing might also be related to more stable dispositions on the part of the respondent.

12.1.3 Why Do Response Styles Matter?

Much of the research interest in response styles has been motivated by concerns about systematic measurement error or bias. For discussion related to comparative research, see, for example, Chun, Campbell, and Yoo (1974), Cheung and Rensvold (2000), Baumgartner and Steenkamp (2001), and Van Herk, Poortinga, and Verhallen (2004); in the general research context, see a seminal statement in Cronbach (1946, 1950).

Systematic measurement error (bias) results when some factor systematically affects construct measurement. The activity of questioning and answering during a survey can be thought of as a *measurement event.* This measurement event involves a minimum set of components (such as question texts or diagrams and answer scales or options) and the respondent's engagement with the questions. Many factors may be involved, including characteristics of the instrument, the interviewer, and the interview context as well as characteristics of a respondent, such as a latent tendency to respond in a certain way.

A response tendency can systematically bias answers, since differences on scores at individual or group level may not reflect true differences on a given target construct. True patterns of relationships could remain undetected or differences observed be actually spurious. In multipopulation research, if differences in response tendencies exist across the populations, inferences derived from country or cultural comparisons will be compromised.

12.1.4 Reducing the Effect of Response Styles on Data

Response styles can be an object of research or can be seen as nuisance factors, resulting in bias. When response styles are seen as nuisance factors, the goal must be to reduce or control their impact on surveys results. Multiple approaches can be taken to limit their effect on data and interpretations. Section 12.3 also discusses approaches within the context of existing literature. It is not the brief of this chapter to elaborate in detail on these methods but we can point to three general approaches. Each has advantages and constraints not considered here. For further discussion on related topics see, for example, Baumgartner and Steenkamp (2006) and Podsakoff, MacKenzie, Lee, and Podsakoff (2003).

One way of reducing response style bias is to improve the design and implementation of the instrument: the questions, the instructions, the answer scales, and the administration procedures. For example, survey questions should be made as easy to answer as possible. The rationale is that although respondents' response tendencies (response styles) cannot be fundamentally changed, the design and implementation can reduce the likelihood of a response style manifesting.

Second, the observed data can be rescored. Various forms of standardization or re-calibrating of the data can be used (e.g., Fischer, 2004; King, Murray, Salomon, & Tandon, 2004). Prerequisites are, however, a theoretical justification for why standardization or re-calibration is appropriate and for the choice of method chosen. Because this involves rescaling scores, the resulting scores and associated statistics must be interpreted with caution; if rescoring involves standardization based on country means and standard deviations, the resulting scores may not be appropriate for cross-country comparisons.

Third, the design can deliberately include items that measure the presence of stylistic response. Balanced scales and scales with items on diverse topics (cf. Greenleaf, 1992a) may be useful in trying to quantify stylistic response in this way. Then, the impact of stylistic response can be partialled out, using a regression or a latent structure model.

The impact of response styles for cross-cultural research can be very complex. Exactly how a response style affects a survey statistic depends not only on the direction and intensity of the response style but also on the direction and intensity of the true relationship and also whether all the variables are affected to the same extent by a given response style. It is essential to evaluate the impact of response styles for a specific situation, rather than rely on broad and general estimates.

Although research interest to date has focused on trying to understand and deal with possible measurement bias, researchers have also been interested in response styles as a reflection of culturally determined perceptions and response. This is reflected in the literature we discuss in the remaining sections.

12.2 RESPONSE STYLES AND CULTURAL DIFFERENCES

The literature discussing cultural differences in acquiescence, extreme, and middle category response tendencies is most diverse. This literature is also often too sparse on documentation of the methods used, the procedures followed, the questions and response scales used, and the populations investigated to conduct the meta-analysis originally intended. Instead we have drawn together the most commonly cited studies discussing cultural differences in acquiescence, extreme, and middle category response tendencies from the late 1960s on. Our comments below indicate the general direction of findings. In Section 12.3 we turn then to critical commentary. Studies discussing cultural differences in acquiescence, extreme, and middle category response tendencies are presented in Table 12.1.

Studies discussing cultural differences in acquiescence, extreme, and middle category response tendencies are presented in Table 12.1.

12.2.1 Acquiescence

Research into acquiescence has suggested that certain ethnic and national groups exhibit greater acquiescence than others. In research on the American continent, these have been found to be Hispanic or Latino populations and African American respondents. In terms of European research, some researchers have suggested a north-south divide in Western Europe, with greater acquiescence in southern regions. Other research interprets differences found in terms of Hofstede's categorization of countries, such as individualistic or collectivist, with collectivist populations displaying greater acquiescence. In each case, as outlined below, findings are contradictory or inconclusive.

The Americas: Respondents of (variously defined) Hispanic or Latino origin have been found to exhibit acquiescence: Aday, Chiu, and Andersen (1980) found that Spanish-heritage samples in the U.S. showed a stronger acquiescence tendency than U.S. non-whites and non-Hispanic whites; Ross and Mirowsky (1984) report that Mexicans in Mexico acquiesced more than Mexican Americans, and that the latter acquiesced more than Anglo Americans. Marín and Marín (1991) and Marín, Gamba, and Marín (1992) examine response tendency differences among Hispanic groups in the U.S. with various countries of origin, as well as non-Hispanic U.S. Whites and argue that Hispanic cultural values might promote an acquiescence responding tendency.

Research has also associated African American populations with acquiescence response tendencies. Bachman and O'Malley (1984b) find that African Americans displayed a stronger tendency to acquiesce than U.S. whites across questions about a broad range of topics. Studying response tendencies using samples of African Americans, Mexicans, Puerto Ricans, and non-Hispanic whites in Chicago. Johnson et al. (1997) find that African Americans and Mexican Americans acquiesced more than non-Hispanic whites.

Some studies considered factors beyond ethnicity/heritage; Marín, Gamba, and Marín (1992) and Ross and Mirowsky (1984), for instance, note that acculturation might play a role in acquiescing behavior. Some of their comparisons indicated that Hispanics less acculturated to the U.S. culture acquiesced more than did more acculturated Hispanics. Marín, Gamba, and Marín (1992) also found that in some cases education matters: Hispanics with lower formal education had more pronounced acquiescent responding than those with higher education. Meanwhile, Johnson, O'Rourke, Chavez, Sudman, Warnecke, and colleagues (1997) find no evidence for acculturation reducing acquiescence among Hispanics but do find higher acquiescence among people with lower levels of formal education. In sum, U.S. Hispanic populations and African American populations are linked to greater acquiescence, although the factors contributing to this may be multiple.

TABLE 12.1. Literature on Cross-Cultural Response Tendency Differences

Groups Involved in Comparison	Acquiescence	Extreme	Middle
Cultural Groups in the Americas			
US: Hispanics and non-Hispanic whites	Aday, Chiu & Anderson (1980) Marín, Gamba, & Marín (1992)	Marín, Gamba, & Marín (1992)	
US: Hispanics and non-Hispanics		Hui & Triandis (1989)	
US: African American, Mexican American, Puerto Rican, and non-Hispanic whites	Johnson, O'Rourke, Chavez, Sudman, Warnecke, & Lacey (1997)	Johnson, O'Rourke, Chavez, Sudman, Warnecke, & Lacey (1997)	
US and Mexico: Hispanics and whites	Ross & Mirowsky (1984)	Clarke (2000a)	
US: African American and whites	Bachman & O'Malley (1984b)	Bachman & O'Malley (1984a, 1984b)	
Cultural Groups in Europe			
11 countries: Belgium, Denmark, France, Germany, the UK, Greece, Ireland, Italy, the Netherlands, Portugal, Spain	Baumgartner & Steenkamp (2001)	Baumgartner & Steenkamp (2001)	
9 countries: Austria, Germany, Ireland, Italy, the Netherlands, Norway, Spain, Sweden, the UK	Welkenhuysen-Gybels, Billiet, & Cambré (2003)		
6 countries: France, Germany, Greece, Italy, Spain, the UK	van Herk, Poortinga, & Verhallen (2004)	van Herk, Poortinga, & Verhallen (2004)	
Cultural Groups across the Globe			
Australia: Asian nationalities and Australian nationals		Dolnicar & Grün (2007)	
Kazakhstan: Kazakhs and Russians	Javeline (1999)		
US: Japanese, Chinese, "American"		Lee, Jones, Mineyama, Zhang (2002)	Lee, Jones, Mine-yama, Zhang (2002)
Belgium: Turkish and Moroccan		Moors (2003, 2004)	
US and Japan		Zax & Takahashi (1967)	Zax & Takahashi (1967)
US and Korea		Chun, Campbell, & Yoo (1974)	
US and China		Culpepper, Zhao, & Lowery (2002)	Culpepper, Zhao, & Lowery (2002)
Canada and Japan		Shiomi & Loo (1999)	Shiomi & Loo (1999)
4 countries: Japan, Taiwan, Canada, U.S.		Chen, Lee, & Stevenson (1995)	Chen, Lee, & Stevenson (1995)
5 countries: Australia, China, Nepal, Nigeria, the Philippines	Watkins & Cheung (1995)		

TABLE 12.1. *Continued*

Groups Involved in Comparison	Acquiescence	Extreme	Middle
Cultural Groups across the Globe			
5 countries: U.S., Mexico, Australia, France, Singapore		Clarke (2000b, 2001)	
9 nationalities: Japan, Singapore, Hong Kong, Indonesia, Malaysia, the Philippines, Thailand, U.S., UK		Stening & Everett (1984)	Stening & Everett (1984)
10 countries and regions: Belgium, Czech, Germany, Hong Kong, Hungary, Japan, Poland, Portugal, Singapore, Turkey	Johnson, Kulesa, Cho, & Shavitt (2005)		
19 countries and regions: Australia, Belgium, Brazil, Czech, Germany, Hungary, India, Japan, Malaysia, Portugal, Turkey, the UK, Mexico, the Philippines, Poland, Singapore, Hong Kong, France, Italy		Johnson, Kulesa, Cho, & Shavitt (2005)	
26 countries and regions: Denmark, Finland, Sweden, Austria, Germany, the Netherlands, the UK, France, Greece, Portugal, Spain, Turkey, Bulgaria, Lithuania, Poland, Russia, Brazil, Chile, Mexico, the U.S., China, Hong Kong, India, Japan, Malaysia, Taiwan	Harzing (2006)	Harzing (2006)	Harzing (2006)
26 countries and regions: Denmark, Norway, Belgium, Austria, Germany, the Netherlands, the UK, Ireland, Italy, France, Hungary, Switzerland, Portugal, Spain, Czech Republic, Poland, Romania, Russia, Slovakia, Argentina, Brazil, , the U.S., China, Japan, Taiwan, Thailand	De Jong, Steenkamp, Fox, & Baumgartner (2008)		
Many countries	Smith (2004)		
Many countries	Smith & Fischer (2008)	Smith & Fischer (2008)	

Other Regions: Research in Europe and elsewhere on acquiescence has produced conflicting results, with some evidence of a relation between acquiescence and Hofstede's (2001) cultural dimensions. Baumgartner and Steenkamp (2001), for example, report negligible differences across 11 European countries. Looking at 10 countries, Johnson, Kulesa, Cho, and Shavitt (2005), on the other hand, find that respondents from countries usually classified as low on Hofstede's cultural dimensions of individualism, uncertainty avoidance, power distance, or masculinity showed a higher tendency to acquiesce. Smith and Fischer (2008) also find that, at country-level, acquiescence was negatively related to individualism. In addition, Smith (2004) suggests cultures high on family collectivism and uncertainty avoidance may tend more to acquiesce on personally relevant questions, while cultures low on uncertainty avoidance may acquiesce more on questions asking them to take positions on their society. Within the small Eastern European region of Kazakhstan, Javeline (1999) found that ethnic Kazakhs had a stronger tendency to acquiesce than did ethnic Russians there.

Using data from 26 countries, Harzing (2006) reports extraversion is positively related to acquiescence at country level, and also reports partial support that power distance, collectivism, and uncertainty avoidance are positively associated with acquiescence. Van Herk, Poortinga, and Verhallen (2004) propose a north-south divide interpretation, finding that people in Italy, Spain, and in particular Greece had stronger acquiescence response tendencies than did respondents from the United Kingdom, Germany, and France.

12.2.2 Extreme Responding

Research on extreme responding has been based on studies that compare different sets of countries or compare cultures within a country (see Table 12.1). As was the case with acquiescence, U.S. research has suggested that Hispanics/Latinos and African Americans display stronger extreme responding tendencies than other groups. Limited research in Europe suggests the same; populations tending to acquiesce are also those displaying more extreme responses.

Bachman and O'Malley (1984a, 1984b), for example, find that African Americans tended more toward extreme responding than U.S. whites. Studying a sample of U.S. navy recruits, Hui and Triandis (1989) find Hispanics made more extreme ratings than non-Hispanics on a 5-point scale, but not when using a 10-point scale; Marín, Gamba, and Marín (1992) report that Hispanics showed a stronger extreme responding tendency than non-Hispanic whites, and Johnson, O'Rourke, Chavez and colleagues (1997) find the same in comparison to U.S. whites for African Americans, Mexican Americans, and Puerto Ricans. Clarke (2000a) again finds that African Americans in the United States and Mexicans show stronger extreme response tendencies than U.S. whites.

For Europe, Van Herk, Poortinga, and Verhallen (2004) find the populations that acquiesce also tend toward extreme responding; they find extreme responding is more prevalent in Mediterranean countries (Italy, Spain, and especially Greek) than in the United Kingdom, Germany, and France. Baumgartner and Steenkamp (2001) again report negligible differences for Europe.

Western-Nonwestern Comparisons: Research on extreme responding often compares western and non-western cultures, considering the possible effect of cultural values and discourse norms. The older and more recent research reported here provides some evidence that U.S. respondents and some other Western populations were more extreme in their responding than were the Asian populations investigated. Cultural dimensions (such as those defined by Hofstede) again seem to be related to differences found.

Comparing U.S. and Japanese college students on a semantic differential scale, Zax and Takahashi (1967), for example, find the Japanese used the endpoints less than the Americans. They also report U.S. females used the endpoints more often than U.S. males. Chun, Campbell, and Yoo (1974) report U.S. students responding more extremely than students in Korea. Chen, Lee, and Stevenson (1995) report greater extreme responding among U.S. school pupils than among Japanese, Taiwan Chinese, and Canadian pupils. Shiomi and Lo

(1999), however, find no significant difference in the use of endpoints between Canadian and Japanese students.

Clarke (2000b) provides further support for ERS in a U.S. population, reporting more extreme responding for U.S. respondents compared to Singaporean and Australian respondents. In keeping with this, Lee, Jones, Mineyama, and Zhang (2002) find that ethnic Japanese and Chinese respondents in southern California were less "extreme" in reporting positive feelings (used the positive answer scale endpoint less) than were "Americans."[2] In addition, when expressing negative feelings, Chinese respondents also showed a tendency to use the positive end of answer scales more than the other samples (i.e., were less extreme in expressing negation). Comparing Australian nationals with Australian residents of Asian origin, Dolnicar and Grün (2007) in turn find that these Asians used endpoints less often than Australian nationals. Clarke (2000a), comparing French and Australian college students, finds French respondents showed stronger extreme responding than Australians. Thus a variety of studies find different degrees of ERS across a number of populations (the studies admittedly also using different samples).

Investigating data from 19 countries, Johnson, Kulesa, Cho, et al., (2005) find that countries high on Hofstede's dimensions of power distance or masculinity tend to have higher extreme response tendencies. In keeping with this, Harzing (2006) finds extraversion is positively related to extreme responding. In like vein, Smith and Fischer (2008) report that, at the country level, extreme responding was positively related to affective autonomy (as defined by Schwartz, 2004), but negatively associated with intellectual autonomy (as defined by Schwartz, 2004).

To complicate the picture somewhat, Stening and Everett (1984) find more extreme responding (endpoint use) among Indonesians, Malaysians, Filipinos, and Thais than for other nationalities in a study of expatriate and local managers from Japanese, British, and American companies in Singapore, and from Japanese companies in the United Kingdom, Hong Kong, Indonesia, Malaysia, the Philippines, and Thailand. They also find that education is relevant: in the Malaysian, Indonesian, Hong Kong, and U.K. samples, respondents with lower educational levels used the endpoints more.

12.2.3 Middle Category Responding

Research on differences across countries in preferences for the middle point or portion of a scale is less extensive. Little is available on middle responding tendency differences among ethnic groups within the Americas or Europe. Some research reports differences among Asian samples, such as the Stening and Everett (1984) study just mentioned. Interest has mainly focused on contrasting Western groups—not considered to exhibit a midpoint preference—with respondents from various Asian countries. A modest body of research indicates that certain Asian

---

[2] Americans were defined as Caucasians whose primary language was English and who identified themselves as primarily American or bicultural.

groups display midpoint preference. There is also some indication that this can be related to Hofstede's cultural dimensions.

Thus, Zax and Takahashi (1967) find that Japanese respondents use the neutral response on a semantic differential scale more than Americans. Chen, Lee, and Stevenson (1995) also find that Japanese and Taiwanese respondents are more likely to use the midpoint than Americans and Canadians, and Shiomi and Loo (1999) also report Japanese respondents used the middle response category more than Canadian respondents in their sample. In addition, Lee, Jones, Mineyama, and Zhang (2002) find that Chinese and Japanese respondents answering questions about positive feelings use the midpoint more often than Americans. In keeping with these results, Harzing (2006) suggests that Hofstede's measure of power distance, individualism, and uncertainty avoidance correlates negatively with a middle responding tendency.

12.3 CRITICAL REVIEW OF THE LITERATURE

Studies reviewed in Section 12.2 are often cited as evidence for the presence or absence of cultural differences in response styles. However, before deciding in any given instance that culturally based response styles are present, several things must be clarified.

First, it is necessary to determine whether an observed response preference indeed involves response bias. This means we must be able to distinguish between possible nuisance factor effects, such as response styles, and responses driven by a respondent's or a group's true values or traits on the target variable.

Second, possible effects of the measurement event must also be controlled for. Moreover, even if a response style tendency seems likely, it could be motivated by a variety of factors other than culture, each of which then needs to be ruled out. Ultimately, convincing theoretical arguments are needed before attributing assumed response style variation to cultural factors. Our review, however, reveals a number of limitations in the existing research, ranging from how response styles were conceptualized to design to implementation limitations that weaken the conclusions variously drawn. We consider a number of these issues below.

12.3.1 Distinguishing between Response Bias and Substantive Responses

Many studies about response styles quantify response bias on the basis of straightforward descriptive statistics derived from observed responses to a set of questions or items. Baumgartner and Steenkamp (2001, 2006) provide reviews of these approaches. Thus, for example, using descriptive statistics, acquiescence has typically been quantified by the mean of item responses (before recoding of reverse items), the proportion or number of responses using one side of a response scale, or the net difference between the count or proportion of using one versus the other side of a response scale. An extreme responding pattern has typically been quantified on the basis of the proportion or number of endpoint responses.

Response range, as measured by the standard deviations across a set of items, is also sometimes used to approximate extreme responding. A middle responding pattern has been measured on the basis of the proportion of responses using the middle points.

However, inferences about response bias based on descriptive indices must be made with caution. Commenting on the nature of observed extremeness, Peabody (1962, p. 72) notes it is important to consider both the "actual differences in intensity" and the "differences in using the response scale." All other things being equal, if a person is observed to be acquiescent, extreme, or middle-of-the-road on a single item or on a set of highly related items, it is not certain whether those responses reflect true opinions, response bias, or combinations of the two. Generally speaking, a distinction cannot be made between substantive and stylistic responses on the basis of a single item or a set of items measuring essentially the same construct.

Heterogeneous Items: A special set of evaluative questions are one way to try to disentangle substantive variance from biasing variance. Since it is unlikely that a respondent will be genuinely extreme on a number of theoretically unrelated constructs, repeatedly observed extremity across heterogeneous items is thus more likely to be a reflection of bias than a matter of true trait. Greenleaf (1992b) therefore argues that a set of heterogeneous items involving diverse constructs are needed to measure extreme response well. Similarly, heterogeneous item sets may also help quantify acquiescence and middle response deviations (Couch & Keniston, 1960; Baumgartner & Steenkamp, 2006). We suggest that content heterogeneity is best viewed as a matter of degree. For instance, items used to assess different domains of consumer attitudes may be viewed as heterogeneous in one study, but be viewed as a homogeneous set in a questionnaire that mainly focuses on well being or political attitudes. Nevertheless, the larger the number of items included and the more certain we can be that they are theoretically unrelated, the more confidence we can have in indices based on descriptive statistics as measures of response bias possibly related to response styles.

In connection with this, a review of the existing literature shows that some studies used heterogeneous item sets (e.g., Bachman & O'Malley, 1984b), whereas others appear to have used items with, at times, rather homogeneous contents. For example, Stening and Everett's (1984) data were collected from surveys about stereotypes of managers, Hui and Triandis (1989) used questions involving descriptions of different types of supervisors, and Lee, Jones, Mineyama, and Zhang (2002) used a set of questions designed to measure sense of coherence.

Balanced Scales: A balanced scale as intended here is a scale consisting of pairs of logically opposite items. Balanced scales can be useful in identifying acquiescence. This is because if respondents simultaneously endorse both items in a pair, they endorse conceptually conflicting responses. If repeated, this could be strong evidence for acquiescence bias. A few studies discussed in Section 12.2.1 have used this approach to study acquiescence (e.g., Aday, Chiu, & Andersen, 1980; Watkins & Cheung, 1995; Johnson, Kulesa, Cho et al., 2005). Javeline (1999) used six pairs of items but did not use a within-subject design. Instead, she randomly assigned the positively worded items to one sample and the negatively worded items to another. Other studies considering acquiescence, such as Harzing

(2006), Marín, Gamba, and Marín (1992), and van Herk, Poortinga, and Verhallen (2004), do not appear to have used balanced scales.

Although conceptually balanced scales can help distinguish between substantive and acquiescent responses, it is not always easy to construct a good balanced scale. In addition, if a statement or item is vague or not salient for a respondent, the intended logical pairing of items will not be relevant for that person. Moreover, in cross-cultural studies, creating balanced scales might encounter difficulties when adaptation or translation of the balanced scale is needed. If items are not successfully reversed, then simultaneous agreement with a pair of oppositely coded items may not indicate logical inconsistency. As a result, endorsement of the items may not be solely attributable to acquiescence. Some studies discussed in Section 12.2.1 provide examples of the pairs of items on a balanced scale, whereas some do not. Additionally, there is no discussion about how translation and adaptation might have affected the balanced scales. It is then hard to judge the quality of the balanced scales used in the literature for the purpose of studying response styles.

Latent Structure Models: Aside from descriptive statistics based approaches, some studies have attempted to conceptualize and quantify response bias using latent structure models. Rost, Carstensen, and von Davier (1997) apply a mixed Rasch model to study responses to personality questionnaires. They find two latent classes, each class requiring different item parameters to describe the response patterns. In one, respondents were more attracted to extreme response categories, in the other, respondents appeared to be avoiding extreme response categories.

Cheung and Rensvold (2000) apply multiple-group mean and covariance structure models to study response styles across cultures. Differential extreme or acquiescence response patterns between groups are operationalized as non-equivalence in the measurement model that relates the observed responses to latent substantive constructs. Under this multiple-group measurement invariance framework, nonequivalent factor loadings across groups would provide some evidence for group differences on extreme response style. Group differences in measurement intercepts can indicate either acquiescence or extreme response style or both. One limitation of this approach is its neglect of within-group differences in response deviations, because the response styles are examined as group characteristics within the multiple-group measurement invariance framework. Thus, although it can detect group-differences in some response style effects, it does not permit quantification at the individual level (Weijters, Schillewaert, & Geuens, 2008). A more serious drawback of the invariance approach is that it is not possible to identify a uniform style effect that pervasively influences all items (Little, 2000). In order to do this, extreme or acquiescence response styles need to be measured independent of the constructs of interests. Weijters, Schillewaert, and Geuens (2008) demonstrate that satisfying the invariance condition in Cheung and Rensvold's (2000) study is a necessary, but not a sufficient, condition for ruling out response style biases.

Weijters, Schillewaert, and Geuens (2008) propose a "representative indicators response style means and covariance structure" (RIRSMACS) model. In this model, response styles are viewed as latent constructs that are measured by multiple indicators that contain measurement errors. The observed response style indicators are summary statistics derived from sets of heterogeneous items which are independent from the items measuring the substantive traits. The authors emphasize the importance of studying the complete response style profile (e.g., acquiescence, extreme responding, middle responding in order to account for potential interdependency among the response style measures). The measurement model for the response styles can be combined with that for the construct of interest and estimated simultaneously. The resulting factor score of the substantive construct is then a "purified" score in that the effects of response styles have been removed.

Billiet and McClendon (2000) conceptualize that acquiescence has constant impact on all items and propose formulating acquiescence as a latent factor. In their model, the observed item scores regress on a common acquiescence stylistic factor with equal loadings, in addition to regressing on the content factors. By including the common style factor, the impact of response style can be partialled out from that of the latent content factor. Welkenhuysen-Gybels, Billiet, and Cambré (2003) applied this approach to study cross-national construct equivalence and found that the effects of acquiescence on item scores were similar across the seven countries investigated. Moors (2003, 2004, 2008) uses a similar approach to study response styles. We note, however, that if the response style actually did have differential influence on items, then a model assuming constant loadings on the common style factor would be misspecified.

De Jong, Steenkamp, Fox, and Baumgartner (2008) propose a multi-level item response theory (IRT) model to measure extreme response style. The model is flexible in three respects. First, items used to measure ERS are not assumed to have the same utility. This allows the latent response style tendency to have differential impact across items used to measure it. Second, the effect of the response style on an item is allowed to be different across groups (e.g., country, language). Finally, the multi-level nature of the model allows investigation of individual- and group-level characteristics.

External Records: Record check studies are one of the most common methods of assessing measurement error (Groves, 1989). When records of the target behaviors are available and can be considered reasonably accurate measures of the variable of interest, they can be contrasted against reported values to help identify the presence of bias or variance. However, response styles are mostly studied in relation to attitude measurement, and records for the "true value" of an attitude are not available. Different models of estimation of error are therefore needed (see, for example, Biemer & Stokes, 1991; Forsman & Schreiner, 1991; Saris & Gallhofer, 2007c).

Van Herk, Poortinga, and Verhallen (2004) use country-level estimates of behaviors as external records—the behaviors are assumed to be correlated with the attitudes measured. That is, if respondents in one country report they enjoy cooking more than do respondents elsewhere, these respondents are expected to cook more often, too. The authors assume that if behaviors are *not* in line with attitudes reported, this may point to response style differences between countries. They found that cross-country differences in attitudinal statements about washing and shaving did not align well with records on related behaviors, and concluded that country differences in responses were contaminated by country differences in response styles.

12.3.2 Confounds of Measurement Event Characteristics

As discussed in Section 12.1.3, the procedure of posing and answering a survey question can be viewed as a measurement event. Any of a number of characteristics of the measurement event can affect a respondent's behavior in intended or unintended ways. If the measurement event affects respondent behavior in unintended ways, it can result in bias responses. Thus the effects of various characteristics of a measurement event also need to be separated out or controlled for in any discussion of response styles

Answer Scale Design: Language issues and method of administration are among the factors which may affect response behavior. For instance, the design, presentation, and wording of an answer scale can affect how respondents understand the questions and how they answer; research has shown how different answer scale formats and wordings can result in different responses to the same questions from the same population (Schuman & Presser, 1981; Sudman, Bradburn, & Schwarz, 1996). Moreover, different questions can affect how respondents perceive and use the same answer scale (Tourangeau, Rips, & Rasinski, 2000). In addition, a number of the studies investigating response styles involve translated questionnaires. As a result, differences in answer scale designs connected with translation and possible effects resulting from these would need to be taken into account (cf. Harkness, 2003; Harkness, Pennell, & Schoua-Glusberg, 2004). The context, implementation, and mode of an interview or self-completion survey have also been discussed in connection with response styles. Dillman, Phelps, Tortora, Swift, Kohrell, and colleagues (2009) investigate administration modes in relation to extremity of response distributions on rating scales. Interviewer effects on measurement error have also been studied (Kish, 1962; Schober & Conrad, 1997; Biemer & Lyberg, 2003). Schuman and Presser (1981) and Carr (1969) consider the potential impact of interviewer effects on acquiescence, although no direct inspection of such effect was made. Instead, deference toward interviewers was hypothesized to drive the tendency of respondents to agree with statements. Bachman and O'Malley (1984a) find that black-white differences in acquiescence were present in the five self-administered questionnaires they examined, but not in the one that was implemented face-to-face. This potential effect has not been further explored; however, Weijters, Schillewaert, and Geuens (2008) find slightly higher levels of acquiescence in telephone interviews than for self-administered questionnaires (both paper-and-pencil and online) in a Dutch sample.

Review of the cross-culture literature on response style differences indicates a general lack of systematic control for, or manipulation of, potential confounding effects posed by answer scale format or the administration methods. Although studies involved multiple languages, they did not usually present details of their translation process and products or of adaptation of questions and answer scales. At the same time, going on the little information available, it did seem possible that translated versions used were indeed problematic. We elaborate these issues in details below.

Answer Scale Length: A few studies intentionally manipulated the length of the scale. Hui and Triandis (1989) find differences between Hispanic and non-

Hispanic respondents using a 5-point scale but not when using a 10-point scale. Extreme responding decreased for the Hispanic population when using the 10-point scale; the non-Hispanic sample showed no difference in this response pattern. Clarke (2000a, 2000b, 2001) reports that extreme responding decreased significantly if the number of scale points increased from a low number to 5 or 7 points. Thereafter, however, effects become small; none of these articles reports clear interactions between a country and scale format. Lee, Jones, Mineyama, and Zhang (2002) study three different lengths of scale (4, 5, and 7 points) and find no statistically significant effect related to length. Most of the remaining studies used 5-point scales, although a few used either 4-, 6-, or 7-point scales. Some studies include scales with different numbers of points but not as a manipulation (e.g., Marín, Gamba, & Marín, 1992; Van Herk, Poortinga, & Verhallen, 2004); sometimes the number of scale points used is unclear (e.g., Ross & Mirowsky, 1984).

Answer Scales—Other Features: A considerable literature explores the effects of answer scale design for monolingual implementation of surveys. The main aspects investigated include the number of scale points (see Krosnick & Fabrigar, 1997, for a review), whether or not to include a midpoint (e.g. Schuman & Presser, 1981), how to label and number scale points (e.g., Schwarz, Knäuper, Hippler, Noelle-Neumann, & Clark, 1991), or color, spacing, and other visual features (e.g., Tourangeau, Couper, & Conrad, 2007). This literature includes little experimentation of answer scale features, with the exception of number of scale points (Hui & Triandis, 1989; Clarke, 2000a, 2000b, 2001) and answer scale format—such as the use of unfolding questions (Albaum, Roster, Yu, & Rogers, 2007). Other aspects of answer scales may also vary within and across studies, without these being intentional manipulations. For example, although many of the studies used agreement scales, some used importance, semantic differentials, or true-false scales. Scale points were sometimes fully and sometimes partially labeled. With regard to the types of labels, some studies appear to have used numerical labels but in many cases it is impossible to tell. The type of verbal labels used also differs across studies. In agreement scales, for example, endpoint wordings included *strongly agree–strongly disagree, disagree–agree, definitely agree–definitely disagree*, and *do not agree at all–agree completely*. Labels used for the midpoint include *uncertain, neither, neither agree nor disagree, agree and disagree equally*, and *unsure*. The directionality of scales is usually not specified but seems to have varied across studies.

Administration Methods: The studies considered also use different administration modes and methods, including face-to-face (in field and laboratory settings), mail, and group administration. Additionally, answer scales were presented orally, orally with show cards, or visually. Finally, the layout of answer scales presented visually is often not clearly described in many studies; nonetheless, differences appear to exist.

The great variation in answer scale designs and administration methods used in these studies provides potentially rich detail for understanding the complex nature of response styles. A meta-analysis synthesizing the effect of response styles across data collected under different measurement event characteristics, for example, could be most enlightening. However, two obstacles currently stand in the way of such a meta-analysis. First, as discussed in Section 12.3.1, response tendencies are quantified differently across the various studies and in addition distinctions attempted between biasing and substantive response variance are frequently unsatisfactory. A meta-analysis of effects, however, requires that studies included have adopted comparable and valid quantification of response bias; analysis is otherwise difficult. Secondly, a number of studies provided too little information about the answer scale design and administration methods to understand the procedures followed. This makes it difficult to integrate their findings into the literature. As it is, we find ourselves frequently forced to list rather than synthesize findings.

Translation and Language Issues: A possibly more critical problem in the existing literature lies in how language and translation issues are addressed. Although a number of studies about cultural differences in response styles used respondents with diverse language backgrounds and appeared to have used translated questionnaires, details about translation procedures and quality are often lacking or are somewhat perfunctory. For example, Culpepper, Zhao, and Lowery (2002) refer merely to using forward and back translations; Aday, Chiu, and Andersen (1980) and Harzing (2006) simply to using forward and back translation with reviews; Chen, Lee, and Stevenson (1995) to using "simultaneous development". Some studies incorporated pretesting (e.g., Aday, Chiu, & Andersen, 1980; Lee, Jones, Mineyama, & Zhang., 2002), others did not. Bilingual interviewers were sometimes used to conduct the interviews; Aday, Chiu, and Andersen (1980), for example, had bilingual interviewers working with questionnaires in English and Spanish. Sometimes, as in the one study reported by Marín, Gamba, and Marín (1992), respondents' relatives or friends were asked to translate on-the-fly (for risks associated with this, see Harkness, Schoebi, Joye, Mohler, Faass, et al., 2007). In other instances, it is not clear how interviews were conducted.

Importantly, these studies lack information on how the answer scales were translated or adapted across languages, making it impossible to evaluate whether any of the group differences observed in responses in a multi-language study might be related to answer scale adaptation issues.

12.3.3 A Cultural Explanation?

Over the past 50 years or so, a number of origins and explanations have been proposed for response styles (for recent reviews see Baumgartner and Steenkamp, 2006; Harzing, 2006). Plausible alternative explanations to cultural background include individual traits and features (e.g., personalities, intelligence, gender, age, and education, social economic class), as well as characteristics of the measurement event. As mentioned earlier, other possible causes need to be discounted before assuming that cultural factors are responsible for what seems to be a response bias. A number of studies suggest a respondent's motivation (Cannell, Miller, & Oksenberg, 1981; Krosnick, 1999) and cognitive ability (Narayan & Smith, 1996a; Krosnick, 1999; Zhou & McClendon, 1999) can influence his/her cognitive processing and resultant response. At the same time, the level of cognitive processing a respondent engages in may encourage the appearance of a response style. Wyer (1969) suggests that less engaged respondents may invest less effort and thus use fewer of the available answer categories than would more engaged respondents. They might, for example, tend to endorse endpoints or mid-point categories. These tendencies are now generally referred to as "satisficing" (e.g. Krosnick, 1999). Clarity of formulation and the perceived meaningfulness and salience of questions may also affect the level of cognitive processing that respondents are willing to engage in. Cronbach (1946), for example, suggests that acquiescence is common when questions or response formats are ambiguous or unclear to respondents.

The possible influence of a variety of factors on cognitive processing should be addressed before attributing response behaviors to culture; one would have to establish, for example, that respondents from different groups are equally willing and able to engage in a given level of cognitive processing or that any differences found in willingness or ability were controlled for. Studies involving multiple languages and translated questionnaires, for instance, would require to demonstrate that a translated questionnaire was as meaningful to a target population as the source questionnaire to another population and that translation has not introduced error or unintended ambiguity.

Attempts to Link Cultural Factors with Response Style Tendencies: A plausible cultural explanation of response style also needs a sound theoretical base. Existing literature on cultural differences in response styles is characterized by a range of explanations with respect to cultural-related factors. Several researchers have speculated that differences in communication styles or norms of responding are associated with cultural differences in response styles. Johnson, O'Rourke, Chavez, and colleagues (1997), for instance, propose that cultural norms, values, and experiences may come into play in each of the four cognitive phases involved in survey responding. Specifically, they suggest that extreme responding might be an outcome of cultural factors interfering at the response formatting phase, and acquiescence a possible outcome of these factors interfering at the response editing phase. Bachman and O'Malley (1984a) suggest black-white differences in extreme and acquiescence response tendency might be explained by cultural heritage manifested as communication styles. Ross and Mirowsky (1984) argue that ethnic differences in acquiescence response tendencies might result from deference or conforming strategies used by minorities to "adapt" in a society. Marín and Marín (1991) and Marín, Gamba, and Marín (1992) interpret the relationship they find between educational level and acquiescent responses in terms of a similar social class explanation. On the other hand, they also find that acquiescence in Hispanic populations was related to acculturation levels and conclude that acquiescence might be related to cultural features.

In addition, cultural norms of responding in ordinary discourse, such as the preference for modesty (Hui & Triandis, 1985) or social conformity (Shiomi & Loo, 1999) among East Asians cultures, sincerity among Mediterranean cultures (Hui & Triandis, 1989), and the customs of hospitality and avoiding offense among Kazakhs (Javeline, 1999) have all been proposed as possible explanations for observed response tendencies. Hui and Triandis (1989) offer a further explanation, arguing that people from different cultures differ in how they map judgment categories onto a given set of response categories.

Some researchers have attempted to connect these behavioral norms with historically rooted cultural heritage. This explanation is particularly favored in studies involving East Asian cultures. Chen, Lee, and Stevenson (1995) and Dolnicar and Grün (2007), for example, argue that acquiescence or middle response tendencies reflect the influence of Confucian philosophy, which prizes moderation, modesty, and cautiousness.

More recently, attempts have been made to connect response styles empirically with cultural values or dimensions by explicitly testing the relationship between cultural dimensions and response tendencies at the individual level (Chen, Lee, and Stevenson, 1995), at the country-level (Harzing, 2006; Smith, 2004b), or within a multi-level framework (Johnson, Kulesa, Cho, et al., 2005; Smith & Fischer, 2008; de Jong, Steenkamp, Fox, et al., 2008).

These studies find acquiescence is negatively associated with Hofstede's individualism (Johnson, Kulesa, Cho, et al., 2005; Smith & Fischer, 2008; Harzing, 2006), power distance (Johnson, Kulesa, Cho, et al., 2005), uncertainty avoidance and masculinity (Johnson, Kulesa, Cho, et al., 2005). Harzing (2006) finds acquiescence positively associated with uncertainty avoidance.

Extreme responding has been positively related to Hofstede's individualism (Chen, Lee, & Stevenson, 1995; Harzing, 2006; de Jong, Steenkamp, Fox, et al. 2008), to power distance (Johnson, Kulesa, Cho, et al., 2005), uncertainty avoidance (Harzing, 2006; de Jong, Steenkamp, Fox, et al. 2008), and to masculinity (Johnson, Kulesa, Cho, et al., 2005). It has also been positively associated with Schwartz's (2004) affective autonomy, but negatively with Schwartz's intellectual autonomy (Smith & Fischer, 2008). Middle responding has been negatively associated with individualism (Chen, Lee, & Stevenson, 1995; Harzing, 2006).

Attempts to Link Cultural Factors with Response Style Tendencies: The empirical research that attempts to link cultural factors with response style differences has limitations with regard to both substantive issues and the methodological and design weaknesses noted earlier. Various conceptualizations of response styles differ, for example, on whether an acquiescence tendency reflects hospitality, submissiveness, or conformity and whether extreme responses result from trying to be clear reflect sincerity.

Second, the operationalization of *culture* poses numerous challenges. It is still quite common to operationalize culture as country or ethnic group. Sometimes this is poorly executed or country is not a suitable defining unit. In many instances, ethnic groups include people with vastly different backgrounds. For instance, "Hispanics" are often presented as one cultural group, regardless of the country of origin, level of acculturation, or educational background of given groups or individuals. The complex nature of culture is seldom addressed in such approaches, making it difficult to pinpoint the sources of cross-cultural differences.

Approaches that empirically link specific cultural factors, such as value dimensions, to response tendencies, provide more precise analysis. However, these approaches are not devoid of challenges. There is, for instance, no consensus on what might constitute a set of stable and generalizable dimensions to explain cultural values. As a result, different studies adopt different frameworks and measures of cultural dimensions; this complicates the interpretation and comparison of findings. For example, Harzing (2006), Smith (2004), and Smith and Fischer (2008) all show inconsistent findings when cultural values are defined and measured using different models. Scores on cultural dimensions are also derived from measures that may not be free of response bias. As Van Herk, Poortinga, and Verhallen (2004) note, it is not always clear whether observed relationships between country-level response style estimates and cultural dimension scores indicate that response styles can be explained by cultural dimensions or whether the cultural dimension scores themselves are subject to the impact of response styles.

The dynamic nature of culture must also be addressed. Values subscribed to by a society or segments within a society can change (Allen, Ng, Ikeda, Jawan, Sufi, et al., 2007; Zhang, 2007). As a result, caution is necessary when associating country-level cultural dimension scores and response style estimates obtained at different points in time. Studies sometimes use samples with homogeneous demographic characteristics (e.g., college students). These samples may not represent the range of values seen in the entire population to whom a study is meant to generalize.

As noted, the literature shows that response styles may vary along with other explanatory factors, such as gender, age, education, and acculturation. Moreover, what people experience within a "culture" can vary greatly, resulting in individual or group differences in the values subscribed to in theory and practice. Different cohorts within a culture oftentimes use language differently and follow different discourse norms. Aggregated analysis at country or ethnic group level may overlook the impact of within-culture variations and lead to spurious conclusions.

If the goal is to consider the impact of contextual factors such as national culture on individual response tendencies, multi-level analysis is a more suitable approach. Johnson, Kulesa, Cho, et al. (2005) show that controlling for background variables at the individual level affects the relationships found between response styles and country-level cultural values. In addition, including individual-level indicators of cultural values can potentially permit more complete analysis of the connection between cultural values and response tendencies. For instance, Smith and Fischer (2008) use multi-level modeling to consider individual-level value measures (i.e., interdependency) and country-level value indicators (e.g., individualism). They show response style differences can not only be explained by country-level or individual-level values, but also by cross-level interactions (i.e. contextual effect). The multi-level approach is promising because it can reveal the complex impact of cultural dimensions on response styles. However, this approach is still rare in the literature.

Finally, some recent studies about the interplay between culture and survey responses undermine sweeping claims about cultural inferences without considering question content. Culpepper, Zhao, and Lowery (2002) show that, depending on the rating tasks presented, Chinese respondents may endorse either more extreme or more moderate responses than American respondents. The authors find that, when faced with factual or nomothetic matters, Chinese exhibit a stronger extreme responding tendency. The explanation offered is that Confucian influence in their culture makes them less likely to weigh different views and more likely to respond without qualification in the direction of what they believe is consistent with the accepted wisdom. In contrast, when questions are about idiographical matters, Chinese respondents can no longer draw upon accepted wisdom. In this situation, the authors suggest, the Confucian tradition of modesty encourages them to respond with caution or deference, which would result in a middle response tendency. The theory of culture as situated cognition, which challenges the assumption that all members of a culture will always follow predominant cultural scripts (Oyserman & Lee, 2007) is also of relevance here. Chapters 10 and 11, this volume, apply this to the relationship between culture and survey responses. They show that response tendencies can result from the interplay between the cultural value dimensions accessible *at the moment of the task* and the pragmatic inferences made by respondents from question features. Such research further emphasize the complex and interactive relationship between culture, response tendencies, and measurement event characteristics, and point to the need for future investigation.

12.4 SUMMARY AND OUTLOOK

Our chapter began with a summary review of the literature on cross-cultural differences in response styles and reported on various arguments about how cultural factors may relate to response styles. We pointed to a number of limitations in the analyses made; studies vary in terms of how well they disentangle biasing and substantive responses; there is a lack of systematic control or manipulations of measurement event characteristics; in addition, other factors apart from culture, which may affect response often cannot properly be discounted.

At the same time, plausible arguments have been advanced for why response styles may indeed be affected by cultural factors, even if details of study design and implementation leave other explanations open. Differently organized research is needed therefore before we can predict with confidence how a given population might respond in a given context to specific questions and answer scales.

It has been well established that multiple factors may contribute to the response selections that people make. The *displayed responses*, that is, the response categories a respondent chooses, need to be considered as the outcome from of a *response process* (Yang et al., 2008). During the response process, predispositions that respondents bring to a measurement event interact with characteristics contributed by the features of a specific measurement event. Respondent predispositions include whatever attitudes, values, or abilities, are targeted by the measurement (target attributes) and any other predispositions, such as nuisance attributes, including response style tendencies. As a result, a displayed response might be a good approximation of a respondent's true standing on targeted attributes. At the same time, it might also be primarily driven by nuisance attributes, including a response style propensity, or be largely determined by factors closely related to the measurement event.

Whether a respondent actually has a response style propensity is determined by individual characteristics/traits and by factors related to his/her experience. At the level of experience, cultural factors might play an important role. At the same time, whether a displayed response actually reflects an existing response style tendency depends upon the *interaction* between (a) what the individual brings to the measurement event and (b) the details of the event itself. These last include the construct of interest and how this is presented (instrument and application), as well as other aspects of the event, such as interviewer effects, third party presence, respondent understanding of their role in surveys, and so forth.

This interactive view thus envisages that a respondent may have a response style propensity but that mediating or moderating factors can mean it is not displayed in a given response. In other words, the manifestation of response style dispositions, which may or may not be related to cultural factors, depends on the interaction of such existing dispositions with other factors within the framework of the measurement event.

If, for example, a respondent has a cognitive representation of the construct of interest that is activated in the measurement event, he/she may proceed to selecting a response choice that accurately reflects this construct of interest (cf. response theory models in Cannell, Miller, & Oksenberg, 1981; Sudman, Bradburn, & Schwarz, 1996; or Tourangeau, Rips, & Rasinski, 2000). If, however, a respondent does not have a cognitive representation of the construct of interest, he/she may adopt any of several courses of action, including strategies which tap into a response style tendency. In this case, the response chosen might actually reflect an existing response style tendency.

We feel that this interactive view of the response process is the most promising avenue toward a better understanding of response styles and their cultural aspects; such an interactive view is also consistent with the theory of culture as situated cognition. Some general discussions of measurement bias (e.g., Cronbach, 1946, 1950; Podsakoff, MacKenzie, Lee, & Podsakoff, 2003) mention an interactive view.

However, much remains to be investigated. The literature reviewed reflects the continuing need for carefully conceptualized and executed studies to tease out potential confounds and to systematically compare alternative interpretations. Greater attention will need to be paid to the proper application of different statistical indices as measures of response styles. The potential of applications of latent structure models to account for an interactive view, such as proposed by de Jong, Steenkamp, Fox, et al. (2008), need further exploration. Given the complex nature of response style phenomena, research must also test specific hypotheses about culture-related factors instead of simply using information on ethnicity or nationality as a representation of culture. In considering cultural factors, the diverse and dynamic nature of culture needs to be taken into account at both micro- and macro-level (Allen, Ng, Ikeda, et al., 2007; Smith & Fischer, 2008). More research is needed from a multi-level perspective. Finally, a consistent protocol of reporting research designs, implementations, and results must be developed and adhered to in future research. Of particular importance for better understanding response styles is the need to report details on answer scale designs, translation, sampling, and survey administration procedures.

PART V

KEY PROCESS COMPONENTS AND

QUALITY

13

Quality Assurance and Quality Control in Cross-National Comparative Studies

Lars Lyberg and Diana Maria Stukel

13.1 INTRODUCTION

Cross-national comparative studies aim at contrasting economic, social, or cultural aspects of different countries or regions. Thus, the underlying statistical problem associated with cross-national studies is part of the larger field of design and implementation of multipopulation surveys (see Kish, 1994). Kish identified five types of multipopulation survey designs, namely periodic surveys such as panels, comparisons of distinct domains from the same survey, multinational comparisons, combinations and accumulations of separate samples, and controlled observations. The last of these is a technique that allows for greater probability of a balanced sample than standard stratification permits, while still retaining probability sampling of each unit. Thus, the common element is the departure of these designs from the classical single population framework.

Kish's view was that the classical theory of survey sampling should be extended to include multiple populations and that such a development had already taken place by 1994, but more from a practical than a theoretical standpoint. Kish envisioned a rapid development in the field of multinational comparisons and he emphasized the need for deliberate rather than ad hoc designs for such comparative studies. Typically, a deliberate design is a mixture between standardization of some design aspects such as definitions, methods, and measurements, and flexibility regarding sample design and sample size. Today this mixture is widely accepted as the best practice, although a general design framework does not exist, and in any case, practical constraints would make rigid standardization difficult to implement.

Kish was right. Cross-national comparative studies are becoming increasingly important but they are still very difficult to design and control. Organizations such as the Organisation for Economic Co-operation and Development (OECD), the

[1] Survey Methods in Multinational, Multiregional, and Multicultural Contexts, edited by Harkness et al.
Copyright © 2010 John Wiley & Sons, Inc.

United Nations, the World Bank, the International Monetary Fund, and Eurostat sponsor surveys across countries and regions on a continuing basis. The purpose of these data collections is usually to produce statistics on economic indicators, welfare, health, labor, literacy, education, and other social and economic phenomena. Examples of surveys of this kind are the International Adult Literacy Survey (IALS), Trends in International Mathematics and Science Study (TIMSS), Progress in International Reading and Literacy Study (PIRLS), World Education Indicators Survey of Primary Schools (WEI-SPS), Demographic and Health Surveys (DHS), Multiple Indicator Cluster Surveys (MICS), and official statistics produced within the European Statistical System. In the social science realm, surveys on topics such as values, time use, happiness, and opinions usually start as research proposals covering a small number of countries and then grow to cover other countries over time. Examples of surveys of this kind are the European Social Survey (ESS), the World Values Survey (WVS), the International Social Survey Programme (ISSP), the Latinobarometer, and the Multinational Time Use Study. Finally, there are studies conducted by marketing firms on multinational comparisons of identification of brands, customer satisfaction, market shares, potential of investments, and expenditures. Typically, however, cross-national surveys are conducted with varying degrees of monitoring and control over the quality of the data collection. The purpose of this chapter is to shed some light on issues related to the assurance and control of various quality aspects of cross-national surveys.

The survey industry has become increasingly competitive and user-oriented. As a result, survey organizations are pressed to produce higher-quality data for increasingly lower costs. In response to this demand, many survey organizations try to apply quality frameworks and continuously improve their processes and make them more cost-efficient. There is, of course, a great deal of variation regarding the extent to which these activities are conducted among survey organizations (see Biemer & Lyberg, 2003), but a reputable organization should have a program for quality assurance that delivers product characteristics in accordance with users' and clients' demands. Furthermore, it should have a quality control program that checks if the quality assurance program works as intended. This means that for each important process step, there is a control function that can decide if the process outcome is in line with specifications.

Quality has become a buzzword in society with a number of different meanings. It can be defined simply as "fitness for use" (see Juran & Gryna, 1980) or "fitness for purpose" (see Deming, 1944). In the context of a survey, quality often means that results must have a total error that is small enough to match the intended use. However, the total error is not the only quality component to consider; other components might come into play. In order to be useful, results must be relevant, easily accessible, and delivered on time. There are also other product characteristics that can be included depending on the needs of the user or client, for instance, a specified wealth of detail in results.

Quality assurance and quality control programs are, in general, less prominent and visible in cross-national comparative studies than in national surveys. The reasons for this are not obvious, but the very size of even moderate international surveys puts enormous demands on conducting even the basic process steps. The endeavor can be overwhelming, leaving little room for quality assurance and quality control, especially in countries and organizations with limited financial and methodological resources. Having said that, it is important to realize that cross-national comparative studies also need these kinds of quality programs, especially since quality problems are magnified compared to national surveys (see Lynn, Japec, & Lyberg, 2006). The literature on the application of quality assurance and quality control programs in cross-national comparative studies is very thin. This chapter, therefore, attempts to discuss some rudimentary issues, in the sequence described below.

Sections 13.2 and 13.3 provide discussions about quality assurance and quality control in a general survey setting and in a comparative survey setting, respectively. In Section 13.2 we also introduce the three-level concept of quality: product, process, and organizational. Section 13.4 describes the notion of product quality and how it can be measured and controlled. Section 13.5 emphasizes the fact that product quality must be based on process quality for which specific tools can be used to check process variation and stability. Section 13.6 discusses what is called organizational quality. Good organizational quality is a prerequisite for good process quality and can be assessed by business excellence frameworks and other measures. In all sections we provide examples from current international studies to illustrate concepts and promising approaches. Section 13.7 gives an overview of survey evaluation mechanisms, and Section 13.8 provides a number of suggestions on ingredients that should be considered in developing more formal programs for quality assurance and quality control in cross-national comparative studies.

13.2 QUALITY ASSURANCE AND QUALITY CONTROL IN SAMPLE SURVEYS

Quality has many definitions. It can be defined as,

- "Fitness for use" (see Juran & Gryna, 1980). In the context of a survey, this translates to a requirement for survey data to be as accurate as necessary to achieve their intended purposes. As discussed above, most of the time there are budget constraints, implying that we have to settle for the maximum accuracy possible given these constraints. In a cross-national comparative study context, accuracy at the national level is not sufficient. Estimates from different countries and regions must also be comparable.

- A multidimensional concept. Users are not only interested in accurate data. They also want data that are timely, easily accessible, that have richness of detail, and so on. These features are product characteristics and we can see them as constraints vis-à-vis the accuracy criterion. There are a number of such quality frameworks available. One example is the framework developed by Eurostat (Eurostat, 2003a) which has six dimensions, namely relevance, accuracy, timeliness and punctuality, accessibility and clarity, comparability, and coherence. For an exact definition of these dimensions, the reader is referred to the Eurostat report.

Generally speaking, the survey process can be illustrated by Figure 13.1. The boxes in the figure represent the main survey steps, and it is important to note that the design of the process has a distinct iterative element. The design involves a number of trade-offs as a function of the planning criterion used. The planning criterion could be maximum precision (minimum variance) given a fixed budget, maximum accuracy (minimum mean squared error) given a fixed budget, or one of the above given a fixed budget, including costs for attaining other product characteristics decided by the user, such as a certain timeliness or a specific documentation system. A rare variant of this criterion is when the precision or mean squared error is fixed and the budget is allowed to be flexible.

Figure 13.1 is intended as a generic description, and most survey process descriptions need to be more detailed to be really useful. In an international survey setting, some of the boxes become very complex, most notably the development of concepts, the development and pretesting of questionnaires including translation, and analysis. Once the parameters for the different process steps are fixed, we need to have procedures in place that can help us achieve the desired quality.

Simply put, quality means delivering all the product characteristics on which the user (sometimes called the customer or the client) and the producer have agreed. To be able to do this, the survey process needs two things. First, there is need for a *quality assurance* (QA) program. By that we mean mechanisms we put in place in relation to the different boxes in Figure 13.1 so that our quality goals are achieved. Examples of such quality assurance ingredients are pretesting of questionnaires, a set of operational specifications, interviewer training, probability sampling design, call scheduling algorithms, formulas for calculating base weights, analytical methods, documentation systems, user communication strategies and channels, etc.

To check if all these quality assurance measures deliver what they are supposed to, there is need for a *quality control* (QC) program. The QC program checks if the QA works. If a process or system part does not deliver as expected, it has to be checked so that the source of the problem is identified. Examples of quality control activities are verification procedures where error rates are recorded for staff, such as coders and interviewers, and for equipment such as scanning devices and software for automated coding. Other examples are recording of nonresponse rates and nonresponse distributions across subgroups, as well as costs and customer reactions. Such data are called process data or paradata (see Couper & Lyberg, 2005; Morganstein & Marker, 1997) and serve the purpose of diagnosing the processes that generate the products or deliverables. Paradata can be analyzed by means of simple tools such as histograms and scatterplots or by using methods such as control charts from the field of statistical process control (see Ryan, 2000). The topic of paradata will be discussed more thoroughly in Section 13.5.1 of this chapter.

A common approach to quality entails a three-level framework that has components of product quality, process quality, and organizational quality; these are described in detail below.

The *product quality*, which is the expected quality of the deliverables in terms of product characteristics, is decided by the client or the main user. The product quality is monitored through customer satisfaction surveys or more direct user contacts and by comparing the resulting product characteristics with the specifica-

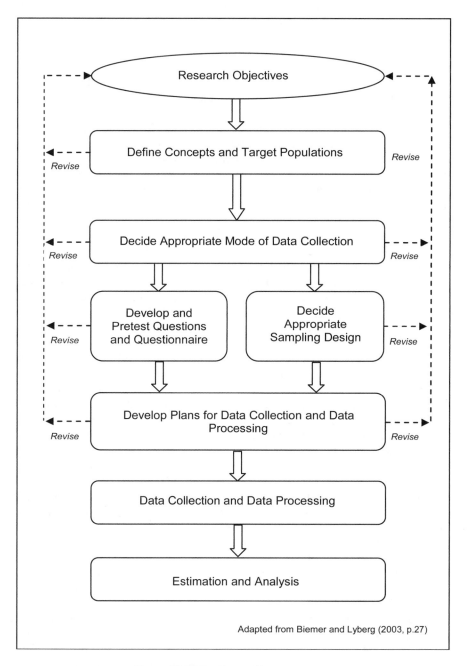

Figure 13.1. The Survey Process

tions in service level agreements (SLAs) or similar documents. The product quality depends on the quality of the underlying processes that generate the deliverables. An underlying process that shows large variability will result in product specifications not being met. Costly and frustrating redoing of work is the only way to fix this problem. For instance, if coders are not properly trained and supervised, the average outgoing coding error rate might be larger than that which can be tolerated, and parts of the material might have to be recoded using more experienced staff.

This *process quality* is controlled by choosing, measuring, and analyzing proper process variables. A process variable is one whose variation can affect the product characteristics. It is essential to check if a process is out of control. For instance, if one of the product characteristics is a specific maximum occupation coding error rate, an analysis of a control chart containing plotted coding error rates for individual coders or entire countries will generate one of two possible types of variation.

One type of variation is such that when the error rates are plotted, some fall outside the limits of the coding process' natural variation. Such variation is an example of "special cause variation" and should be eliminated. It could be that some coders within a country are performing poorly and need retraining or that the average error rates for some countries are too high, and this must be investigated to ascertain the root cause. For instance, were the coding instructions correctly translated and were the coders properly trained? Once this special cause variation is eliminated, the process is said to be in control.

In contrast to special cause variation, a second type of variation is natural variation, which is typified by error rates that are considered high, even though the plotted error rates fall inside the limits that define the process' natural variation. It would be a mistake to try to change the pattern relating to natural variation without first analyzing the process itself. Using the above example of coding, we should not tamper with individual coders in a natural variation situation, since this kind of variation stems from the process itself. Every process has a natural variation, the limits of which are determined by the efforts put into developing the coding nomenclature, the hiring and training of coders, and the coding instructions. However, if we think that the natural variation for this specific process is too large, we must change the process itself. This means that we might first investigate the factors that contribute most to this large natural variation. We might find that coder training was too short and did not cover some of the difficult coding situations that occur, and that this was the reason for the large natural variation. We would then try to improve the training, thereby decreasing the variation. Paradata, such as individual coder error rates, country error rates, and error rates by coding category, are used to control the processes.

The quality of the processes depends on the *organizational quality*. An international survey is usually an enormous undertaking, and the machinery that has to be in place to plan and lead it resembles that of a business. If the organizational quality is not sufficient, the processes will suffer accordingly. It goes without saying that very few of the components in the big survey picture can be allowed to vary if we want comparability. This means that we must have consistent approaches toward areas such as leadership, development of competence, results, user relations, and survey processes themselves. Experience shows that a good organization is absolutely essential for achieving results that are comparable. The typical set of specifications for a multinational study sometimes include hundreds of requirements that imply access to experienced survey managers, methodological know-how, and a general capacity to work according to the specifications. Without good organizational quality, there may be insufficient resources to apply processes as specified, which might lead to deviations from the specifications. Organizations deviate from specifications for a number of reasons: they do not have the experience necessary, they do not understand the justifications for some requirements, or other national interests take over. The result is that a specific requirement is simply not followed, or a variant of a specification is used in its place.

The organizational quality can be measured by means of so-called business excellence models such as the European Foundation for Quality Management (EFQM) model and the Malcolm Baldrige Award (see Lyberg & Biemer, 2008). An excellence model is a tool for self-assessment of the organization regarding a number of criteria. EFQM, for instance, has nine criteria. Five of them (leadership, people, policies and strategies, partnerships and resources, and processes) cover what the organization does, and the remaining four (people results, customer results, society results, and key performance results) cover what the organization achieves. For each criterion, the organization describes what approaches are in place, how widely they are applied across the organization, and to what extent these approaches are evaluated. Based on the description, the organization can position itself compared to world-class performance. Obviously, an organization that suffers from a lack of approaches in certain areas, or displays a large variation in approaches, and that does not evaluate work processes on a continuing basis has a lesser chance of handling requirements of the kind we find in multinational surveys. The maximum score that can be achieved using these models is 1,000. It is very common for organizations to score between 100 and 200 points on the first assessment.

Table 13.1 summarizes the control situation in general survey work. This framework can be extended to a multinational survey setting.

13.3 QUALITY ASSURANCE AND QUALITY CONTROL IN CROSS-NATIONAL COMPARATIVE STUDIES

Currently, there is little existing literature on the topic of quality in relation to cross-national comparative studies. However, one such paper by Lynn, Japec, and Lyberg (2006) describes seven unique aspects of cross-national comparative studies with respect to quality issues.

- The objective is to compare, and in order to do so, one must take country-specific circumstances into account so that equivalence is achieved. There are more than 80 definitions of equivalence in the literature; at a minimum, so-called procedural equivalence should be observed. That is to say, it is essential that operational aspects be conducted in equivalent ways across countries to achieve comparability.

TABLE 13.1. Three-Tiered Framework for Assuring and Controlling Quality

Quality Level	Main Stakeholders	Assurance and Control instruments	Measures and Indicators
Product	User, client	Product specs, service level agreements	Frameworks, compliance to specs, estimates of mean squared error, user surveys
Process	Survey designer	Process variables, current best methods, standards, checklists, verification	Control charts and analysis of variation, other paradata analysis
Organization	Agency, firm, owner, society	Business excellence models, code of practice, standards, reviews, audits, self-assessments	Scores, identification of strong and weak points, follow-up of improvement activities

Source: Adapted From Lyberg and Biemer (2008).

- Concepts and items must be relevant across nations and cultures. This means that some concepts that work in a national context might not work in an international context. For instance, in the International Adult Literacy and Lifeskills (ALL) Survey, the concept of "teamwork" was not uniformly interpreted or even understood across cultures, and, therefore, differences in cross-national analytical results based on this concept may be confounded by the differing meanings understood. In general, conceptualization (definitions, developing questions, adaptation and translation, and comparison) in the context of international surveys is more complex than in national survey design steps.
- Cross-national comparative studies are enormous undertakings and demand considerable resources and strong, innovative leadership.
- The emphasis on comparability might suggest a resource allocation that differs from the one used in a national survey. In cross-national comparative studies there are operations that must be given special care. Otherwise comparability or equivalence is not achieved. One such operation is translation and adaptation of survey materials, including the questionnaire. Therefore considerable resources must be allocated to these operations in accordance with risk management principles.
- The financial resources and methodological sophistication may differ between countries and organizations within countries. A stark example of differing budgetary circumstances is illustrated by the fact that the cost of participating in the European Social Survey can, according to ESS planning staff, vary by as much as a factor of 20 between countries. Methodological differences may show themselves in terms of level of expertise and the number of appropriately skilled methodologists allocated to the survey, the viability of certain design options, the level of

access to frames and auxiliary information, and the types of survey organizations involved.

- Many international surveys have multiple objectives. For almost all international surveys there are also national interests in terms of data needs, and those needs might clash with the international goals. As an example, a cross-national comparative study design might conflict with a national optimal design of the same study. Sometimes the international aspect is seen as a by-product of the national study, which is not surprising since the national interests sometimes are used as the incentive to participate. In addition, if the cross-national comparative study concentrates its efforts on producing rankings of results of composite indicators more than it does on producing in-depth analytical results, national pride may cause countries to focus on the former aspect exclusively. This is unfortunate since comparisons should be analytical in nature, rather than reduced to mere rankings.
- The error structures might differ between countries due to the use of technology, methodological know-how, respondents' perceptions of response scales, and the inclusion of sensitive information.

A number of authors have discussed specific challenges in relation to quality in the context of cross-national comparative studies (see Jowell, 1998, 2008; Harkness, 2008a; Harkness, Mohler, & van de Vijver, 2003; Lievesley, 2001; Lynn, Japec, & Lyberg, 2006; Verma, 2002). Jowell, Harkness, and Lievesley have separately contributed to a listing of a number of issues that might be problematic across countries and cultures:

- So far, there has been very little research associated with controlling the quality in cross-national surveys. One possible explanation is that most resources have gone toward developing concepts, survey materials and in targeting equivalence. As more knowledge is gained regarding these essential design steps, more concerted efforts should be put on transforming general survey methodology to the international multi-population context.
- Attitudes, traditions, and infrastructures vary. The levels of literacy and use of technology are different across countries, as is the general willingness to participate in surveys. As such, standardization can be difficult. A prescribed standard method may not be the one considered best national practice or may be considered infeasible because of cultural norms. This is a problem that the European Statistical System has tried to solve by introducing different levels of harmonization to achieve comparability. Input harmonization in the European Statistical System is a concerted effort to have comparability achieved through central design and support by Eurostat, Europe's Official Statistics Office. All the main survey steps are conducted in similar ways across participating countries. To avoid clashes with national norms, output harmonization is an option. With this kind of harmonization, product characteristics are fixed, while

survey design and methods to achieve them are flexible. This span between input and output harmonization reflects the conflict between demands for strict standardization and national standards that are above the standard design chosen (see Clemenceau & Museaux, 2008).

- The complexity of design and control efforts is affected by factors such as topics covered by the survey, number of countries, and degree of similarity between countries.
- Quality control indicators must be realistic and make sense universally.
- Some countries experience difficulties achieving even the minimum quality standards. Therefore, there must be an element of national methodological capacity building in international surveys to ensure that more than just a minimum standard is achieved and to enable countries to conduct future surveys more independently.

Quality issues are important in any national survey, but in a cross-national comparative study the quality problems are magnified due to the problems listed above and may threaten the validity of comparisons. Therefore a cross-national comparative study must have a program for quality assurance and quality control. In the three sections that follow (Sections 13.4–13.6), we use the product and process organizational levels as a framework for the discussion on quality and give examples of the practices of some cross-national comparative studies in relation to these.

13.4 PRODUCT QUALITY IN CROSS-NATIONAL COMPARATIVE STUDIES

As indicated in Table 13.1, the main instrument used to control product quality is a series of specifications. There seems to be a general agreement that the quality assurance approach should consist of a fair amount of standardization and harmonization with proper monitoring. Standardization generally means that specifications concerning specific survey steps are defined and countries are urged to follow them. The exceptions are the steps of sampling and estimation, where considerable flexibility can be allowed and should, in fact, be encouraged in order to adapt to different national contexts (see Chapter 14, this volume). Adaptations of questions to different languages and cultural contexts are further aspects in cross-national surveys that demand flexibility in specifications.

In general survey work, the product quality is decided by the producer or designer, preferably assisted by the user. In multinational surveys, the distance between the users and the survey organizations is much greater and the implications of this fact must be addressed. In some surveys, meetings with stakeholders are conducted, creating room to discuss design options and product characteristics so that the entire exercise is much more collective than in other surveys. Also, there is a need to widen the user base and to develop means to improve communication between the survey organizations involved and the users and sponsors. Since users need to be informed on quality indicators for each survey step, indicators to monitor quality and innovative documentation approaches must be developed to effectively communicate the large body of results that come from studies of many countries simultaneously. For instance, sometimes it is not practical to have a single document containing all information, so a "document of documents" is needed. Interactive user contact and feedback are another underdeveloped area in multinational studies.

13.4.1 Approaches to Setting Standards

Without clear standards, there will be unnecessary variation, a situation that should be avoided. Lynn (2001) has defined five approaches to setting standards.

- Approach 1 aims at achieving maximum quality, which means using best possible practices in each country. This approach is difficult to justify since it is expensive and makes comparability difficult to obtain. In effect, with major variations in error rates and design characteristics across countries, one has to create a separate standard for each country.
- Approach 2 aims at achieving consistent quality across countries. That is accomplished by adjusting the standard to the lowest common denominator. This is not a good strategy since some countries would be forced to use methods that are inferior to those used in their national surveys.
- Approach 3 is the constrained maximum quality approach and is a compromise between Approaches 1 and 2. A standard is prescribed for some key design aspects and, within those constraints, maximum quality is adopted. The key design aspects chosen are those that affect comparability most, according to some risk assessment protocol. Traditional aspects of this kind are question development, translation, and interviewer training. This is a good strategy because countries with less sophisticated survey cultures have a chance to gradually improve their capacities.
- Approach 4 aims for a specified target quality. With this approach, countries with more developed survey cultures set the standard and it is clear at the outset that some countries will not be able to adhere to them in full. Again, the hope is that the less developed survey cultures will be stimulated to improve, but it is also clear that comparability will be compromised in some instances.
- Approach 5 is the constrained target quality approach. Here, too, within a few constraints, challenging targets are set. The targets can be viewed as minimum quality standards.

Among the approaches mentioned, Approaches 1 and 2 are typically ruled out, depending on the set of countries involved and the degree of heterogeneity between them. A standard for a survey involving three countries that are geographically close (or countries at the same level of development) might be very different from one involving 26 countries spread all over the world (or countries at varying levels of economic development and survey-taking sophistication). What is important, as Lynn (2001) points out, is the structured team-thinking around the decision on how realistic various approaches are.

Input harmonization and output harmonization are two extremes among the many options for approaches to standardization. Input harmonization is used by statistical organizations such as Eurostat and the European Statistical System, whose job it is to coordinate surveys. It means that comparability is to be achieved via regulations regarding concepts, definitions, classifications, and technical requirements. Although the regulations are usually very rigid, sanctions are relatively rare. Input harmonization is a more general notion than standardization and includes features such as consistency and similarity. It should be pointed out, though, that many EU surveys lack standardization, and in these cases comparability is achieved by other means, such as output harmonization. This kind of harmonization consists of specifying only the statistical outputs, leaving it to the individual countries to decide how to collect and process the data necessary for achieving the outputs. It is clear that output harmonization is preferable in cases where countries are heterogeneous in terms of level of economic development and statistical sophistication.

On the topic of harmonization, a general topic-independent standard is the recent ISO 20252 standard on marketing, opinion, and social surveys (International Standards Organization, 2006). This standard is not primarily aimed at international comparisons, even though such comparisons are made easier if survey organizations follow the standard. A distinction is made between standards for surveys and standards for organizations.

The general approach to reach comparability is standardization through specifications on as many processes as reasonably possible. However, it should be noted that standardization can be detrimental to the quality of the survey process in some circumstances, so when enforced, it should be appropriate (see Harkness, 2008a). As an example, strategies for contact attempts can be allowed to vary, but countries should be aware of how different strategies work. Typically, any contact scheduling algorithm starts with a prescribed number of minimum number of attempts, while allocation of time can be decided locally. For instance, an allocation based on Swedish work hours would not function well in Spain or Italy, where other work patterns are prevalent. So the prescriptive part for this process would be the number of contact attempts, while the spread of the attempts would be allowed to vary. Collection of information about local constraints and other specific knowledge regarding local circumstances is very important but time-consuming. A methodology should be put in place to collect such information more systematically.

Finally, specifications can sometimes be allowed to be loose. Examples of this could be "control of coding should be performed on 5% of the cases" or "data capture errors should be estimated as prescribed in ISO 20252, with the assumption that the local organization strives toward using known reliable methods to do this." When specifications are kept loose in this manner, it is usually a concession made because other specifications were kept more rigid, based on some kind of risk analysis. Examples of more rigid specifications would be "probability sampling must be used" and "translation of survey materials should be performed by teams according to the following procedure…" Translation is, in fact, an example of a process where research teams have worked together to establish ground rules that can be continuously improved. One ground rule is to use team translation. The ESS uses the variant of the ground rule called TRAPD (Translation, Review, Adjudication, Pretesting, and Documentation) (Harkness, 2002, 2003, 2007). Despite the extensive methodological research that has taken place in the field of translation, it is still viewed by some as a task that can be performed by anyone with basic knowledge in two languages, translation software, and a dictionary. The Swedish language guru Fredrik Lindström once said that whenever Swedes tell him they are fluent in English, he asks them to provide the English words for all the utensils in a kitchen cupboard. The outcome of the test is typically eye-opening!

13.4.2 Examples of Specifications in Cross-National Comparative Studies

In the International Life Skills Survey (ILSS), which later became the Adult Literacy and Life Skills (ALL) Survey, a management group provided over 100 specifications to which countries were instructed to adhere. Each specification consisted of the headings presented in Table 13.2.

The vast number of specifications implied that countries were not given much flexibility, at least not in theory. Of course, countries had opportunities to perform tasks in different ways, but the specifications had to be addressed. The specifica-

---

TABLE 13.2. Headings for ILSS Specifications

Automated Coding

Rationale
Some countries have the option to utilize software for automated coding. For the portion of the cases that are coded by means of this software, coding consistency is achieved.

Accompanying Procedures
Coding should be performed by means of software that has been successfully used in statistics production. For the portion that cannot be coded automatically, parts of 1.1 and 1.2 *(referring to other specifications)* apply with the exception that number of coders can be reduced and that training can be relaxed, provided that the manual coding needed is not that extensive. In the latter case, coding can be performed by trainers, and independent verification is used as control.

Input from ILSS Management Group
Description of how process data should be collected.

Key Process Variables / Supporting Documentation
Portion that is coded automatically, data on types of descriptions that are referred to manual coding, number of codings based on exact matches, number of codings based on inexact matches, and data on key process variables for the manual part (same as specification 2.4). Data are delivered after coding is terminated.

Dates and Timing
During main survey.

---

tions cover the survey processes step by step, and it is easy to see that, as the number of participating countries grows, this becomes a very extensive exercise. Below is one example of a specification from ILSS. The entire set is provided in Darcovich, de Heer, Foy, Jones, Lyberg, et al. (1999).

All the specifications strive to explain to country institutes and to other local stakeholders the purpose of the specifications, the reasons for their existence, how support can be obtained from the central team, and how adherence should be checked. At the same time, these specifications serve as a standard for the specific survey. They should cover all known error sources relating to the frame, sampling, nonresponse, measurement, processing, and analysis.

A second example is the World Education Indicators Survey of Primary Schools (WEI-SPS). This is a school survey sponsored jointly by the OECD and UNESCO-UIS on fourth-grade students and teachers, and conducted in 11 countries. In the case of this survey, six specification documents were issued to instruct countries on acceptable procedures for conducting various aspects of the survey.

- Survey Participation Protocol Document (or Service Level Agreement): This was an agreement that laid out the responsibilities of each of the parties involved in the survey.
- Translation and Adaptation Guidelines: This provided the guidelines to the countries for translation of the English-based questionnaires. It also gave tools to keep track of the question adaptations they were undertaking. A database of translations and question adaptations that were observed was developed during the survey. This was maintained to assist the experts in their analyses, as some of the results would be influenced by the adaptations.
- Sampling Guidelines: This elaborated the sampling rules that would guarantee an agreed-upon level of accuracy.
- Operations Manual: This provided the countries with the implementation guidelines to follow in order to maintain a certain degree of consistency between countries and to guarantee some rigor in the implementation of the survey in the various countries.
- Data Entry Manual: This provided the countries with the software and documentation to limit the data capture errors.
- Analytical Framework, Indicator Descriptions, and Questionnaire Items Document: This document set the dimensions that would be researched using the survey data, and the theoretical frameworks upon which some of the outputs would be based.

13.5 PROCESS QUALITY IN CROSS-NATIONAL COMPARATIVE STUDIES

Quality assurance involves all the procedures, approaches, and methods put in place in an organization to ensure minimal errors and cross-national comparability as perfectly as possible. On the other hand, quality control encompasses many different, but not necessarily all, activities in a survey. Quality assurance is necessary but not sufficient to achieve comparability. It is also necessary to check whether the mechanisms that are in place work as intended; we must, therefore, perform quality control.

An example of quality assurance is the following: Coding is a specific error source that can generate substantial error rates. To reduce these error rates as much as possible, various assurance measures are implemented, such as accessible instructions, training, a verification system, a certain degree of automation, analysis of the outcome of verification, and possible improvements. Ideally, the collection of these measures should constitute current best practices or standard operating procedures, as we know them. The quality control checks whether these practices and procedures work as intended, and if they do not, then corrective and improvement measures are applied. In this case, the quality control system may consist of error rate estimation; analysis of error structures by coding category, coder, and country; and suggestions for improvements of relevant quality assurance components so that they remain the best. Similar approaches should be developed for other operations that constitute high-risk areas from a comparability standpoint. Examples include questionnaire development, translation of survey materials, frame construction, data collection, reducing and adjusting for nonresponse, and analysis (see Morganstein & Marker, 1997; Lyberg & Biemer, 2008).

The tools for quality control (QC) depend on the purposes of the quality control.

A. The simplest form of QC is to check whether or not the specifications have been followed. The outcome is basically "yes" or "no."
B. A more advanced QC is to analyze variability patterns of processes through the use of process data or paradata. A common tool is the control chart. Results can be used to adjust the ongoing process.
C. The analysis of outcomes from A and B might suggest process improvements to future applications of the process or future rounds of the survey.
D. Evaluations and reviews might also be part of the QC toolbox. Typically, these activities take place after the process or survey is terminated.

In international surveys, unfortunately, type A is much more common than the other three. There is need for more advanced procedures, one of which would be the systematic use of paradata.

13.5.1 Paradata

One way of performing QC is to select, measure, and analyze key variables about the survey process. The resulting data are called *paradata*. Paradata in the data collection process might include nonresponse rates broken down by interviewer, country, and type of nonresponse. These data can be plotted on a control chart and used as a tool in statistical process control.

Figure 13.2 shows a control chart that can be used for generic processes within a usual survey setting. The control chart has upper and lower control limits (UCL and LCL, respectively). The limits are defined separately for each process being monitored. The control limits are a multiple of the standard deviation of the data points on the chart, in this case and most commonly $+/-3\sigma$. As long as data points fall inside the control limits, the process behaves normally. This is the case of the process in Figure 13.2, and, thus, we say that the process variation has common causes. If there had been data points falling outside the limits, this would have been indicative of special cause variation. As indicated earlier, such variation must be eliminated. For example, in the case of coding, if one specific coder has an unacceptable error rate (or at least an error rate not within that which the process normally delivers), the coder would then be retrained. When the process is stable again with all data points back within the limits, we can decide whether we find acceptable the normal variation associated with the process. If we think that the normal variation is too large, then we have to change the process itself. In the case of coder error rates, we might want to make changes in the training program or the instructions in order to narrow the gap between the UCL and the LCL (see Ryan, 2000).

Figure 13.3 presents an example of a control chart excerpted from Japec (2005). The chart shows fluctuations in response rates among interviewers in the Swedish results of the European Social Survey. In this control chart, the control limits do not form straight lines due to variations in workloads between interviewers. In this particular case, survey managers would only intervene with regards to special cause variation for the three interviewers whose response rates have fallen below the LCL. An intervention might include retraining or other corrective measures aimed at the three interviewers, which hopefully puts them within the control chart limits in future plotting. However, in this case, the survey managers would also have to deal with the issue of natural variation because it is obviously very large, ranging from more than 80% down to 50–60%. (Note that

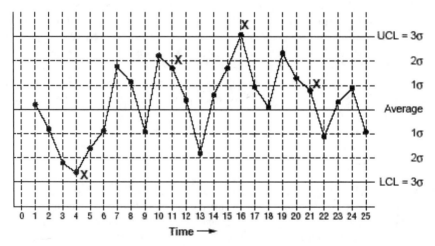

Figure 13.2. Generic Control Chart

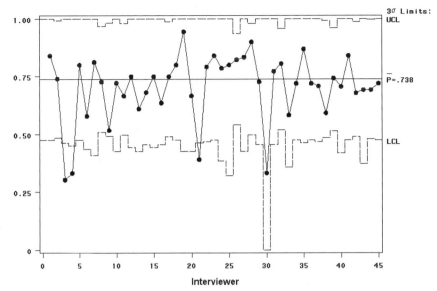

Figure 13.3. Control Chart for Monitoring Interviewer Response Rates (*Y*-Axis) in the Swedish European Social Survey, ESS. (Excerpt from Japec, 2005)

the lowest response rates that were part of the natural variation were much lower than 50%, but, in this instance, the interviewer had too small a workload to be classified as being a special cause.) Given the large remaining natural variation, the survey manager might contemplate measures that could narrow the gap between UCL and LCL. Such measures are similar to those in the coding example and could include training sessions with clarifications, improved instructions, and improved supervisor feedback.

It has become increasingly common that paradata are collected and analyzed in international surveys, but, unfortunately, because these data are not often analyzed using methods and theories from statistical process control, their full utility is not exploited.

13.5.2 Examples of QC Procedures and Initiatives in Cross-National Comparative Studies

The following are some examples of QC measures from three surveys: the 1994 IALS (see U.S. Department of Education, 1998), TIMSS, and ESS:

In IALS, the number of interviewers varied more than specified for the study, resulting in an average interviewer workload between 6 and 30 across countries. To elaborate on the significance of this, average workload is one of the components that determine the interviewer design effect, which, in turn, is a component of confidence intervals used in analytical work. Varying design effects mean varying effective sample sizes that are typically smaller than the actual sample sizes. Furthermore, the effect of interviewers is likely to be larger for the background questionnaire (in which basic socio-demographic information on households and individuals is collected) than for administering the substantive literacy items because the interviewer is typically more passive when eliciting responses for the latter. Nevertheless, the cognitive situation and the interviewer's role in surveys involving tests have not been investigated in detail.

The 1994 IALS had a special procedure for controlling the scoring of the literacy assessments. A second scorer rescored 10% of all assessments, and 50% of these were conducted early in the survey process to improve the learning curve. In general, rescoring is analyzed by comparing the original score with the rescore, and the agreement rate is calculated. Acceptance tolerances are set in advance. In the case of IALS, the scoring reliability was high. Most countries had more than 97% agreement rates. There was also a cross-national exchange of scorers for at least 300 randomly selected booklets in those countries where another country shared the same test language; the results were equally good. In meetings between chief scorers and national study managers, any clarifications were added to the scoring rubrics. In addition, Statistics Canada and the Educational Testing Service in the United States had a scoring hotline during the scoring process, where unusual responses were discussed.

In IALS, 100% keystroke validation was a given specification. Two countries did not comply. One chose 20% and the other 10%. For some other error sources, unjustified assumptions about random error distributions were made instead of actually performing the QC. It is not clear why some countries chose to deviate.

TIMSS and some other surveys have used unannounced site visits to check on test conditions and the test leader activities. It is not clear how information from these visits has been used, however.

The ESS checks such indicators as interviewer reports on respondent reactions, frequencies of "don't know" responses, as well as item nonresponse and nonresponse rate comparisons. Call records have also been informative. Finally, the ESS conducts quality enhancement meetings on design effects and interviewer effects.

Finally, in some surveys only very limited quality control efforts are made, often restricted to a limited number of domains, such as translation and back-translation or interviewer-respondent matching.

13.6 ORGANIZATIONAL QUALITY IN CROSS-NATIONAL COMPARA-TIVE STUDIES

13.6.1 Choice of National Survey Organizations

One important element of quality assurance in international surveys is the choice of survey organization in each country. It turns out that in many countries there is a severe shortage of organizations that can undertake all required operations for a cross-national survey. The experience in international work and the methodologies involved, as well as the experience in dealing with other cultures, are all sadly lacking.

During the work on upgrading IALS surveys in the European context that took place as a result of France's challenging the methodology behind the 1994

IALS, a study was conducted of survey organization resources in France, Germany, Greece, Italy, The Netherlands, Portugal, Sweden, and the United Kingdom (Carey, 2000). The purpose of the study was to investigate the extent to which the subject matter and methodological resources to carry out a literacy survey in the countries studied was up to the level required. More specifically, the review of survey practices in the eight European countries undertaken as part of the project was a necessary preliminary step to understanding the issue of comparability. By documenting survey practices across a number of European countries, the range of practices that could be offered at the time was identified. The review centered on four aspects of survey practice:

- Sample design and sampling procedures
- Survey experience
- Field work organization and strategies
- Survey processing

The main conclusion was that there seemed to be no institute in any of the eight countries studied that had sufficient experience and expertise in all aspects of the survey process in relation to literacy surveys.

A second important conclusion and recommendation was that it seemed to be necessary to form consortia of institutes in each country to combine all necessary expertise and experience in order to ensure the best possible survey practices and to avoid unnecessary variation. The study is documented in full in Carey (2000). It is difficult to tell whether the situation has improved since then; new studies are needed to see whether improvements have been made.

Survey sponsors tend to be more interested in the capabilities of potential research providers than before. Specific studies are not available to confirm this change in attitude, but examples of such indications in general survey work include the development of the ISO standard 20252 on Market, Opinion, and Social Research released in 2006 and the fact that some sponsors ask research providers to present proof of an existing quality assurance or business excellence model when submitting proposals. Some sponsors ask for proof of certification or the result of organizational evaluations conducted by external bodies. Similar tendencies are not as prevalent in international surveys, but there is certainly pressure from large sponsors and stakeholders such as Eurostat and the OECD for organizations involved in surveys to be reputable. For instance, OECD's recommendation is that countries participating in the current Programme for the International Assessment of Adult Competencies (PIAAC) choose national statistical offices as data collectors.

13.6.2 Building an Infrastructure of Coordination

For cross-national comparative studies, there is a general agreement on the need for an overarching body whose mission would be to plan, coordinate, support, and improve international survey endeavors. Most international surveys are conducted on an ongoing basis, which is a prerequisite for investing in such infrastructure. The European Social Survey has an elaborate infrastructure with a Central Coordinating Team (CCT) at the center. This team is supported by a Scientific Advisory Board, as well as four Specialist Advisory Groups on question module design, methods, sampling, and translation. In addition, the CCT receives feedback from national coordinators, survey institutes, and other stakeholders in participating countries. There is also a Funders' Forum. Within this model, each participating country ensures the majority of the funding of its own survey, as well as the survey organization and implementation.

The coordinating infrastructure is part of the quality assurance plan and provides the infrastructure for tasks such as general planning, development of specifications for processes and participating survey organizations, design of quality control procedures, and suggestions for future improvements. High-profile international surveys such as ALL, IALS, and PIAAC have developed such functions. Surveys without such infrastructure are at risk of experiencing large local deviations from the specifications. However, the actual formation of the infrastructure can be a complicated matter, and no organization, as far as we know, has evaluated the effect of the infrastructure model used. The recent evaluation of the European Social Survey (Groves, Bethlehem, Medrano, Gundelach, & Norris, 2008) mentions the survey's infrastructure and provides some suggestions about the tasks of various groups, but no in-depth review of this aspect is carried out. It is likely that evaluation studies could eventually result in more efficient infrastructure formations. The real quality problem, however, is that many international surveys still have very primitive and weak infrastructures with very little central coordination and monitoring, resulting in poor comparability.

13.6.3 Quality Assurance Approaches of Coordinating Units

The models for quality assurance adopted by coordinating units vary depending on know-how, ambition, and funding. There exist two extreme models, as well as a "middle ground" variant (see Lynn, Japec, & Lyberg, 2006). The three variants have a common element, namely some kind of unit (even if just one part-time person) that communicates with local national representatives responsible for conducting the survey.

At one extreme, a coordinator issues written instructions on how various countries should implement design phases and quality control activities. Involvement of a single coordinator is a high-risk approach since experience tells us that large deviations from instructions can take place due to ignorance or misunderstanding. For instance, in the 1994 IALS, a survey that, at the time, used written instructions as the main coordination vehicle, one country opted to tell the respondents that the IALS was a test rather than the real survey. Another country administered an incentive to boost the response rate, a measure that was not part of the written instructions. Still another country was not able to calculate the base weights correctly. Even sophisticated systems, such as the official statistics run by

Eurostat, rely quite heavily on instructions and regulations to achieve comparability.

At the other extreme, a large central coordination team is set up. The team's mission is to liaise closely with, and monitor the activities of, each national team throughout the implementation of the survey. The ESS, PIAAC, Programme for International Student Assessment (PISA), TIMSS, as well as more recent rounds of the IALS, are examples of this model.

13.7 EVALUATIONS

Unfortunately, in cross-national comparative studies, there is generally a lack of effort invested in evaluating survey results. Two kinds of evaluations are used to provide estimates of bias components or to provide recommendations for improvements. First, comparisons with gold standards, experiments, or other methods may allow estimation of bias components or identification of improved methodological solutions. Second, peer reviews or audits may allow the determination of whether a survey has been conducted according to specifications and/or whether areas needing improvement remain. Both the ESS and IALS have made strides toward evaluating aspects of survey results.

For instance, Billiet, Philippens, Fitzgerald, and Stoop (2007b) compared cooperative and reluctant respondents in the ESS, and analyzed contact forms without finding severe signs of nonresponse bias. There are, of course, many similar studies related to such aspects of the survey process as questionnaire design, translation, coverage issues, and scoring as examples in other chapters in this volume demonstrate.

More extensive evaluations are rare. In the ESS review already mentioned, Groves, Bethlehem, Medrano, Gundelach, and Norris (2008) recommended the development of quantitative indicators for all process steps, standardized contact forms, boundaries set on effective sample sizes, as well as the conducting of several rounds to improve capacity building and to expand the number of users.

An early evaluation of the peer review type in relation to the 1994 IALS was conducted by Kalton, Lyberg, and Rempp (1998). The main recommendation was that future rounds should be monitored much more closely, because many specifications were not followed by some countries. Coordinators of many larger international surveys have since worked hard to improve central support, monitoring, and infrastructures.

Apart from the ESS and IALS, PISA and TIMSS let the national project managers submit self-reports on process implementation, but it is not totally clear how these are used for improvement activities or whether they are useful at all. These surveys allowed quality monitors to study and report on assessments, but participant feedback did not seem to be part of the process. PIRLS, a survey studying comprehension among fourth-grade children, also allowed national managers to complete an activities questionnaire regarding their experiences.

13.8 RECOMMENDATIONS

In this final section we summarize our thoughts on how QA and QC approaches can be improved in cross-national comparative studies.

1. General experience seems to indicate that a multinational study must have a strong coordinating body responsible for adhering to user demands, study design, communicating with national bodies, implementing and supervising survey operations, and documenting results. Approaches assuming an automatic understanding of disseminated instructions and other survey materials tend to fail for various reasons, resulting in comparative measures that are difficult to assess (see Kalton, Lyberg, & Rempp, 1998; Carey, 2000; Lynn, Japec, & Lyberg, 2006). The ideal composition, tasks, and competence mix of an efficient coordinating body need to be investigated further. Such studies should also address more subtle issues, such as how to secure participation from countries with methodological know-how, but, at the same time, prevent those stronger countries from dominating the weaker ones. Few evaluation attempts have been performed so far. Kalton, Lyberg, and Rempp (1998), Carey (2000) and Groves, Bethlehem, Medrano, Gundelach, and Norris (2008) are three rare cases.

2. Typically multinational studies have national coordinators whose role is to ensure that prescribed operations are executed according to specifications. The national coordinator is frequently a subject matter specialist with minimal survey experience. Since subject matter experience could be secured through existing expert groups, it would be preferable if the national coordinator were a more experienced survey manager rather than a subject matter expert. It is important for national coordinators to appreciate the need for QA and QC and to be familiar with error structures associated with different survey operations.

3. Most multinational surveys focus on product quality. More attention to process quality is needed to avoid rework and to keep variation to a minimum. Basic concepts associated with controlling and adjusting processes should be part of every multinational survey's toolbox. Unfortunately, even many national surveys have not recognized educational efforts regarding the role of quality (see Morganstein & Marker, 1997; Biemer & Lyberg, 2003; Groves & Heeringa, 2006).

4. Thought must be given regarding how to motivate national bodies to accurately report quality. What is the reward for conducting effective interviewer training? In whose interest is it to report low response rates if they occur? Are public ranking and external reviews the main motivators? More positive motivators to stimulate quality work and quality reporting are needed.

5. Methodology should be put in place to systematically collect information on local constraints and circumstances. Such information could be used to adjust both design and implementation instructions.

6. Studies of evaluation should not be confined to development of coordinating bodies. Evaluation is needed for the effect of mixed-mode data collection, constraints on nonresponse bias, and correlated response error with effective sample size. Such evaluation is needed particularly in the case of relaxed standardizations. One might speculate that with more knowledge on these issues, it would be possible for a national survey institute with good interviewer training and supervision to decrease its sample size somewhat, since nonsampling errors are reduced. There is also need for periodic review of quality development in continuing multi-national studies. It is very important to identify reasons for deviations from specifications. Root cause problem analysis must serve as input to continuous improvement.

7. QC methods clearly need considerable upgrading. The use of quality monitors is an inefficient use of resources. More modern verification and analysis using control charts and other tools from the statistical process control theory are urgently needed (Ryan, 2000).

8. There is a need for more careful design approaches taking into account various trade-off situations based on risk analysis. Most designs currently used are very limited in the sense that design resources have not taken total survey error into account. Although some surveys spend a lot of resources on developing concepts, questionnaires, and a translation procedure which is in line with a proper risk analysis, many other surveys do not. Multinational survey stakeholders should be much more aware of the trade-offs that exist at the design stage. For instance, a fixed maximum nonresponse rate does not seem rational when other design decisions, such as maximum interviewer workload, are much looser.

9. Thought should be given to the need for national capacity building in some developing countries. It is possible that, in some extreme cases, external entities might conduct the survey in the short run, although this scenario is not preferable in the long run.

10. Documentation is a necessary part of any QA and QC effort. However, this is problematic for most cross-national surveys, in that the documentation is often of low quality or key aspects are missing altogether. Documentation needs to be improved (see Chapter 16, this volume; Harkness, 1999).

11. Related to documentation is the issue of confidentiality regarding the release of microdata. This is a problem for international studies, since many participating countries have limited or no legal frameworks, such as Statistical Acts, in place. Procedures for handling confidentiality issues in an international comparative setting must, therefore, be developed.

12. There are a number of quality frameworks for national surveys (see Biemer & Lyberg, 2003). Similar initiatives have been suggested by Groves, Bethlehem, Medrano, Gundelach, and Norris (2008) for the European Social Survey, but the general framework is applicable to all cross-national studies. Such a framework should have quantitative indicators associated with various dimensions of quality.

14

Sample Design for Cross-Cultural and Cross-National Survey Programs

Steven G. Heeringa and Colm O'muircheartaigh

14.1 INTRODUCTION

Over the past 40 years, successful multinational, multicultural programs of population survey research have employed probability sample designs and procedures to select nationally representative samples to which to administer the common survey interview (Verma, Scott, & O'Muircheartaigh, 1980; Heeringa et al., 2008; Lynn, Häder, Gabler, & Laaksonen, 2007). This chapter blends established probability sampling principles and the experience and empirical outcomes gained from these past programs to define a model for collaborative sample design efforts that will be required in future multinational, multicultural survey programs.

In the strictest sense, the standardization of the sample designs across nations and cultures is really more a coordination of the multiple sample design efforts that draws on the principles of sampling and survey program objectives than a standardization of the specific probability sample design features (e.g., frames, strata definitions, clusters). A coordinated focus on the design objectives and target populations for the overall survey program is essential, but strict uniformity of designs across countries or cultures is often not possible, nor is it necessary to the scientific integrity of the overall project.

14.1.1 Process Complexity, Management Models

In thinking about sample designs for such survey programs, it is important to recognize what Lynn, Japec, and Lyberg (2006) call the additional "layer of complexity" that enters the management of design decisions. Optimal sample design for a single population survey is itself a sufficiently complex undertaking, incorporating multiple domains of measurement (e.g., demographic, health-

[1] Survey Methods in Multinational, Multiregional, and Multicultural Contexts, edited by Harkness et al.
Copyright © 2010 John Wiley & Sons, Inc.

related, economic) as well as within-population heterogeneity in the variances of these measures and the cost of reliably collecting the survey data. A strictly *statistical perspective* on designing samples for multinational, multicultural surveys sees the process as addressing an extension of the single population sample design problem with increased inter-country or inter-cultural heterogeneity in survey errors and costs. The standpoint then is that we solve the more complex statistical optimization problem, everything else will fall neatly into line. This view of the problem is often short-sighted. It can, for example, fail to recognize that while many of the survey program objectives are universal, there may be local objectives that must be met to obtain funding or to gain the cooperation of local agencies or stakeholders.

These issues are not resolved even in cases such as the World Fertility Survey, where the coordinating organization is also funding the costs of the program surveys. Obviously, survey programs in which the central coordinators hold the purse strings can exert more influence over local decision making than might be possible in a voluntary consortium of countries or cultural groups, when each entity is thus expected to fund its specific data collection. However, outside money alone is not sufficient to ensure the full and enthusiastic cooperation of local experts, especially when local investigators and technical staff may have concerns that the central coordinators are not sensitive enough to local knowledge and essential local survey conditions. Just as countries and cultures are diverse in their demographic composition, language, social norms, attitudes, and beliefs, there can be substantial differences in the accepted survey practices and basic survey conditions and infrastructure. The coordinated design effort will require a patient, collaborative approach involving fact-finding, discussion, and, ultimately, agreement on a best possible design that is consistent with the objectives of the broader survey program.

Just as dictatorial centralized control over local sample designs is bound to fail, so will a management model that lacks any centralized technical expertise or control over the sample designs employed by participating entities. Based on our experience and that reported by others, we are convinced that coordination of a multinational survey design effort requires centralized leadership and a commitment to basic probability sample design principles (see also Chapter 13, this volume, on quality control). The centrally coordinated design effort will require teamwork with local survey statisticians and other practitioners to ensure that participant designs meet the basic standards established for the overall program but are also efficiently adapted to local resources and conditions.

14.1.2 Chapter Objective

The objective of this chapter is therefore to address the full complexity of designing samples for cross-national, cross-cultural survey programs—the basic principles and procedures that apply universally, as well as required flexibility and adaptations that may be needed to match local conditions and resources.

We avoid extensive references to texts on sampling and instead make reference to a selection of basic texts and also to some more advanced readings which have particularly relevant observations to offer. Our partiality to the classic works is clear in our choice, but does not imply that more recent work is not valuable to the practitioner.

We also choose to focus primarily on sample designs for programs of household survey research. There are a number of important international survey programs, including cross-national education surveys carried out by the International Association for the Evaluation of Educational Achievement (IEA), that utilize school-based sampling or sampling of other nonhousehold populations. In general, the principles and experiences that we describe for household population surveys extend to coordinated probability sampling of such nonhousehold populations.

14.2 EXAMPLES OF MULTINATIONAL, MULTICULTURAL SURVEY PROGRAMS

Three major multinational survey programs are used as exemplars in this chapter: the World Fertility Survey (WFS), the World Mental Health (WMH) Initiative, and the European Social Survey (ESS). Each of these three survey programs used a team approach to coordinate the sample design efforts of the participant countries. In addition to the survey staff in each of the participating countries, the central program team included one or more sampling experts who had responsibility for determining the sample design in collaboration with the national researchers.

14.2.1 The World Fertility Survey (WFS)

The WFS was a program of surveys carried out on behalf of the United Nations Fund for Population Activities (UNFPA) between 1974 and 1982. Comparable surveys, with a questionnaire covering marital and birth histories and knowledge and use of contraception, were carried out in 44 developing countries with a common organizing principle and design. Members of a central sample design team traveled to the participating countries and designed the samples in collaboration with national (statistics office) staff. The central WFS sampling staff worked with the participating countries to develop a sample design appropriate for the country. In some cases an existing sample design already in use in the country was adopted for the WFS survey. The general principles of the WFS designs are described in Verma, Scott, and O'Muircheartaigh (1980). Surveys were carried out also in a small number of developed countries, but without this coordination. There may be a lesson to be learned from the fact that sampling errors were computed and published for all the developing countries, and for none of the developed countries.

Although the WFS was by design a set of standardized surveys, "(s)tandardization does not, however, extend to the sample designs; these are worked out individually to suit each country's situation—though in practice similar problems often lead to similar solutions" (Verma, Scott, & O'Muircheartaigh, 1980, p. 31). The approach adopted by WFS emphasized the requirement of probability sampling but provided flexibility in country-specific designs.

We have chosen the WFS as one of our exemplars for five reasons. First, it covers all continents, many languages, religions, and ethnicities. Second, it included the whole range of situations, from countries with a history of survey sampling to those for which this was the first probability sample survey. Third, the processes and procedures were consolidated and published in a sampling manual. Fourth, the appropriate analysis of the data (including design effects) was carried out and published. Fifth, it is a landmark study that illustrates all the challenges of cross-national and cultural research, and also some of the solutions.

14.2.2 The WHO World Mental Health (WMH) Surveys

The WMH initiative is an ongoing international program of psychiatric epidemiology surveys in over 30 countries. The primary objective of each WMH survey is to estimate the national prevalence of DSM-IV (*Diagnostic and Statistical Manual of Mental Disorders Fourth Edition*) mental health disorders (mood, anxiety, and substance disorders) as well to measure potential risk factors for these conditions. Heeringa et al. (2008) provide a description of the coordinated sample designs for the first 18 WMH countries to complete data collection. Each prospective participant in the WMH initiative was provided with a common set of requirements and performance standards that their probability sample design would be expected to meet (Heeringa et al., 2008). Each country that sought to join the WMH initiative submitted a sample design and research plan for the preferred design for their country's survey to a panel of technical experts for a critical review and recommendations. The WMH coordinating centers and technical experts provided consultation on request throughout the survey process, but the day-to-day oversight of the sampling and other survey activities was the responsibility of the local research team.

14.2.3 European Social Survey (ESS)

The ESS comprises an academically oriented program of research designed to measure and profile change in the attitudes, beliefs, and behaviors across the span of European populations and cultures. ESS Round 1, conducted in 2002–2003, studied household populations in 22 nations. Lynn et al. (2007) provide a description of the procedures employed by the ESS to establish productive, collaborative relationships with survey coordinators and statisticians in each of the 22 participating countries and to guarantee that participant country surveys met ESS design standards. These included statistical precision requirements (effective sample sizes) for key survey estimates. The ESS employed a management model similar to that of WFS in which a technical panel of four sample design experts worked individually and collectively with the participant countries to gather required planning data, review design options, and finally agree upon an approved ESS sample design. Post-survey evaluations of the full ESS collaborative effort highlighted the effectiveness of the technical panel approach to coordination of the multinational sample design efforts.

14.3 BUILDING THE FOUNDATIONS FOR MULTINATIONAL, MULTI-CULTURAL SAMPLE SURVEYS: SURVEY OBJECTIVES, TARGET POPULATIONS

The WFS, WMH, and ESS exemplify how sample designs can be harmonized without being identical. The amount of flexibility that can be permitted will actually vary across the sequence of sample design steps. For obvious reasons, specification of the survey objectives and definition of the target population must be highly standardized across participating surveys, while choice of the sample frame, the number of sampling stages, stratification, and clustering can and should be optimized on a country-by-country basis. In this section, we explore the process of laying a foundation for building a coordinated set of probability samples for the survey program—a clear specification of the program-wide survey objectives and the target population for survey inference.

14.3.1 Specification of Survey Objectives

At first glance, the specification of the primary analysis objectives for a program of multinational or multicultural research should be a simple task. However, even in cases such as the ESS where there was broad consensus among the participants that the survey program was "designed to chart and explain the attitudes, beliefs and behaviour patterns of Europe's diverse populations" (Lynn, Häder, Gabler, & Laaksonen, 2007, p. 108), there are important details that must be considered before participating countries can proceed to the specification of a sample design for their population. Some of the issues that must be addressed in the centralized planning process include those below.

Country-Specific Estimates? Comparisons? Or Pooled Analyses? Consensus on the analytic objectives of the survey program is critical to the coordinated planning of sample size determination and sample allocation across the participating entities. Although participation in a multinational program implies that cross-national analyses are important, it is essential that the analytic goals be carefully and explicitly defined in advance of the sample design planning stage. Almost certainly, individual participants will want to conduct stand-alone analysis of the data for their particular country or cultural group. Specific participants may want to ensure that the sample allocation plan and sample size determination provides sufficient statistical precision to contrast their population sample with one or more of the other populations or cultural groups. Although it is less common, the program's central planners may also envision a pooled analysis of the data from the multiple country surveys.

Local Survey Objectives. Most multinational and multicultural survey programs rely on local or locally allocated funding (i.e., through country ministries, targeted programs of international agencies, or nongovernmental organizations). Although the survey may be conducted under the umbrella of the multinational program, it is rare that the local survey team will not also have objectives that are specific to their country. These may include special analysis requirements for geographic domains, and ethnically and culturally distinct subpopulations. Individual countries may also want to add local content to the core survey questionnaire or nest the multinational program survey content in an ongoing national survey program.

Subpopulations. Do the overarching or local objectives mandate estimation for major subpopulations such as men and women, the young and old? Survey planners often neglect to plan for critical subpopulation analysis requirements. If subpopulation estimates are important, they should be considered in determining key features of the country-specific design including overall sample allocation, total sample size, and stratification criteria.

Impact on Sample Design. The paragraphs above illustrate three of the basic maxims of sample design (O'Muircheartaigh, 2007) whose relevance is likely to be neglected in multinational, multiregional, and multicultural (3M) studies, where the need for coordination and expertise may submerge the recognition of subject matter expertise. The natural tendency toward centralization and standardization must be tempered by an understanding of the special circumstances of each individual study. First, we must understand and accept that survey sampling is an applied discipline and not a branch of mathematical statistics. The subject matter experts in each country/domain must therefore have input to the shape of the sample design; it cannot be left to the statisticians in the country. Only by relating the sample design to the particular circumstances can good decisions be made. Second, as a corollary, good sample design (which in this case is country/domain specific) is based on knowing the population, the frame, and the objectives. Visiting (or remote) experts will not be able to design the most appropriate sample. Third, optimal design does not minimize cost per case completed; it maximizes information per dollar/euro/peso/yuan spent.

14.3.2 The Target Population

As investigators set out to define a common target population for a multinational survey program they face a number of questions. How can the simple definition of the target population be operationalized in the design of the sample survey and survey procedures for household screening and respondent selection? What about persons who are temporary residents, guest workers, or those who have legal claim to medical treatment or services? Is it reasonable to expect that the government and nonprofit agencies who support the surveys in participating countries will wish to focus survey resources on de jure or de facto populations? What about adults in the target population who are incapable of participating in the survey—institutionalized populations or persons with cognitive limitations or other impairments that make a survey interview impossible? How about population elements in remote places that require disproportionate amounts of survey resources to sample and interview? The answers to all of these questions will differ from one country to the next and will need to be recognized in the centralized planning process.

The survey population is defined as the subset of the target population that is truly eligible for sampling under the survey design. Restrictions in the survey population definition for each country survey can include geographic scope limitations, language restrictions, citizenship requirements, and whether to include special populations such as persons living in military barracks and group quarters or persons who were institutionalized at the time of the survey (e.g., hospital patients, prison inmates).

In the WFS, consideration was given to possible formal geographic exclusions in coverage for each country. There were no exclusions in 20 countries; seven countries excluded geographic areas containing less than 5% of the national population; six countries excluded areas with between 5 and 10%; and only two, Indonesia and Malaysia, excluded sparsely populated islands or interior regions containing more than 10% of the target population. After extensive negotiation by its 21-country steering group, the ESS chose to define its target population as persons aged 15 and older residing in a private household within the borders of the survey country.

Under the ESS target population definition, there were to be no exclusions based on nationality, citizenship, language, or legal status and formal translations of the ESS questionnaire were mandated whenever a minority language group exceeded 5% of a country's survey population.

In the WMH, specific arrangements were made to accommodate special circumstances in a number of countries including restriction to urban areas only (Mexico, 75% of population; Colombia, 73% of population), restriction to the 48 contiguous states (United States, 98% of population), and restriction to telephone subscriber households only (France). Specific language fluency was an eligibility criterion in 13 of 18 survey population definitions, with 6 of 18 requiring that respondents be citizens of the country.

Subject to obvious constraints on the survey budget, the necessary steps to maximize coverage of the target populations in multinational studies include: (1) avoid major geographic exclusions; (2) design survey procedures and interviewer training to maximize coverage of hard-to-identify or difficult-to-access population elements; (3) translate questionnaires to accommodate major ethnic minorities and language groups in the target population; and (4) schedule the survey data collection to avoid seasonal absences due to vacation, work patterns, or religious or national events (Ramadan, for example).

14.4 GENERAL STRATEGIES FOR OPTIMIZING DESIGNS FOR INDIVIDUAL COUNTRIES: SAMPLE FRAMES, CLUSTERING, STRATIFICATION

Section 14.3 addressed the two elements of the sample design process—specification of survey objectives and definition of target populations—that must be highly standardized across the multiple surveys in a multinational survey program. This section considers features of the probability sample designs that can and should be specifically adapted to each participating country to optimize the statistical and cost efficiency for its survey sample.

14.4.1 Frames for Sample Selection

Probability sampling requires a sampling frame that provides a high level of coverage for the defined target population. The technical panel responsible for the coordination of the sample development must work with each participating country in a multinational survey program to carefully review the available choices and select the best sample frame for that country's survey population.

There is one question that should be asked at the start of this exercise. Is there already an existing frame for a current or recent survey that could provide a suitable sample for the survey planned, possibly through augmentation, or by subsampling? There was a clear contrast here between the strategies used by WFS, WMH, and ESS. For the ESS, in all but two countries in Round 1 there was an existing, reliable frame of addresses available for social research. Existing probability sample frames were available and used in approximately half of the initial set of 18 WMH countries. For the WFS, none of the developing countries had an existing frame for probability sampling of households and new (generally area probability) sample frames needed to be developed.

Sampling frames must contain information that is up-to-date and accurate (e.g., addresses, stratifying variables, size measures). It is rare that the information included on existing sample frames is completely current. The fact that an existing frame is out-of-date does not mean that it cannot be used. In some instances, it is possible to update the frame information before the sample is selected. When the survey team has the option of using an existing sample frame, the costs of updating that frame must be weighed against the substantial costs of building a completely new frame for the target population.

For surveys of household populations, the range of sample frame choices commonly includes: population registries, new or existing area probability sampling frames, postal address lists, voter registration lists, and telephone subscriber lists. Many countries, regions, or states maintain population registers for administrative or other purposes, although the coverage and quality of these registers vary widely depending on the country and the purpose for which they are developed and maintained. For registers maintained on a local level or in cases where a stratified random sample from a national register would not permit a cost effective survey, a primary stage sample of local administrative units could initially be selected. A second stage cluster of eligible individuals could then be selected from the population registers for selected sample localities. Before selecting an administrative list as a sampling frame for a survey, the survey team should carefully evaluate its quality. The managers and owners of lists almost invariably claim a higher degree of coverage and timeliness for their lists than is warranted. Unless there has been some independent evaluation of the lists, caution is recommended in accepting these claims.

In the absence of population registries or administrative lists that provide a high degree of coverage of the target population, an area probability sample design is the logical choice. Efficient procedures for area probability sampling of households are covered in detail in a number of texts and guidance documents (Kish, 1965; United Nations, 2005). In today's world we expect that most national statistical agencies and nongovernmental organization (NGO) researchers world-wide have extensive experience in using area frames for demographic and epidemiologic surveys.

The WMH collaborating countries generally chose one of three types of frames: (1) a database of individual or household contact information provided in the form of national population registries, voter registration lists, postal address lists, or household telephone directories; (2) a conventional multistage area probability sample frame, or (3) a hybrid multistage frame that combined area probability methods in the initial stages and a registry or population list in the penultimate and/or final stages of sample selection.

14.4.2 Clustering, Sampling Stages

Once an appropriate sampling frame has been chosen, the next step is to decide whether a multistage sample design is necessary. In many cases, two or more stages of sampling will be employed. The first or primary stage is usually an area stage and will always be necessary unless the country is geographically compact and a reliable list (of addresses, dwellings, households, or individuals) is available for the whole country.

In the WFS, there were no countries with a reliable register of addresses or individuals. Consequently, all the samples involved selection of a sample of areas as a first stage. In the ESS, most (20) countries had at least one of the lists above; even in the ESS, however, 17 of the 22 nations in Round 1 used a multistage sample.

The first decision in the multistage sample design process is to determine the area unit within which households or individuals will be selected for interview; we designate this as the Ultimate Area Unit (UAU). Typically the UAU will be considerably smaller if listing needs to be carried out; if a list already exists, the size of the UAU is determined primarily by the cost of travel for interviewers. A common UAU is the Census Enumeration Area (EA) for which boundaries are clearly defined and generally available.

The number of households to be selected within each UAU (the cluster "take") will be roughly constant. The second decision in the multistage design process is to decide the average size of the cluster "take." The choice of "take" will be determined by a combination of statistical considerations (the relative heterogeneity within and among UAUs) and practical considerations (the number of interviewers in each UAU, the time necessary to carry out the requisite number of call-backs, etc.).

The third decision is whether to introduce an earlier stage of clustering into the design. Again this will be determined by geographic distribution of the country population as well as interviewer travel or supervision costs. Somewhat surprisingly, in the WFS there were very few countries in which this additional (earlier) stage of clustering was found necessary; we also emphasize that even then, this additional stage was necessary only in the more rural parts of these countries. Unless travel costs are prohibitive, a two-stage design that employs many smaller geographic primary stage units (enumeration areas) and smaller subsamples of eligible subjects per geographic cluster is preferred. A common multistage sample design mistake is to draw a very small primary stage sample of very large geographic areas (e.g., 11 of 13 states, 12 of 30 health regions), followed by additional stages of geographic subsampling within the large area. Such designs have all the properties of a multistage probability sample but their statistical efficiency is poor. If the geographic area spanned by the survey population is large and a three- or four-stage design is required, one should choose primary sampling units (PSU) that are many in number and select a primary stage sample with no fewer than 30–40 sample PSUs.

The fourth decision is how to select the sample within the UAUs. To achieve the highest level of sample quality, the preferred method for sampling households within UAUs is to begin with an enumerative list of the dwelling units in the selected EA or local area unit. Kish (1965) describes procedures for housing unit enumeration or listing that can be applied in most every setting, including densely settled urban neighborhoods or sparsely settled villages and rural areas. A simple random sample or systematic random sample of the housing unit addresses is then selected from the enumerative listing and provided to the field interviewers for contact, screen, and interview of eligible sample individuals. Strict control is exercised over the original sample selection. Interviewers contact predesignated sample households and are not permitted to substitute a new address for a sample household that cannot be contacted or refuses to participate.

14.4.3 Stratification of the Sample

In our view, the primary purpose of stratification in national sample designs is to provide design flexibility in domains of the population where different cost and precision arguments apply. However, the most commonly presented purpose—increase in precision (reduction of variance)—is also relevant. In particular, stratification permits the incorporation into the sample design of much of the structural knowledge of the population we may have. We may wish to mirror these population characteristics in the sample, either to increase precision or to increase the credibility of the sample, or both. Common stratifiers at the cluster stage are geographic region, population density, and ecological zone.

In stratification, essentially we divide the population into subpopulations and then select a sample independently from within each stratum. Given this independence between strata, there is no need to have a uniform stratification plan across countries or even across domains within a single country.

14.5 DESIGN EFFECTS AND EFFECTIVE SAMPLE SIZE FOR COMPLEX SAMPLE DESIGNS

In multinational, multicultural surveys, the precision of sample estimates for single-survey analyses and for comparison of estimates between country surveys will be a function of the survey sample size and the "efficiency" of the specific sample design. Due to differences in stratification, clustering, and weighting, the efficiency of the individual sample designs will vary from one country to the next. Following the approach adopted by the ESS technical sampling group (Lynn, Häder, Gabler, & Laaksonen, 2007), we recommend that precision standards for participating country designs be standardized or that minimum precision goals be set for key statistics that will be estimated and compared from the participating country survey datasets.

As described in the previous sections, probability sample design features such as stratification and clustering for the individual country surveys will not be highly standardized. This flexibility in design across participants in the multinational program complicates comparison of the effective precision of the resulting samples—a sample of $n=5,000$ from country A may have very different precision than a sample of identical size from country B. The concept of effective sample size provides a common language for discussing precision within the Babel of differing probability sample designs that may be encountered in the multinational or multicultural survey program. "Effective sample size" for a design is specific to the survey variable(s) and statistic of interest and is interpreted as the size of the simple random sample, n_{srs}, that is required to yield sample precision equivalent equal to the actual design sample size, n_{des}.

The design effect statistic is used to estimate the effective sample size. The following section describes the design effects for probability samples that include stratification, clustering, or disproportionate sampling (i.e., weighting) and how design effects may be used to set targeted precision standards for the survey program.

14.5.1 Design Effects: Definitions and Sources

For convenience we designate as a complex sample design any design that is not a simple random sample (SRS). Practical sample designs employed in multinational survey programs are generally stratified multistage designs. Stratification is introduced to increase the statistical and administrative efficiency of the sample and to provide flexibility in the design. Clusters of elements are selected at the initial stage (or stages) in multistage designs to make list construction possible and/or to reduce travel costs and improve interviewing efficiency. Disproportionate sampling of population elements may be used to increase the sample sizes for subpopulations of special interest, resulting in the need to employ weighting in the estimation of population prevalence or other descriptive statistics. Often the lack of a frame of individuals leads to the selection of a single individual in each selected household in an epsem (Equal Probability of Selection Method) sample of households, similarly requiring weighting in the analysis (this was typically the case in the WMH and ESS). Relative to simple random sampling, each of these complex sample design features influences the size of standard errors for survey estimates. Figure 14.1 illustrates the effects of these design features on standard errors of estimates. The curve plotted in this figure represents the SRS standard error of an estimate of a proportion as a function of sample size. At any chosen sample size, the effect of sample stratification is generally a reduction in standard errors relative to SRS. Clustering of sample elements and designs that require weighting for unbiased estimation generally have larger standard errors than an SRS sample of equal size. Relative to an SRS of equal size, the complex effects of stratification, clustering, and weighting on the standard errors of estimates are weighting for unbiased estimation generally have larger standard errors than an SRS sample of equal size. Relative to an SRS of equal size, the complex effects of stratification, clustering, and weighting on the standard errors of estimates are termed the design effect and are measured by the following ratio:

$$\text{Deff}(\hat{\theta}) = \frac{\text{Var}(\hat{\theta})_{\text{complex}}}{\text{Var}(\hat{\theta})_{\text{srs}}}$$

where :

$\text{Deff}(\hat{\theta})$ = the design effect for the sample estimate, $\hat{\theta}$;

$\text{Var}(\hat{\theta})_{\text{complex}}$ = the complex sample design variance of $\hat{\theta}$; and

$\text{Var}(\hat{\theta})_{\text{srs}}$ = the simple random sample variance of $\hat{\theta}$.

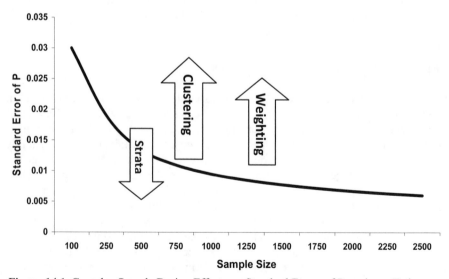

Figure 14.1. Complex Sample Design Effects on Standard Errors of Prevalence Estimates

Survey designers make extensive use of design effects to translate between the simple analytical computations of sampling variance for SRS designs and the approximate variances expected from a specific complex design alternative. Survey statisticians have developed models for the design effects attributable to stratification, clustering, and weighting (Kalton, 1977; Skinner, Holt, & Smith, 1989). Lynn, Häder, Gabler, and Laaksonen (2007) demonstrate how these models may be applied in the planning stage of a multinational survey program to predict design effects and thus the relative efficiency of proposed design alternatives for participating countries.

14.5.2 Effective Sample Size: A Common Language for Specifying Sample Precision of Probability Sample Designs

In working with collaborators in a multinational or multicultural survey program, the sampling team may choose the related measure of design efficiency termed the effective sample size:

$$n_{\text{eff}} = n_{\text{complex}} / \text{Deff}(\hat{\theta})$$

where :

n_{eff} = the effective sample size, or the number of SRS cases required to achieve the same sample precision as the actual complex sample design;

n_{complex} = the actual or "nominal" sample size selected under the complex sample design.

The design effect ratio and effective sample size are therefore two means of expressing the precision of a complex sample design relative to an SRS of equal size. For a fixed sample size, the statements "The design effect for the proposed complex sample is 1.5" and "The complex sample of $n = 1,000$ has an effective sample size of $n_{\text{eff}} = 667$" are equivalent statements of the precision loss expected from the complex sample design.

14.5.3 Design Effects in the WMH, ESS, and WFS Programs

To illustrate the typical magnitude of the design effect ratios, Table 14.1 provides the design effects for estimates of the prevalence of major classes of mental health disorders for each WMH survey. The design effect estimates provided in this table are based on diagnoses from Part II of the World Mental Health Composite International Diagnostic Interview (WMH CIDI) questionnaire. With several exceptions, design effects for prevalence estimates tend to be larger than 1.0, the average across countries being 1.7 for anxiety disorders, 1.1 for mood disorders, 1.9 for substance disorders, and 1.8 for the prevalence of any disorder. Within individual WMH country samples, the magnitude of the estimated design effects

TABLE 14.1. Design Effects for Estimates of the 12-Month Prevalence of Major Classes of DSM-IV Disorders

| Country | Mental Health Disorder | | | | Part 2 Sample Size |
	Anxiety Disorders	Mood Disorders	Substance Disorders	Any Disorder	
Colombia	2.19	1.0	1.40	1.44	2381
Mexico	1.55	1.32	1.17	2.40	2362
United States	1.80	1.03	2.49	2.34	5692
Nigeria	2.82	<1.0	<1.0	2.03	2143
South Africa	2.01	2.39	3.03	2.90	4315
Lebanon	1.37	1.06	5.14	2.04	1031
Belgium	2.57	1.26	1.30	2.38	1043
France	1.45	1.23	1.63	1.61	1436
Germany	2.10	<1.0	1.79	2.32	1323
Israel	1.36	1.27	<1.0	1.20	4859
Italy	1.04	<1.0	1.78	1.12	1779
Netherlands	1.07	1.07	1.05	<1.0	1094
Spain	2.75	1.35	2.84	1.61	2121
Ukraine	1.25	1.22	1.84	1.73	1720
New Zealand	1.45	1.63	1.98	1.63	7435
People's Republic of China	1.35	1.21	1.65	2.00	1628
Average	1.70	1.08	1.89	1.81	—

Adapted from: Heeringa et al. (2008). World Mental Health Survey Program.

shows a consistent trend Deff($p_{substance}$)>Deff($p_{anxiety}$)>Deff(p_{mood}). The lower values of Deff(p) for mood disorders could reflect smaller intra-cluster correlations in the survey populations for depression, generalized anxiety disorder (GAD), etc. particularly in relation to substance abuse disorders (which may be more geographically clustered in urban areas or other localized areas of the survey population). Across WMH surveys, design effects are smaller in countries such as Israel and the Netherlands that employed only limited clustering of observations in the survey sample design. Design effects for prevalences were highest in larger countries such as the United States, Mexico, and South Africa, where distances between sample points necessitated greater primary stage clustering of the sample selection.

Kish, Groves, and Krotki (1976) and Verma, Scott, and O'Muircheartaigh (1980) provide in-depth analyses of the design effects for estimates based on the

WFS data; they report a consistency of patterns of design effects across countries and variable types that parallel those found in the WMH. For ESS Round 1, Lynn, Häder, Gabler, and Laaksonen (2007) report average design effects for a selected set of estimates from the 22-country surveys that range from 1.0 (no effect) to just under 2.0 (a 50% loss in effective sample size).

14.6 DETERMINING SAMPLE ALLOCATION AND SAMPLE SIZES FOR PARTICIPANT COUNTRIES

The goal of all survey sample designs is to minimize sampling variance and bias for a fixed total cost. In multinational survey programs, there is no single path to this goal. Survey cost structures and the total financial resources available to conduct the survey play a key role in determining optimal sample design. Survey cost structures are highly variable from one country to another, depending on factors such as available and accessible survey infrastructure (government or commercial survey organizations), availability and costs for databases and map materials required to develop sample frames, labor rates for field interviewers and team leaders, and transportation costs for getting trained interviewers to distributed samples of households. Total funding for the survey may also vary widely across the participating countries; in many cases funding restrictions not only limit the total size of the sample but also limit the definition of survey populations or the use of preferred, but more costly sample design options.

Consequently, ensuring that participating countries meet minimum precision standards in the design of their probability sample while, at the same time, honoring the difference in funding and local expertise are major challenges for the sampling team. Cross-national, cross-cultural surveys occupy a position somewhere between two extremes. They are not equivalent to a single study carried out at a world (or regional) level; neither are they at the opposite extreme: a collection of independent national surveys linked only by a common subject matter. Programs like WFS, WMH, and ESS needed to formulate criteria to provide some balance between the samples for the participating countries. We describe the criteria for WFS and ESS below.

In the WFS funding was provided to the countries by the WFS program [primarily from UNFPA and the United States Agency for International Development (USAID)], and thus the countries themselves had no particular incentive to use small sample sizes. WFS used a set of rough criteria, as follows:

1. A minimum of 2,000 women, to allow for sufficiently detailed tabulation.
2. A maximum of 10,000 women, in order to limit the organizational burden, but also to preserve an approximate equity between countries.
3. Within these limits, larger samples were generally allocated to (a) countries with larger populations; (b) countries with greater ethnic or regional heterogeneity; and (c) countries which could argue that they had already carried out a similar survey and the main justification for another one could only be the need for more detailed analyses, requiring a larger sample. This

is an example of how a country's own special objectives can lead to a departure from a general rule.

For the ESS, the investigators in each country raised their own funds, independently of the central secretariat. Consequently, ESS defined only minima, and not maxima, for the sample sizes.

1. A minimum sample size (completed interviews) of 2,000 (with an exception for countries with a population below 2 million; for those the minimum was 1,000).
2. A minimum effective sample size ($n_{eff} = n/\text{Deff}$) of 1,500 (800 for the small countries described above).

The general strategy for determining country-specific sample sizes in multi-national survey programs should follow a sequence of six steps:

1. Specify the primary analysis objectives for the multiprogram including desired cross-country comparisons, pooled analyses for multicountry regions, etc.
2. Determine the precision requirements for each participating country based on the objectives established in step 1. Consider using the ESS approach of establishing a minimum precision standard expressed as a minimum effective sample size.
3. Use standard formulas to determine sample size for each participating coun-try under simple random sampling (SRS) assumptions (Cochran, 1977).
4. Use simple models for clustering and weighting effects to evaluate the potential size of design effects for country-specific sample design options.
5. Adjust the SRS sample size calculation for the expected design effect for the country-specific design: $n_{final} = n_{srs} \times \text{Deff}$.
6. Re-evaluate steps 1, 2, and 4, based on implied costs of the chosen design and required sample sizes. Recompute the final sample size as needed in step 5.

14.7 OTHER CONSIDERATIONS IN MULTINATIONAL, MULTI-CULTURAL SAMPLE DESIGN

The goal of this chapter has been to describe design principles and management models that have proved to be successful in the coordinated sample design programs for multinational, multicultural survey programs. The focus has been on design strategies and methods that minimize survey errors that originate directly from the properties of the chosen sample design, specifically sampling variances of estimates and sampling biases due primarily to sample noncoverage that arises in the survey population definition, the choice of a sampling frame, or the procedures employed in sample screening and respondent selection.

In reality, the design team responsible for developing sampling plans for the component surveys in a multinational, multicultural survey program must reach beyond these errors in sampling to consider the implications of the chosen design for other sources of error in survey data.

For example, work on the WFS (O'Muircheartaigh, 1984) showed that the same variables tended to be subject to correlated interviewer effects across countries. The impact of these effects on the variance of estimates is similar in form to that of clustering in the sample design, with the variance being multiplied by a factor approximately equal to $(1 + \rho (m-1))$, where m is the interviewer workload. For variables sensitive to correlated interviewer effect, the size of the interviewer workload can have a critical impact. This is rarely considered in cross-national (or even national) sample design, and is certainly an area in which a great deal more empirical information (and eventually evidence) is needed. We recommend that this information be collected by survey coordinators and recorded in the description of cross-national surveys.

Features of the sample design can also influence noncontact rates and other nonresponse outcomes of the survey process. A statistically efficient sample design with many highly distributed clusters of small average size will result in increased travel costs and effort, reducing the opportunities for interviewers to make call backs to sample addresses or to follow up with appointments to conduct the survey interview.

These and other aspects of the survey that impact the quality of the data are the subject of other chapters in this volume. As these other chapters are read, we encourage the reader to think critically as to how sample design choices can impact these other important dimensions of survey quality and survey error.

15

Challenges in Cross-National Data Collection

Beth-Ellen Pennell, Janet A. Harkness,
Rachel Levenstein, and Martine Quaglia

"For other aspects of survey design and implementation, such as field work practices, almost nothing is known." Lynn, Japec, and Lyberg, (2006, p. 15).

15.1 INTRODUCTION

This chapter discusses issues and considerations in the data collection phase of cross-national and cross-cultural surveys. Although the extensive literature on comparative research frequently discusses the complexity of cross-national designs and how these may threaten data comparability (Chapter 2, this volume), there is little investigation of the contribution that data collection plays with regard to comparability. This is particularly surprising given the considerable cost of data collection relative to all other phases of the survey life cycle. The comparative literature also lacks detail on the design and implementation of data collection, and how decisions with regard to these dimensions may challenge the goal of equivalence. This chapter explores the various phases of data collection and alerts the reader to issues especially relevant to cross-national and cross-cultural survey research.

The material that follows is organized into seven sections, including this introduction. Section 15.2 sets the stage for later discussion by addressing the complexity of cross-national survey research. Section 15.3 discusses organizational structures in large-scale projects, the importance of achieving standardization at the appropriate level of implementation, and the impact and relevance of country-specific variations in research infrastructure and traditions. Section 15.4 focuses on data collection modes, considering the options and constraints in cross-national data collection efforts. Section 15.5 deals with the numerous considerations directly related to fielding a survey. Section 15.5.1

---

[1] Survey Methods in Multinational, Multiregional, and Multicultural Contexts, edited by Harkness et al.
Copyright © 2010 John Wiley & Sons, Inc.

addresses field staffing, the languages to be accommodated, and interviewer recruitment, training, and remuneration. In Section 15.5.2, study protocols and procedures relevant for the data collection phase are considered; topics covered include ethics reviews, community engagement models, household contact protocols, privacy issues, nonresponse bias reduction, and the increasing collection of biomarker and other physical measures. Section 15.5.3 deals with field structures and with interviewer supervision and management during fielding. In Section 15.6, quality control and documentation are revisited, illustrating the importance of monitoring, documenting, and evaluating survey outcomes. The final section, Section 15.7, highlights areas of future research and considers the development of best practices for cross-national or cross-cultural survey data collection.

15.2 COMMON CHALLENGES

As discussed in the introduction to this volume, potential challenges to comparability are myriad in cross-national data collection. The "common ground" and shared understandings with regard to the purpose of surveys or procedural approaches that may be present in a single country survey stand in contrast to the obvious heterogeneity of populations, languages, contexts, and perspectives found in multinational studies. In addition, challenges that are atypical in a one-country context often become the norm in cross-national contexts.

In the following we briefly revisit the defining features of cross-national surveys that make such undertakings particularly demanding. Examples include the need to field in multiple languages (some possibly unwritten); locating and engaging respondents who may live on remote islands, mountainous regions, or other rural areas; seasonal conditions (e.g., winter or monsoon seasons); and varying holiday or vacation periods where respondents are away from their usual place of residence. Respondents may also be inaccessible because of migration (e.g., nomadic populations) or because of cultural norms (e.g., women) or other barriers (e.g., miners in camps). Refugees, undocumented workers, and internally displaced populations may also be difficult to locate and interview. Countries undergoing periods of political unrest or war may increase the risk for both respondents and interviewers.

In minority countries,[2] access impediments such as gated communities or locked high-rise buildings may be present. Likewise, access to villages or communities in majority countries may need the permission of village elders or other "gatekeepers." Wide variations in response bias within and between countries are to be anticipated and declining response rates are well-documented in minority countries (Billiet, Philippens, Fitzgerald, & Stoop, 2007; Couper & de Leeuw, 2003; de Leeuw & de Heer, 2002).

---

[2] In this chapter, we refer to "transitional" and "developing" countries as *majority* countries, since these are where a majority of the world's population resides. Accordingly, "developed" nations are referred to as *minority* countries.

The ability to reduce social desirability bias will vary across populations and countries because of wide variations in literacy which limits the use of self-completion modes (Johnson & van de Vijver, 2003; World Bank, 2003).

It has also long been recognized that interviewer-related bias may differ widely across countries. Literature of vintage date has described such considerations well (Bulmer, 1998; Bulmer & Warwick, 1983; Ralis, Suchman, & Goldsten, 1958; Warwick & Osherson, 1973).

It has also long been recognized that interviewer-related bias may differ widely across countries. Literature of vintage date has described such considerations well (Bulmer, 1998; Bulmer & Warwick, 1983; Ralis, Suchman, & Goldsten, 1958; Warwick & Osherson, 1973).

Countries vary widely both in official requirements and in unwritten rules and customs pertaining to data collection and data access. Heath, Fisher, and Smith (2005) note that countries such as the Democratic People's Republic of Korea/North Korea and Burma/Myanmar officially prohibit survey research, while others severely restrict data collection on certain topics or allow collection but restrict the publication of results (e.g., Iran). Regulations pertaining to informed consent also vary greatly.

Individuals working at the level of project development and data collection management in different locations can have varying degrees of proficiency in the *lingua franca* in which the project chooses to communicate. Collaborators may also be located across many time zones, further complicating communication.

Projects vary greatly in the level of coordination and standardization across the many phases of the survey life cycle, in their transparency and documentation of methods, and, of special interest in this chapter, in their data collection requirements and approaches.

Close examination of large-scale projects also reveals that the research traditions of individual countries and/or data collection agencies can differ greatly in methodological approach and rigor. For example, some countries have research traditions that routinely employ quota sampling or sample-unit substitution at the last stage of selection rather than probability sampling methods at all stages.

Country differences in tradition notwithstanding, many studies officially target equivalence by trying to keep the design and implementation the same across countries. Harkness (2008a), Kalgraff Skjåk and Harkness (2003), Harkness, Mohler, and van de Vijver (2003), and Lynn, Japec, and Lyberg (2006) discuss some of the issues involved in trying to keep things the same, suggesting that attempting to follow a "one size fits all" approach can ultimately threaten data comparability and that appropriate standardization and appropriate variation in approach are essential. However, the challenges to identifying and setting suitable implementation standards and procedures in cross-national research are many. This holds as much for a country trying to adhere to centrally-required procedures as it does for a central coordinating organization aiming to determine the level at which standardization of procedures is practicable.

In sum, the challenges to requiring similar implementation strategies across very diverse populations and socioeconomic situations are considerable. In what follows, therefore, we explore the ground between standardization and local accommodation, acknowledging that with many aspects of data collection, the appropriate balance remains to be discovered.

Throughout this chapter reference is made to documentation as a basic requirement for monitoring the quality of data collection processes and understanding survey outcomes. At the same time, although awareness of the importance of survey documentation seems to be growing (Mohler, Pennell, & Hubbard, 2008; Chapter 16, this volume), very few cross-national studies provide the detail needed to undertake an assessment of the outcomes and quality of the survey.

15.3 ORGANIZATIONAL STRUCTURES

The organizational structure of a cross-national survey has a major impact on how the study is designed and implemented. The structure is often influenced by two factors: the source and flow of funds and the experience and research infrastructure at both central and local locations. In many cases, if the funding has been obtained through a central source, the organizational structure is then determined by the organization in receipt and control of those funds. The control of funds also often prescribes the level of influence an organizing body may have across countries to set standards and to specify various aspects of the study implementation. Centralized funding, therefore, may make standardization more likely and more consistent. Standardizing aspects of design and implementation may also be more likely if the countries involved are minority countries with similar research infrastructures, such as many parts of the European Union. At the same time, decisions about standardization cannot simply be based on what counts as good or best practice in one location or one region of the world.

Large survey efforts such as those highlighted in this chapter can be funded in a variety of ways. Multiple sources of funding and the organizational aspects of such endeavors generally mean that many individuals and organizations will be involved in a project. The Study of Health and Ageing in Europe (SHARE), for example, has one central source of funding and a centralized administrative unit, with 130 researchers working in 11 countries (Börsch-Supan, Jürges, & Lipps, 2003; Chapter 25, this volume). The European Social Survey (ESS) receives partial funding from the European Union to provide overall coordination, but each of its 24 countries funds and organizes its data collection activities (Jowell, Kaase, Fitzgerald, & Eva, 2007a; Chapter 26, this volume). The International Social Survey Programme (ISSP), currently involving 45 countries, has volunteer working groups that coordinate components of the survey (e.g., archiving, quality monitoring and reporting, and the Secretariat coordination). The ISSP has no central funds (cf. Chapter 27, this volume). Thus, funding is local, even for central operations; organization is partially centralized but undertaken by volunteer members; and specifications on quality are agreed to by the annual general assembly.

If funding is locally controlled, standardization of implementation may be less likely and more difficult to achieve. If standardization is, nonetheless, perceived as a project goal, other mechanisms must be found to achieve sufficient levels of influence on local implementation. Appropriate mechanisms could include encouragements for collaborators to cooperate, such as access to standardized data collection instruments they do not need to produce themselves, offers of specialized training, assured access to the cross-national dataset, and organized and supported involvement in joint publications based on the cross-national data. In practice, most large-scale cross-national survey projects have a mix of central and local funding and also vary in their levels of centralization and control. For example, the ESS (European Social Survey, 2005), the World Mental Health (WMH) survey (Kessler & Üstün, 2008), and SHARE (Börsch-Supan, Jürges, & Lipps, 2003) all require the following: probability samples at all stages of selection, minimum sample sizes and a minimum response rate, shared core survey content, a highly specified translation protocol, and a minimum set of quality control measures. Because of their highly complex survey instruments, WMH and SHARE require specific interviewer training protocols and use of the same data collection software across all countries. ESS and WMH require pretesting of final translated instruments in every country. Because of the institutional affiliations and the sensitive topic of the WMH, this study also requires that each country carry out an ethics review.

These projects recognize that it is neither practical nor good practice to standardize every aspect of the project. For example, setting rigid contact protocols (number of attempts across days and time of day) across countries can be unwise. At-home patterns may vary significantly, as will the ways the sample is assigned to interviewers. In the WMH studies, guidelines specified the minimum number of contact attempts and strategies that might be used to minimize nonresponse (the use of incentives, for example). However, countries adapted these protocols on the basis of known at-home patterns and the appropriateness of incentive offers in the local context. Contact attempt protocols were also adapted to account for sample assignments. Thus, for example, interviewers in Colombia traveled in teams from place to place, which restricted the number of contact attempts possible. In the U.S. implementation, interviewers were assigned by geographic area. Many more contact attempts were necessary in the United States over a much longer period of time to achieve a comparable response rate outcome (Pennell, Mneimneh, Bowers, Chardoul, Wells, et al., 2008).

Decisions over what aspects of study implementation to standardize, what aspects to leave to local control, and where responsibility and accountability for these aspects will lie must be made for every phase of the survey life cycle. For data collection, these include: deciding the mode of survey administration; preparing the sample; determining in what languages the survey will be administered; setting translation or interpreting guidelines; pretesting, determining ethics review procedures, developing interviewer training content and protocols; selecting and training interviewers; determining the timing and length of data collection; and setting standards for data collection monitoring, as well as data transfer and review. The configuration and composition of the data collection teams must also be decided, including whether interviewers will travel in teams or be locally assigned, and whether interviewers will be matched on key characteristics beyond language, such as ethnicity, race, gender, or religion.

Country infrastructure must also be assessed in terms of resources available for transportation, communication, accommodation, as well as technological support such as programming resources and computer and Web access. Finally, quality control procedures must be clearly defined, procedures agreed upon, and decisions taken on sanctions for failure to fulfill requirements. None of these can be properly addressed on an ad hoc basis. Three examples illustrate different approaches to the implementation of standards.

The ISSP is one of the largest and longest running academic cross-national surveys. It is comprised of a federation of researchers who are largely self-funded. New members to the program are vetted in terms of their ability to comply with the ISSP Working Principles (http://www.issp.org/organisational.shtml/). These principles, in turn, have come under review and elaboration as new countries with new challenges join the program. They determine general sample requirements but say nothing, for example, about contact procedures, pretesting, or translation requirements. Although survey module content is coordinated across countries, most details of the actual data collection implementation are left to the local agencies and their sponsors. Research on features of ISSP data collection relevant to nonresponse (Smith, 2007c) highlights differences across countries regarding, for example, the use of incentives.

SHARE, on the other hand, is largely centrally funded by the European Union, with further support from U.S. agencies. As noted above, SHARE has stipulated procedures for many aspects of the study design (Börsch-Supan, Jürges, & Lipps, 2003).

The WMH model lies between that of the ISSP and SHARE. In this study, some centralized funding is used to develop and maintain the data collection instruments; conduct interviewer (trainer) trainings; monitor aspects of data collection, quality control, and data processing; and to conduct cross-national analysis. Country-level funding of data collection and its survey management is undertaken by the collaborators in each country. Centrally provided benefits to collaborators in the WMH project include having access to the standardized and programmed instruments and to interviewer training sessions and training materials, as well as receiving analysis support for joint publications (Kessler & Üstün, 2008; Pennell et al., 2008). The ISSP, SHARE, ESS, and WMH all also benefit from at least yearly face-to-face meetings of the collaborators and staff.

The second organizational consideration that is central to deciding how studies are structured is the research infrastructure at central and local levels, including technical and human resources. Many centrally organized projects will not commit to data collection in a country without a local collaborator. The rationale is that local knowledge can be critical to understanding traditions and customs, to recognizing possible limitations, and to being able to gauge the feasibility of the research design. In addition, some countries strictly control access and will not authorize a study without local collaboration (cf. Nigerian Federal Ministry of Health, 2007). Ethics boards in minority countries may also require local collaboration (Council for International Organizations of Medical Sciences, 2002; National Bioethics Advisory Commission, 2001; Sugarman, Popkin, Fortney, & Rivera, 2001).

Researchers contemplating a cross-national survey project will often first engage their own professional networks to identify appropriate local collaborators. Funding sources may also require certain oversight committees or suggest collaborators. Longer-term projects may have numerous boards and committees. The ESS, for example, has a Funders Forum, an independent Methodology Committee, and a Scientific Advisory Board, each with different focus and functions. Aspects to be considered at the country level include whether substantive and methodological expertise are available locally for the project at hand. When, as in the ongoing research program of the ISSP, the topic of the project changes annually, different experts may be required for each round of the study.

Typically, a collaborator is needed who is based in the country, has undertaken similar research, understands the local context, and can oversee the project. Because local conditions can vary greatly, it is unrealistic to expect central staff of a large cross-national project to have sufficient knowledge of country-specific conditions to be able to make appropriate decisions about implementation. However, local knowledge or collaborators are not a panacea for avoiding potential problems. Limited local experience with best practice research methods or a weak local research infrastructure may present obstacles. There may be resistance to trying unfamiliar methods or to changing existing approaches. Conducting exploratory research (focus groups, in-depth interviews) and pretesting can be valuable tools in such cases, providing evidence and information critical for a successful survey implementation (Bloch, 2007).

Questions to be answered in organizing the project at the country level include: Is there a survey (or other) research infrastructure in place in the country? If yes, what are its capabilities? Is there access to up-to-date sampling frames? In some countries, only the government's statistical agency has access to area-based sampling frames or current registries of the population. Do local data collection agencies have experience with probability sampling, or various modes of data collection (including computerized methods)? Is there knowledge of the potential sources of survey error across survey administration modes in the country (e.g., phone coverage rates or literacy levels)?

Organizations employed to collect data at the local level are often selected through a tender and bidding process. Local collaborators may also belong to a university with a survey research capacity. Alternatively, the country's statistical agency may be employed. In the case of a 15-country project on stigma and mental health (White, 2008), the central organization, based in the United States, tapped the ISSP network to organize local data collection efforts. In some cases, if there is no local agency or no local agency qualified to conduct the study, an interviewing team might need to be assembled from scratch. In all cases, just as with the overall project structure, roles and responsibilities must also be specified in relation to the supervision and oversight of these data collection organizations.

As soon as multiple countries or regions are involved, a well-articulated organizational structure is essential in order to effectively manage many sub-groups and sub-tasks. Unambiguous procedures for decision making, clear lines of authority, and assigned responsibility across all aspects of the study are critical organizing principles. However, project planning and organizational and contract negotiations can take considerable time, depending on how the project is funded, organized, and whether local collaborators need to be sought and assessed. For example, the ESS took five full years of planning before the first round of fielding.

15.4 MODE OF DATA COLLECTION

The choice of data collection mode(s) is a key decision that affects survey errors, survey costs, and the management of all aspects of the survey life cycle. Selecting a mode also involves striving for a balance among a number of potentially competing goals that will have differing effects on error sources and the costs of the survey. Although this balancing applies to a single country survey (Biemer & Lyberg, 2003), it becomes a more complex choice in a cross-national survey project where there are wide variations in such aspects as sample coverage and population access, nonresponse trends, the number of languages involved, literacy levels, social desirability, as well as labor availability and cost structures. If standardization of mode(s) is a goal, differences in individual countries with regard to technical infrastructures may also reduce the mode options available.

There are many ways in which surveys can be conducted, such as using Interactive Voice Response (IVR), text messaging, and Web-based options, that are not discussed here. Instead, we focus on interviewer-mediated modes, in part because these newer modes are currently not viable in all areas of the world. We note, however, that the use of technology is also rapidly increasing. For example, Eng, Wolkon, Frolov, Terlouw, Eliades, et al. (2007) successfully implemented a data collection project using handheld computers in Niger and Togo. Because electricity was not always available, the devices were charged by using extra battery packs, solar chargers, and car chargers. These handheld computers provided the survey with the advantages of computer-assisted interviewing, including the ability to automatically navigate complicated skip patterns, and also provided Global Positioning Systems (GPS) to assist in sampling procedures. In another example, Bloch (2007) reports using the Web to access highly educated "forced migrants" from Zimbabwe. In any event, the generic issues addressed below will frequently apply, by extension, to these emerging mode options.

The current practice in many cross-national studies is to try to standardize mode across locations, even if this results in sub-optimal choices for some locations (Kalgraff Skjåk & Harkness, 2003). The ESS, for example, requires all countries to collect data with face-to-face interviews. The ISSP requires a questionnaire in a format suitable for self-completion. This can take the form of a mail survey, paper and pencil, Audio computer-assisted self-interview (ACASI) self-administered with interviewer attending, self-completion as drop-off, or, if necessary, an interviewer administered interview. The motivation for this variety lies in the beginnings of the ISSP when the survey was added to four established national surveys (see Chapter 27, this volume), each administered somewhat differently.

Recent deliberations about Web-based implementation in the ISSP have led to the re-establishment of the modes work group which investigated data collection modes in the late 1990s (Kalgraff Skjåk & Harkness, 2003). In the ESS, too, a group was formed to explore telephone data collection as a second mode in some countries. In a preliminary analysis, Jäckle, Roberts, and Lynn (2006) found little evidence that using two modes on the ESS significantly affects survey estimates.

In contrast to the paucity of literature on mode effects in cross-national studies (Fu & Chu, 2008; Jäckle, Roberts, & Lynn, 2006; Kalgraff Skjåk & Harkness,

2003), the literature on mode effects in general survey research is considerable (see reviews in de Leeuw, 2008; Groves et al., 2009). This literature also considers mode with respect to religious, racial, or ethnic populations.

15.4.1 Coverage Error

The type of sampling frame available will often determine which mode is used. If a sampling frame must be created because none exists or available frames are out-of-date or have known coverage problems, an area-based frame may have to be constructed (Häder & Gabler, 2003). Large-scale cross-national projects that decide to keep the mode of administration constant across all countries are usually forced to choose face-to-face surveys (Fu & Chu, 2008). Even if some members of the target population are difficult to reach (Iachan & Dennis, 1993; Kalsbeek & Cross, 1982), face-to-face surveys generally have better coverage properties. At the same time, face-to-face surveys are labor-intensive and, if labor costs are high, this is also the most expensive mode (see Groves, 1989; Biemer & Lyberg, 2003) and may not be viable in practical terms, especially in minority countries. In majority countries, the opposite generally holds. For example, the Afrobarometer surveys (http://www.afrobarometer.org/surveys.html/) all use face-to-face data collection. However, the face-to-face mode may also take longer to plan and implement than other modes, especially if the sampling frame must also be constructed. Telephone surveys tend to be both cheaper and faster because less labor is required to complete an interview, so fewer staff are needed to carry out data collection. Travel expenses are also avoided. However, there is also very wide variation in telephone coverage across countries (World Bank, 2006); thus, telephone surveys are often not possible as a uniform mode across countries. Telephone penetration rates also vary widely within countries, with urban areas generally having higher coverage (International Telecommunications Union, 2007; Minges & Simkhada, 2002).

A solution sometimes adopted is to implement a mixed-mode approach. For example, the European Crime and Safety Survey was a cross-national study conducted in the European Union (van Dijk, Manchin, van Kesteren, Nevala, & Hideg, 2005). In 16 countries considered to have good telephone coverage, telephone surveys were conducted. Face-to-face interviewing was used in Poland and Estonia, where telephone penetration was relatively low. In countries where there was great variation in telephone coverage within the country, the country was stratified by telephone coverage and then a mix of telephone and face-to-face was used. The Gallup World Poll currently takes a similar approach. Here, data are collected by telephone in countries with at least 80% estimated telephone coverage with face-to-face interviewing is used in all other contexts (Gallup, 2007a).

15.4.2 Sampling Error

The mode of administration may also affect sampling error—that is, the error associated with the fact the survey is conducted with only a subset of the target population (Kish, 1965). In face-to-face surveys, cluster sampling is often used to reduce interviewer time, travel, and related expenses in order to reduce costs. Clustering, in turn, can increase sampling variance due to the heterogeneity between clusters. Because face-to-face surveys are expensive, the overall sample size may have to be reduced, which also increases sampling variance. Extensive substitution practices in some countries are a further factor to be considered.

15.4.3 Nonresponse

Unit nonresponse results from failing to include a sampled unit in the final dataset because of household or respondent noncontact, refusal, or because selected persons are unable to participate (e.g., due to language problems, health issues, etc). Nonresponse in minority countries has been studied extensively (e.g., Groves & Couper, 1998; Groves, Dillman, Eltinge, & Little, 2002; Singer, 2006) with response rates generally higher in face-to-face surveys than in telephone and mail surveys (Biemer & Lyberg, 2003) or Web (Vehovar, Batagelj, Manfreda, & Zaletel, 2002). Exceptions to general rules are also found. For example, in a nonresponse study in 16 minority countries in North America and Europe, de Heer (1999) found that in Finland and Sweden response was higher in the telephone surveys than in face-to-face surveys. De Heer (1999) speculates that field procedures such as the use of differential incentives may have affected the response rates, but also notes that cultural aspects relating to refusals and noncontact may have been present.

Nonresponse in cross-national studies has not been extensively researched (Couper & de Leeuw, 2003). The effect of mode on nonresponse across countries or in individual majority countries is, thus, much less clear. Kalgraff Skjåk and Harkness (2003) note that in minority countries, higher nonresponse in urban areas is more common than in rural areas. However, in majority countries, rural populations may be more cautious about outsiders asking questions and, therefore, more likely to refuse than urban respondents (Johnson, O'Rourke, Burris, & Owens, 2002).

In general, however, response rates in majority countries tend to be higher than in minority countries (Couper & de Leeuw, 2003). Multiple explanations can be advanced for this. For example, requests for participation in minority countries far exceed requests for participation in many majority countries (Groves & Couper, 1998), and in some locations, over-surveying may decrease participation. In locations where the survey process is novel and interesting, response rates may be higher (Chikwanha, 2005; Kuechler, 1998). Couper and de Leeuw (2003) also suggest high response rates could reflect that in some countries respondents are afraid to decline to participate.

Johnson, O'Rourke, Burris, and Owens (2002) outline potential reasons for differential nonresponse rates across cultures. They suggest that individuals in more collectivist cultures are more likely to view out-group members with suspicion; thus, the use of local (in-group) interviewers can greatly increase participation. They also suggest that some cultures may be more likely than others to base decisions to participate on factors such as the interviewer's nonverbal behavior or racial or ethnic background.

Nonresponse can also result when the language of the sample person does not match that of the interviewer. This may happen frequently in some locations, if many languages are spoken in one region. In Zimbabwe, for example, interviewers for the Afrobarometer encounter 20 different languages and many more dialects (Gordon, 2005). Each region is not homogeneous with respect to language or dialect, making interviewer assignments to areas challenging. Chikwanha (2005) recommends increased efforts to recruit multilingual interviewers to help reduce this source of nonresponse.

Finally, response rates can differ by at-home patterns and the level of effort that is made to contact households, as well as the restrictions that may be placed on how reluctant or refusing respondents can be addressed. For example, can an initial refusal be contacted again in order to encourage participation? Local laws, regulations, or norms may disallow re-contact of an initial refusal to encourage participation.

Item nonresponse occurs when respondents choose not to answer specific questions or if questions are omitted through technical problems in the instrument or mistakes on the part of the respondent or the interviewer. Item nonresponse can be related to mode. For example, well-tested computerized questionnaires can greatly reduce questionnaire navigation mistakes that result in item nonresponse. At the same time, it is widely accepted that respondents may refuse to answer sensitive questions in interviewer-administered questionnaires (e.g., Tourangeau, Rips, & Rasinski, 2000). In minority countries, techniques such as self-completed sections and ACASI are, therefore, used to reduce item nonresponse on questions considered sensitive. However, these approaches assume a degree of literacy which may not be present in many majority countries. In addition, Owens, Johnson, and O'Rourke (1999) provide within-country evidence for cultural differences with regard to sensitive questions. While this literature sheds some light on item nonresponse across cultures, it is still unclear how this may manifest across countries. Few cross-national comparisons of item nonresponse patterns have been undertaken. However, Pilmis (2006) found no relationship in the ESS between sociodemographic variables and increased item nonresponse to political questions.

15.4.4 Measurement Error

The choice of mode has repeatedly been shown to affect measurement error in studies conducted in minority countries. In a meta-analysis of studies largely carried out in the United States, de Leeuw and van der Zouwen (1988) found small differences between telephone and face-to-face modes in item nonresponse and in social desirability bias. Also in the United States, Tourangeau and Smith (1996) found that responses to interviewer-administered questions were more susceptible to social desirability bias than in self-administered modes. We can expect that measurement error differences across modes will also differ across cultures for sensitive questions.

The choice of mode may be affected by other factors such as infrastructure, including such aspects as availability of private and public transportation, sample access routes (roads, rivers, trails, etc.), electricity or sufficient battery power to run computers, or the cost of duplicating and printing. If laptops or other data collection devices such as handheld personal digital assistants are to be used, human and technical resources to support these computerized methods are essential.

Natural disasters may affect mode and fielding decision. For example, Nielsen Media Research conducts surveys of television viewing behaviors in the United States and typically recruits by telephone. In 2005, Hurricane Katrina devastated much of the U.S. Gulf Coast of Louisiana and Mississippi, and many residents there lost their landline telephones. As a result, the Nielsen media surveys could no longer recruit respondents by telephone Random Digit Dialing. Nielsen utilized area probability sampling instead and modified their definition of a household for the Gulf Coast region. Damaged homes that could be rebuilt without a construction permit were counted as housing units, as were temporary housing trailers provided by the Federal Emergency Management Agency (FEMA). As a result, households remaining in the Gulf Coast after Katrina could be represented in a probability sample (Palutis, 2008).

Sometimes surveys will not be fielded in areas that have experienced a disaster; the Afrobarometer has followed this policy (Institute for Democracy in South Africa, Center for Democracy and Development—Ghana, & Michigan State University, 2005–2006). Concerns for interviewer safety may also determine which mode is chosen. Thus, although interviewers do conduct interviews in dangerous regions (cf. Burnham, Lafta, Doocy, & Roberts, 2006; Iraq Family Health Survey Study Group, 2008), at times it may not be acceptable to send interviewers into the field. Varughese (2007) reports on areas in Afghanistan that were considered too dangerous and excluded from his study on attitudes toward democracy, the government, media, and the role of women in society. In Israel, some regions are regularly excluded in the ISSP and ESS surveys due to continuing conflict between cultural and religious groups.

Whether an area is dangerous may also depend on the interviewer's own group membership (e.g., race, or ethnicity) or the organization for which the interviewer is working. The equipment interviewers are using might also put them at risk if, for instance, it has a resale market value (laptops, cell phones). Projects have successfully used strategies such as hiring local escorts for protection and negotiating with relevant parties to gain safe access to sampled areas. The local context will determine which parties are the relevant gatekeepers in these cases.

Interviewers also sometimes travel in teams for reasons of safety and/or ease of supervision. These teams can take many forms and use various techniques to communicate. For example, in the ISSP, interviewers in the Philippines travel in teams with a supervisor to sampled areas. A similar procedure is used in South Africa to field the ISSP. Here the supervisor is also the driver. The interviewers make chalk marks on or near the entrance to a house to indicate where they are interviewing; this way their colleagues know where each interviewer is located at all times.

15.5 FIELD IMPLEMENTATION

The choice of survey mode affects aspects of field implementation, such as the timing and length of data collection, interviewer assignment to the sample, the interviewer skills needed, and whether characteristics of interviewers and respondents will be matched. Related decisions include how interviewers will be recruited, supervised, and remunerated. In a sobering historical overview of the challenges of acquiring accurate data in "less-developed" countries, Cleland (1996) remarks: "The quality of the demographic data depends more on the skills, training and supervision of field staff than on the design of the data instrument" (p. 446). These issues are discussed below.

15.5.1 Field Staffing

Constraints on data collection timing and the local research infrastructure dictate many aspects of field staffing.

Timing. The timing and length of a data collection may depend on a number of factors, including the topic, funding sources, research infrastructure, and attributes of the target population. If the topic is time-sensitive in nature, e.g., a national election, then the timing of the data collection will be constrained. This may mean that more interviewers are needed to cover the sample or that travel in teams may not be feasible in order to cover the population of interest in a short period. Since elections occur at very different times across countries, the timing of data collection in these studies may even vary across years (Shively, 2005). On the other hand, political and social events may constitute times to avoid data collection. For example, some countries may become unstable around an election. Requests for survey participation may also be mistaken for recruitment for political parties. In order to minimize this possibility, the Afrobarometer, for example, tries to avoid data collection in the six months before and six months after an election (Chikwanha, 2005).

The timing of data collection may also be affected by weather and geography. Travel during certain parts of the year may be impossible or difficult. If it is impor-tant that the timing of the data collection is the same across countries (as when gauging response to a global event or issue), timing issues will be more difficult to resolve since the northern and southern hemispheres will have different challenges depending on the time of the year. Budget concerns may also constrain timing. If interviewer training is extensive or there is a limited pool of skilled interviewers to conduct a study, a smaller team of interviewers may be necessary. This, in turn, will lead to a longer field period and an increase in interviewer-related variance (Kish, 1962). We consider further interviewer assignment factors below.

Assignment to Sample. In one-country surveys, interviewer and respondent matching on one or more demographic characteristics is sometimes used to lessen certain forms of bias related to interviewers (cf. Davis, 1997; Dotinga, van den Eijnden, Bosveld, & Garretsen, 2005; Kane & Macaulay, 1993; Schuman & Converse, 1971). However, the possibilities for matching to reduce interviewer-related bias may not be equally viable across all countries or populations.

If many languages need to be accommodated, then the assignment of interviewers to sampled areas assumes some a priori knowledge of the distribution of those languages in the population. Interviewer matching may be quite complex and can become costly. Depending on the topic under study and the range of respondents (head of household, randomly selected adult, child, etc.) and/or social and cultural norms regarding who may appropriately question household members, interviewers may have to be matched on gender, age, race, religion, or other factors (caste, for example). In some Muslim countries, matching on gender is considered essential but can introduce complexity where female interviewers may not travel without a male relative as an escort (Feld & Mohsini, 2008).

Third-party presence may be a further source of bias, especially if the head of the community, one or more family members, or a chaperone are required and/or expected to be present during an interview. The permission of multiple third parties (gatekeepers) may be required before beginning to collect data. Interviewers may also be sent out in teams if more than one person in a household needs to be interviewed or if it is necessary to distract household members with other tasks so an interviewer can conduct an interview in private. Such considerations will all affect how many interviewers may be needed and how they will be assigned to the sample. Here again, there is a balancing of potentially competing pressures. What may be optimal for one context may not be for another; matching interviewer skills and characteristics appropriately, for example, could vary both across and within countries. Team interviewing also increases costs and, thus, may be too expensive in high-wage contexts (see also Section 15.5.3).

Interviewer Recruitment. Interviewer recruitment needs and protocols will largely depend on the depth and breadth of the research infrastructure present in the country and the design and content of the survey. Where a research infrastructure is present, finding experienced field staff is a somewhat easier task. The local survey organization chosen to conduct the study may have a pool of experienced interviewers available to work on the project. At the same time, however, design features of the new study may be unfamiliar and meet with skepticism at several levels (Pennell et al., 2008). Matching interviewer and respondent characteristics may strain resources if previous studies were staffed differently or if interviewers have to be matched on new attributes.

Where there is no research infrastructure, local contacts become critical to building an interviewing staff from scratch. Here, studies and countries have used various strategies to address these issues. Examples include the use of traveling nurses, college students, teachers, or other professionals. Depending on the context, these strategies can be optimal or potentially problematic. For example, nurses and students may have other priorities and demands on their time that deter them from attempting interviews during optimal times when respondents can be found at home. Because of cultural norms, students may not be able to interview older people; the role of a nurse, on the other hand, may be confused and confounded with his or her role as an interviewer. Previous interviewing experience is valuable, assuming the new set of tasks do not differ significantly from local research traditions or, if they do, that these traditions are not difficult to "unlearn." In the latter case, it may be simpler to recruit and train new interviewers than to change the habits of interviewers trained in other traditions.

No matter how interviewer recruitment is approached, certain required attributes of potential interviewers are universal. These include good organizational skills (handling forms, keeping track of paperwork), an interest in research and people, attention to detail, and the ability to follow complex instructions. Also important are the ability to work independently since they are often working on their own; familiarity with the norms of the culture in which they will be working to be able to engender culturally suitable rapport; and, depending on the mode of interview and geography, the ability to travel by public transport, car, bicycle, or even to walk long distances.

Recruiting interviewers with adequate language skills may be particularly important in a multilanguage context. Here, some means of assessment may need to be developed. Interviewers need to have adequate conversational skills including comprehension and speech level and speed in the languages in which they will be interviewing. Clearly, the ability to read aloud and to be clearly understood is critical. Interviewers' accents need to be acceptable and intelligible to the target population(s). Assessment of writing skills is also important and may include evaluating the correct use of grammar, noun use (gender), sentence structure, and spelling.

If a country has little or no research infrastructure, recruitment and training can be expected to take longer. As a general guide to conducting surveys in countries without a research tradition, Glewwe (2005) recommends up to a month to train inexperienced interviewers. Even where survey research is common, local research traditions may question new training protocols. For example, the WMH study specified a 5-day interviewer training. Research agencies in many minority countries challenged the need for such lengthy training; other studies they had conducted had required little or no training (Pennell et al., 2008). Ongena, Dijkstra, and Smit (2008), in contrast, report that a 3-day training was too short to teach interviewers in minority countries to use an event history calendar questionnaire.

Training models also need to take into consideration the language that will be used in training since most cross-national studies will be conducted in many languages. One possible approach is to implement a "train-the-trainer" model. Here, the initial training is conducted in a *lingua franca*, with each country sending one or more individuals to the central training session who can understand and work in the language of the trainers (see Chapter 3 on *lingua franca*). The newly trained individuals then return to their own country, adapt, and translate the training materials as needed, and in turn train their interviewers. This model helps to ensure that all trainers are approaching training content and delivery in a similar fashion, but it also allows for tailoring at the country level. Of course, such a two-step process increases the time needed for training (Alcser, Kirgis, & Guyer, 2008; Chapter 28, this volume; Pennell et al., 2008).

Another interviewer training approach uses the training center model. Here, as in the train-the-trainer model, a centralized training is held, but instead of each country being represented, language "regions" are represented. This model is effective when it is not possible for every country to send trainers who are functional in the central trainer's language. As an example, a central training might be held in English with regional representation from Spanish-speaking or Arabic-speaking nations. The regional trainers would then, in turn, return to their country and hold trainings in Spanish or Arabic for other nations. This model is used in the World Health Organization's Composite International Diagnostic Interview trainings. Here, for example, trainers from Lebanon were trained in the United States and they subsequently trained the trainers in Oman and Iraq.

Training activities generally include a thorough review of the techniques used in interviewing, the survey and its content, implementation protocols and procedures, aspects of quality monitoring and quality control, and various administrative issues such as submitting for time and expense reimbursement. Informed consent, respondent privacy, and confidentiality are also usually addressed, as well as some form of test or certification process (written or mock interview) to ensure that interviewers understand the study protocols before they are permitted to begin interviewing (Alcser, Kirgis, & Guyer, 2008).

Interviewer Remuneration. This can also be expected to vary and may depend on local research traditions, the mode of the survey, local labor laws, or contractual obligations. Typically, interviewers are either paid by the hour or by completed interview. Each of these approaches has advantages and disadvantages. If each completed interview takes approximately the same amount of interviewer effort (more likely in a short telephone survey), payment by interview may make the most sense. Such an approach also makes it much easier to predict and control data collection costs. However, equal effort across interviews is rare. In a face-to-face survey, depending on assignment of cases, some interviewers may have to travel further or make more attempts to reach some respondents. Interview adminis-tration times also can vary widely depending on the respondent's characteristics or behavior. If the effort to complete an interview varies widely, it is generally recommended that interviewers be paid by the hour. Failure to do so may induce interviewers to take short cuts in the interviewing process, such as only interviewing respondents who are easy to reach, or in the worst case, falsifying interviews. At the same time, paying interviewers by the hour means that mechanisms need to be in place to monitor and control interviewer costs.

Other Logistical Issues. Where road and transportation infrastructure are limited, it may be necessary to secure transportation for field staff. For example, vehicles, fuel, oil, and maintenance may need to be secured. Some means of backup transportation may also be needed in case of emergencies. Accommodations may need to be arranged as well as communication (satellite phones, for example). If computers or other equipment will be used and electrical power is not widely accessible, solar backup or extra batteries will need to be obtained.

15.5.2 Study Protocols and Procedures

In this section, we discuss the design and implementation of the field procedures and protocols including ethics reviews, community engagement, methods and timing of contacting respondents, privacy, use of incentives, nonresponse reduction techniques, and collecting physical measures and biomarkers.

It should be clear from the foregoing that knowledge of local conditions is critical to successful survey implementation. Every survey will have unique aspects and what may work for a given survey in a given context may not work for another, even in the same context. Therefore, preliminary exploratory work is nearly always necessary to test assumptions and develop alternative approaches. This is particularly important when determining the contact protocols that will be used. Contact protocols include the procedures for making contact with the community, sometimes referred to as community engagement, as well as the details for making initial and subsequent contact with households or sampled persons.

Ethics Review. The protocols for contacting respondents will often start with a review by an independent ethics committee, board, or governmental entity, which may also review the actual content of the survey. These committees may also regulate data access and dissemination but we limit our discussion here to those aspects of the review that affect data collection.

Although there is surprising consensus across countries in the principles contained in ethical codes of conduct and the types of issues with which ethical boards will concern themselves, regulations can and will vary by country and often, by organizations within countries (Dawson & Kass, 2005). The variation in the interpretation of regulations by Institutional Review Boards in the United States is a good example of how even well-established regulations can be interpreted very differently depending on the organization and composition of these boards (Jansen, 2005).

Many of the current ethical standards and principles used in survey research are derived from the Declaration of Helsinki (World Medical Association, 1964). These widely accepted principles include the protection of free will, privacy, confidentiality, and well-being of research participants while also minimizing the burden the study may place on them. For a comprehensive review of ethical principles and standards in cross-national survey research, see Singer (2008) and Bowers, Benson, Alcser, Clemens, and Orlowski (2008).

Vulnerable populations such as children, pregnant women, the elderly, prisoners, the mentally impaired, and members of economically and politically disadvantaged groups may also require extra review and protections. Sensitive topics may also be given special attention by review committees and, as mentioned previously, what constitutes sensitive questions can be expected to vary across countries. Finally, many ethics committees will review issues of coercion, including the use of incentives to participate, that may interfere with the voluntary nature of the research.

At a minimum, the survey request will typically describe the research, the affiliations of the researchers, how the respondent was/will be chosen, and indicate the voluntary nature of participation. Statements about promises of confidentiality or anonymity are also frequently included as are the risks and benefits of participation. This process will often also involve informed consent, by which a sample member voluntarily confirms his or her willingness to participate in a study after having been informed of all aspects. Informed consent can be obtained with a written form or orally (or implied if the respondent returns a mail survey).

Whether such consent is secured in writing or verbally will vary depending on the mode of the survey, the research tradition in the country, the literacy level of the population, and local laws and regulations. Asking a respondent to sign a document in some countries and in some contexts can be potentially threatening (Dawson & Kass, 2005; Marshall, 2001). Here, too, local knowledge is critical in deciding procedures to follow. Indeed, institutions in the United States that receive federal funding for cross-national research are required to have "sufficient knowledge of the local research context" and that knowledge must have "been obtained through extended, direct experience with the research institution, its subject populations, and surrounding community" (Puglisi, 2000).

Community Engagement. Community engagement can take many forms depending on the context, sample design, and survey topic. Activities may include communication or meetings with local officials or stakeholders to either inform them or seek their permission to work in an area. Media outlets might also be used. In many minority countries with well-developed research traditions, a letter to local authorities may suffice. In majority countries, it may take a whole series of one-on-one meetings with village or tribal leaders or other stakeholders or constituents to secure cooperation and permission (e.g., Bloch, 2007; Christopher, McCormick, Smith, & Christopher, 2005; Hershfield, Rohling, Kerr, & Hursh-César, 1983; Twumasi, 2001). Where there is no local experience in a specific region, visiting the area or consulting local authorities about such issues as interviewer safety or other logistical challenges is essential.

Household Contact. The first contact with a household or respondent may be by a letter received in the mail or hand-delivered by an interviewer (de Leeuw, Callegaro, Hox, Korendijk, & Lensvelt-Mulders, 2007). Such a letter may describe the purpose of the survey, establish the legitimacy of the organization doing the data collection, explain the voluntary nature of the survey, and promise confidentiality. These forms of advance information are not feasible where there is poor mail delivery or where literacy rates are low. In all cases, however, interviewers should carry personal identification and any permits or other legal documents that might be required. At-home patterns may vary greatly both within and across countries. Although a project may choose to set a minimum number of contact attempts, the number of contacts needed to reach a household can have a wide range.

Nonresponse Reduction Techniques. Research in minority countries (primarily the United States) has shown that incentives are an effective means to increase response rates (Groves & Couper 1998; Groves et al., 2009; Singer, van Hoewyk, Gebler, Raghunathan, & McGonagle, 1999). Monetary incentives are generally more effective than gifts and pre-payment more effective than promised payment.

There also appears to be a linear relationship between the amount of the incentive and the response rate (Singer et al., 1999), and incentives can have differing effects on subgroups of a given population (e.g., Groves, Singer, & Corning, 2000).

There is little comparative literature on the effect of incentives, however. In the absence of such evidence, since the components of nonresponse and principles of reciprocity will vary across countries and possibly across cultures within countries, this suggests that decisions about the use of incentives should be locally determined. The WMH surveys took this approach. In China and Japan, for example, incentives were considered culturally inappropriate. In countries that did use incentives, these ranged from cash, to gift certificates (for food or gas), to bath towels and alarm clocks. If an incentive is offered, the type and amount should be indicated in the context of the request. Incentives should always be in proportion to the request and in line with cultural norms. A large monetary incentive in a majority country could be seen as coercive. If it is not appropriate to give incentives to individuals, it may be appropriate to give one to an organization (a school, for example) or the village. Even in these cases, coercion must be avoided, lest pressure is put on sample units to participate.

Interviewers' efforts to secure an interview also play an important role in nonresponse reduction (Groves & McGonagle, 2001; Hox & de Leeuw, 2002). Techniques to address respondent concerns to increase cooperation are usually covered in interviewer training. The range of viable approaches may differ across surveys and locations depending on survey topic, cultural appropriateness, and possible legal constraints. Approaches need to be sensitive to and directly address respondent concerns (Groves & McGonagle, 2001). Concerns might include interviewer characteristics, and, here, matching respondent and interviewer by gender, race, ethnicity or some other characteristic may be necessary. Often, an ethics committee review will address what is both appropriate and legal in the local context.

In order to monitor the field production and response rates (within and across countries), it is essential to set a standard for determining the final status of every sampled unit or member. For example, it is important to be able to distinguish completed interviews with eligible respondents from vacant units, noninterviews with known eligible respondents, and noninterviews where eligibility is unknown. Reasons for noninterviews are also essential information. Consistency in definitions across sample types is also very important for other comparative purposes. For example, knowing whether nonresponse is comprised primarily of noncontacts or refusals may inform different strategies to mediate these outcomes. Standards for determining response rate calculations are well-established (for example, see American Association of Public Opinion Research, 2006). Strategies for monitoring sample outcomes are further discussed below.

Privacy of Interview. Many surveys are intended to be conducted in a private setting in which survey questions and responses cannot be overheard. This is particularly important when the survey asks about sensitive topics and where self-administered techniques are not feasible. In many contexts, it may be difficult to achieve this privacy. For example, people may be living in close and crowded quarters, weather or some characteristic of the interview may preclude conducting the interview outside, or it may simply be inappropriate for the respondent to be alone with a stranger. In some cultures, young women might require a chaperone or other family member present. It may also be that the interview process itself is such a unique event that others want to be involved or watch the process. Small children may also be present and distract from the interview. Sometimes respondents will be more readily found at the village market or other public area than at home, and the challenge here, again, is to find a quiet spot to conduct the interview. Studies have used various techniques for addressing these issues.

Biomarkers and Physical Measurements. Increasingly both biomarkers and physical measures are being collected in large-scale cross-national studies, although for a variety of cost and logistical reasons these are generally focused on cross-national studies in minority countries [with exceptions: see DHS (http://www.measuredhs.com/topics/biomarkers/start.cfm/)]. These include simple measures such as height, weight, waist and hip circumference, and walking speed; measures involving special equipment such as blood pressure, grip strength, and lung capacity; and biomarkers such as saliva and blood. The rapid development of easy-to-administer biomarker kits at increasingly lower costs now places these forms of collection more readily within reach of organizations on modest budgets and without highly trained interviewers (Bentley & Muttukrishna, 2007). At the same time, local understanding is necessary to devise appropriate strategies to make certain requests of groups culturally apprehensive about giving samples of blood, hair, or saliva.

Where specialized equipment is needed, frequent calibration of these may be necessary. Since these measures are often taken in homes instead of clinics, the logistical issues regarding proper handling, transportation, and processing of the samples can be considerable. Interviewer training is very important, for instance, with regard to handling blood. Informed consent procedures may also be more complex for these studies and, if DNA is to be extracted, there may be considerable restrictions on the use of such data.

From this discussion, it should be apparent that although past fielding experiences in a given setting may be a good guide to issues that will be encountered with a new study, a thorough exploration and testing of alternative approaches should be undertaken. These may include focus groups, one-on-one interviews; and consultations with experienced researchers, interviewers, or key stakeholders familiar with the topic or the population under study. Generally, a mix of these activities will provide the widest range of information and possible options, but none of these will substitute for a field test of the protocols and procedures. In addition, because these procedures are context-specific, testing them in one or two locations will not be sufficient. Ideally, they should be tested in every country or culture where they will be implemented. The time it may take to develop, test, revise, and implement these procedures and protocols as well as the ultimate benefits of doing so should not be underestimated.

15.5.3 Field Structure, Interviewer Supervision, Production, and Data Monitoring

This section discusses field structure, interviewer supervision, and their effect on production and data monitoring. The structure of the field interviewing will depend on the timing of data collection, the number of languages that need to be accommodated and how those languages are distributed in the population, whether interviewers and respondents will be matched on other characteristics, and how unique the interviewer skills are that will be needed (familiarity with computers, for example). The mode of interviewing and the size of the interviewing team will also affect the field structure. More interviewers generally necessitate more supervisory levels. Although the names of roles will differ, the field staff will generally be comprised of interviewers and supervisors who may, in turn, be supervised by production managers or field directors. Supervisors can play many roles and perform such tasks as recruitment and training of interviewers, production monitoring and quality control, interviewing, data entry, and communication with the rest of the project team.

The ratio of number of interviewers to supervisors will vary by the experience of the field staff, the difficulty of the task, and the mode. Generally, fewer supervisors are needed for telephone surveys (Biemer & Lyberg, 2003) and when interviewers are experienced (Couper, Holland, & Groves, 1992). If travel is over a great distance or the administration time of the survey is particularly long, more interviewers may be needed. Typically, interviewers will be assigned to geographic areas or will travel in teams to sample locations. The advantage to traveling teams, in addition to providing a greater measure of safety, is that they can be accompanied by a supervisor and feedback on performance can be immediate and ongoing. Team interviewing generally limits the number of calls that can be made over a given time frame partly because teams usually travel over large areas and cannot afford to stay in one area too long. It may also be difficult to cover a wide range of languages or match on other characteristics when employing a traveling team, since these aspects will often vary by region.

If interviewers are not in teams, supervision and quality monitoring must take other forms. Depending on the country's infrastructure, supervisors may have to travel extensively to monitor interviewers' work. Where good communication and postal services exist, supervisors can keep in touch with interviewers by phone and review completed work through the mail. Clearly, where the sample management and survey administration is computerized and detailed information on the disposition of the sample and completed interviews can be reviewed daily, more can be done from a central location and much less field supervision is needed (Nicholls & Kindel, 1993). This is discussed below.

15.6 QUALITY ASSURANCE, CONTROL, AND DOCUMENTATION

15.6.1 The Main Dimensions of a Quality Framework

In the survey context, "quality" can be assessed by the overall usefulness of the data, sometimes described as the "fitness for use" (Biemer & Lyberg, 2003; Juran & Gryna, 1980), the total survey error (Groves, 1989; Groves et al., 2009), or the survey process quality (Lyberg, Biemer, Collins, de Leeuw, Dippo, et al., 1997). A number of frameworks have been used to describe the dimensions of quality (see also Chapters 13 and 16, this volume). Below, we directly relate the main dimensions to cross-national survey research:

Relevance—Do the data meet the needs of the client or users? As noted earlier, in a cross-national study, where more than 100 collaborators and multiple funding sources—with possibly competing goals—may be involved, the dimension of relevance becomes more challenging to fulfill.

Accuracy—Are the data describing the phenomena they were designed to measure? Accuracy refers to the distance between the estimate and the (often unknown) "true" value and is usually measured by the mean squared error. Wide variations in the level and sources of error can be expected both across and within countries.

Timeliness—How much time has elapsed between the end of the data collection, and when the data are available for analysis? Here, we note the timing of national elections across nations as one example of the challenges of meeting the timeliness dimension across nations (see Chapter 30, this volume; Lagos, 2008).

Accessibility—The ease with which data may be obtained by users. In the cross-national context, data access can mean more than simply making data publicly available. Particularly in majority countries, making the data truly accessible to local populations may need to include capacity building or training activities. Country-level data access laws and regulations will also come into play.

Interpretability—Are supplementary data available to analysts that describe the major characteristics and structure of the data (metadata) as well as data about the survey processes (paradata)? As elaborated below, the lack of transparency in cross-national data collection is long-standing and persistent.

Coherence—Are the data available for further recombination with other statistical information for various, secondary purposes?

Comparability—To what extent are observed data differences due to genuine variation as opposed to other factors? The quality dimensions of coherence and comparability are *the raison d'être* for cross-national and cross-cultural survey research and are discussed throughout this monograph.

Professionalism—Are staff provided with clear behavioral guidelines and trained appropriately and adequately? Are there adequate provisions to ensure compliance with relevant laws? Laws and regulations will vary by country and often by organizations within countries, as will local research traditions and procedures. Professionalism also includes impartiality in data analysis and report writing.

Cost and Respondent Burden—To what extent did cost play a factor in implementation decisions? Were the concerns of respondents adequately considered? Local laws, regulations, or traditions will shape what constitutes undue respondent burden. Balancing costs and burden, and decisions regarding these, will have to be made at local levels, given the wide variation in cost structures across and within counties.

Design Constraints—Were there context-specific constraints on study design that may have affected quality? As discussed, the examples are myriad in cross-national survey research: multiple languages, physical barriers such as islands, mountains, seasonal conditions, natural disasters, wars, populations that are displaced or nomadic, undocumented workers, and hidden populations, to name a few.

In order to provide end-users with sufficient information to assess the overall quality of the end product, quality assurance, quality control, and documentation should form an integrated set of tasks. The above framework serves as the basis of developing a quality assurance plan, implementing quality control and quality monitoring, and ultimately publishing a quality profile. In brief, these terms can be defined as follows:

Quality Assurance Plan—Developing and applying planned, systematic activities to ensure that the project meets or exceeds expected goals. The outcome of these efforts may or may not be measurable.

Quality Control—Monitoring specific project results against a predetermined baseline to ensure that standards are met or exceeded. The results should be quantifiable, but again, may not cover all aspects of the project.

Quality Profile—Publishing a document summarizing the quality assurance plan and including the indicators collected during the quality control effort. Quality profiles synthesize information from other sources, documenting all aspects of the survey, providing indicators of process quality, sources of sampling and nonsampling error, and recommendations for improvement and further research. They allow analysts to make an informed judgment about the overall quality and usability of the data.

15.6.2 Implementing a Quality Framework for Data Collection

A minimum set of quality control processes should be built into every step of the survey life cycle, starting generally with detailed specifications. In order to be able to adjust procedures and rectify problems, quality monitoring should be done in real time. All too often, quality assessment or documentation of quality controls takes place too late to permit corrective action. Even documentation delivered after the event needs to be recorded close to the phase or event being documented; it is not possible to accurately reconstruct implementation details or outcomes on the basis of recollections or impressions.

If corrective interventions are to be possible, documentation and quality monitoring must take place during data collection and be available to those deciding the corrective action to be taken. Centrally organized quality supervision faces challenges in this respect, as does quality assurance at the local level. If supervision and interventions take place centrally, shared systems will be required to deliver the necessary monitoring data quickly. Without such systems, supervision will necessarily be at the local level. In either case, managers need to be informed about options and required standards in order to devise appropriate interventions. If interventions are decided centrally but information collected only locally, time lags may hamper successful intervention. In many such cases, a mixed model may be most viable. Clear specifications, documentation tools, examples, timely and regular local monitoring, and assessment of the collected data can be partnered with assistance from an accessible centrally organized team. For example, the ESS does not have a shared system but does have a range of tools, specified requirements, and reporting deadlines throughout fielding; various steps are also monitored by the Central Coordinating Team. Once fielding is completed, countries submit detailed prespecified documentation in required formats to the ESS archive.

In the ISSP, members know what they will be required to report after fielding. Quality monitoring during fielding is the local agency's responsibility. Activities of the ISSP methodology committee (research projects, requests for country details, circulars on problems) and discussion of outcomes and any problems found are addressed at the annual ISSP meetings and help keep quality improvement on the ISSP agenda. The ISSP's annual reporting document for each module is a Web-based questionnaire completed after fielding. It is based on a paper version developed in the mid-1990s and refined over time. Countries new to the program are informed about documentation requirements before they conduct a study. This is necessary because the ISSP does not have a detailed set of implementation specifications. The European Values Survey (EVS) based its reporting form on the older ISSP paper version and added some questions relevant to the EVS survey.

The WMH took a similar approach to the ISSP, using an online modular form that performed much as a Web survey (Mohler, Pennell, & Hubbard, 2008; Pennell et al., 2008). The WMH modules covered general study information (goals, contact information, etc.), sample design, ethics review, interviewers and interviewer training, pretesting, data collection, quality control, data preparation, and final report (with specified study and sample outcomes). The advantage of the modular approach was that different people could complete different modules. For example, it is not uncommon that the individual who has designed and chosen the sample is different from the person who managed the data collection.

Documentation and monitoring again raise the question of language. Local research managers or collaborators may also differ in their ability in the chosen *lingua franca*. This may lead to misunderstandings on both sides; instructions to the local collaborators may be misunderstood and, in turn, requests or questions from the local collaborators also misunderstood and inadequately answered. A modularized question format also ensures that the detail obtained from each country is comparable. Asking for long narratives written in a *lingua franca* about technical aspects of the survey will generally produce documents of varying quality and detail. No matter the approach to collecting such monitoring information, however, it is important to clearly define all terms to ensure a common understanding of the information that is being sought.

In the past several years, the level of detail sought in such monitoring systems has improved. For example, the ESS now collects call-record-level data, as well as many other details about the conduct of the study (cf. Billiet, Koch, & Philippens, 2007; Chapter 18, this volume), and makes these data available to analysts (/http://ess.nsd.uib.no/). Since these data are not collected in real-time, central-level intervention can only be from round to round. The SHARE project has systems in place to achieve active monitoring and has increased the intensity of such monitoring in each subsequent wave of the project. SHARE uses the same data collection systems across all participating countries; call records are collected using a custom-built sample management system and all data are collected using the same CAPI software (de Luca & Lipps, 2005). As a result, call attempts and outcomes, response rates, interview length, and data quality can be monitored centrally throughout data collection. SHARE has also recently implemented a certification program modeled after approaches used to monitor the World Mental Health Surveys (Pennell et al., 2008) and intends to publish these outcomes in a quality profile. Examples of items collected include details on the process and results of verification, aspects of interviewer recruitment and training, and on other phases of the survey life cycle, such as sampling. The monitoring data are collected using a modularized, online system.

Although such approaches, in combination with site visits, comprise a comprehensive cross-national quality monitoring program, they still fall short of the "active management" and real-time quality monitoring undertaken in a number of single minority country surveys. Techniques such as "responsive design" rely on daily measures and paradata across many dimensions of the survey. For example, many of these monitoring systems track interviewer activities and costs in addition to call records. This is especially important if interviewers are paid by the hour instead of per interview. Some of these sample management systems also provide a mechanism to collect household observations. In combination with records of call and contact attempts, these can help survey managers guide interviewers to maximize household contact, increase efficiency, or balance workloads across interviewers. Such process data or paradata (Couper, 2005) are increasingly being used in combination with the interview data to minimize nonresponse bias (Groves & Heeringa, 2006; Wagner, 2008).

The rapid diffusion of technologies to majority countries is changing methodology and quality control. For example, in the pilot phase of the Chinese Family Panel Study, Peking University's Institute for Social Survey (iSSS) fielded CAPI questionnaires on state-of-the-art netbooks, transmitting data daily via wireless Internet cards (air cards) and using the University of Michigan Survey Research Center's sample management system. The system has been adapted and translated for the Chinese context. This allowed the iSSS to monitor more than 50 indicators of field operation status on a daily basis. It will not be long before quality monitoring of this kind can be accomplished in large-scale cross-national studies.

As noted above, organizations and projects will vary in the cost-quality tradeoffs that are made, as well as items that will be monitored for quality. In any study, the cost and error relationship may be complex or unknown. In a cross-national context, however, where the optimal design will vary across countries and little is known about the sources of error in all contexts, the decisions regarding cost and error tradeoffs are much more difficult. In these situations, investments might need to be devoted to discovering the relationships among various error sources and costs to inform later design decisions. The dimension of timeliness is also a limiting factor in the evaluation of tradeoffs. If timeliness is critical, some quality dimensions may be assigned lower priority, for example, a smaller sample size might be chosen or some populations might be eliminated to avoid the need for multiple translations. These issues are discussed further below.

15.6.3 Pretesting

Ideally all components of a survey should be tested before it is fielded including: preparation and management of the sample, respondent selection procedures, contact protocols (including use of incentives and informed consent), administration of the questionnaire, and any additional features such as collecting biomarkers, as well as administrative aspects such as recording interviewer time and expenses. An effective pretest should also include testing the protocols for monitoring and documenting field progress and quality control (see also Chapter 3, this volume). If more than one mode of data collection is involved, pretesting protocols should take this into account. Obviously, these protocols and materials must also be translated (or developed) in the languages to be used and these documents should also be reviewed for language adequacy.

15.6.4 Ethics Review

It may be necessary to document that all local laws and regulations have been followed. Generally, there will be some kind of official notification or letter from the authorizing organization conferring permission to conduct the research. Of course, these will generally be in the local language and may need to be translated into the *lingua franca* if the central organization needs to verify the content of these documents. If written consent is required, a mechanism (such as an ID number) is needed to link the consent form with the interview. Contract specifications may also stipulate how long these documents must be kept. Interviewers and project staff should be required to complete some form of a confidentiality pledge, which may need to be tailored to local employment laws.

15.6.5 Interviewers and Interviewer Training

If specific criteria have been set for the recruitment and training of interviewers, details of how these were implemented locally may be important to understand survey outcomes. At a minimum, the number of interviewers, their characteristics, and how they were trained should be recorded as should the results of certification tests. These records should be appended to data files along with the interviewer's unique identification number (i.e., linking each sample unit to the interviewer of record) so that interviewer effects can be studied. In the SHARE project, which uses a train-the-trainer approach (described above), site visits to the interviewer training were added to the quality control procedures to ensure that the centrally specified training protocols were being implemented in each country. Obviously, such site visits must be made by people versant in the language of the local training.

15.6.6 Sample Management and Production Monitoring

At a minimum, a system will be needed to determine the final disposition of every sampled unit or member (see Section 15.5.2) in order to calculate response rates across sample designs and countries. Although this system will allow a comparison of sample outcomes, it will not be sufficient to monitor the sample during data collection. For that, a system to receive detailed reports or data from the participating countries is needed. The reports should provide information on progress by individual interviewers, interviewing teams, or region. They may state only the status of all completed work or include the status of work in progress. Reporting might consist simply of updates by phone or e-mail or be made through more complex sample management systems. Finally, production monitoring should include a timely and ongoing review of the actual data to detect any anomalies. Information such as length of interview and any response patterns to key questions (especially any questions that determine whether sets of questions are asked or not) should be systematically checked for each interviewer over the course of the entire field period.

15.6.7 Household and Respondent Selection and Interviewing Protocols

The frequency of mistakes in both household and respondent selection will largely depend on the sample design. It is preferable, whenever possible, to have someone other than the interviewer select the units to be interviewed. A quality check should be built in to assure that selected units were indeed where the interviews took place. Respondent selection protocols should be systematically checked for random or systematic interviewer-based error, such as misunderstood procedures or systematic exclusion of some household types or unit members. A sample of each interviewer's work should be reviewed in a timely fashion to detect systematic errors and provide feedback as soon as feasible. One model frequently used in traveling teams is for the supervisor to review the work of the team daily. This may also enable the team to collect any missing data before they move on to another sample location. In the WMH study in Colombia, the supervisor entered the completed paper surveys into a laptop computer daily and was able to run diagnostics on the resulting data on an ongoing basis. This also facilitated the review of data over time to detect errors possibly missed in a manual review of the paper instruments.

Interviewers may inadvertently or deliberately deviate from required procedures (Biemer, Groves, Lyberg, Mathiowetz, & Sudman, 2004; Biemer & Stokes, 1989; Harrison & Krauss, 2002). A sample of the interviewer's work needs to be selected for re-interview to ensure that the interviews took place and that there were no systematic errors. This procedure can be costly if the re-interview requires another face-to-face visit. It may be possible to have interviewers audio record their interviews for monitoring; how feasible this is will depend on the sensitivity of the survey topic and on local infrastructure. Such interviewer monitoring can be expensive but is a critical investment.

15.7 CONCLUSIONS

This chapter demonstrates that a wide range of knowledge, skills, cooperation, and capacity-building at all levels of a project (both horizontally and vertically) are essential for planning and executing successful data collections in cross-national surveys. This concluding section revisits how the range of skills and the cooperation needed are interlinked, before turning to consideration of promising new developments and research that is urgently needed.

15.7.1 Know-How, Skills, and Documenting Lessons Learned

The range of knowledge and skills needed at the different stages of planning and implementing high quality cross-national data collection efforts is quite considerable, given the number of languages, varying geographical and social conditions, and personnel of diverse experience and training that will be encountered. Everyone involved in such an effort needs to have a complete understanding of the larger goals of the study and how the study's processes and procedures contribute to these goals. This includes demonstrating the importance of documentation and transparency as also emphasized in other chapters in this volume. Criteria and recommendations such as advocated in the Strengthening the Reporting of Observational Studies in Epidemiology initiative (STROBE) (von Elm, Altman, Egger, Pocock, Gøtzsche et al., 2008) may well influence general expectations and specifications for reporting on studies of different kinds.

Collaboration and Multilateral Capacity Building. Cross-national studies bring together researchers from various locations with different experiences, needs, and perceptions. To optimize the research project, multilateral collaboration is essential. One team or set of teams may have more technical know-how or survey experience than another team or set of teams. They, on the other hand, might have more insight into substantive issues or more cross-national experience. Whatever the situation, learning on all sides will always be required. Informed decisions can only result from collaboration across research teams and a multilateral sharing of knowledge and experience. Without a mutual understanding of reasons and motivations for decisions, agreed-upon strategies may be low on viability and even if "do-able," they may not be properly implemented.

Capacity-building requires training. With some notable exceptions, such as the Atlantida initiative of the 1960s (United States Bureau of the Census, 1965), very few international programs directly address survey methods. However, with a declared goal of improving social science research in Europe, the European Science Foundation funds programs of seminars and workshops on qualitative methods in the social sciences (http://www.esf.org/index.php? id=4858cw/). Aided by EU funds for that purpose, the ESS has also organized quality enhancement meetings and workshops for countries participating in the ESS and for young researchers across Europe and beyond. A number of cross-national projects investigating pretesting strategies for comparative research are also under development (Edwards, 2009; Fitzgerald & Miller 2009; Miller, Fitzgerald, Caspar, Dimov, Gray et al., 2008; Miller, Mont, Maitland, Altman, & Madans, 2008; Willis, 2009). Wide dissemination of data and increasingly accessible tools also helps change official attitudes toward data publishing more generally. The Central Statistical Agency of the Federal Democratic Republic of Ethiopia, for example, began in 2005 to distribute a wide variety of microdata products for the first time (http://www.csa.gov. et/). The Data Documentation Initiative (DDI) is an effort to establish an international XML-based standard for the content, presentation, transport, and preservation of documentation for datasets in the social and behavioral sciences (see Chapters 16 and 17, this volume).

15.7.2 Amassing and Disseminating Knowledge

Each cross-cultural project leaves its researchers with a wealth of new understanding and lessons learned. However, much of this knowledge and expertise often remains with the project team, either because there are insufficient time, funds, or human resources to document this knowledge or because it is viewed as proprietary, with commercial worth (for future competitive bids, for example). Whatever the reasons, lessons learned are rarely documented for use for the benefit of the larger scientific community. Activities in the Guidelines Initiative of the International Workshop on Comparative Survey Design and Implementation (http://www.csdiworkshop.org/) are of note here. The members of the Initiative have developed guidelines for key phases of the survey life cycle (http://www.ccsg.isr.umich.edu/). While acknowledging the fledgling aspects of their work, the project is a deliberate attempt to draw together knowledge, research, and experience on all aspects of comparative survey design and implementation, including data collection.

15.7.3 Research Needed

Gathering information about solutions to challenges in various contexts is the key to being better able to plan for research in contexts about which information is lacking. German sociologist Erwin Scheuch (1989) had this to say about comparative research:

> The real problem is not the methodology per se, but it is methodological in its consequences: what can be done to make methodological advances

and practical experiences in comparative research more cumulative? Or phrasing the question both more realistically and more depressingly: how can we make knowledge in this area cumulative at all? (p. 147)

Two decades later we still face the same question. Recent prominent publications on data collection methods and modes have virtually nothing to say about comparative issues (Dillman, Smyth, & Christian, 2008; de Leeuw, 2008). The vintage literature on data collection, although instructive about the past contexts and populations described, also does not reflect current options, contexts, and needs.

At the same time, we are perhaps closer to some resolution than ever before. The quality revolution in survey research has begun to alter expectations about comparative research design and outputs. With the proper training at local levels, new developments in survey data collection methods can enable majority countries to collect and publish their data.

In commenting on the little progress made in comparative methodology in the nineteen sixties, Scheuch (1968) felt that, "The major progress has been to increase the awareness of the real sources of difficulties" (p. 176). It is precisely through learning about difficulties that some progress has recently been made. As indicated, the detailed documentation now provided by some studies and available on the Web can be tapped as a valuable source of information and can initiate new research. The fact that some surveys are collecting and providing information about interviewer performance, for example, sets standards which will, we expect, ultimately affect other surveys and research domains.

The idea that awareness can promote change is a realizable route to accelerate progress. By making research sponsors aware of the sources of difficulties and their consequences and by demonstrating the importance of transparency and documentation and the potential for improvement, we are a lot more likely to gain the funding needed to record and disseminate information for the general good of the research community.

16

A Survey Process Quality Perspective on Documentation

Peter Ph. Mohler, Sue Ellen Hansen, Beth-Ellen Pennell,
Wendy Thomas, Joachim Wackerow, and Frost Hubbard

16.1 INTRODUCTION

This chapter discusses the documentation of multinational, multilingual, and multi-regional (hereafter 3M) surveys from a survey process quality perspective. In this introduction we outline our understanding of the needs and purpose of documentation in such comparative contexts. Section 16.2 relates documentation to a survey quality framework and the "total survey error" (TSE) paradigm; Section 16.3 introduces the notion of structured systems of data documentation, focusing on the Data Documentation Initiative; Section 16.4 provides examples of documentation in multinational surveys; and Section 16.5 offers summary conclusions.

Documentation serves two main purposes in surveys: First, it provides data users and researchers with information on how a study was designed and implemented, permitting them to test and assess various aspects of design, implementation, and findings on the basis of either benchmarks or replication of components of the study. Differences found between populations can be differences in "true values" but could also be the result of bias stemming from incomparable sampling designs, faulty translations or adaptations, or improper interviewer behavior, to name but a few possibilities. Detailed documentation of a study helps clarify what might be involved. Numerous multipopulation surveys and programs are designed to repeat measures over time as well as across populations (see Chapters 25–31, this volume). Proper documentation of the survey design and production process is essential also for primary researchers in order to know how to replicate measures accurately either across countries or at

[1] Survey Methods in Multinational, Multiregional, and Multicultural Contexts, edited by Harkness et al.
Copyright © 2010 John Wiley & Sons, Inc.

different points in time. In each instance, the information needed to make decisions and judgments is recorded in "documents."

Second, documentation is an indispensable tool for quality assurance and quality monitoring for any and all stages in the survey life cycle. Documentation on sample allocation, for example, can inform quality assessment of fielding progress and permit timely interventions to address any fielding problems. Documentation of adaptations to instrument design could be needed at the stage of harmonization or during translation, as well as at the stage of data analysis. Thus documentation plays a crucial role in the development, testing, and use (analysis) of any survey.

However, much in the way that a pilot's handbook does not enable everyone to fly a plane, those working with documentation need to understand the processes and outcomes that are recorded in the documentation. Technical know-how and training and a thorough understanding of the subject matter are also essential. In addition, standards such as officially recognized specifications, best practice definitions and protocols, or accepted guidelines are also critical, since it is in terms of these that research is assessed and described. If any of these three sources of knowledge and expertise are lacking or deficient—training and knowledge, standards, and pertinent and sufficient documentation—quality assessment and data analysis will be hard to accomplish.

Variety may be "the spice of life," but in comparative survey research variation can threaten comparability. Nonetheless, variation is the usual state of affairs. Even within one country, survey organizations differ in many respects regarding, for example, their preferred instrument and sample designs, the standards and benchmarks they use, and their field organization and quality assurance frameworks. Research disciplines also differ in their preferred designs, testing and analysis, and quality assurance and quality monitoring (QA and QM). They also differ in their typical (standard) definitions of such common concepts as "income," "household," or "education;" in procedural requirements (e.g., requirements for ethical board reviews, choice of sample management); in the type of training players receive (e.g., interviewer training); in the detail and form of the documentation typically required on or for procedures (e.g., contact protocols, interviewer manuals); and, importantly, in researcher perceptions of the relevance of all these to assuring and monitoring survey quality.

In any situation where training, standards, and documentation approaches differ, it becomes challenging to evaluate or compare designs and research undertaken in different contexts. However, such comparisons are exactly what is necessary to plan, conduct, and assess multinational research. Survey documentation for comparative research must obviously document the survey data products but it must also record details of how the survey was designed and implemented in the various locations and languages. Well planned and implemented documentation thus plays an essential role in dealing with the challenges involved in multinational survey research.

Marked changes have taken place in the last decade in survey documentation standards and procedures. International survey programs such as the International Social Survey Programme (ISSP) and the World Mental Health Survey (WMH) Initiative have developed Web-based documentation, for example. Increasingly, too, survey projects or the archives holding data are making access to their data and documentation free or affordable to many. The Data Documentation Initiative, in collaboration with researchers active in comparative research, has also developed a comprehensive specification framework for documenting survey metadata (data about data), with a focus on the data life cycle across countries and time (see Section 16.3.1). Cross-cultural survey guidelines (Cross-Cultural Survey Guidelines, 2008) have also been developed that include a documentation and dissemination module (http://www.ccsg.isr.umich.edu/).

16.2 TOTAL SURVEY ERROR, COMPARABILITY AND PROCESS QUALITY DOCUMENTATION

The total survey error (TSE) paradigm provides a widely accepted conceptual framework for evaluating the quality of survey data (Groves, 1989). Quality in TSE is defined as the maximal reduction of the mean square error, usually taking into consideration error related to representation (coverage, sampling, non-response, and statistical adjustment), measurement (questionnaire, mode, interviewer, and respondent), and processing errors (Groves 1989; Groves et al., 2009). Multinational surveys, however, have a further essential quality require-ment—comparability. There is to date no ready formula that defines compara-bility. However, numerous chapters in this volume discuss how it might be achieved and tested.

To assess data quality in terms of both the TSE paradigm and with regard to comparability, survey producers and data users need detailed documentation of the survey production process. This should include access to information (data) behind the data (metadata), such as the contact protocol information that may lie behind nonresponse documentation. It should also include information on other factors re-levant for the survey process and outcomes, such as interviewer profiles. Such data are termed paradata (see discussion in Couper, 1998; Couper & Lyberg, 2005).

Computer-assisted interviewing (CAI) makes it possible to collect a multitude of diverse metadata and paradata, all of which can potentially facilitate quality assessment in terms of TSE and comparability. However, the usefulness of the information is largely decided by the adequacy of the documentation structure and the details recorded: These should be specified to meet the needs of different user groups.

16.2.1 Process Quality and Continuous Process Improvement

Older approaches to quality tend to focus on the quality of outputs—the resulting product of a procedure. One disadvantage of this is that faults are discovered only "after the fact" and their cause may not be easily identified. Searching for causes of faults, where possible, then becomes costly and time-consuming. In practice, what often happens is that: (a) more outputs are rejected; (b) quality standards are lowered in order to reduce the number of rejections; (c) ad hoc, spur-of-the-moment "improvement" procedures are implemented; or (d) in order to "fix" at

least something, all the known sources of faults and errors are addressed, even if
the cost-benefit gain for the overall quality of a product is quite small.

Modern quality assurance and monitoring principles, on the other hand, are
multidimensional and focus on "fitness for use" (cf. Juran & Gryna, 1980; Chapter 13, this volume), aspects which are not addressed in the older TSE literature
(cf. Biemer & Lyberg, 2003). Modern QA and QM frameworks address each of
the steps involved in assembling a product (in our context, a survey) or providing
a service, from planning, to completion, and distribution. Today's quality management also defines quality benchmarks for production steps and takes care to
implement quality controls to achieve each desired outcome (see Figure 16.1).
Outcomes are thus accepted or rejected at each step. For instance, if the outcome
of a component in questionnaire design, say an item scale, fails to meet a targeted
quality benchmark, the scale is looped back into the development process. If, after
several iterations, a satisfactory scale quality has not been achieved, the scale is
discarded, like a faulty product in a production line. Agreed quality standards
and procedures also make it possible to concentrate on output quality at each
given step and to reduce or eliminate input assessment efforts. In other words,
each production step is undertaken in accordance with QA/QM specifications and
thus can accept the approved deliveries from the foregoing step without
repeating QM controls. In this kind of framework, each production step can be
thought of as a comprehensive unit accompanied with full documentation of its
process quality. This strategy makes it possible to narrow down both the sources
of error/faults and the kinds of possible faults within a step, thus also making it
easier to target remedies and to reduce the costs of doing so. Such models call for
refined documentation schemes, such as that described in Section 16.3.1.

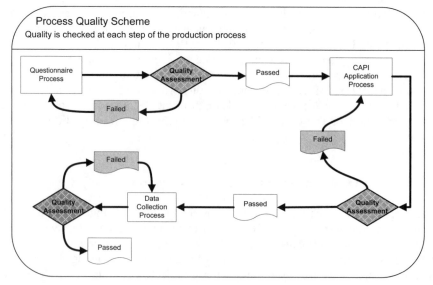

Figure 16.1. A Simplified Process Quality Scheme for Some Production Steps in a
Social Survey

16.2.2 The Survey Life Cycle of Comparative Surveys

The "survey life cycle" can be seen as a series of interlinked and often iterative processes, including survey production, publication, and secondary analysis. Figure 16.2 represents the survey life cycle for a monocultural study.

However, the picture is much more complex for a 3M survey life-cycle model. The simplified diagram in Figure 16.3 suggests something of the layers of design and implementation to be carried out and documented. For more detailed discussion regarding documentation, see Mohler and Uher (2003).

Traditional codebook documentation formats for one-culture/nation surveys present question text, answer categories, and interviewer instructions alongside frequency counts of each question/variable. For 3M surveys following this format, cross-tabulations of questions by country are usually added to the above. This approach reflects a linear conception of comparative projects, seeing them as comprised of such steps as presented on the following page.

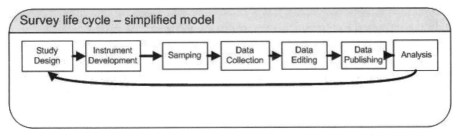

Figure 16.2. The Survey Life Cycle for a Monocultural Study

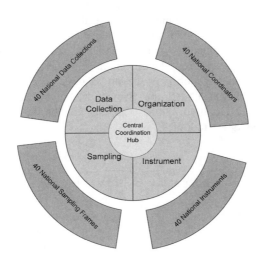

Figure 16.3. Simplified Model of 3M Documentation Complexity

1. Source questionnaire production
2. Local questionnaire version production
3. Local data collection
4. Data editing of individual data sets
5. Central merging of individual data sets
6. Tabular comparisons

This linear view, however, does not capture the true multilayered production process of modern 3M surveys developed within a QA and QM framework. It therefore fails to reflect the documentation actually needed. In order to monitor quality in comparative surveys, we need to be able to navigate easily back and forth between (possibly ongoing) procedural steps and between local and centralized levels of data and of documentation. The complex web of data, paradata, and metadata needed for this calls for detailed planning in accordance with agreed documentation requirements and rules. The next section considers tools with which such complex documentation become viable.

16.3 DOCUMENTATION FRAMEWORKS FOR METADATA

This section first describes an existing tool to record information about the survey life-cycle process and its outputs and then outlines a survey metadata documentation framework which supplies definitions to be used in future versions of this tool, also for comparative surveys.

16.3.1 The Survey Metadata Documentation System

Documentation should be as rich as necessary for stated purposes but also as economical as possible. To achieve this for the survey production process, a documentation framework and protocol need to be created that cover the components to be documented in the detail and specificity required. In 2004, colleagues at Survey Research Operations at ISR, University of Michigan, and then ZUMA, Mannheim, Germany, began to develop a tool to collect survey metadata. The tool was called the Survey Metadata Documentation System (Mohler, Pennell, & Hubbard, 2008). The first version developed comprised the following documentation modules:

- General Project Information
- Ethics Review
- Sample Design
- Questionnaire Development
- Translation Process
- CAI Programming/Systems Development

- Pretesting
- Interviewer Recruitment and Training
- Data Collection
- Quality Control
- Dataset Preparation/Final Report Information

The Survey Metadata Documentation System (SMDS) looks and feels like a Web-based survey. It has closed and open-ended questions which cover all standardized metadata on all aspects of the survey life cycle. From the start, some comparative elements were envisaged (e.g., translation). The tool is organized in modules; each leads the person compiling the documentation through the documentation process for that topic. The data collected can be used to produce a wide array of reports. Different people can complete different modules, allowing whoever is best informed on a given topic to complete the module for it. A beta version of SMDS was used to document the WMH surveys. A newer version of the tool is currently under development. Longer term, the aim is to be able to convert metadata collected with SMDS into a form that complies with the Data Documentation Initiative (DDI) specifications described below.

16.3.2 The Data Documentation Initiative

In the mid-1990s, researchers affiliated to social science data archives began to address the need for standardized survey metadata. This group later became the Data Documentation Alliance. They created a set of standardized tags for survey metadata written in eXtended Markup Language (XML), calling this system of descriptions the Data Documentation Initiative (Data Documentation Initiative, 2003). Using XML for DDI permitted easy sharing of tagged text files between different software programs and offered researchers and documenters the flexibility to create many forms of documentation from just one XML data source. XML-based text and data structures are indeed quite common in modern computing; office tools such as Microsoft Office 2007 and Open Office, as well as websites, often format text that has XML as its underlying structure.

DDI is not a software tool; It is a nonproprietary (open source) model for documentation or, as it is sometimes called, a "data grammar." The model is designed to structure internal and external metadata content and relationships. It serves as an underlying systematic structure that can be used to organize stringent survey descriptions (i.e., documentation). Users do not need to be aware of the underlying structure; they can simply use their higher level tools, such as browsers, editors, and spreadsheets, much as they do with XML-based office systems. An early version of DDI appeared in 2000. Early versions were used primarily by archives in North America and Europe to record basic study and survey variable metadata and as support for data access systems. DDI has served in this way as the metadata structure for over 24 projects, ranging from metadata creation and dissemination software for single files, such as in the NESSTAR system (http://www.nesstar.com/) and the Microdata Management Toolkit of the International Household Survey Network (IHSN; http://www.surveynetwork. org/), to the development of multistudy information dissemination systems, such as the Center for Comparative European Survey Data Information System (CCESD; http://www.ccesd.ac.uk/), the National Historic Geographic Information System (NHGIS; http://www.nhgis.org/), and the Dataverse Network system (http://www.thedata.org/).

However, these early versions of DDI did not provide either the detailed coverage needed by data producers or the stringent format control required by software programs. The early DDI standard was primarily a structured electronic version of a printed codebook. We pointed earlier to limitations of these. It assumed the existence of a dataset and did not support documentation of early stages of study design and survey development. A structure was therefore needed that could better support more complex surveys and files, better cover the survey development process, and also support software development. By 2005, proposals were on the table to change DDI radically so as to address these needs.

This shift in perspective also prompted the Alliance to review the overall structure of DDI and ultimately to move toward a flexible structure based on modules and schemes that could be used and published at any point in the survey process (for details, see http://www.icpsr.umich.edu/DDI/ddi3/). Metadata could be captured at the point of creation, be reused or referenced at later stages, and could also be used to inform and drive subsequent steps in the data collection process.

The new design made it easier to establish repositories of shared information (concept banks, question banks, variable banks, and other shared materials) in formats that could be easily moved into or out of DDI metadata files. The new structure treated multiple data products of a single data collection (e.g., the original dataset and subsets or aggregated files) as part of the same study, thus reducing duplication of data collection information across products. Studies that belonged together could be grouped as a series, and questions shared or changed across the series could be easily identified. Harmonization procedures (see Chapter 17, this volume) could be captured and information provided for coding individual studies into a harmonized data structure using a format that could drive data analysis software. DDI Version 3.0, based on this structure, was released in 2008 (Data Documentation Initiative, 2008). It followed the data life-cycle model, from the initial research proposal, through the survey development process, into data collection, storage, eventual analysis, and reuse. Figure 16.4 shows how this version was conceptualized (see also Figure 16.2 in Section 16.2.2).

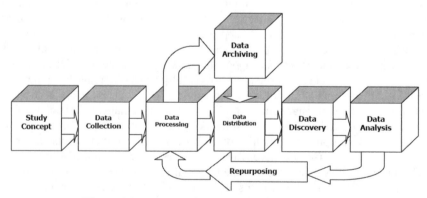

Figure 16.4. DDI Version 3.0 Data Life-Cycle Model

DDI Version 3.0 was intended to meet expressed needs of the survey research, archive, and software development communities. These needs included:

- Supporting computer-assisted survey instruments (e.g., Blaise) through expanded description of the questionnaire content and question flow
- Supporting the description of data series, such as longitudinal surveys
- Supporting comparison and both input and output harmonization
- Improving support for describing complex data files, such as record and file linkages
- Providing improved support for geographic content to facilitate linking to geographic files, such as shape and boundary files
- Ensuring a consistent, reliable information structure and also the content needed for programming and software systems which transform how data are collected, organized, and published

The key strengths of this 2008 version of DDI directly address issues important in documenting survey development, process quality, and, of special relevance in this chapter, multipopulation surveys. Version 3.0 allows publication of information independent of any specific study. Examples are concept definitions and question structures that might occur in multiple studies. The metadata are structured to separate such independent content from "within-study" content. For example, the independent information relevant to a question includes the question text, the response option(s), and possibly some instructions. The information on the same question as used in a given survey, on the other hand, would include its position on the questionnaire and any relevant filter (universe) information. Independent "publications," as they are sometimes called, provide standardized structured metadata that can be searched, reused, modified, and documented over time. They thus facilitate the replication of studies and survey instruments. As such, they are a valuable resource for designing and replicating question structures and content.

Revisions in questionnaires across versions or within a round of development can also be recorded, along with the final textual outcome. Sections of metadata that are often reused, such as the purpose of an ongoing study program or any core questions, can be entered (stated) once, as can features such as concept definitions. Once stated, they can be reused by reference, providing explicit replication of that content. Further enhancements to DDI are underway to provide information on survey modes and methods used, record the survey design process, and also provide structured information on quality control processes.

As it currently stands, DDI can be understood as a data model for the documentation of surveys and other forms of data collection. The metadata are broken up into small pieces, each referenced by an ID. Any institution can create a version with a unique ID and maintain major sets of these pieces for publication or reuse. In this way, DDI has become a viable foundation for developing tools to document multipopulation surveys. In Section 16.4.1 we consider tools designed to take advantage of DDI-compliant structured metadata.

16.4 EXAMPLES OF SURVEY LIFE CYCLE AND PROCESS DOCUMENTATION

Lyberg and Stukel (Chapter 13) point to reasons for a slower growth in QA and QM activities in 3M contexts. Nonetheless, a change of awareness is underway and considerable progress in process and output documentation necessary for QA and QM has been made. Standardized specifications and accompanying tools can be expected to make important contributions to quality assurance and monitoring in large comparative surveys.

Two advances in large-scale comparative documentation were pioneered by the ISSP; the first was access to the target language questionnaires for all cultures/nations; and the second was the introduction of a standardized monitoring instrument (Park & Jowell, 1997a) to collect metadata. These data are now obligatory in ISSP study reporting (see Chapter 26, this volume). A major step toward extensive comparative documentation was also taken in the European Social Survey (ESS). The ESS website provides information about its procedural protocols and specifications, instrument development, the national questionnaires, a comprehensive data report, national datasets, national additions and deviations, questionnaire images, and the like. To date, the ESS provides the most comprehensive publicly available survey documentation on a large-scale, cross-national project. Figure 16.5 is a screen shot from the ESS website on the survey documents available for download.

At present, each category of documentation is presented separately. Users must thus open each of the relevant files to access all the information available. To see source question text associated with variables listed in the data protocol, for example, they need to open the main questionnaire or the variables and questions document (see Figure 16.2). Target language texts are stored in separate

ESS Round 3 – 2006

The ESS3-2006 Edition 3.2 was released on 16 October 2008. Please see Version notes for complete information.

The ESS3-2006 main questionnaire is made up of the core module and two rotating modules: Timing of life (section D in the questionnaire) and Personal and social well-being (section E). See introductions to the modules.

A full Technical Report on the 2006/2007 survey is also available on the ESS Home Site.

Survey Documentation
ESS3-2006 Documentation Report
App1: Population statistics
App2: Classifications and coding...
App3: Variables and questions
App4: Variable lists
App5: Other country specific documents
Weighting ESS Data
Response Based Quality Assessment
Data Protocol
More...

Deviations and Fieldwork Summary
Overview of anomalies in data, fieldwork dates, response rates and sample sizes.

Classifications and indices
Human Values Scale by Shalom Schwarz

Fieldwork Documents
Main questionnaire
Supplementary questionnaires
Administration of supplementary quest...
Showcards
Fieldwork instructions
Contact forms
Country specific documents
More...

Data Download
ESS3 Integrated file [SPSS] [SAS]
More files...

Online Analysis
ESS3 Integrated file - Open in Nesstar

Figure 16.5. ESS Round 3 Survey Documentation (http://www.europeansocialsurvey.org/)

files. Hyperlinks across documentation files would therefore be a valuable addition to the current form of ESS data presentation.

Comprehensive documentation must include metadata and paradata on survey quality. The ease or burden involved in managing and documenting such data depends to a considerable extent on the tools available for collecting and publishing them. In the next section we consider such tools.

16.4.1 Examples of Survey Data and Instrument Documentation Tools

The early focus of DDI, as said, was on providing very basic after-the-event documentation on surveys and on variables in datasets. The framework and standard tags were not tailored to be compatible with computer-assisted survey instruments. At the same time, computer-assisted survey implementations (CAI) were becoming increasingly common. These made it possible to produce a great deal of metadata about survey instruments and data collection efforts (e.g., question text, interviewer instructions, and flow logic, as well as aspects such as data entry times).

As a result, organizations began to create their own tools to help solve their metadata documentation needs. One of the earliest was the Tool for the Analysis and Documentation of Electronic Questionnaires (TADEQ; Bethlehem & Hundepool, 2004). This tool was designed for use with instruments developed by different computer-assisted survey software to provide printable and electronic textual and graphical documentation. TADEQ provided an XML-based Questionnaire Definition Language to structure instrument metadata.

A second initiative was led by the Survey Research Center (SRC) at the University of Michigan's Institute for Social Research (ISR) in the early 2000s. Prompted by the need to provide comparative documentation for the World Mental Health (WMH) surveys, a program was developed to document the surveys that used Blaise for computer-assisted personal interviewing. The program generated XML-based metadata that documented variations in WMH surveys across countries. The system was later enhanced to become the Blaise Documentation System and, later, the Michigan Questionnaire Documentation System (MQDS). This system was used to document and compare major U.S. and cross-national surveys, such as the Collaborative Psychiatric Epidemiology Surveys (CPES) study. Figure 16.6 shows the MQDS instrument documentation for CPES, including question text as it appeared in the interviewer's screen.

In 2008, Survey Research Operations (SRO), ISR, and the Inter-University Consortium for Political and Social Research (ICPSR), both University of Michigan, collaborated to enhance MQDS to provide a metadata system that was DDI Version 3.0-compliant and that structured metadata in both XML and relational database formats (Guyer & Cheung, 2007). Much of the MQDS metadata content regarding question text, flow logic, and instructions was actually incorporated into the 2008 DDI 3.0 structure. The purpose was to make it easier for SRC to test and document large and complex Blaise instruments and for ICPSR to document and archive survey data collected using Blaise, as well as to use this new DDI-compliant database structure to store metadata captured from other systems.

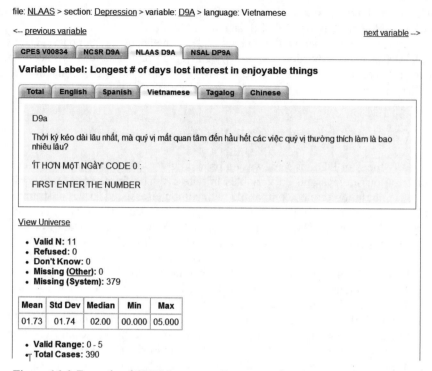

file: NLAAS > section: Depression > variable: D9A > language: Vietnamese

<-- previous variable   next variable -->

| CPES V00834 | NCSR D9A | NLAAS D9A | NSAL DP9A |

Variable Label: Longest # of days lost interest in enjoyable things

| Total | English | Spanish | Vietnamese | Tagalog | Chinese |

D9a

Thời kỳ kéo dài lâu nhất, mà quý vị mất quan tâm đến hầu hết các việc quý vị thường thích làm là bao nhiêu lâu?

'ÍT HƠN MỘT NGÀY' CODE 0 :

FIRST ENTER THE NUMBER

View Universe

- **Valid N:** 11
- **Refused:** 0
- **Don't Know:** 0
- **Missing (Other):** 0
- **Missing (System):** 379

Mean	Std Dev	Median	Min	Max
01.73	01.74	02.00	00.000	05.000

- **Valid Range:** 0 - 5
- **Total Cases:** 390

Figure 16.6. Example of CPES Instrument Documentation Created Using MQDS—Vietnamese Form

16.4.2 Toward Optimally Enhanced 3M Documentation

National surveys may certainly require extensive documentation efforts; nonetheless, in comparison to 3M surveys, they are relatively straightforward. In the 3M case, the increased complexity of documentation called for at every step ultimately results in procedures that seem qualitatively different, not just more complex. The complexity results from the need to document aspects such as culture-specific collateral information, culture-specific questionnaires, and culture-specific data collection implementation practices and protocols. Each requires general documentation on the topic, plus specific details for each individual culture or nation (Mohler, 2007, 2008; Mohler & Uher, 2003).

The goals of such documentation are threefold: to provide timely, quality-relevant information on the ongoing survey, to provide data users with essential information, and third, to provide information relevant to improving the quality of future surveys. In the case of a 40-nation study, for instance, this means 40 "documents" recording culture-specific data collection implementations and their respective assessments. To be useful in a comparative context, the documents must be comparable, have well-defined terms, and be compiled (merged) in a common information base which provides access to quality characteristics for each individual culture or nation, as well as for the whole survey.

The process quality perspective provides an optimal framework for effective documentation of complex comparative surveys and for dissemination of survey data from these. The documentation should provide users with information on aspects such as:

- How each question/item relates to the substantive research goals and to analysis
- How indicators are defined
- Reliability and validity reports on item performance
- Outcomes of any earlier research using items in the study literature references
- Independent and within-instrument information on questions used (origin, wording, as well as universe and sequence) and information on the various language versions and adaptation
- Data collection mode(s)
- Respondent and interviewer profiles
- Details of respondent selection and briefing (e.g., informed consent)
- Details of interviewer selection and training
- Field sample management and field implementation
- Quality assurance and monitoring specifications for process
- Quality assurance and monitoring specifications for outputs (e.g., missing data rates, response rates, measure of nonresponse bias)

The kind, degree, and organization of data documentation will vary by the kind of comparative study involved—in terms of whether it is a one-country cross-cultural study, a one-country panel study, a one-time multinational cross-sectional study, repeated multinational cross-sectional studies, or a multinational panel study. In the remainder of this section, we provide examples of various types of documentation that point the way toward addressing such aspects.

Minimal Comparative Documentation. Minimal documentation for a comparative survey includes a labeled data file ready for input into statistical software. Figure 16.7 provides an example taken from the Round 3 Data Protocol of the European Social Survey (ESS, 2006).

Basic Comparative Documentation. Going one step further than minimal documentation, basic documentation, such as found in traditional paper forms of codebooks, often includes the full question text (source language) and frequencies. Figure 16.8 provides an example taken from the codebook for the 2000 module in the International Social Survey Programme.

Data protocols and codebooks such as shown here do not provide information on such quality aspects as listed in Section 16.3. Indeed, standard codebooks such

B24	STFLIFE	HOW SATISFIED WITH LIFE AS A WHOLE	F2.0	00	Extremely dissatisfied
B25	STFECO	HOW SATISFIED WITH PRESENT STATE OF ECONOMY IN COUNTRY		01 02	
B26	STFGOV	HOW SATISFIED WITH THE NATIONAL GOVERNMENT		03 04	
B27	STFDEM	HOW SATISFIED WITH THE WAY DEMOCRACY WORKS IN COUNTRY		05 06 07 08 09	
				10	Extremely satisfied
				77	Refusal
				88	Don't know
				99	No answer

B24, 25B25 etc: question number in source questionnaire; STFLIF, STFECO etc.: variable name in data file; text field: abbreviated question text; F2.0: FORTRAN type format description (Fixed, 2 digits, 0 decimals); 00, 01 etc.: value codes; text field: value labels.

Figure 16.7. Minimal Comparative Data File Documentation with Variable Names/Labels and Value Labels

```
V11   Worry: about future environment

Location:   24    MD1: 9
Width:       1    MD2: 8

Q.4  And how much do you agree or disagree with each of
these statements?
(Please tick one box only)

Q.4a  We worry too much about the future of the environment
and not enough about prices and jobs today.
_____

     1. Strongly agree
     2. Agree
     3. Neither agree nor disagree
     4. Disagree
     5. Strongly disagree

     8. Can't choose, don't know
     9. NA, refused
```

	D-W	D-E	GB	NIRL	USA	A	IRL	NL	N	S	CZ	SLO
1	81	96	70	83	127	89	40	71	80	42	186	129
%	8.6	18.9	7.4	11.6	10.7	9.4	3.4	4.5	5.7	4.1	15.4	12.5
2	212	130	280	239	328	248	296	401	292	152	312	391
%	22.5	25.6	29.5	33.4	27.6	26.1	24.9	25.6	20.7	15.0	25.8	37.8
3	155	73	129	124	213	175	171	332	230	237	203	143
%	16.5	14.4	13.6	17.3	17.9	18.4	14.4	21.2	16.3	23.3	16.8	13.8
4	357	133	386	213	398	313	598	657	616	387	339	300
%	37.9	26.2	40.7	29.7	33.5	32.9	50.4	41.9	43.7	38.1	28.1	29.0
5	136	75	83	57	123	126	82	108	192	198	167	72
%	14.5	14.8	8.8	8.0	10.3	13.2	6.9	6.9	13.6	19.5	13.8	7.0
8	29M	18M	6M	20M	20M	60M	37M	35M	15M	31M	37M	42M
9	4M	2M	18M	9M	67M		8M	5M	27M	20M		
Sum	974	527	972	745	1276	1011	1232	1609	1452	1067	1244	1077

Figure 16.8. Example of Codebook with Question Text and per Country Frequencies ISSP 2000 (http://www.issp.org/data.shtml)

as currently provided by data archives do not cover central aspects of the comparative survey production process across the cultures or countries involved. Some features are relatively simple to document—PDF files of questionnaires in each target language, for example—but even these are not always available, in part for proprietary reasons. Study documentation on contextual and background information is not always linked to the relevant codebook, leaving secondary analysts to their own devices to find the information pertinent for analysis.

Locating even such basic information as gender of the interviewer, number of contacts, region involved, and any country-specific codes can be time-consuming, costly, and a haphazard undertaking (cf. van Deth, 2003).

Enhanced Comparative Documentation. Web-based documentation provides much enhanced navigation options. For example, the U.S. cross-cultural CPES (http://www.icpsr.umich.edu/CPES/) provides access to interactive documentation of three harmonized mental health survey datasets. Users can browse variables within and across surveys, by questionnaire section, major diagnosis, and language, and can access publications based on primary and secondary analysis. The longitudinal U.S. Health and Retirement Study (HRS; http://hrsonline.isr. umich.edu/) takes another route; users can follow an interactive "data collection path" across years to select specific instruments (see Figure 16.9) or go to a Web page with links to the data products and documentation for each instrument.

16.5 OUTLOOK AND CONCLUSIONS

Proper survey documentation is indispensable for producers and users. It provides information needed to asses survey quality and to guide analyses. Standard docu-

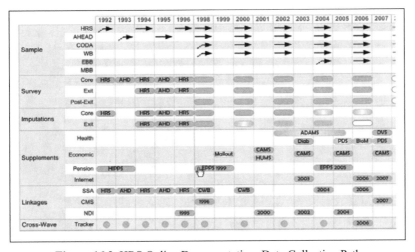

Figure 16.9. HRS Online Documentation: Data Collection Path

mentation such as provided in a codebook does not provide navigable, comprehensive, transparent, and accurate information on metadata and paradata for 3M surveys. This can only be achieved using complex database systems such as presented in the chapter. DDI 3.0 offers a structure for standardized documentation of the survey life cycle, but this in turn requires documentation tools along the lines of the Michigan Questionnaire Design System. For comprehensive 3M documentation to prosper, such tools need to become widely available and affordable.

To date, 3M survey research is still a long way from producing standardized well-structured metadata captured at the time of occurrence at a given step of the survey production process. However, as suggested here, the examples set by a number of multinational surveys show the way forward and initiatives such as DDI will doubtless help pave the way.

17

Harmonizing Survey Data

Peter Granda, Christof Wolf, and Reto Hadorn

17.1 INTRODUCTION

The ability to use population surveys to study social phenomena across time and across countries is a great achievement of the social sciences. To enable valid conclusions from this type of research, data from the respective surveys must be comparable. To ensure comparability, it is often necessary to harmonize data.

Harmonization is a generic term for procedures aimed at achieving or at least improving the comparability of different surveys. These procedures may be connected to any part of the survey life cycle, e.g., choice of indicators, question wording, questionnaire designs, sampling, field work, data coding, or data editing. The need to harmonize survey data always arises when data from different surveys are to be compared—be they surveys across time or surveys from different countries. This is particularly true if the goal is, as we assume here, to combine the data into a single integrated dataset.

Over the last decades in particular, research interests and the needs for information by governments and administrative bodies have focused increasingly on international comparisons. The need for harmonized cross-national data has thus grown apace, as have efforts to supply harmonized data.

In the field of social sciences, regular programs of research such as the International Social Survey Programme (ISSP) and the European Social Survey (ESS) have increased interest and potential concern about the comparability of data. In the area of official statistics, the European experience of integrating numerous nation states into a supranational structure has raised the issue of comparability of statistics to a question of paramount importance. While Eurostat, the statistical office of the European Communities (the legal name of the European Union; EU), does not itself collect data; it does integrate micro- and macrodata from the EU member states. Eurostat has developed different strategies of harmonization

[1] Survey Methods in Multinational, Multiregional, and Multicultural Contexts, edited by Harkness et al.
Copyright © 2010 John Wiley & Sons, Inc.

specifically with a view toward improving data quality, especially with regard to their comparability. A number of these are presented below.

The chapter is organized as follows: in Section 17.2, we distinguish three approaches to harmonizing survey data. In Section 17.3, we discuss methods to assess the quality of harmonized data which also lead to specific requirements for the documentation and distribution of harmonized data. In Section 17.4, we discuss these requirements in detail. In the concluding section of the chapter, we outline recommendations for producers of harmonized datasets.

17.2 APPROACHES TO HARMONIZATION

A prerequisite for using data for comparative purposes is that the data are indeed comparable. This may seem to be a truism, but in actuality, assessing and ensuring comparability is far from easy. Before we describe different approaches to arrive at harmonized data, we present examples of two situations in which data harmonization is necessary.

Example 17.1 Comparisons across Countries/Cultures

Assume we want to investigate job satisfaction in Britain, France, and Germany. We, therefore, turn to the ISSP because the Programme fielded a survey on "work orientations" in the year 2005. The Programme follows an "ask-the-same-question" design format (see Chapter 3, this volume). This survey asks about job satisfaction. This question was devised following ISSP procedures by a multicultural drafting group and ultimately accepted by the ISSP General Assembly. ISSP developmental procedures aim to ensure that questions can be meaningfully asked in all countries in whatever language is used for a population (translations). However, the example reveals (for the polyglot) slight variations among the three versions:

Britain	France	Germany
How satisfied are you in your main job?	Etes-vous satisfait ou insatisfait de votre emploi principal? *(Entourer seulement un chiffre)*	Wie zufrieden sind Sie im allgemeinen in Ihrem Beruf? *Nur EINE Markierung möglich!*
1 Completely satisfied	1 complètement satisfait	1 Völlig zufrieden
2 Very satisfied	2 très satisfait	2 Sehr zufrieden
3 Fairly satisfied	3 assez satisfait	3 Ziemlich zufrieden
4 Neither satisfied nor dissatisfied	4 ni satisfait, ni insatisfait	4 Weder zufrieden noch unzufrieden
5 Fairly dissatisfied	5 assez insatisfait	5 Ziemlich unzufrieden
6 Very dissatisfied	6 très insatisfait	6 Sehr unzufrieden
7 Completely dissatisfied	7 complètement insatisfait	7 Völlig unzufrieden
8 Can't choose	8 ne peut pas dire	*8 Kann ich nicht sagen*

1. The British and the Germans ask "How satisfied are you …," while the French ask "Are you satisfied or dissatisfied …."
2. The British and French ask about satisfaction with "your main job," while the Germans ask about satisfaction "in general with your job/profession."

The extent to which these differences affect comparability of the data is not easy to determine. Van Deth (1998b) suggests one approach: to investigate whether the items in question have comparable covariance structures with other items. This, however, assumes that these "other" items are (more) comparable across datasets than the questions under observation.

Example 17.2 Using International Standard Classifications

Assume we want to investigate educational achievement in Europe. One option might be to analyze the highest educational level attained by respondents in the European Social Survey (ESS).

Because education depends heavily on country-specific institutional arrangements, the ESS asks participating countries to collect this information according to national practices and then to recode these data into (a simplified version of) the International Standard Classification of Education (ISCED-97; see UNESCO, 2003). By applying an international standard for the assessment of education, the ESS team hoped to guarantee comparability of the resulting measures. However, standards such as ISCED have to be applied with care and on the basis of expert knowledge. As Schneider (2007) shows, many countries in the ESS did not strictly follow UNESCO coding rules or the recommendations given by the Organisation for Economic Co-operation and Development (OECD) (1999). The resulting data produced ISCED scores that cannot be compared.

These two examples show that obstacles in the quest for harmonized data arise for different reasons, in different circumstances, and at different stages of the survey life cycle. Two important distinctions can be made between three different approaches to harmonization (cf. Figure 17.1).

The first refers to whether we want to compare on the basis of existing, but not yet comparative, data or whether we are designing a study to collect data for comparative purposes.

If we aim to compare on the basis of existing data not originally collected for comparative purposes, the data must be carefully examined to find similar or identical questions and to identify possibilities to construct comparable indicators. This approach is called *ex-post harmonization.*

If the data have not yet been collected we have two further options. First, we can design a comparable study with its different surveys as if it were one study with a single survey, i.e., one questionnaire, one set of rules for data coding and data editing, etc. This approach is referred to as *input harmonization.* Alternatively, we could agree on a set of predefined target variables and their categories but leave the decision on how these data are collected to the different substudies. This last approach is called *output harmonization.* We will discuss these three approaches in more detail in the sections that follow.

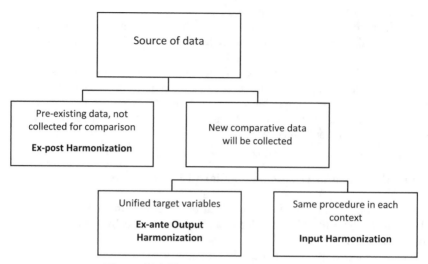

Figure 17.1. Three Approaches to Harmonization

17.2.1 Strategies to Harmonize Existing Data: Ex-post Harmonization

We first turn our attention to the harmonization of existing data that were not collected for comparative purposes. Researchers are often interested in the analysis of social change. Have the spending priorities of the U.S. public changed over the last 30 years? How does the social-structural basis of party identification change? Can the declining religiosity in Western Europe be attributed to a cohort or period effect? Does post-materialism decline with growing unemployment rates and rising inflation? Questions like these can only be answered on the basis of comparable data that cover relatively long time periods. However, because such data were typically not collected in order to study social change, they have to be harmonized *ex-post*, after the event.

After identifying the datasets containing the data to be harmonized, the first step is to define the variables of interest. Then a databank is constructed containing these target variables as cases and the relevant variables from the selected surveys as their attributes. The final step provides mapping routines, recoding the original variables to the predefined target variable. Table 17.1 provides an example. For this example, we assume we want to create a harmonized variable reflecting the religious affiliation of individuals, distinguishing between "Catholic," "Protestant," "Other," and "None." We look through the surveys to be harmonized (in the example here there are only two) and list the relevant variables together with the target variable. Finally, we recode the original variables into the target variable "RELIG."

A procedure like this was used in a project investigating whether, as hypothesized by individualization theory, voting behavior had individualized over time (Kohler, 1995; Schnell & Kohler, 1995). To test whether this was the case, Kohler

TABLE 17.1. Target Variables and Corresponding Variables from Surveys to Be Harmonized

Target variable: RELIG	Survey 1, variable: V32	Survey 2, variable: V433
Religion	Are you …	What is your religion?
1 Catholic	1 Catholic	2 Catholic
2 Protestant	2 Protestant	1 Protestant
3 Other	3 Muslim 4 Jew 5 Other	3 Other Christian 4 Other Non-Christian
4 None	6 I do not belong to a religion	5 None
9 Missing	9 Missing	

selected 37 surveys conducted between 1953 and 1992 from the holdings of the German Social Science Infrastructure Services (GESIS) Archive in Cologne, Germany. From these surveys, he harmonized 15 variables that played an important role in his theoretical model. On the basis of these data, Kohler was able to show that the impact of structural conditions on voting behavior has indeed diminished considerably over the 40-year period studied. Kohler's work has inspired a number of further projects investigating social change on the basis of *ex-post* harmonized data collections.

If the goal of a harmonization effort is to provide a general resource to the scientific community to permit observation of social change, then a different approach to coding harmonized variables is preferable. This approach, developed at the University of Minnesota, is usually referred to as *hierarchical coding* (see Section 17.4.2 below). In this approach, the harmonized variable is coded into multidigit codes where the first digit signifies the most general category (e.g., Protestant), the second digit a more specific subtype of the first digit (e.g., Lutheran), and so forth. At the end of the procedure, all the harmonized datasets should be comparable at the first digit, while only a subset will be comparable at more detailed levels. The advantage of this approach is that it largely preserves the original information and is, thus, less restrictive with respect to the research questions that can be addressed using these harmonized data.

A valuable tool for this kind of harmonization would be a databank containing all the survey questions for which data exist in the public domain. Such a resource would greatly simplify a search for identical or similar questions and facilitate the creation of harmonized trend data. The Question Bank, an academic project initiated by the department of sociology at the University of Surrey in the UK, is one such databank (http://qb.soc.surrey.ac.uk/). Commercial question databanks are also available, such as Roper's iPOLL (http://www.ropercenter.uconn.edu/data_access/ipoll/ipoll.html/). As described in Section 17.4, question banks of this kind could well be a byproduct of comprehensive data management tools of the future.

17.2.2 Strategies to Collect Harmonized Data: Input and Ex-ante Output Harmonization

The situation is completely different when we plan a longitudinal or cross-national comparative survey. In these instances, two strategies are available to assure that the resulting data are comparable. We can decide to collect the data using the same procedures [questionnaire, data collection mode(s), sampling frames, fielding instructions, etc.] in each context. Alternatively, we can define target variables that each participating survey must attain, while the actual questions used for data collection might vary between surveys. These two approaches are called *input harmonization* and *ex-ante output harmonization*, respectively.

Comparative surveys usually apply a mix of these approaches (Scholz, 2005). For instance, the ISSP follows a strict input harmonization approach with respect to the thematic modules (see Example 17.1). This means the participating countries agree on one common source questionnaire—written in English—defining the exact question order, question wording, response categories, filtering rules, and interviewer instructions. As Example 17.1 shows, minor deviations of the national questionnaires from the source questionnaire may, nonetheless, occur. In some cases, source questionnaires deliberately leave room for interpretation/adaptation at the national level. In the 2008 ISSP module on Religion, for example, the source questionnaire contains the following question: "For religious reasons, do you have in your home a shrine, altar, or a religious object on display such as a (COUNTRY-SPECIFIC LIST icon, retablos, mezuzah, menorah, or crucifix)?"

Because religions differ widely with respect to their objects of worship, the ISSP survey designers decided it was impossible to provide one meaningful list of such objects that would yield equally valid data across the world. Instead, it was recommended that each country provide its own list, as indicated in the instruction "COUNTRY-SPECIFIC LIST" in parentheses in the source question.

With respect to what in the ISSP are called "background variables," that is, demographic and socioeconomic data, the ISSP originally used an *ex-post* harmonization approach as described in the previous section. However, as ISSP membership grew, this created enormous problems, eventually leading to a change in ISSP policy regarding background variables. Since 2001, the ISSP has adopted an *ex-ante* output harmonization approach. Thus, today each participating country is required to deliver their data on background variables in accordance with a predefined list of target variables (Braun & Uher, 2003; Scholz, 2005). Education, for instance, has to be coded according to ISCED-97. As Example 17.2 shows, the agreement on a given standard classification (or some other target variable) does not automatically guarantee comparable data. Those coding the data according to the standard classification have to be trained on how to do this. Moreover, both data collection and data coding have to be carefully monitored and documented.

Learning from the ISSP experience and from other comparative survey programs, the European Social Survey planners opted for strict input harmonization; some variables, such as education, are *ex-ante* output harmonized, however (see Example 17.2). Although this might seem to ensure the highest level of comparability, a number of pitfalls have been identified in the meantime. One problem relates to the construction of the source questionnaire. If it is not jointly developed with equal input from experts representing all the countries in which the survey is conducted, questions may be culturally biased. The ESS instrument for measuring income provides an example. Warner and Hoffmeyer-Zlotnik (2005) show that the income questions and categories used in the ESS work well for the UK but are much less appropriate for Luxembourg and Portugal. One may speculate that this, in part, reflects the dominance of a UK frame-of-reference among the ESS team.

Following institutionalized rules for collaboration and appropriate methods of questionnaire development and translation helps to minimize cultural biases in the development of questionnaires for cross-national research (Harkness, 2003; Harkness, van de Vijver, & Johnson, 2003). Thus, if applied properly, input harmonization as the *main* mode of designing harmonized surveys is the method of choice for cross-cultural studies. But input harmonization will not be possible in all cases. For example, in the context of cross-cultural surveys, as indicated already, education has to be measured in a country-specific way and, thus, can only be *ex-ante* output-harmonized. In such cases, it is critical to provide the original country-specific data and to document the mapping procedure by which these measures are converted into the harmonized measure (see also Section 17.3).

17.2.3 The Role of International Standard Classifications

International (statistical) standard classifications are usually developed or propagated by international organizations such as the United Nations (UN), the International Labour Organization (ILO), or the Organisation for Economic Co-operation and Development (OECD). Examples are the International Classification of Status in Employment (ICSE) (Hoffmann, 2003), the International Standard Classification of Occupations (ISCO) (ILO, 1990), and the International Standard Classification of Education (ISCED) (UNESCO, 2003). Of course, standard classifications are compromises between different national approaches and differing substantive interests. Additionally, their application often is difficult and those having to code or work with the data have to be specifically trained. Particularly, coding, that is, the assignment of code numbers to survey responses, can be a serious source of error. If coding is done manually, coders have to be trained intensively. If coding is done automatically, the coding procedures have to be evaluated and constantly improved. In both cases, extensive quality controls should be applied (for more detail, see Biemer & Lyberg, 2003).

Nonetheless, international standard classifications are of great value because they provide unambiguous definitions plus instructions for coding country-specific data into the international standard classification. Thus, at least in principle, researchers conducting a comparative survey only have to agree that they want to measure occupation according to ISCO-88 and do not have to discuss the detailed questions that yield the information necessary to code occupational data into this standard.

Recently, Eurostat's Task Force on Core Social Variables (Eurostat, 2007) proposed to take the development of international standard classifications one step further by presenting a list of 16 "core variables" that should be included in all relevant statistics of the European Union, such as the Labour Force Survey and the

Survey on Income and Living Conditions. This set of core variables contains demographic, regional, and socioeconomic characteristics, including occupation measured according to ISCO-88 (COM), the ISCO variant for Europe, and education using the ISCED-97 schema. Should this set of core variables be adopted by the European Union and its member states and used in relevant official surveys, it would boost adoption into other surveys as well, ultimately leading to easier cross-national and cross-survey comparison of the respective variables.

Typically, the standards cited so far are presented in an output-harmonized format, meaning that definitions and descriptions for variables and categories are provided but with no questions, interviewer instructions, and the like. This is different for standardized measures usually found in psychological and health research. One example of such a standard is the Short-Form 36 Health Survey or SF-36. This instrument, developed by Ware, Kosinski, and Keller (1994), consists of 36 items capturing different aspects of respondents' health. The instrument has been translated into more than 50 languages and extensive research was conducted to assure cross-country comparability of results. The "International Quality of Life Assessment" working group has carefully reviewed the difficulties that arose in translating the SF-36. In their view, comparability of translations was greatly improved by explicating the meaning of the original American English item wording and the provision of American English synonyms (Wagner, Gandek, Aaronson, Acquadro, Alonso et al., 1998).

In as far as standards are accepted and widely used, they play an important role in collecting input or *ex-ante* output harmonized data and thereby facilitate the collection of comparative data.

17.3 ASSESSING THE QUALITY OF HARMONIZED DATA

Statistical data in general, and harmonized data in particular, can be assessed against a variety of quality criteria. Among these, consistency, completeness, and comparability seem to be the most relevant in the current context (see Ehling, 2003, for the criteria proposed by Eurostat). The consistency of a harmonized variable can be assessed by comparing the results from multiple independent efforts of harmonizing this variable.

The degree of completeness—here in the sense of the degree to which the original information is preserved in the harmonized data—can be assessed numerically by comparing the variability of original and harmonized variables. Assume we have k surveys, each including a measure for some attribute X. Furthermore, let us assume these measures are not identical. They may, for instance, have a differing number of categories and/or partly differing definitions of categories. If X is measured on an interval or ratio scale, there is no problem; the original values can be retained. If necessary, a conversion of units has to take place, e.g., from dollar to euro, but this will not affect the amount of differentiation captured in the measures. Bigger problems are posed by ordinal or nominal variables. If the definition of such variables differs between surveys, a harmonized measure will almost always be less differentiated than the original variables. In these cases, it would be desirable to know how much is lost through harmonization. For this purpose the following general measure is proposed:

$$Q_{X^{h_i}} = \frac{disp^{h_i}}{disp^{o_i}}$$

The quality Q of a harmonized measure X^h is equated with the amount of dispersion that X^h has in dataset i $(disp^{h_i})$, relative to the dispersion that the original measure of X has in dataset i $(disp^{o_i})$. For $disp$ any suitable dispersion measure may be inserted. Thus, for ordinal data, one may use the concentration measure d^2 proposed by Blair and Lacy (2000):

$$d^2 = \sum_{j=1}^{k-1} \left(cp_{ij} - \frac{1}{2} \right)^2 \qquad \text{with:} \qquad 0 \le d^2 \le (k-1)/4$$

cp_{ij}:   cumulated relative frequency of X for the j-th category.

For nominal variables, dispersion can be assessed by the index of diversity D described by Agresti and Agresti (1977):

$$D = 1 - \sum_{i=1}^{k} p_i^2 \qquad \text{with:} \qquad 0 \le D \le (k-1)/k$$

k: number of categories of X
p_i: proportion of observations in category x_i

If metric variables are modified in the course of harmonization, the same logic can be applied by defining $disp^h$ and $disp^o$ as the sum of squares of harmonized and original variable, respectively.

An example of this approach is given in Table 17.1. Here we have listed the variables from two different surveys asking for religious affiliation. Both surveys use slightly different categories of which only two are identical (Catholic, None). With diversity indexes of 0.70 and 0.71, respectively, both surveys show a similar level of religious diversity. In order to integrate the two variables into a single dataset, they have to be harmonized. One way to do this is presented in Table 17.2. The harmonized measure distinguishes between the categories Catholic, Protestants, Others, and None. It captures 81% of the religious diversity of sample 1 and 76 % of religious diversity of sample 2 (see values for Quality Index in Table 17.1). Thus, being able to jointly analyze the data by using this harmonized measure comes at a cost, namely losing approximately 20%–25% of the original dispersion.

An alternative way to assess completeness of a harmonized measure is to analyze it *relative to a criterion variable*, that is, to study its predictive validity or power. In this case, the quality of a harmonized variable, X^{h_i}, of a measure, X, can be assessed by regressing a criterion variable, Y, on the harmonized and the original variable and comparing the ratio of the explained variance (for non-metric

TABLE 17.2. Quality Assessment of Harmonized Nominal Variable, Frequencies in Percent

	Original Variables		Harmonized Variables		Target Variable
	Survey 1	Survey 2	Survey 1	Survey 2	
Catholic	42	34	42	34	Catholic
Mainline Protestant	33	—			
Protestant	—	36	37	36	Protestant
Other Protestant	4	—			
Other Christian	5	—			
Muslim	—	3	10	9	Other
Other Non-Christian	5	—			
Other	—	6			
None	11	21	11	21	None
Total	100	100	100	100	Total
Diversity Index D	0.70	0.71	0.57	0.54	
Quality Index Q			0.81	0.76	

— category not used in survey.

variables X^h and X^o can be split into an appropriate number of dummy variables). Thus the quality measure is:

$$Q_{(X^{h_i}|Y)} = \frac{\Sigma(\hat{Y}^{h_i} - \bar{Y})^2}{\Sigma(\hat{Y}^{o_i} - \bar{Y})^2}$$

where $\hat{Y}^{h_i}$ is the predicted value of Y in dataset i using the harmonized measure of X^{h_i}, i.e., $\hat{Y}^{h_i} = a + bX^{h_i}$ and $\hat{Y}^{o_i}$ is the predicted value of Y using the original variable X^{o_i}, i.e., $\hat{Y}^{o_i} = a + bX^{o_i}$. Again this coefficient reaches its maximum value of one when the original variability of X relative to Y is retained by the harmonized measure. As the indices of Q imply, this approach assesses the quality of a harmonized variable, X^{h_i}, *relative* to a criterion variable, Y, in a given dataset i. Thus, this method assumes that the criterion variable Y is comparative. Using this approach, Schneider (2008) assesses the quality of different measures of education in the European Social Survey. She examines the following question:

> What is the effect of reclassifying country-specific educational attainment variables into the levels-only ISCED–97 and ES–ISCED [a variant proposed by Schneider for European surveys], and how do years of education compare with the country-specific variables on the one hand and the two comparable categorical variables on the other in terms of predictive power [in models with socioeconomic status as dependent variable] (p. 18; material in square brackets added).

Put another way, Schneider asks how much explanatory power one loses when using different harmonized measures for education in a status attainment model, that is, linking education to socioeconomic status.

Finally, in the special case of multiple item measurement, the degree of comparability or equivalence across surveys may be assessed by comparing their covariance structure (for a good example, see van Deth, 1998b). Expanding this approach, Davidov (2008) demonstrates how multigroup confirmatory factor analysis can be used to test for "configural invariance" (i.e., same pattern of factor loadings across countries), "metric invariance" (i.e., same factor loadings across countries), and "scalar invariance" (i.e., identical intercepts of items when regressed on a common factor). As Davidov (2008) points out:

> … in order to conduct a comparison of factor means across countries and over time and interpret it meaningfully, three levels of invariance are required, configural, metric and scalar. Only if all three types of invariance are supported can we confidently carry out mean comparisons (p. 37).

What can be said about the advantages and disadvantages of these methods? If the data are harmonized to serve multiple research needs, not all possible uses of the data can be foreseen. In this situation, it is not clear against which criterion predictive validity of harmonized measures should be evaluated. Instead, the proportion of variability of the original measure preserved in the harmonized measure is a suitable quality indicator. If, however, a particular phenomenon is studied, assessing and possibly optimizing predictive validity of a given harmonized measure relative to the variable of interest could be useful. In the case of multiple measures, the approach described by Davidov (2008) will result in the strictest test of measurement equivalence. However, social science surveys mostly contain single item measurements.

The proportion of dispersion retained by a harmonized variable or the degree to which it has predictive power is important but by no means the only quality criteria against which a harmonization effort should be judged. Equally important are a close reading of the original questions and answer categories, an inspection of the placement of the question in the context of the respective questionnaire, and the acknowledgment of interview modes, sampling frames, field procedures, nonresponse rates, etc. To be able to attain all this information, it is essential that the harmonized data are carefully and comprehensively documented.

17.4 ENCOURAGING SECONDARY ANALYSIS: DOCUMENTATION AND DISSEMINATION

As cross-cultural and cross-national harmonization efforts continue to expand, it is increasingly important to provide researchers with all of the information they need to take advantage of the enormous effort and expense in collecting such data. After all, the greatest value of any harmonized dataset may be as a long-term resource: one used by policymakers to study specific social issues, by social scientists to advance knowledge in their fields of interest, and by students in their coursework. All must have access to these resources and need to know how to use them.

While the tasks involved in preparing and disseminating public-use files are ubiquitous to all survey research, harmonized datasets present data producers with some unique issues to consider.

17.4.1 Documenting the Harmonization Process: A Call for New Software Tools

Complete documentation of the harmonization process presumes that all necessary information is available to do the job. But, if the documentation is created only at the end of the process, decisions made at the beginning will be forgotten or incompletely reported. Documentation must start ideally at the beginning of the project, since multiple decisions occur throughout the data life cycle. This is ideal, but not easy to follow in practice when all efforts are often concentrated on data collection (Mohler & Uher, 2003).

Increasingly, producers of harmonized files advocate the inclusion of specific materials and the use of recognized standards when providing documentation to end users [Krejci, 2008; Cross-Cultural Survey Guidelines (CCSG), 2008]. The United Nations Economic Commission for Europe's (UNECE) Population Activities Unit recommends creating detailed data description tables for each harmonized variable in their Generations and Gender Surveys, as well as a data availability report which will inform users when variables could not be fully harmonized (UNECE Population Activities Unit, 2008).

Since documentation is generated throughout the data life cycle, one of the main problems facing coordinators, producers, and editors is the lack of integration of the tools in use. Descriptions of original concepts and questionnaires are often developed using text editors, modifications are communicated over e-mail, and subsequent translations generate a wide collection of text files often without any relationship to the original materials. Data collection programs use a wide range of software, data processing relies on multiple programs as well, and at the end of the process comes data dissemination and display software, which should present the user with all useful information. At all stages, information is lost, in part because the involved actors do not recognize the importance of the information, in part also because it would be much too cumbersome to re-collect and re-capture all that information for further dissemination.

What we propose, therefore, is a standardized process to alleviate these problems, whereby a software tool would support both the harmonization and documentation processes. To attain this goal, two challenges have to be met: the elaboration of a general model, which would collect documentation on all activities involved in the harmonization process, and the development of applications based on this model, which would meet the requirements and the organizational complexities of cross-national survey studies.

All of the documentation which describes the harmonization process, from initial concept to a full description of each harmonized variable, falls under the term "metadata" (National Information Standards Organization, 2004). Used extensively by the library and electronic information communities, metadata ("data about data") have a critical function in the social science survey world. Metadata comprise not only traditional catalog records to discover a particular dataset, but also encompass all of the information needed to understand the history, creation, and content of a dataset so that users can effectively use the resource. Formerly, social science metadata were available in static documents ("codebooks"), but now the same content can be categorized and searched on the Web. Such searches locate individual datasets and can also facilitate linkages between datasets in ways never before possible.

How can producers enrich the amount and quality of the metadata they collect and provide to users? Harmonization of data from various sources or from various datasets collected under a similar data schema is currently undertaken using statistical packages like the Statistical Package for the Social Sciences (SPSS) and the Statistical Analysis Software (SAS). If appropriately used, these data processing and analysis programs are capable of some—very limited—self-documentation. However, the processing information stored there is likely to be scattered across many files that are usually not retained. There are also some dedicated tools such as the ISSP Data Wizard. Developed by the German Social Science Infrastructure Services (GESIS), the Data Wizard supports harmonization work by comparing individual country-specific files with the expected standard of the merged, harmonized file. It provides a robust user interface that permits more efficient data processing during the harmonization process (Strotgen & Uher, 2000). The integrated nature of this tool is a significant feature. At the same time, the metadata capture process starts only once the data are already available and is focused on variable definitions, so a lot of the remaining metadata have to be handled by other tools. Since the ISSP is based on *ex ante* output harmonization, the processes programmed in the tool are limited to the requirements of that approach. These are nonetheless good first steps in organizing all of these diverse materials.

What should come next? We envision a computer-assisted harmonization tool (CAHT) that automates all harmonization work except for the steps that call for human decisions and interventions. The tool would also document those human actions, including explanations for decisions, assumptions surrounding the decisions, rejection of alternative solutions, and so forth. The following characteristics are key to a successful harmonization tool:

- Metadata must be captured throughout the whole process beginning when the data producer develops a plan to harmonize data and ending when the file product becomes available to researchers.
- Metadata must be stored in a structure that allows continued use by future data producers.
- Metadata must cover the complexities of the organization of survey programs.
- The harmonization tool should be just one component within a larger system supporting the whole of the survey process.

How would all of this work? The storage structure for that huge volume of information is obviously a database or a set of coordinated databases. Access to this tool must be available throughout the planning and construction phases of harmonization projects. All staff involved in the project (planners, producers, translators, data processors, archives, and publishers) should work closely with this tool to input, edit, and output content. Each user must be able to generate the metadata needed at any stage of the project. For example, producers can use the tool to review decisions they made about harmonization, particularly at the variable level, while data archives would be able to extract general descriptions of the entire process to document public-use files.

It is clear that such an ambitious goal is not attainable in one single development step, and, thus, we must define a development strategy. Just as the standardization of procedures to prepare and field comparative surveys is an area that has long dominated survey research, equal time must also be given to creating a tool for the end user to comprehend the entire process. What are the mechanisms for such a system? This is a topic the comparative research community must address.

17.4.2 Documenting the Harmonization Process: The Final Result

Data producers should provide users with documentation for the two key areas of the harmonization process: the variable level and the file level. At the variable level, they can explain similarities and differences in how individual variables change in the harmonization process and include discussion of how they handled variable universe, question wording, coding schemes, and missing data definitions in the material provided to secondary analysts.

At the file level, they should describe the steps they took to merge a number of individual, often national-level, data files into a single output file. In this case, producers must document the decisions they made regarding weights, imputation procedures, variance estimation, and differences in key substantive and demographic concepts in the original files. Let us look at a few examples of how certain projects have handled these issues.

To preserve as much information as possible from original sources, the University of Minnesota has developed a multilevel system when harmonizing the public-use microdata samples produced over the decades by the United States Census Bureau and the Health Interview Surveys conducted by the U.S. National Center for Health Statistics. This system uses multidigit coding employing both a *general* code, which captures information in all sampled surveys, and a *detail* code, which follows the general code and preserves any information that is available only in some of the sampled surveys. For example, adopted children were recorded in the U.S. Census enumerations for 1920 but not for 1940. The variable that summarizes and harmonizes the relationship of each individual to the head of the household includes a general code of "03" to designate "child" in all Census years but also a detail code of "02" to designate "adopted child" for those years when the Census Bureau collected that information. Thus, the coding is "0301" for "child" and "0302" for "adopted child" in the harmonized data file. This allows users to analyze children both as a single category and adopted children as a subcategory (see a general discussion of the Integrated Public Use Microdata Series at http://usa.ipums.org/usa/intro.shtml).

The Eurobarometers, conducted since the early 1970s, present a case study of file harmonization effort and the complexity of these surveys over time. Using weighting as an example, the early surveys included 9 countries and contained a single weight variable that produced a representative sample of the entire European Community at that time. More recent surveys sample some 30 countries and include multiple national and European weights as well as those for the Euro zone, countries which recently joined the European Union, and those that are candidates to do so. The creation of these weights permits users to analyze these surveys from many different perspectives. The public-use documentation includes descriptions of how these weights were generated. All producers must provide similar resources for users in describing how their harmonized files differ from the original source inputs. On the other hand, the World Values Survey, while providing a wide array of documentation and data access tools on its project Web site, could include more information on the weights used in its harmonized data files (see http://www.worldvaluessurvey.org/).

As we have described, the entire harmonization process involves a steady stream of decisions throughout the entire survey life cycle. It imposes greater responsibilities on those who prepare documentation since their goal should be to disseminate the harmonized data files as soon as is practical. It is too often the case that secondary analysts see only the final fruits of harmonization efforts that may have taken many months to create. To conduct informed research on harmonized data, analysts should have the opportunity to work with the original materials, test and assess the harmonization strategies employed, and make alternative decisions if they deem them appropriate for their research projects.

What then should producers provide to users at the end of the harmonization process? They should provide the widest range of data and documentation products available. Users should have access not only to the harmonized end result but also to detailed information about all steps taken by the producers so that they can fully understand what decisions were made during the entire process. The following are key elements that producers should consider when they disseminate the results of harmonization efforts (for other lists of such elements see van Deth, 2003, a "schematic overview of process-oriented documentation" in Mohler & Uher, 2003, and, on cross-cultural survey guidelines, CCSG, 2008):

- Document each target variable with information from all source variables, transformation algorithms, and deviations from the intended harmonized approach if known.
- Provide users with the original data files used in producing the harmonized file.
- Supply the code or syntax used in creating new variables for the harmonized file.
- Create a complete set of documentation including crosswalks that describe all the relationships between variables in individual data files in connection with their counterparts in the harmonized file.

- Include original questionnaires and information about the data collection process whenever possible.
- Produce a User's Guide that describes all of the idiosyncrasies and special characteristics of harmonized files and their component parts.
- Summarize all data processing steps in a series of processing notes.
- Distribute full specifications for any recoded variables.

Much of the metadata which describe these elements can be captured within the emerging international standard for social science documentation: the Data Documentation Initiative (DDI). Use of the DDI provides a well-defined structure for all of these metadata because it is encoded in Extensible Markup Language (XML), stored as plain text for easy exchange between organizations, and encourages reuse by others (Hoyle & Wackerow, 2008).

17.4.3 Distribution of Harmonized Datasets

After all of the necessary materials are produced, what is the best way to disseminate them to the research community? While this may seem a simple question, the choice can affect how easy the data are to use and how long they are available to secondary analysts. The Web has presented a host of new opportunities for the dissemination of public-use files that appear to make things easier but also pose significant long-term questions.

Data access via the Web permits data producers and archives to update and correct files whenever the need arises. If, however, they do not establish a set criterion for distinguishing between different versions of their data files, secondary analysts may be unaware that the file they are using may have been modified since they downloaded it. Similarly, someone who wants to replicate an analysis must be sure that the data file they use is the same one as that used for the original analysis.

Putting files on the Web does not guarantee their long-term availability or permanent preservation. Web sites must be constantly updated and enhanced and data producers must also take steps to make backup copies of all versions of their data and documentation files. They must also have a system in place that allows them to retrieve these files a week, a year, or 20 years later.

Data producers have two choices. They can disseminate the data themselves, most often through a Web site that they create. This is often the choice of large national or international projects such as the European Social Survey (ESS). These projects have the financial and methodological resources to provide the social science research community with a high level of service, including full access to all final data and documentation files, detailed information about all aspects of the data collection project, and user support for those who need assistance in using project materials. Alternatively, data producers can deposit their data in a social science data archive which can provide many of the services just mentioned, as well as a permanent preservation site for all of their materials. These archives normally update data and documentation files when necessary, and maintain a system of version control so that users know which particular files they have used.

They may also provide new tools for researchers such as the capability to read the data into new statistical software packages, perform online analyses through a special Web interface, and subset and download sections of both the data and documentation files. Of all of these factors, version control and long-term preservation may be the most important and are also those least discussed.

Because of the need to monitor dissemination processes long after the completion of data collection, it is not uncommon for producers of large harmonization projects to partner with data archives in the dissemination function. Both the ESS and ISSP collect data from many countries, use a common questionnaire, and employ the services of national social science data archives which play a key role in the creation and distribution of the public-use data and documentation files. The Norwegian Social Science Data Services (NSD) in Bergen performs the final ESS processing work, while archivists and data managers at GESIS–Data Archive and Data Analysis in Cologne, Germany, perform similar tasks for the ISSP.

Partnering with a data archive is often a good choice since harmonized files may require more complex dissemination tools and a more robust technological infrastructure than data producers can support on their own. For example, users should be able to see not only a harmonized variable but also how questions were asked in different countries in their original languages. Related questions should be easy to find and special notes about the construction of complex variables should be readily available. Making all of this information available to the researcher often requires complex metadata storage systems, sophisticated Web programming, and customized interfaces that only a larger organization can provide.

Whatever the mode of dissemination chosen, producers and archivers must anticipate the need to modify and update harmonized datasets after public release based on comments from the research community. They should also consider facilitating the proper use of their new harmonized data products through presentations at social science research conferences in which they describe their objectives and provide practical examples and practice exercises for interested researchers. Activities like this widen the user base for important datasets.

The costs of creating all of these metadata for complex, harmonized datasets can be considerable. But data producers must realize that the utility of the dataset depends almost entirely on the level and quality of the information that they provide to the end user. Obtaining advice from data archives, using standard tools, and capturing information routinely during the entire data collection and processing stages will keep these costs to a minimum and produce a dataset that will provide researchers indefinitely with a valuable resource.

17.5 CONCLUSION AND RECOMMENDATIONS

Data harmonization is an integral part of any multisample data collection and analysis project (time series or cross-national survey). In this chapter we distinguished between the harmonization of existing data (*ex-post* harmonization) and planned collection of harmonized data either by input harmonization or by *ex-ante* output harmonization. In this connection we also highlighted the importance of international standard classifications. All approaches can be jointly used within the same project, although the weight given to each approach depends on researcher experience within a project and the availability of standards and resources.

Whatever the solution selected, comprehensive and structured documentation of all steps in the survey life cycle and all decisions made in the process of harmonizing data is essential. Dissemination of harmonized datasets has to include not only the harmonized data but also the original data and well structured documentation (metadata). We urge producers to follow a standardized procedure to create their harmonized data which will enable users to validate the work done in the harmonization process. In short they should:

- Make each harmonization decision explicit as Eurostat has done, for example, with the creation of the Consumer Price Index (Makaronidis, 2008).
- Document the survey and harmonization decisions comprehensively according to an established standard, e.g., DDI.
- Always provide original data and structured metadata alongside the harmonized data.

These requirements cannot be met fully with the tools available today because data and documentation are scattered across too many document types, formats, and owners. New tools are necessary to achieve better integration of data and documentation from the original and the harmonized files. They should be suitable for any kind of survey, provide containers for most information, and provide linkages between related pieces of information (e.g., original country questions, variables, and harmonized variables). To exploit their proper potential, new harmonization tools must be integrated into a framework that supports not only the harmonization process but the entire life cycle from data definition through data production to ultimate publication.

Steps in this direction have been undertaken by several organizations. The Data Documentation Initiative is working on a new version of its metadata model, covering the data life cycle with a special module for comparative data (see information at http://www.ddialliance.org/). Networked Social Science Tools and Resources (NESSTAR) has created an infrastructure for many social science data archives to process, disseminate, and provide online analysis of datasets. It continues to extend its data publisher to include earlier stages of data production (http://www.nesstar.com/). It is not necessary for all roads to lead to Rome; we will be happy if just one of them makes Rome a reality.

ACKNOWLEDGMENT

We thank Uwe Warner, Dominique Joye, and participants of workshops at which we presented preliminary versions of this chapter for their insightful comments and recommendations.

PART VI

NONRESPONSE

18

The Use of Contact Data in Understanding Cross-National Differences in Unit Nonresponse

Annelies G. Blom, Annette Jäckle, and Peter Lynn

18.1 INTRODUCTION

Cross-national studies of nonresponse are important for several reasons. Researchers want to know whether and why response rates differ between countries and whether and why nonresponse bias differs. In particular, they want to know to what extent and in what circumstances methodological findings regarding nonresponse in one country are likely to be generalizable to another country. The factors that might lead to differences between countries in either response rates or nonresponse bias can be broadly classified as (1) differences in survey implementation (e.g., the survey organization, skills of interviewers, times, days, and number of contact attempts); (2) differences in the population distribution of characteristics associated with propensity to respond (e.g., household size, economic activity status, urbanicity); and (3) differences in the association between factors in categories (1) or (2) and propensity to respond (e.g., making contact attempts on Sundays may be very helpful in one country but have no effect in another country).

This chapter reviews the use of auxiliary data in studies of cross-national variations in the above factors and the impacts upon survey nonresponse. We focus on contact data, also known as call-record data. Contact data contain information about the contact attempts made at each sample unit, such as the number, timing, and outcomes of calls. We review how contact data have been used to understand cross-national differences in nonresponse, and discuss ways in which the usefulness of contact data could be enhanced in future. We concentrate on face-to-face interview surveys but many of the issues we address have close analogues in the case of telephone surveys. Finally, we deal only with *unit* nonresponse (see Groves et al., 2004, Chapter 6).

---

[1] Survey Methods in Multinational, Multiregional, and Multicultural Contexts, edited by Harkness et al. Copyright © 2010 John Wiley & Sons, Inc.

We note that the comparative study of survey nonresponse is of central interest to researchers using cross-national survey data for substantive analysis; they need reassurance that their cross-national substantive comparisons are not biased by differential nonresponse error. It is also of interest to those responsible for carrying out cross-national surveys: for example, whether standardized field practices produce more comparable outcomes in terms of nonresponse bias than country-specific practices is an open question. But the comparative study of survey nonresponse is also of interest to other researchers; for example, those who want to know whether they should use methodological findings from other countries to guide survey design in their own country.

The extent of nonresponse bias in any given estimate depends on a combination of the nonresponse rate and differences between respondents and nonrespondents. An unadjusted design-based estimate based on respondents will be unbiased only if nonrespondents do not differ systematically from respondents in terms of the survey items that contribute to the estimate. This situation is referred to by Little and Rubin (2002) and others as *missing completely at random* (MCAR). If nonrespondents differ systematically from respondents (*not missing at random*—NMAR), then estimates will be biased unless these differences can be fully explained (in a statistical sense) by other available data items, in which case unbiased adjusted estimates can be produced (*missing at random*—MAR). That is, nonresponse bias can be removed from estimates if data items are available that correlate both with the survey estimate and with propensity to respond.

In comparative research key survey estimates are estimates of differences between countries. Unbiased comparisons require one of three conditions to hold: (a) nonresponse in each country is MCAR, (b) NMAR processes introduce equivalent bias to each national estimate, or (c) analysts must identify—and use appropriately—a set of additional items that turn the process from NMAR to MAR.

The first key concern for research on nonresponse in cross-national surveys is therefore measuring non-response bias across countries, in order to identify the implications for analysis. This requires information about both components of bias: nonresponse rates and differences between respondents and nonrespondents. Response rates vary vastly between countries. Even for harmonized surveys (surveys that implement procedures designed to achieve a considerable level of standardization of design and implementation between countries) such as the European Social Survey (ESS; see Chapter 26, this volume) response rates in round 1 ranged from 33% in Switzerland to 80% in Greece (Billiet et al., 2007a). Identifying differences between respondents and nonrespondents is a more difficult exercise, since it requires information about nonrespondents, about whom by definition usually not much, if anything, is known. For cross-national surveys, comparable information on nonrespondents is especially hard to obtain.

The second key concern is to identify ways of reducing (differential) nonresponse bias. This requires information about the survey participation process in each country, in order to identify potential fieldwork actions to reduce nonresponse and to identify correlates of nonresponse that can be used to turn NMAR processes into MAR processes. Note that different additional information may be needed for different countries, to statistically explain differences between respondents and nonrespondents.

In this chapter we review existing studies that have attempted to shed light on these questions, in particular highlighting the insights gained from micro-level contact data. As background, we first discuss expected sources of cross-national differences in survey participation processes (Section 18.2). We then classify potential sources of auxiliary data, focusing on contact data (Section 18.3). In Section 18.4 we review existing cross-national studies of nonresponse and summarize findings with respect to the key questions outlined above. We then discuss the main limitations of currently available contact data (Section 18.5) and suggest three quality criteria which would improve the usefulness of contact data for the study of cross-national nonresponse: equivalence of design of contact forms, of implementation, and of the coding of outcomes (Section 18.6).

18.2 THE SURVEY PARTICIPATION PROCESS

The survey participation process is typically complex, but usually involves three main stages: *location*, *contact*, and *cooperation*. These stages may have rather different impacts on bias and hence different implications for bias treatment. The location stage involves the survey organization locating each sample unit. For example, for a face-to-face interview survey this involves both obtaining a correct address and successfully locating it in the field. Generally, the contact stage involves making verbal contact with the sample unit. Contact with a person at the sample address does not necessarily constitute contact with the sample unit, depending on the survey definition of a sample unit. In some surveys, it is possible to attempt to contact the sample member in mediums other than that of the survey interview. For example, many face-to-face surveys allow telephone contacting to set up an appointment for the interviewer to visit. Since in a face-to-face survey cooperation of the sample person is not possible on the phone, we consider that a personal visit by the interviewer is necessary to complete the contacting stage. Similarly, advance letters will not count as successful contact if the survey is not conducted by mail. In sum, the contact stage involves making contact with the sample unit in the same medium as the survey interview. The cooperation stage involves successfully obtaining the desired survey data, usually either through an interview or a self-completion questionnaire.

Response is only obtained if the location, contact, and cooperation stages are all passed successfully. It should be obvious that the processes leading to success are different at each stage. Consequently, the correlates of success are likely to be different at each stage. These correlates will include both characteristics of the sample units (see Lynn et al., 2002) and characteristics of the fieldwork strategies adopted by the survey. For example, the propensity to make contact will (in the case of a face-to-face survey) depend on characteristics of sample members such as the amount and distribution of time spent at home, which will in turn be influenced by employment patterns, leisure activities, and so on, but also on characteristics of the fieldwork strategy such as the number and timing of call attempts (Groves & Couper, 1998, Chapter 4). Cooperation may be influenced by a different set of factors, such as social integration, attitudes toward authority, and disposition toward the interviewer on the one hand and interviewer introductions (including whether these are scripted introductions or flexibly tailored by the interviewer), survey materials, incentives, etc., on the other (Groves et al., 2000).

Cross-national surveys differ from national or sub-national surveys in a number of important ways (Lynn et al., 2006). There are likely to be design differences between countries that impact on the survey participation process: for example, the availability of different kinds of sampling frames in different countries (Lynn et al., 2006; Lipps & Benson, 2005). The implementation is also likely to vary, since data collection is typically organized at the national level, with a different survey organization involved in each country. This potentially introduces *house effects* (Smith, 1978, 1982) that are confounded with country effects. There may be differences between survey organizations in the types of persons recruited to work as interviewers, in the interviewer payment structure, in the reputation of the survey organization, and so on. This, and cultural differences, may lead to different interviewer practices in different countries, such as differences in working hours and days (see Lipps & Benson, 2005). To some extent house effects may be eliminated (and their confounding with country effects reduced) if the cross-national survey has a detailed specification of how the survey should be implemented and, for example, prescribes call schedules, interviewer training, and the length of fieldwork. There are also likely to be differences in the *population distribution* of characteristics associated with nonresponse (for example, the proportion of persons living in households where all adults have a full-time job). Finally the *relationship* between those characteristics and the propensity to respond may differ between countries. This can happen, for example, because full-time jobs have different implications for at-home patterns in different countries (e.g., average commuting times can be very different) or because interviewers have different typical working hours in different countries.

18.3 AUXILIARY DATA FOR THE STUDY OF CROSS-NATIONAL NONRESPONSE

Studying the survey participation process and identifying correlates of response that can be used to adjust for bias requires auxiliary data about both respondents and non-respondents. Auxiliary data that directly measure aspects of the survey data collection process have become known as paradata (Couper, 2005). Couper and Lyberg (2005) cite Scheuren (2001) as the source of the distinction between macro- and micro-level paradata. Macro-level paradata (also referred to as meta-data) are survey process indicators measured at the level of the sample (e.g., response rate or length of fieldwork period), while micro-level paradata are measured for each sample member (e.g., the time and date of each attempt to contact a sample member or the length of time it took a respondent to answer a specific question).

Micro-level auxiliary data that are available for both respondents and non-respondents can be more powerful in explaining nonresponse than survey variables, as the latter are only available for respondents. The key is to find "variables that are related to both the probability of response as well as to key survey outcomes" (Kreuter et al., 2007, p. 3,143). Potential sources of auxiliary data include:

- The sampling frame (e.g., sex and age if the frame is a population register)
- Linked micro-level data (e.g., publicly available administrative data that can be linked using identifiers that are on the frame, such as full names)
- Linked geographical or other aggregate-level data (e.g., Census small area data or other data that can be linked via grid reference or postal code)
- Systematic interviewer observations (e.g., regarding the sampled dwelling or the neighborhood)
- Interviewer characteristics (e.g., sex, age, years of experience)
- Contact data
- Other survey process data

The latter four sources of auxiliary data, i.e., interviewer observations, interviewer characteristics, contact data, and other survey process data, are especially relevant in a cross-national nonresponse context, since they can be collected in most countries and, if care is taken, can adhere to a high level of comparability. To be effective in adjusting for nonresponse bias, auxiliary data will need to explain all three stages of the survey participation process.

Contact data are measurements of key aspects of the process that lead to a fieldwork outcome. They provide information about the contact attempts made to all sample units—respondents and nonrespondents—in the quest to obtain response. The data are generated to monitor and optimize different stages of the data collection process (see Section 18.5.1 below) but, as we argue here, can also be of considerable interest to researchers. We can broadly classify contact data as being of one of two types: *case-level* or *call-level* contact data. Case-level contact data consist of data items defined at the level of the sample unit, summarizing the call attempts made to the unit. Such items might include the total number of call attempts made and the total elapsed time between the first and final call attempts. Call-level contact data consist of data items defined at the level of the call attempt. These typically include the date, time, mode, interviewer number, and outcome (no contact, interview, refusal, unable, ineligible, appointment, etc.) of the attempt (Section 18.5.2 below discusses how contact data are collected) (Blom, 2008, p. 2). Contact data may be particularly well-suited to explain the location and contact stages of the survey participation process. In addition, contact data measure key aspects of fieldwork which can be manipulated to increase response rates.

The following section reviews studies using contact data to analyze cross-national nonresponse. All of these studies use data from one of three surveys: the European Social Survey (ESS), the Survey of Health Ageing and Retirement in Europe (SHARE), and the European Community Household Panel (ECHP). While the findings are a great step forward in the state of the art of cross-national nonresponse research, we argue that some analyses are hampered by incomplete or inconsistent contact data, by a lack of clarity regarding data definitions or by a lack of further auxiliary information to complement the contact data. These limitations are discussed in Section 18.5 below.

18.4 CROSS-NATIONAL STUDIES OF NONRESPONSE

Research on cross-national comparisons of nonresponse is still in its infancy. To date the existing studies can be distinguished by whether they describe (1) cross-national differences in response rates, (2) the processes leading to cross-national differences in response rates, or (3) cross-national differences in potential non-response bias.

The earliest studies, starting in the 1990s, were of the first type. These studies used country-level data to document differences in nonresponse rates and trends (see Section 18.4.1). The information from these studies provided some insight into differences in the potential risk of nonresponse bias across countries.

More recent studies have used contact data to explain differences in non-response rates and to describe nonresponse bias. Studies of the second type aim to make causal inferences and focus on differences in fieldwork characteristics which are in the control of the researcher and could be manipulated in order to achieve more comparable outcomes across countries (see Section 18.4.2).

The third type of studies examines correlations between auxiliary information about the sample unit and fieldwork characteristics with response propensities and substantive survey variables (see Section 18.4.3). These studies are not necessarily interested in the causal processes, but instead in identifying correlates of response (whether manipulable or not) to be used in nonresponse adjustments.

18.4.1 Studies Using Survey-Level Data

Already in the early 1980s Goyder (1985) was concerned about the reportedly lower response rates in Canada compared to the United States. In a comparative meta-analysis of response rates in probability sample surveys in the two countries he found that the social structure and cultural values, "plus the sparing use of call-backs, has entailed the persistence over some two decades of low response on national Canadian interview surveys" (p. 246). Lyberg and Dean (1992) first coined the notion of different *survey climates* across countries and changes in survey climate within countries that affect response rates. Although only based on anecdotal evidence from the Netherlands, Germany, and Sweden, the notion of survey climates has remained an important concept. Lyberg and Dean showed that a country's survey climate can abruptly change, for example, when people lose trust in surveys and official statistics, as was the case in the 1980s with the Metropolit study in Sweden, and the public debates regarding the censuses in Germany and the Netherlands. (Metropolit was a research project that collected register information for a sample of 15,000 persons born in 1953 without their consent and knowledge and which, on becoming public, caused response rates in all Swedish surveys to plummet. The census debates in Germany and the Netherlands emanated from World War II scars and the fear of Big Brother, resulting in enormous public and media attention.) For the United States, Harris-Kojetin and Tucker (1999) found that indicators of macro-level political and economic conditions were related to refusal rates in the Current Population Survey: "Over the entire period 1960–1988, higher presidential approval was associated with lower refusal rates... However, periods of decreasing unemployment and increasing consumer expectations for the economy were associated with periods of increasing refusal rates" (Harris-Kojetin & Tucker, 1999, p. 180). Groves and Couper (1998) called for assembling *time trends* in response rates across countries. Supplemented with metadata on "social environmental correlates of survey participation (e.g., degree of urbanization, level of political participation, alienation, education, crime rates, etc.), a database could be built to permit comparative analysis of response rates within and across countries" (p. 173). These kinds of data they hoped "could reveal important differences in response to comparable surveys across countries that could be explained by variation in survey-taking climate" (pp. 172–173).

De Heer (1999) published the first comprehensive study of cross-national nonresponse. Comparing surveys in 16 countries, he found that there were "large differences in response rates and response trends between countries for official statistics" (p. 140). For example, published response rates on labor force surveys in the mid-1990s ranged from 58% in Netherlands to 99% in Germany, while noncontact rates ranged from 1% in Slovenia to 15% in Denmark. Couper and de Leeuw (2003) examined response rates for three cross-national surveys (the International Social Survey Programme, International Adult Literacy Survey, and the Trends in International Mathematics and Science Study) and concluded that the differences in response rates and trends across countries and surveys indicated "differences in survey design and effort as well as societal differences" (p. 165). In detailed multilevel analyses of de Heer's (1999) data, de Leeuw and de Heer (2002) found that survey-management and socio-economic metadata on the survey and country level were associated with country differences in outcomes. Factors that were positively associated with contact rates included larger average household size, higher proportion of young children, panel as opposed to cross-sectional surveys, and more lenient rules for sampling and respondent selection. Factors positively associated with cooperation rates were higher unemployment rates, lower inflation rates, and the mandatory status of surveys (pp. 52–53).

Despite what we have learned from the studies describing cross-national differences in response rates and trends using country-level data, they inherited two main problems. First, the country-level data do not allow causal inference about survey participation processes in different countries. For example, although de Leeuw and de Heer found that contact rates were higher in countries with larger average household sizes, they could not infer from this that a household of a larger size would have a higher contact propensity than a household of a smaller size. Such individual-level inference would require individual-level data. Second, neither the surveys nor the calculations of response outcomes were necessarily comparable across countries. The studies relied on reports of survey outcomes over which the authors had no control. As a result, even descriptive inferences about differences between countries in response rates and trends were limited. As de Heer (1999) pointed out, "without a detailed description of the response, it is impossible to evaluate the quality of a survey. Without comparable response rates it is difficult, to say the least, to compare or integrate data from different sources or countries" (p. 141).

18.4.2 Studies Using Contact Data to Explain Differences in Response Rates

Some more recent studies have used individual-level response outcomes in combination with contact data in an attempt to explain differences between countries in response rates (primarily contact rates). The underlying question of these studies is whether optimal contact strategies, defined as the most efficient number and timing of calls to achieve a good response rate, are country specific or whether previous results from the United States and some European countries (see, for example, Japec, 2005; Bennett & Steel, 2000; Campanelli et al., 1997; Swires-Hennessy & Drake, 1992; Weeks et al., 1987) also apply to other countries. A related question is whether differences are explained by differences in the composition of national populations and of fieldwork characteristics, or by differences in response behaviors between countries. The answers to these questions have implications for how best to achieve equivalent fieldwork outcomes across countries: whether by standardizing procedures, for example, by requiring a specific minimum number of calls at specific times and days, or by allowing each country to adapt strategies to the situation it faces. All the studies reviewed here and in Section 18.4.3 below use data from rounds 1–3 of the ESS, wave 1 of SHARE, or the 1995/6 waves of the ECHP.

Several studies have shown that there is considerable variation between countries in fieldwork procedures. Billiet and Philippens (2004), Billiet and Pleysier (2007), and Symons and colleagues (2008) investigated the fieldwork strategies in countries participating in the ESS rounds 1, 2, and 3, respectively. These studies documented that some ESS countries had high proportions of noncontacted sample members who were called fewer than the required minimum of four attempts. However, they could not find any cross-country association between the mean number of calls to noncontacted sample units and contact rates. The timing of calls, whether during a weekday daytime, evening, or weekend, also varied hugely across countries. Overall, the studies found that for most countries contact attempts on weekday evenings were more effective than attempts during the day. Nicoletti and Buck (2004) compared two independent surveys in Germany and Britain (the German Socio-Economic Panel, SOEP, and the British Household Panel, BHPS) and two surveys in these countries that were part of the harmonized European Household Community Panel (ECHP). They found that the distribution of fieldwork characteristics, conditional on individual and household characteristics, was slightly more similar between the independent BHPS and SOEP surveys than between the ECHP surveys. This suggests that the attempted harmonization of fieldwork procedures in the ECHP was not successful.

Regarding cooperation, Billiet and Philippens (2004) and Beullens and colleagues (2007) showed that both the implementation and the results of refusal conversion procedures varied hugely across countries in rounds 1 and 2 of the ESS. While some countries issued nearly all refusers to the conversion stage, other countries issued hardly any, or only those that appeared to be "soft" refusers. As a result, success rates at the conversion stage also varied widely. Beullens and colleagues (2008) further showed that in all countries examined, conversion probabilities were higher the more time elapsed between a refusal and a conversion attempt. The length of elapsed time was, however, determined by two different factors. In the Netherlands, the interval seemed to be chosen strategically, with sample units rated as being unlikely to respond having longer "cool-off" periods than apparent soft refusals. In Switzerland the time interval seemed mainly to be determined by time pressures: those who refused earlier in the fieldwork period had longer cool-off periods than those who refused later.

A further group of studies has examined cross-national differences in the association between the propensity to respond and both population characteristics and fieldwork efforts. Conclusions about whether these associations differ across countries are, however, mixed.

Philippens and colleagues (2003), Nicoletti and Buck (2004), and Blom and colleagues (2007) found country differences in the contact probability and optimal calling strategies for the general population, while Lipps and Benson (2005) found no country effects for the population aged 50+. This may suggest that contact processes for the general population differ more across countries than contact processes for the population aged 50+. The results should, however, be interpreted with care, since the different conclusions may also be the result of the different analysis methods used in the studies. Philippens and colleagues (2003) showed that countries differed in the ease of making contact. In addition, the probability of contact decreased more strongly with additional calls in countries which had a high contact rate at the first call than for countries with populations that were harder to contact. The benefits of focusing on evening or weekend calls, and of increasing the minimum number of calls, also varied across countries: the number of previous contact attempts was negatively associated with the conditional probability of success in some countries, but had no effect in other countries. The authors concluded that "the 'optimal' timing of calls was country-dependent and illustrate[s] the importance of tailoring fieldwork strategies towards specific national contexts" (p. 9). Lipps and Benson (2005) found hardly any country-level variation in the probability of contact, but instead large interviewer-level effects. The interviewer variation was not explained by differences in the timing of a calls, the mode of contact or interviewer observations about the physical state of the sampled address, the state of the neighborhood, and the existence of barriers to access of the housing. The results suggested that differences in outcomes between interviewers were not explained by differences in their contact strategies or characteristics of the assigned sample points. Instead, explaining interviewer effects would possibly require additional auxiliary information about the interviewers' characteristics.

Nicoletti and Buck (2004) and Blom and colleagues (2007) examined whether differences in the distributions of fieldwork variables and population characteristics explained differences in response outcomes between countries. Nicoletti and Buck (2004) suggest that differences in contact and cooperation rates between Britain and Germany were mainly due to differences in response behaviors (i.e., the coefficients of the probit models) rather than differences in population characteristics and that data-collection variables were more important than individual and household characteristics. They concluded that harmonizing fieldwork procedures would not necessarily produce comparable response outcomes across countries: "Even if the explanatory variables distribution were equal between two surveys running in two different countries, the contact and the cooperation rates would not be equal because of a different impact of the variables. In other words the ease of contact and the propensity to cooperate, every explanatory variable being equal, are different across surveys running in different countries" (Nicoletti & Buck, 2004, p. 14). The comparison of 11 countries by Blom and colleagues (2007) further suggested that the reasons for differences depend on the countries compared and vary between countries with register and household or address based sampling frames.

18.4.3 Studies Using Contact Data to Explain Differences in Nonresponse Error

The third type of studies examines associations between contact data and both response and survey outcomes, which would be indicative of nonresponse bias and could inform weighting adjustments. The associations between contact information and survey variables tend to differ between countries, but are generally small. Billiet and colleagues (2007b) tested for associations between survey variables and respondent cooperativeness and showed that differences in summed attitude scores were larger between cooperative and hard-to-convert respondents (identified by the number of attempts required to persuade them to participate after an initial refusal), than between cooperative and easy-to-convert respondents. After controlling for differences in sample composition, by adding background variables thought to predict the attitudinal score, the associations between attitude scores and cooperativeness were reduced but not removed. This suggests that nonresponse bias could be adjusted further by including information about cooperativeness in addition to the standard socio-demographic background variables used in nonresponse weighting models. The authors also concluded that the nonresponse bias was likely to be small, since removing the indicator of cooperativeness from the predictive model neither affected the explained variance nor the size of coefficients of the substantive covariates. Overall, they concluded that "the relationship between the type of respondent (cooperative, reluctant) and the attitudinal and background variables was not all in the same direction in all countries. This needs further research and discussion because it creates a serious challenge to any scholar who believes there is a theory of nonresponse that applies cross-nationally" (p. 159). The inconsistency of biases across countries was corroborated by Beullens and colleagues (2007), who compared the socio-demographic characteristics and survey variables for cooperative respondents with those of reluctant respondents, who had been converted after an initial refusal.

Kreuter and colleagues (2007, 2008) tested whether interviewer observations were correlated with survey variables. In addition, the authors tested whether the interviewer observations were correlated with response outcomes and what the effect of using them to construct nonresponse weights would be. The results suggested that correlations were low and varied by interviewer-observation item, survey item, and country. Weighting hardly changed point estimates. For the countries analyzed the patterns were very similar, but exploratory analyses of other countries had apparently shown that this was not necessarily the case. The auxiliary variables tested were interviewer observations about the type of housing (whether multi-unit), signs of litter and vandalism in the neighborhood, and whether the sampled address had an alarm system. The survey variables were indicators of social involvement, fear of crime, general health, and activities in the home. The approach is novel in that contact data and interviewer observations have mainly been used to predict response probabilities. Correlations with survey variables have rarely been tested. Correspondingly, contact data and interviewer observations have rarely been used for nonresponse adjustment.

Kreuter and Kohler (2007) used contact data to derive contact sequences and test hypotheses that these might be correlated with contactability, cooperativeness, interviewer behavior and fieldwork regulations in different countries. If this holds, the authors propose using contact sequences for nonresponse adjustment. They define a contact sequence as a series of calls, which may either lead to no contact, contact with someone other than the sample person, contact with the sample person but no interview, or an interview. The contact sequence is composed of elements (each call attempt) and episodes (a subsequence of calls with the same outcome). The results indicate that the number of contact attempts is correlated with indicators of time spent at home (labor force status, time spent watching television). Similarly, countries which allow calls on Sundays produce contact sequences with fewer no-contact calls.

In sum, various studies have documented that there are vast differences across countries in contact and cooperation rates, in fieldwork practices and the effectiveness of different fieldwork actions, and in the direction and extent of potential nonresponse biases. The results imply that survey participation processes vary across countries and that the standardization of fieldwork characteristics would not necessarily achieve equivalent outcomes. Some of the studies, and hence their conclusions, are, however, restricted by limitations of the cross-national contact data and other auxiliary data.

18.5 LIMITATIONS OF CONTACT DATA FOR ANALYZING CROSS-NATIONAL NONRESPONSE

In their analyses of the cross-national decline in response rates, Couper and de Leeuw (2003, p. 174) noted that "for valid cross-cultural and cross-national comparisons, it is of utmost importance that the various sources of nonresponse are reported." Nowadays we can take this a step further and state that access to comparable call-level contact data is of utmost importance. These, in combination with other types of paradata on survey implementation, enable researchers to code response outcomes and conduct response analyses according to strategies that they regard as optimal. So far the ESS is, however, the only cross-national survey to make such contact data readily accessible. It is therefore no coincidence that most studies that have used cross-national call-level contact data are based on this survey.

However, the review in Section 18.4 has shown that the use of contact data to aid understanding of cross-national differences in nonresponse is hampered by missing data, inconsistencies, and a lack of equivalence. Kreuter and colleagues (2007), for example, only considered countries with near-complete interviewer observation data, with the result that only three of the 21 ESS round 1 countries were included in their analysis. Similarly, Blom and colleagues (2007) excluded 11 countries from their analyses due to item nonresponse. These limitations arise from the ways in which contact data are usually collected, which in turn are driven by the uses to which they are put in the survey process. We here review those features in order to provide context for a discussion (Section 18.6) of how the quality of contact data could be improved.

18.5.1 Uses of Contact Data in the Survey Process

Contact data are typically used at three different levels as a routine part of the survey process. At the case level, the contact data are used, typically by the interviewer, to decide when and how to make further calls. At the interviewer level, contact data can be used by the survey organization to monitor interviewer performance. At the survey level, contact data can be used to monitor where field-work progress lags behind schedule, which types of sample units are underrepresented in the achieved sample and should receive more effort, or whether refusal conversion strategies are effective. For the purpose of survey management, contact data are often analyzed after the completion of fieldwork if the data are not available to the survey organization electronically and in real time. The lessons learnt are then applied to subsequent surveys.

Some organizations use computerized case-management systems that allow automatic edit checks, call scheduling (especially in telephone surveys but also in face-to-face surveys) and real-time analysis of the fieldwork progress. Increasingly, researchers also use real-time analysis of contact and interview data for so-called responsive designs. A decision to adjust or keep the planned survey design is made during fieldwork, based on real-time information about fieldwork and survey outcomes which affect costs and errors (Groves & Heeringa, 2006).

18.5.2 The Collection of Contact Data

Contact data are collected by the interviewer, usually by means of a contact form. Figure 18.1 illustrates what the relevant part of a contact form may look like. Outcome code schema for contact attempts must be specified by the researcher and will depend on the survey objectives, the population researched, and the sampling frame. [See AAPOR (2006) for a review of outcome codes in various survey situations in the United States and Lynn and colleagues (2001) for outcome codes adapted to survey settings in the United Kingdom.] The schema need not be complex: the ESS uses just eight general outcome codes. Some outcomes codes (for example, some ineligibility codes) will only be assigned in the office, when the final case outcome of a sample unit is defined. Other outcome codes are recorded in the contact forms by the interviewer in the field. The way response outcomes are collected can vary across surveys, survey houses, and countries.

Differences in the design of contact forms reflect differences in the purposes for which survey organizations collect contact data. The purpose, for example, determines whether contact data are recorded as call-level data or summarized and recorded as case-level data. For case management, face-to-face interviewers may use full call-level data about each past call to plan their workload and decide on the next action for each case. In a centralized computer-assisted telephone interviewing (CATI) setting, computerized data about past calls may provide the input for automated calling schedules, which are algorithms that trigger the next call to each case. The software for computerized contact forms "usually includes a report that summarizes the most recent disposition field in the sample database" (McCarty, 2003, pp. 398–399). In some fieldwork management systems the information regarding a call attempt is only stored until the next call attempt is made. At that point, the previous information is overwritten by the outcome of the new call. The final outcome of a series of contact attempts is therefore determined by the outcome of the last contact attempt. No full record of the contacting process is stored and available for later analysis.

Even if full contact histories are initially recorded, survey organizations vary in whether and at which stage the original call-level contact data are reduced to a summarized case-level form: (1) interviewers may return only case-level contact summaries to the survey organization, (2) the survey organization may summarize full contact data before releasing the data, or (3) researchers may derive their own summary measures from call-level contact data. In the first case, it is left to the interviewer to derive the required indicators. Typically the indicators may include only a final case outcome (e.g., interview, refusal, noncontact, ineligible) and the total number of calls/visits made. For each case they were assigned interviewers are only expected to return these summary indicators even though they may well have recorded more information in the course of carrying out the fieldwork. Consequently, no information on intermediate contact attempts is stored and available for later analysis. Even if survey organizations receive call-level contact data from the interviewers, they might decide to derive final case outcomes from the call outcomes. This can happen for a variety of reasons including data protection, commercial sensitivity, or because the sponsor prefers summarized contact data. Finally, the third possible scenario is that researchers may derive final outcome codes at the case level from call-level response outcomes in order to use their own preferred outcome definitions in analyses.

18.5.3 Cross-National Contact Data: ESS, SHARE, and ECHP

Comparative contact data from cross-national surveys are still rare. The studies reviewed in Section 18.4 were all based on one of three surveys: the ESS, SHARE, and the ECHP. The small number of cross-national surveys with available contact data is one severe limitation to cross-national nonresponse research, but each data source also has its own limitations.

The ESS was the first cross-national survey to collect and make publicly available cross-national contact data for both respondents and nonrespondents. Except for a few countries, where data protection laws forbid the publication of all or parts of these data, the data from all countries can be accessed via the ESS data archive website (http://ess.nsd.uib.no/). Most countries use the ESS model contact

Respondent ID number	

Number of contact attempts	Date (DD/MM)	Time (24 hour clock)	Mode of visit (code)	Outcome of visit (code)	Notes
1	/	:			
2	/	:			
3	/	:			
4	/	:			
5	/	:			
6	/	:			
7	/	:			
8	/	:			
9	/	:			
10	/	:			

Outcome: 1=Completed interview; 2=Partial interview; 3=Contact with someone, don't know if target respondent; 4=Contact with target respondent but no interview; 5=Contact with somebody but not target respondent; 6=No contact at all; 7=Address not valid; 8=Other information about sample unit.

Figure 18.1. Example Contact Form

form. Countries may, however, use their own forms, provided that they submit all the compulsory variables to the data archive. Mandatory data collected include date, time, mode, and detailed outcome of each contact attempt, interviewer number, reasons for refusal, and whether a contact attempt is a re-issue. In addition to contact data, the ESS collects information on the housing and neighborhood of the sample unit. Unfortunately, since many countries implement the contact forms on paper and do not carry out edit checks on the data, in some countries the data have many item missings. More recently efforts have been undertaken by the central coordination team to carry out *ex post* edit checks to improve the contact data of the ESS.

SHARE (see Chapter 28, this volume) collects and centrally stores call-level contact data for all sample units through a common computerized case management system. During interviewer training, which is streamlined across countries by means of a train-the-trainer program, interviewers are instructed on operating the system and on how to assign result codes. Unfortunately, the SHARE contact data are not publicly available and therefore not easily accessible to external researchers.

Data from the ECHP are available through the EuroPanel Users Network, epunet (http://epunet.essex.ac.uk). Only case-level contact data are available. For each sample household and for each wave of the survey, the relevant items available are the total number of visits made to the household, an indicator of whether the household was successfully traced, and an indicator of whether the household was successfully interviewed.

18.5.4 Summary of Limitations of Cross-National Contact Data

The review (Section 18.4) of studies using contact data to analyze cross-national nonresponse has highlighted that analysis is still limited by the nature and quality of the available contact data and found four main limitations:

- The lack of publicly available cross-national data. Only few cross-national surveys collect contact data, and of these only the ECHP and the ESS have published theirs, making it possible for external researchers to carry out cross-national nonresponse analyses with these data.
- Missing information due to item nonresponse and lack of consistency. This is primarily a problem of quality control in the collection of contact data and could be improved if completeness and consistency checks were incorporated in the fieldwork process. Such improvements would probably aid fieldwork processes as well as methodological research.
- The lack of equivalence of cross-national contact data. The information collected in contact data, the level of detail, the mode of collecting contact data, as well as the process control and data cleaning procedures can vary across countries of the same cross-national survey, giving rise to criticism regarding the comparability of cross-national contact data. Therefore, even with complete and publicly available data, inferences about cross-national differences in nonresponse are conditional on the equivalence of contact data. Section 18.6 proposes three criteria for equivalence, which should enable high-quality comparative nonresponse analyses based on contact data.
- The absence of other related auxiliary data. Both Blom and colleagues (2007) and Philippens and colleagues (2003) concluded that missing variables hampered the interpretation of their findings with regards to cross-national nonresponse. Philippens and colleagues, for example, speculated that differences between countries in optimal calling strategies may be due to differences in at-home patterns, but were unable to explore this further without further auxiliary data. Blom and colleagues hypothesized that additional micro-level measures for both responding and nonresponding cases—in addition to the limited interviewer observation items—might provide more explanatory power.

18.6 QUALITY STANDARDS FOR CROSS-NATIONAL CONTACT DATA

Although contact data are a valuable source of information, the methodology for defining, collecting, and recording contact data is underdeveloped. We still lack best practice and coherence in (1) the design of contact forms to collect the data, including the technology used for their collection, structure, and content; (2) the implementation of contact forms, i.e., instructions given to interviewers regarding how they should fill in the forms; and (3) the coding of contact data, specifically how best to derive a final outcome for a sample unit from individual outcomes of call attempts. Without a common understanding on these issues comparability of contact data cannot be achieved. This is a major obstacle to their use on cross-national surveys where comparability is a main objective. Drawing from standards generally agreed upon for substantive survey measurement, this section points to issues of comparability in the design and collection of cross-national contact data.

18.6.1 Equivalence of Design

Both the content and structure of the contact forms need to be equivalent across surveys. In terms of the content, at a minimum the code frame for call outcomes should be the same. This also includes the eligibility criteria (and ineligibility codes). For detailed comparative analyses, the more information is collected in the contact forms of each survey, the more comprehensive the analysis can be. In addition to equivalence of the call outcome, equivalence in the collection of date and time, mode of call, and other additional variables is therefore desirable. In terms of structure, not only the information collected, but also the way it is collected should be the same across surveys. From questionnaire design methodology we know that differences in question formulation, question format, and translation can have a serious impact on the data (see, for example, Dillman, 2007; Harkness, 2003). Although no such evidence yet exists for contact forms, keeping the structure of contact forms similar should allow maximum comparability. Care needs to be taken, however, that variations in national constraints, such as differences in available sampling frames, are reflected in the codes available to the interviewer and in the structure of the contact forms.

18.6.2 Equivalence of Implementation

In addition to the design, data comparability is likely to be influenced by the implementation of contact forms in the field. It might therefore be desirable to standardize implementation, including (1) whether contact data are collected on the computer or on paper, (2) how interviewers are trained and briefed on filling in the forms, and (3) what the fieldwork control procedures are for the contact forms.

On any survey, interviewers know the characteristics and outcomes of all calls they have made. Surveys, however, differ in whether and how this information is captured. Face-to-face surveys often use *paper-based systems*, where interviewers record the information about the contact process using pen and paper, even if the interview is carried out as a computer-assisted personal interview (CAPI). This may be because the paper technology offers greater speed and flexibility in the field, where contact data are often recorded on the doorstep or in other public places and where the interviewer may be moving rapidly from one case to another. Some organizations still use paper-based systems for telephone surveys, though this may be more likely when the interviewing is decentralized than when a centralized facility is used. Centralized CATI surveys, where cases are worked by a number of different interviewers, typically use *computerized systems*, which can vary greatly in capability and complexity. Relational database systems are most frequently used, where interviewers enter a call outcome code for each contact attempt according to a code frame specified by the researcher. Date and time are recorded automatically by the system. Computerized systems are increasingly also used for face-to-face surveys, though the data entry by the interviewer may not always be instantaneous. This may take place some time after a call attempt is made, or even just once at the end of each field trip. Instantaneous entry of contact data is less important for face-to-face surveys as only one interviewer is assigned to work on each sample case at any one time.

Interviewer training and *fieldwork control* are crucial, regardless of the technology (see Morton-Williams, 1993; Loosveldt et al., 2004). Some survey organizations implementing computerized contact forms in face-to-face surveys, for example, report that interviewers can be resistant to filling in accurate contact protocols. In addition, frames of disposition codes are often difficult to apply to real fieldwork situations. On the doorstep it might, for example, be difficult to decide whether a person saying that he/she has no time is refusing to participate or inviting the interviewer to come back another day. Likewise, in cases with language problems a rejection might be a refusal or a genuine inability to participate because of the language. In addition, interviewers often regard contact forms as unimportant, since they do not concern the main questionnaire and they may perceive that they are not being paid for completing them. Without consistent quality checks contact data are likely to suffer from negligence. Unfortunately, detailed interviewer training on contact forms and quality control during field-work are still the exception. As a result some contact data suffer from high levels of item nonresponse and are excluded from cross-national comparisons by analysts. Finally, the lenient implementation of data collection can lead to interviewer effects, reducing comparability even within a survey. Detailed interviewer training and close monitoring of interviewer performance, in combination with an *ex post* analysis of interviewer effects are key to achieving low levels of variation. Improving the contact data through better interviewer training and data checking can also benefit the fieldwork process itself and is therefore a worthwhile quality improvement project. And after all, one cannot but wonder how good the main survey data can be, if the process data are incomplete or inconsistent.

18.6.3 Equivalence of Coding Fieldwork Outcomes

Our final quality standard concerns the coding of call outcomes for each sample unit into a final case disposition code for the sample unit. There are three main methods by which the code may be derived: *most recent*, *priority*, and *subjective* coding. With most recent coding the outcome of the last call to a sample unit is defined as the case outcome (see AAPOR, 2000). With priority coding some outcome codes take priority over others. For example, one would define that achieving an interview takes priority over a refusal, which in turn takes priority over a noncontact (for details on priority coding see Lynn et al., 2001). A situation in which an interviewer tries to convert an initial refusal, yet never manages to make contact again, would be coded differently in the two coding systems. If the last call outcome defined the final disposition, this would be a noncontact. According to a priority coding system, it could be a refusal. Finally, subjective coding refers to situations where the rules for deriving a case outcome from call outcomes are not defined. Typically in such situations, only descriptions (which can vary in their precision) of each case outcome code are provided. It is left to the coder to decide how to allocate cases to outcomes. This kind of coding is perhaps most common when interviewers are asked to return case-level codes to the survey organization, though it may also be used by survey organizations carrying out in-office coding.

Comparing the results of different types of coding schemes, McCarty (2003) found only little differences in overall response rates. This study may, however, underestimate the potential impact of different coding schemes, since the definition of eligibility and interview outcomes is less likely to vary by coding method than, for example, the definition of contact or refusal rates. Blom (2008) examined the differential impact of using most recent coding and a priority coding scheme on cross-national differences in outcome distributions and outcome rates in the ESS. Blom found that the choice of coding scheme had a large impact on the number of noncontact and refusal outcomes recorded as the final disposition code, especially in countries where considerable, yet unsuccessful, refusal conversion attempts were undertaken. Furthermore, the analyses showed that response, refusal, and noncontact rates calculated with the nationally derived final dispositions did not correspond to those calculated with the most recent coded or priority coded final dispositions. The findings illustrated the importance of consistently deriving case outcomes from call outcomes, to achieve comparable fieldwork outcomes and outcome rates.

18.6.4 Prospects for Standardization

We have presented three equivalence criteria for cross-national contact data and believe that respecting these high-quality comparative contact data can be achieved. For the first two criteria (equivalence of design and equivalence of implementation) we can only guess the impact inequivalence has on the comparability of contact data. For the last criterion (equivalence of coding fieldwork outcomes), however, analyses exist proving that different coding strategies lead to considerable differences in the distributions of fieldwork outcomes and outcome rates across countries, hindering cross-national comparisons thereof (Blom, 2008).

A note of caution is in place though when analyzing cross-national contact data. Even in the ideal case of perfect equivalence of design, implementation, and coding, cross-national differences in available sampling frames across countries lead to necessary differences in the design of contact data and the distribution of response outcomes. Countries with an individual sampling frame drawn from a register necessarily have some different call outcome codes than countries with a household or address-based sample design. More importantly, with different sample designs the distribution of the outcomes are bound to be different, as are the fieldwork processes leading to them. For cross-national research, differences in sampling frames and designs must therefore be taken into account.

18.7 CONCLUSIONS AND RECOMMENDATIONS

The growing availability of contact data opens up new possibilities for comparative nonresponse research. The earliest cross-national studies of nonresponse documented large differences in nonresponse rates and trends across countries, suggesting that there are high risks of differential nonresponse bias. Only with contact data and other micro-level auxiliary data can researchers explain differences in contact and response rates across countries. The findings suggest that fieldwork actions have different effects in different countries, and that harmonizing fieldwork procedures would not necessarily lead to the same response outcomes in each country though it may be useful in limiting some of the more extreme differences between countries. Studies testing the utility of contact data for nonresponse adjustment have had some, if limited, success in identifying correlates of both response propensities and substantive survey outcomes. So few studies have, however, been carried out to date, that most findings have not been replicated and should therefore be interpreted with care. Traditionally, studies of fieldwork implementation have had little place in cross-national survey projects. Instead cross-national surveys have emphasized other methodological problems, such as translation and the cultural applicability of concepts. Contact data and other auxiliary data can aid understanding of the impacts of fieldwork implementation. Specifically, they can help the researcher to move from a situation of NMAR to MAR—or at least to move some way in that direction—by explaining an additional proportion of the variation in response propensity.

Currently the main restrictions for cross-national studies of nonresponse are data problems. We recommend the following remedies:

- First, all cross-national surveys should make contact data and other auxiliary information publicly available. Currently very few do so.
- Second, efforts should be made to minimize item nonresponse. Levels of item nonresponse in the few available sources of cross-national contact data vary greatly between countries, imposing significant analysis limitations.
- Third, international standards for the collection of contact data should be agreed and adopted. This would aid quality control and would improve comparability between countries.

From a researcher's point of view, some types of contact data are more useful than others. Thus, we recommend that:

- Full call-level data should be made available where possible. [This enables researchers to (a) choose how to define the final outcome code based on the sequence of call outcomes, (b) create other case-level indicators, for example, concerning sequences of call attempts or timings of attempts, and (c) carry out analyses at the call level in addition to analyses at the case level.]

- If call-level contact data are not to be made available, case-level outcome codes should be derived centrally by the survey organization rather than by interviewers. (Coding is likely to be more consistent and the definitions applied to different cases are more likely to be comparable.)
- Computerized systems for recording contact data in the field should be preferred to paper-based systems. (The data are likely to be of better quality, as routing and edit checks can be built into the script, reducing the potential for interviewer errors. In addition, the data are more likely to be available for analysis, as they already exist in electronic form and do not have to be keyed, as would paper-based data. In any case, contact data should be collected in the same mode across all sample units to prevent mode effects.)

Standards for the content, implementation, and coding of contact data should be developed and adhered to. This is likely to improve the quality and coherence of cross-national research into survey nonresponse and thereby to extend our knowledge of the factors driving differences between countries in survey outcomes. As soon as this type of data receives more attention from survey researchers and methodologists, it should be possible to reduce data quality problems. Simple edit checks and interviewer training on the importance of contact data could quickly change their perceived role from a burdensome by-product (see Stoop et al., 2003) into an invaluable data source. The criterion regarding equivalence of outcome coding would become redundant if surveys made available to researchers full call-level contact data that had been collected in a fully standardized way. As with any cross-cultural comparison, researchers will need to be aware of differences in the interpretation of contact data that will remain despite adherence to quality standards. Differences in sampling frames, for example, influence outcome codes in the contact data.

In conclusion, we believe that the availability of high-quality comparative contact data is in sight. We encourage all those with influence over cross-national surveys to bring this about by encouraging developments in the directions that we have outlined.

19

Item Nonresponse and Imputation of Annual Labor Income in Panel Surveys from a Cross-National Perspective

Joachim R. Frick and Markus M. Grabka

19.1 INTRODUCTION

A common phenomenon in population surveys is the failure to collect full information due to various forms of nonresponse behavior (see, e.g., Groves, Dillman,
Eltinge, & Little, 2002). In contrast to a complete refusal, i.e., unit nonresponse
(UNR), respondents may only partly cooperate due to confidentiality and privacy
issues, or the respondent's unwillingness or inability to provide a requested piece
of information. This nonresponse behavior is referred to as item nonresponse
(INR) (see de Leeuw, Hox, & Huisman, 2003). High rates of INR are typical for
questions about wealth and income, which are among the most important outcome
measures used in economic analyses (see, e.g., Riphahn & Serfling, 2005). There
is a growing body of literature dealing with this phenomenon in microeconomic
research as a specific form of measurement error (see Cameron & Trivedi, 2005).
Most importantly, INR on income questions has been found to be selective in
several respects: There is strong empirical evidence that individuals at both ends
of the income distribution as well as those experiencing changes in their labor
market status are less likely to provide the requested income information, thus
yielding downward-biased measures of inequality and mobility (see Frick &
Grabka, 2005; Jarvis & Jenkins, 1998; Watson & Wooden, 2009). Nevertheless,
ignorance about such selectivity is still widespread in empirical research using
only observations with completed interviews and leaving aside all observations
where at least one relevant item is missing ("casewise deletion"), thereby
assuming the underlying missing mechanism to be completely at random.
Approaches that explicitly attempt to compensate for the potential selectivity

[1] Survey Methods in Multinational, Multiregional, and Multicultural Contexts, edited by Harkness et al.

arising from INR usually include weighting and imputation, with the latter approach being used most widely (see Rässler & Riphahn, 2006; Rubin, 1976).

Virtually every national (panel) survey is affected by these problems as well as the need to choose appropriate treatment, which clearly magnifies comparability problems in internationally comparative research. In recent years, a large body of empirical literature has emerged focusing on cross-national comparisons of harmonized microdata including the ECHP (European Community Household Panel; http://circa.europa.eu/irc/dsis/ echpanel/info/data/information.html) and the Cross-National Equivalent File (CNEF; see Frick, Jenkins, Lillard, Lipps, & Wooden, 2007). Not only is the harmonization of the microdata itself crucial for optimal cross-national comparability but so are many methodologically relevant decisions in the pre- and post-data-collection phases such as the definition of the relevant population (e.g., information collected at the household or individual level), the choice of the data collection method (e.g., interview or register data), and the handling of any phenomena around data treatment (e.g., data entry, editing) especially when it is related to unit and item nonresponse.

It is in the latter area where this chapter adds to the literature by providing a comparative investigation of INR in three major national panel datasets, namely, the British Household Panel Study (BHPS), the German Socio-Economic Panel (SOEP), and the Household, Income and Labour Dynamics in Australia Survey (HILDA). We analyze the incidence and relevance of INR on labor income measures from a cross-sectional and longitudinal perspective, as well as the impact of alternative imputation approaches on inequality and mobility. Following the postulates of the "Canberra Group on Household Income Measurement" for harmonized national household income statistics (Canberra Group, 2001), we present evidence that it is important to harmonize not only income measurement but also the procedures for handling INR.

The chapter is organized as follows: Section 19.2 outlines the basic characteristics of the three panel surveys focusing on the incidence of INR with respect to labor income. It demonstrates the selectivity entailed by INR and investigates the time dependence of nonresponse behavior. Section 19.3 describes the imputation methods applied in the three surveys. Based on rather typical empirical research questions using labor income, Section 19.4 demonstrates the effect of imputation on earnings inequality and mobility as well as on wage regressions. Finally, Section 19.5 concludes with recommendations on how to deal with INR and imputation in cross-national comparative research.

19.2 INCIDENCE AND SELECTIVITY OF INR FROM A CROSS-NATIONAL PERSPECTIVE

This section briefly describes the most important features of the underlying panel datasets, all of which are included in the Cross-National Equivalent File (CNEF). The labor income measure and the accompanying information on imputation status which are used in this analysis are included as standard variables in the CNEF.

19.2.1 The Three Panels

The British Household Panel Survey (BHPS) is carried out by the Institute for Social and Economic Research (ISER) at the University of Essex (see Taylor, Brice, Buck, & Prentice-Lane, 2005). It started in 1991 with about 5,500 households and roughly 10,300 individuals (aged 16 and over) in England and has been repeated every year since then. The following analyses are based on this original sample, covering data on about 11,000 individuals participating in waves 1991 through 2004. In 1999, the interview mode was changed entirely for the whole sample from paper and pencil to computer-assisted personal interviewing (CAPI). Annual gross labor income in the BHPS is generally measured by means of a single question asking the amount of the last gross pay including any overtime, bonuses, commission, tips, or tax refund. This measure also includes income from self-employment. Apparently, this kind of "one-shot" question aimed at determining retrospectively a rather complex construct, namely the aggregation of a variety of income sources over a period of 12 months, bears a high risk of measurement error due to understatement, rounding, forgetting, omission of income components, and nonresponse.

The Household, Income and Labour Dynamics in Australia (HILDA) Survey started in 2001 with about 7,700 participating households and almost 14,000 individual interviews (see Watson, 2005b). HILDA, compiled by the Melbourne Institute of Applied Economic and Social Research, provides information on the living conditions of private households in Australia. By and large, the panel design used in HILDA resembles the one used in the BHPS, although respondent age starts at age 15. Interviews are carried out every year, mainly by CAPI with the telephone being used as a mode of last resort. Annual gross labor income in HILDA comes from three sources. First, all respondents are asked for their total wages and salaries from all jobs in the last financial year (July 1 of the previous year to June 30 of the survey year). Second, income from the respondent's own business/farm or from incorporated businesses is added, and to this are added the total shares of profit or loss from unincorporated businesses or farms. One-time payments and irregular payments are not surveyed explicitly. Results presented here use data on about 11,500 respondents from the first five waves, covering the period 2001–2005.

The German SOEP is the longest-running household panel study in Europe (see Wagner, Frick, & Schupp, 2007). Since its inception in West Germany in 1984, all household members aged 17 and over are surveyed individually each year. The initial sample size was about 6,000 private households comprising more than 12,000 respondents. In order to keep the sample representative and to cope with immigration and German unification, various additional subsamples have been incorporated in more recent years. We make use of all available data for unified Germany collected from 1992–2004, which matches fairly well with the BHPS observation period 1991 to 2004. However, as already noted, SOEP started 8 years earlier, in 1984. This raises the question whether long-time panel respondents "behave" differently with respect to INR than do individuals in a rather young panel. A new SOEP sub-sample, started in 2000 with about 5,000 households, provides the basis for checks of robustness (Frick, Goebel, Schechtman, Wagner, & Yitzhaki, 2006). The SOEP sample as of 2004 comprises about 11,800 households and 22,000 respondents. Interviews usually take place face-to-face. Although CAPI was introduced in 1998, paper-and-pencil interviews are still the main mode used to collect data. Information about gross annual labor income is gathered through 10 individual questions. There are separate questions asking about individual labor income in the previous calendar year from paid employment and from self-employment. In each case, the respondent is asked to state his/her average monthly income as well as the number of months in which this income type was received. Additionally, respondents are asked to report one-time or irregular payments such as a 13- or 14-month salary, holiday pay, or bonuses separately, and these are then added (see Frick & Grabka, 2007, for the exact wording of the various income questions in the three surveys).

While the three surveys deviate somewhat from each other with respect to panel duration, number of cross-sectional observations, and interview mode, their longitudinal performance as a panel survey is quite similar: The response rates after five waves vary between 70 and 74% (see Frick, Jenkins, Lillard et al., 2007).

19.2.2 Incidence of INR in Cross-Sectional and Longitudinal Perspective

INR may be a function of various factors such as the respondent's unwillingness to answer questions that are perceived as highly sensitive or in violation of confidentiality and privacy, the fact that the information requested is too complex, or simply the fact that the answer is not known (see Schräpler, 2004). The specific formulation of questions and the complexity of the construct being measured may also play a role (Hill & Willis, 2001). Another strand of research has shown that the interview situation, the survey mode, the presentation of the question with a "don't know" answer option, and possible interviewer effects including a change of interviewers in panel studies, are relevant determinants of INR (see, e.g., Riphahn & Serfling, 2005; Groves, 2006; Watson & Wooden, 2009).

While all such factors have been found to be related to INR in general, cross-national comparability of earnings inequality and mobility may also be affected by cultural and attitudinal differences in response behavior (see de Leeuw & de Heer, 2002) as well as by survey-specific features. Frick, Goebel, Schechtman, and colleagues (2006) find clear indications of a positive correlation between the number of interviews an individual provided in the course of a panel study and the probability of providing complete information. This effect might be driven by both increasing interview experience on behalf of the respondent as well as increasing confidence in the interviewer and the interview situation.

Despite the apparent differences among the three panels in the means used to collect annual labor earnings data, we find surprisingly little cross-national variation in the incidence of INR, defined as the share of persons with at least one missing income component. While for the longest running panel, namely the SOEP, the share of INR is on average about 14% across the entire observation period (1992–2004), the BHPS shows a rate of about 15% (based on data for 1991–2004) and the young HILDA survey is affected by about 16% (over the period 2001–2005).

Obviously, the probability of having missing data in the aggregated annual measure is related to the number of different income items observed (which raises the odds of at least one missing component). In the SOEP for 2004, income information is collected on 10 different components. Here, the overall share of those with INR on aggregated labor income is 12.5%, of which 8.3% need to be completely imputed. In the case of the BHPS, on the other hand, one would expect INR to be less frequent given that only one income question is asked. However, especially for individuals with a rather volatile employment history, the difficulty of recalling their various past income data might make it difficult to provide accurate aggregated income information (see Jürges, 2007). Obviously, there is a tradeoff problem between a simple "one-shot" question and a more detailed set of questions aimed at collecting the same information (see Micklewright & Schnepf, 2007), and this may in turn affect cross-national comparability. Furthermore both the HILDA and BHPS questionnaires offer a "don't know" category, which may also tempt respondents to refrain from giving an income value. Finally, one should note that any seemingly valid observed income information may be affected by measurement error as well (see Abowd & Stinson, 2007). Again, from the perspective of cross-nationally harmonized data, this latter argument may be more relevant in cases where the respondents are forced to provide aggregated information.

Following from these findings and given our substantive analytical interest in comparative inequality and *mobility* analysis, there is an inherent need to control for possible state dependence of INR. Although there is growing awareness of the risk of selectivity inherent in INR (at least since Ferber, 1966), much of the literature on nonresponse behavior in longitudinal studies has focused on unit nonresponse (UNR) and on the possible bias arising from selective attrition in such surveys (see Groves, 2006; Watson & Wooden, 2009). Several studies have argued that these two types of nonresponse, INR and UNR, should be analyzed in a common framework (e.g., Lee, Hu, & Toh, 2004) or have proposed that respondents be arranged on a "cooperation continuum" (see Loosveldt, Pickery, & Billiet, 2002; Schräpler, 2004) ranging from (a) those who will (always) be willing to participate in surveys and also to provide valid answers, (b) those who will be more or less willing to cooperate (i.e., who will take part in the survey as such but who may refuse to answer certain items, causing INR), and finally (c) those who will not take part at all (causing unit nonresponse, UNR). Above and beyond these basic traits, there will most likely also be situational factors that interfere with an individual's basic willingness or ability to cooperate. These may include severe illness, exceptional events such as the death of a relative, or an unpleasant encounter with the interviewer.

The results presented in Figure 19.1 shed some light on the intertemporal link between INR and UNR in the three panels considered here and on the potential selectivity this entails for mobility analyses in a cross-national setting. Separating individual observations by imputation status at time t0 (i.e., observed income [Obs. in t0] vs. missing income [INR in t0]) and thus ignoring observations with zero labor income, we differentiate four potential outcomes at time t1, namely "observed earnings," "INR with subsequent imputation," "not employed, thus zero labor income," and "attrition." In all three panel studies, we not only find indica-

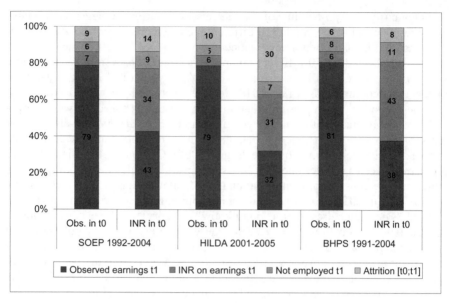

Figure 19.1. INR in Panel Perspective: The Case of Labor Earnings
Source: SOEP 1992–2004; HILDA 2001–2005; BHPS 1991–2004.

tions of state dependence of INR, but also clear support for the "cooperation continuum" hypothesis, according to which INR is a valid predictor of subsequent UNR, namely attrition. Among those who reported their earnings in the previous year, only about 6–7% refuse to do so in the subsequent wave. Individuals with INR in the labor income measure in t0, however, show not only a much higher share of INR in the following year (ranging from 31% in HILDA to 43% in the BHPS), but they are also clearly more likely to suffer attrition. This is a very relevant finding for international comparisons based on panel data focusing on income mobility (see Section 19.4.1): Obviously, the more the datasets at hand differ with respect to scope as well as selectivity of nonresponse (INR and UNR) across time, the greater the uncertainty about the comparability of such mobility analyses.

19.2.3 Selectivity of INR

Thus, for the sake of cross-national comparability, it is of prime importance to control for whether the missing mechanisms are similar in the datasets considered here. For each of the panels separately and utilizing the panel nature of the underlying data, we specify a random effects probit model estimating the probability of INR on our measure of annual labor earnings. The application of this method allows us to control for unobserved heterogeneity in the response behavior of the respondents.

Based on currently employed individuals (including the self-employed) aged 20–65 years, we control for sociodemographic characteristics, the interview situation, the survey experience of the respondent, as well as for the complexity of the income portfolio. The latter is operationalized by various dummy variables indicating changes in an individual's labor market status over the previous (calendar or financial) year by identifying experience of unemployment, the number of months spent in employment, and a possible exit from education (see Table 19.1).

The marginal effects shown in Table 19.1 reflect the change in the probability of INR given a unit change in the independent variable. As a result, an additional month in full-time employment, for example, significantly reduces the probability

TABLE 19.1. Estimating the Probability for INR on Labor Income—Marginal Effects from Random Effects Probit Models

Covariates (D=Dummy)	Germany (SOEP)		Australia (HILDA)		UK (BHPS)	
Age	−0.002	(0.006)	−0.013	(0.011)	0.001	(0.007)
Age squared	0.000	(0.000)	0.000	(0.000)	0.000+	(0.000)
Male (D)	0.067*	(0.021)	−0.237**	(0.034)	0.075*	(0.030)
Educ. level = Low (D)	0.039	(0.026)	−0.104*	(0.042)	−0.073*	(0.033)
Educ. level = Middle (D)	0.043	(0.033)	−0.236**	(0.048)	−0.142**	(0.032)
Educ. level = Univ (D)	−0.033	(0.032)	−0.250**	(0.067)	−0.266**	(0.081)
Disability status (D)	0.045	(0.039)	0.106**	(0.040)	−0.019	(0.030)
Married (D)	−0.010	(0.021)	−0.149**	(0.039)	−0.009	(0.025)
# HH members aged 0–14	0.010	(0.011)	0.026	(0.017)	0.005	(0.009)
Metropolitan area (D)	−0.029	(0.029)	−0.145**	(0.042)	−0.087**	(0.028)
Remote area (D)	0.025	(0.019)	0.092+	(0.051)	−0.107**	(0.037)
Tenure (years)	−0.002	(0.003)	−0.027**	(0.005)	−0.068**	(0.004)
Tenure squared	0.000+	(0.000)	0.001**	(0.000)	0.002**	(0.000)
Foreigner (D)	−0.024	(0.032)	0.058	(0.043)	0.020	(0.040)
Public service (D)	0.019	(0.021)	−0.221**	(0.083)	−0.092**	(0.022)
Firm size: small (D)	−0.018	(0.019)	0.257**	(0.037)	0.015	(0.022)
Firm size: large (D)	0.000	(0.021)	−0.045	(0.062)	−0.012	(0.036)
# Months FT last year	−0.018**	(0.003)	−0.048**	(0.008)	−0.015**	(0.004)
Unemployed last year (D)	0.090**	(0.030)	0.237**	(0.065)	−0.083*	(0.036)
Left educ. last year (D)	0.022	(0.033)	−0.028	(0.047)	−0.127+	(0.075)
Self-employed (D)	0.475**	(0.028)	1.328**	(0.041)	1.053**	(0.029)
Problems during interv. (D)	0.204**	(0.016)	−0.206*	(0.098)	0.117	(0.074)
Two interviews (D)	−0.124+	(0.065)	−0.221**	(0.081)	−0.014	(0.084)
Three and more interv. (D)	−0.364**	(0.047)	−0.385**	(0.062)	−0.294**	(0.060)
Obs.	120 818		35 238		72 696	
N	24 178		10 722		11 134	
−2 Log-likelihood	−36536.506		−5151.03		−24036.69	
Pseudo-R-squared	.1244		.1609		.2120	

Note: Dummies for waves included, but not reported. Reference groups: Education (no vocational education); Area (intermediate); Firm Size/number of employees (intermediate); Standard errors in parentheses; Significance level: + significant at 10%; * significant at 5%; ** significant at 1%.
Source: SOEP 1992–2004; HILDA 2001–2005; BHPS 1991–2004.

of INR by about 1.8% in the case of SOEP, by 4.8% in the HILDA survey, and by 1.5% in the BHPS. In brief, INR on previous year's labor income is clearly more frequent among the self-employed, while it becomes less likely with an increasing number of months in full-time employment. As expected, one finds a higher probability of INR in SOEP and HILDA among those who were unemployed at some point during the last year, but the opposite effect is seen in BHPS. This finding may result from the different wordings used to ask for this information in the BHPS for those currently employed versus those currently unemployed.

Findings are also inconsistent with respect to gender, with SOEP and BHPS showing more INR among men, whereas in HILDA women more often provide what appears to be a valid answer to labor income questions. In HILDA and BHPS, there is a significant negative education effect—i.e., more highly educated individuals are more likely to provide the requested information. We observe a similar tendency among SOEP respondents holding a university degree, but this effect is not statistically significant. Controlling for long-term employment patterns, it appears that INR declines with the number of years spent with an employer, but at a reduced pace. When controlling for education, *ceteris paribus*, there is no significant immigrant/citizenship effect in any of the three surveys. The UK and Australian panels do, however, confirm our expectation of higher response propensity among public servants. The strong and significant results for the INR-reducing effect of survey experience, here measured by the number of interviews conducted in the course of the panel survey, are consistent across all three country datasets. This finding clearly illustrates the need to control for methodological features above and beyond (standard) substantive characteristics.

Summing up the results of this section, we observe a rather similar incidence of INR across surveys, despite the fact that the surveys ask for labor income in different ways. With respect to the selectivity of INR in the three panels, we find some cross-country similarities, but country-specific patterns of INR as well.

19.3 IMPUTATION RULES IN THE THREE SURVEYS

It is beyond the scope of this chapter to provide an exhaustive description of imputation procedures other than those used in SOEP, BHPS, and HILDA. However, it should be noted that even a very sophisticated approach of substituting for nonresponse may not completely eliminate bias (see Nicoletti & Peracchi, 2006). As such, the choice of the adequate imputation technique is a problem in itself. Potential bias due to imputation may creep in due to "regression-to-the-mean effects," and a potential change in total variance—most likely a decline—may occur. See Rubin (1987) for a discussion of imputation methods. For an evaluation of alternative treatments of INR by means of weighting see, for instance, Rässler and Riphahn (2006). Focusing on imputation in panel studies, the findings by Spiess and Goebel (2003) on the basis of matched survey and register data for Finland clearly argue for the use of longitudinal data in the imputation process. In fact, the three panel datasets at hand explicitly take advantage of the existence of longitudinal information in their respective imputation approaches. While all three panels use single imputation routines, one area for improvement is in the application of *multiple* imputation techniques (see Rubin, 1987) which would more adequately reflect the uncertainty embedded in the imputation procedure as such.

Annual individual labor income in the BHPS is imputed using the regression-based predictive mean matching (PMM) procedure proposed by Little (1988), also known as regression hot-deck imputation. The basic idea of PMM is the use of observed predictor variables from a linear regression to predict variables with missing values. The advantage of this method is that a possible real value is imputed and that a random error component is added to preserve variance. The PMM method adopted in the BHPS also considers longitudinal information, but only from 3 years. Depending on the availability of observed information about labor income in previous and subsequent waves as well as possible job changes, either forward or backward imputation is applied (ISER, 2002).

HILDA and SOEP both use a two-step procedure to impute any income information missing due to INR. The main method is based on "row-and-column imputation" as described by Little and Su (1989; hereafter L&S). Row-and-column imputation takes advantage of cross-sectional as well as individual longitudinal information, using income data available from the entire panel duration, by combining row (unit) and column (period/trend) information. It then adds a stochastic component resulting from nearest neighbor matching. While the SOEP applies this L&S procedure to the entire population, HILDA uses a modification of this technique by matching donors and recipients within imputation classes defined by seven age groups (see Starick, 2005; Starick & Watson, 2007).

A secondary method is needed whenever longitudinal information is lacking. This includes not only first-time respondents, but all those observations for whom a given income variable has been surveyed for the very first time. Hence, a purely cross-sectional imputation method needs to be applied. In the case of HILDA, a nearest neighbor regression method (similar to that used by the BHPS) is used. In the SOEP, a hot-deck regression model is employed, supplemented by a residual term retrieved from a randomly chosen donor with observed income information in the regression model.

In an evaluation of various imputation methods, Starick (2005) argues that "in a longitudinal sense, the L&S methods perform much better when compared to the nearest neighbor regression method. Evidence shows that the L&S methods preserve the distribution of income between waves. Furthermore, these methods perform better in maintaining cross-wave relationships and income mobility" (Starick, 2005, p. 31). This finding is confirmed by Frick and Grabka (2005) for the SOEP.

To check for robustness and to control for possible effects of the choice of imputation strategy on income inequality and mobility, we ex-post harmonize the imputation technique across all three surveys by applying the L&S imputation method to the BHPS data as well. It must be noted that we do not impute the single income components but only the aggregated "annual labor earnings" measure used here. About 80% of individuals with missing labor earnings can be imputed with the L&S method, while for the remaining 20% we use the original BHPS regression results. In other words, there are no *longitudinal* earnings data available for the latter group.

In the following, we will compare results obtained from the imputation techniques as given by the various original data providers: SOEP, HILDA, and BHPS. For the latter, we will provide a point of comparison using the alternative imputation method based on Little and Su (1989).

19.4 EMPIRICAL APPLICATION: THE EFFECT OF IMPUTATION

Keeping in mind the above findings on the incidence and selectivity of INR across panels as well as the differences and commonalities in the respective imputation processes, the following analyses focus on the effect of imputation on prototypical applications: earnings inequality and mobility (Section 19.4.1) as well as wage regression (Section 19.4.2).

19.4.1 The Effect of Imputation on Earnings Inequality and Mobility

Accepting the applied imputation strategies, i.e., assuming that these adequately correspond to the underlying missing mechanism, any increase in selectivity of nonresponse will obviously be reflected in the deviation of empirical results based on truly observed cases ("complete case analyses") from those derived on the basis of all observations (i.e., observed plus imputed cases).

The analysis of income inequality is based on pooled, deflated income data for all available years as described in Section 19.2, and thus the results presented in Table 19.2 give the average picture over that period (see Frick & Grabka, 2007, for details on time series of inequality). A comparison of basic statistics of annual gross labor income (Table 19.2) shows income levels (given by mean and median) for "all cases" to be clearly lower than those of the population with observed values in the case of BHPS and HILDA, while in the SOEP this tendency is weakened. For example, the overall median in HILDA based on "observed and imputed cases" is about 2.3% lower than the value resulting from "observed cases" only—the median of the imputed values is 11% lower than that of the observed ones. This is some indication for the selectivity of the missing process as persons who experienced transitions in and out of employment during the previous year (e.g., after leaving the educational system or when re-entering the labor market after some spell of unemployment) are more likely not to report or not to know their often below-average annual labor incomes.

Extending our perspective to cross-sectional measures of *inequality*, a rather robust picture of understated inequality appears when using "complete case" analysis. Using various indicators, we find statistically significant differences after including imputed values. For example, both the 90:10 decile ratio and the MLD (mean logarithmic deviation) for the observed cases in HILDA understate inequality—measured on the basis of all cases—by about 5%, while in Germany the top-sensitive HSCV (half-squared coefficient of variation) increases by almost 4%, and even the change in the rather robust Gini coefficient indicates increased inequality when considering the imputed cases as well. The results obtained from the row-and-column imputation of missing income data in the BHPS instead of the

TABLE 19.2. Income Inequality and Income Mobility by Imputation Status

	Germany (SOEP)		Australia (HILDA)		UK (BHPS)			
					Original Imputation		"L&S" Imputation	
	"All cases"	Deviation: "All" vs. "Obs." (%)	"All cases"	Deviation: "All" vs. "Obs." (%)	"All cases"	Deviation: "All" vs. "Obs." (%)	"All cases"	Deviation: "All" vs. "Obs." (%)
Basic statistics*								
Mean	24408	+0.0	27349	−1.0	13621	−1.8	13849	−0.2
Median	21940	−0.6	23375	−2.3	11360	−2.7	11553	−1.1
Inequality								
Theil 0 (Mean log deviation)	0.4096	+1.0	0.4587	+4.5	0.4425	+8.6	0.4211	+3.4
Gini	0.4141	+1.0	0.4273	+1.9	0.4280	+1.7	0.4268	+1.4
Half-SCV (top-sensitive)	0.3488	+3.5	0.4456	+1.5	0.4709	+4.9	0.4652	+3.6
Decile ratio 90:10	13.71	+0.4	14.92	+5.2	12.69	+6.1	12.43	+3.9
Decile ratio 90:50	2.13	+1.0	2.20	+2.3	2.33	+1.8	2.32	+1.1
Decile ratio 50:10	6.45	−0.6	6.76	+2.7	5.44	+4.3	5.38	+3.0
Average N per cross-section	10773	+13.4	9082	+15.3	5002	+18.1	5002	+18.1
Mobility								
Quintile matrix mobility: Avg. jump	0.448	+19.1	0.530	+15.5	0.456	+30.7	0.457	+31.0
Quintile matrix mobility: Norm. avg. jump	0.179	+19.3	0.212	+15.2	0.183	+30.7	0.183	+30.7
Fields & Ok: Percentage income mobility	24.38	+29.1	28.81	+17.5	25.42	+47.0	24.74	+43.1
Fields & Ok: Non-directional	0.333	+10.6	0.447	+16.4	0.348	+27.5	0.326	+19.4
Shorrocks: Using Gini Coefficient	0.0290	+19.8	0.0445	+30.8	0.0279	+40.3	0.0254	+27.8
Average N per 2-wave balanced panel	9878	+30.8	7474	+19.9	4389	+37.7	4389	+37.7

* Germany in 2000 euros; UK in 1996 GBP; Australia in 1989/90 AUD.
Shaded cells indicate statistically significant deviations at the 95% confidence level (for HILDA and SOEP based on a random group approach; in the case of BHPS bootstrapping with 200 replications was applied).
Source: SOEP 1992–2004; HILDA 2001–2005; BHPS 1991–2004.

originally provided hot-deck imputation yields somewhat higher imputed values, but inequality among the L&S-imputed observations is less pronounced here. Following from this, the deviation between "all" and "observed" in the UK data using the alternative imputation scheme is not significant in any of the measures employed.

As shown above, the missing mechanisms for INR on labor income point toward selectivity with respect to characteristics found more often among attriters. Given that attrition is controlled for in most panel surveys by means of weighting factors that represent the inverse probability of being selected into the sample as well as dropping out, one may assume that the use of population weights in the present context will increase the percentage of the population showing INR. Indeed, the weighted population share containing imputed labor income data is as high as 13% in SOEP, 15% in HILDA, and 18% in BHPS.

With respect to labor income *mobility*, as is true for any longitudinal analysis, one can expect the effect of imputation to be even larger because INR may be an issue in any one of the waves under consideration. For simplification, we only use a series of two-wave balanced panels (pooled across all available waves in each survey), i.e., the effects shown here would be even more pronounced in multi-wave analyses (see Table 19.2).

Above and beyond the general finding of inequality being understated among the "observed cases," clearly more distinct and statistically significant differences can be found for labor income *mobility*—conditional on the imputation techniques applied. Depending on the mobility measure used and on the population share affected by imputation, the results between "observed" and "all cases" (including the imputed ones) deviate in the original BHPS by as much as 27–47%, while the degree of mobility is less pronounced using the alternative imputation (index values between 19 and 43%, respectively). In SOEP, the corresponding shares are between 10 and 30% and in HILDA this range is from 15 to 31%. In other words, having in mind our focus on international comparability, it should be noted that the effect of imputation on substantive research results becomes more similar across countries when using a similar imputation procedure.

Focusing only on "complete cases" would yield an even higher loss in statistical power or efficiency due to the massive reduction in the number of observations. The last row in Table 19.2 indicates that the (weighted) population share containing imputed data in at least one of the two waves considered is as high as 20% in HILDA, 31% in SOEP, and 38% in BHPS.

19.4.2 The Effect of Imputation on Wage Regressions

Obviously, there is convincing evidence for selectivity in INR on labor income questions in all three panel datasets considered. It therefore stands to reason that coefficients derived from (simple) wage regressions based on observed data only will be biased as well. In other words: can correct inferences be drawn from a dataset excluding observations with INR? Such phenomena could potentially be dealt with by estimating a Heckman selection model (Heckman, 1979), where the selection function would focus on the INR and the wage regression would be based only on the "observed" values. Even if this allowed for a perfect correction, there would still remain the problem of a loss in efficiency (caused by the loss in observations).

In the following, we will try to shed some light on this issue by comparing the results of survey-specific fixed-effects wage regressions based on the "observed" cases (Column 1 of Table 19.3a) to those based on the entire population including the imputed ones (Column 2). Finally, in Column 3 we repeat the estimation from Column 2, but add a dummy variable identifying the imputed observations. Table 19.3a gives the results separately for the three panels (as well as for the alternative BHPS imputation), controlling for usual covariates relating to human capital, sociodemographics, regional agglomeration, health status, and (changes in) labor market participation over the last year. We refrain from including covariates focusing on the current employment situation in order to be able to include individuals who are currently not employed but did receive earnings in the observation period (such as those who recently retired or who are currently unemployed). Given the analytical interest of this chapter, the results presented in Table 19.3 focus on the imputation status and strategy (for full results, see Frick & Grabka, 2007). By and large, the findings based on "observed cases" are widely consistent for SOEP and BHPS with respect to the direction and significance of most parameter estimates as well as with respect to the overall degree of explained variance (about 50%). For HILDA, however, the specified model performs rather poorly with an exceptionally low R-squared (approx. 22%) for this kind of analysis (see Watson, 2005a). Nevertheless, the estimated coefficients show the expected direction although they sometimes lack statistical significance.

More importantly for the purposes of this chapter, however, is the effect of the additional consideration of imputed observations (see Column 2): in all three panels, this yields a pronounced reduction in the degree of explained variance. This decline is most prominent for HILDA with an approximately 23% reduction in R-squared to only 0.172. Obviously, this effect is driven by the consideration of a less homogenous group of individuals due to the aforementioned selectivity of INR. This is seen in the fact that "all" observations (see Column 2) include significantly more self-employed people in all three datasets. Other striking differences are seen in BHPS in the under-representation of retired individuals, and in SOEP and HILDA in the under-representation of those who experienced at least one month of unemployment in the previous year. Observations from the first waves of BHPS and HILDA are also underreported among the observed cases, while this is not the case in the more mature panel population in SOEP.

Comparing regression results for the original BHPS imputation to those based on the alternative L&S imputation, it appears that the deviations between the coefficients derived from the observed cases and from the overall sample are not always perfectly in line. Although the two methods do not show any explicit contradictions, the coefficients for "remote area" and "disabled" lose statistical significance when using the L&S method and the consistently significant age effect is reduced in size. Such variations, however, may simply result from the selection of controls in the PMM regression model underlying the original BHPS imputation.

TABLE 19.3a. Results from Fixed-Effects Wage Regressions (Dependent Variable: Log Annual Labor Income)

	Germany (SOEP)			Australia (HILDA)			UK (BHPS) Original Imputation			"L&S" Imputation		
	(1) Observed cases	(2) All cases	(3) All cases	(1) Observed cases	(2) All cases	(3) All cases	(1) Observed cases	(2) All cases	(3) All cases	(1) Observed cases	(2) All cases	(3) All cases
Controls *	yes	yes	yes	yes	yes	yes	yes	yes	yes	yes	yes	yes
Imputation status (1=yes)	—	—	0.064**	—	—	0.052**	—	—	−0.042**	—	—	0.047**
			(0.005)			(0.014)			(0.006)			(0.006)
Constant	7.515**	7.543**	7.533**	5.247**	5.240**	5.223**	5.959**	6.071**	6.093**	5.959**	6.805**	6.781**
	(0.042)	(0.042)	(0.042)	(0.172)	(0.176)	(0.176)	(0.262)	(0.271)	(0.271)	(0.262)	(0.270)	(0.270)
Observations	119 030	134 337	134 337	35 661	38 681	38 681	62 049	72 729	72 729	62 049	72 904	72 904
N (Persons)	24 183	25 487	25 487	11 097	11 522	11 522	10 352	11 137	11 137	10 352	11 138	11 138
R-squared	0.4869	0.4474	0.4484	0.2228	0.1723	0.1727	0.5169	0.4368	0.4372	0.5169	0.3661	0.3666

* In all estimations the following control variables are included but not reported in the table: age; age squared; sex; education; co-residing children; marital status; disability status; regional information and community size; dummy variables for leaving education and for moving into retirement during the last year; number of months in unemployment, in part time employment and in full time employment during the last year; time effects. Full results are available in Frick & Grabka, 2007.

Population of working age: 20–60 (Germany), 20–65 (Australia and UK)

Note: Standard errors in parentheses; Significance level: + significant at 10%; * significant at 5%; ** significant at 1%.

Source: SOEP survey years 1992–2004; HILDA survey years 2001–2005; BHPS survey years 1991–2004.

TABLE 19.3b. Results from Quantile Wage Regressions (Dependent Variable: Log Annual Labor Income, Normalized)

	Germany (SOEP)			Australia (HILDA)			UK (BHPS) Original Imputation			"L&S" Imputation		
	p25	p50	p75	p25	p50	p75	p25	p50	p75	p25	p50	p75
Controls *	yes	yes	yes	yes	yes	yes	yes	yes	yes	yes	yes	yes
Imputation status (1=yes)	-0.0274**	0.004	0.0457**	0.195**	-0.017	0.078**	-0.119**	-0.076**	-0.036**	-0.139**	-0.004	0.066**
	(0.0068)	(0.0054)	(0.0041)	(0.030)	(0.013)	(0.021)	(0.010)	(0.006)	(0.007)	(0.012)	(0.010)	(0.008)
Constant	1.4541**	2.155**	2.6652**	-0.960**	0.446**	1.652**	0.532**	1.141**	1.720**	0.537**	1.251**	1.935**
	(0.0327)	(0.0274)	(0.0284)	(0.115)	(0.086)	(0.071)	(0.055)	(0.043)	(0.042)	(0.073)	(0.055)	(0.043)
Observations	139321			38681			72729			72904		
R-squared	0.477	0.395	0.349	0.264	0.208	0.168	0.331	0.279	0.243	.307	.256	.222

Test on significant differences of imputation effect between the 25th and 75th percentile:

	F(1,139321) = 94.41			F(1, 38661) = 69.15			F(1, 72700) = 46.77			F(1, 72875) = 241.16		
	Prob > F = 0.0000			Prob > F = 0.0000			Prob > F = 0.0000			Prob > F = 0.0000		

* In all estimations the following control variables are included but not reported in the table: age; age squared; sex; education; co-residing children; marital status; disability status; regional information and community size; dummy variables for leaving education and for moving into retirement during the last year; number of months in unemployment, in part time employment and in full time employment during the last year; time effects. Full results are available in Frick & Grabka, 2007.

Population of working age: 20–60 (Germany), 20–65 (Australia and UK)

Note: Standard errors in parentheses; Significance level: + significant at 10%; * significant at 5%; ** significant at 1%.

Source: SOEP survey years 1992–2004; HILDA survey years 2001–2005; BHPS survey years 1991–2004.

Finally, Column 3 contains the repetition of the estimation from Column 2, but controlling for imputation status. The corresponding effect indicates that individuals with imputed incomes, *ceteris paribus*, earn significantly above average in SOEP and HILDA (about 5–6% more), while they earn 4% less than the average in the original BHPS data. However, changing the imputation strategy for the BHPS yields a positive imputation effect of similar size. In other words, applying the "row-and-column" imputation to all three panels, we also find a very similar effect for the imputation dummy in the fixed-effects wage regression model.

For each set of panel data separately, we estimate quantile regressions (at the 25th, 50th, and 75th percentiles), controlling for potential regression-to-the-mean effects emerging from the imputation process across the earnings distribution (see Table 19.3b). Consistently for all estimations, the result for the imputation dummy is smallest at the 25th percentile, intermediate at the median, and finally, strongest at the 75th percentile. A Wald test confirms this effect to be statistically different between the 25th and the 75th percentile. We interpret these findings as an indication that the imputation techniques applied did not produce a relevant regression-to-the-mean effect.

For the BHPS, in line with the changing effect of the imputation flag when changing the imputation strategy in the fixed-effects wage regressions, we find an almost identical effect across the UK earnings distribution. The L&S imputation method also produces a significant negative effect for imputed observations at the 25th percentile, which becomes insignificant at the median, and finally positive and significant at the income threshold to the upper quartile. Again, without even arguing about the pros and cons of the various imputation techniques at hand, it is the decision to apply the same or at least a very similar imputation technique across all surveys considered in the analyses that clearly affects (and in this case improves) the cross-country comparability of substantive research results.

19.5 SUMMARY AND RECOMMENDATIONS

Cross-national comparability of empirical research based on microdata is hampered by deviation among national datasets in the scope and selectivity of item non-response (INR) as well as in the approach chosen to handle this phenomenon. Analyzing annual labor earnings in three large panel surveys (German SOEP, British BHPS, and Australian HILDA), we find only minor differences in the incidence of INR, while at the same time there is considerable cross-country variation with respect to the selectivity of INR.

Longitudinal imputation is the preferred way to handle INR in all three panels, with HILDA and SOEP using the basic strategy suggested by Little and Su (1989), and the BHPS using a hot-deck regression approach. In all three surveys, the selectivity of INR and hence the imputation of such missing data appears to have a significant effect on both the distribution of earnings and earnings mobility: Considering only those cases with observed information significantly understates income inequality as well as variability over time. Moreover, our study provides evidence for a positive intertemporal correlation between INR and any kind of subsequent (item and unit) non-response. Results from multivariate regression models analyzing the determinants of labor income indicate that individuals with imputed incomes, *ceteris paribus*, earn significantly above average in SOEP and HILDA, while this relationship is negative using BHPS data. Applying the imputation approach used in HILDA and SOEP to the BHPS provides an empirical basis for robustness and sensitivity checks with respect to the choice of the imputation technique. Indeed, cross-country variation in the descriptive results on inequality and mobility is reduced and the effect of controlling for imputation status is reversed for the BHPS data when using the same imputation technique.

The most important lesson to be learned from the present study is that the cross-national variation in INR presented here—variations in scope and selectivity, in strategies used, and consequences for prototypical labor income analyses— emphatically confirms the importance of further harmonizing the methods used to handle missing (income) data in (panel) surveys. For panel studies, this includes the choice of an imputation procedure that explicitly considers longitudinal information, if available. In line with our empirical findings, for comparative research the imputation techniques should be as similar as possible for all country datasets involved. In this way, any potential bias arising from the choice of the technique rather than reflecting true cross-national differences will be minimized. A proper imputation is certainly preferable to simply ignoring cases with missing data by assuming the underlying missing mechanism to be completely at random as well as reducing efficiency due to the reduced sample size, but even this may yield biased results. Nevertheless, the question of whether to use imputation for the treatment of missing values and if so, which imputation techniques and control variables, may depend heavily on the specific question under analysis. Thus, no definitive, "one-size-fits-all" imputation method exists, and our evidence under- scores the possible variety of imputation methods that may be used by data providers. Data users should therefore not view the imputations produced by data providers as a panacea but should keep the potential shortcomings of the various methods in mind. When using imputed data, especially in a cross-national context, one should control for imputation status, thus effectively controlling for different characteristics underlying the missing mechanism as well as for any specific treatment of data in the imputation process.

While this chapter dealt with three countries only, it is apparent that even more discussion of ex-ante standardization and harmonization needs to be in place to improve comparability across 20 or more countries, e.g., at the level of the European Union's Statistics on Income and Living Conditions (EU-SILC). As argued elsewhere in this volume, a mix of centralized agreements and national expertise allowing for country-specific applications is necessary at all levels of the data production process, starting from the design stage of the survey instruments, through the fieldwork phase, the post-data-collection treatment (including imputation, its documentation and flagging in the microdata), as well as at the research level. Throughout these processes, institutions and all their staff involved ideally observe quality management procedures (see Lyberg & Biemer, 2008). This is in line with the ISO 20252 standard, which aims to set a common level of quality for market research globally. In so doing, data producers will enable data users to conduct sensitivity tests to determine the impact of imputation, which—as shown in this chapter —may be even more significant in the case of cross-national analyses. In the long run, this kind of methodological feedback from the user community may help to further improve the quality of the imputation methods used by data collection, production, and dissemination agencies.

ENDNOTE

This chapter uses confidentialized unit record data from the Household, Income, and Labour Dynamics in Australia (HILDA) survey. The HILDA Project was initiated and is funded by the Commonwealth Department of Families, Community Services and Indigenous Affairs (FaCSIA) and is managed by the Melbourne Institute of Applied Economic and Social Research (MIAESR). The findings and views reported in this chapter, however, are those of the authors and should not be attributed to either data producers or financing agencies.

PART VII

ANALYZING DATA

20

An Illustrative Review of Techniques for Detecting Inequivalences

Michael Braun and Timothy P. Johnson

There are currently several analytic approaches available that have been proposed as useful tools for assessing measurement comparability in cross-cultural research. This introduction gives a brief overview of the main issues involved in statistically dealing with comparability problems, describes the main approaches used in the literature (including those not covered in Chapters 21–24, this volume), and applies them all to the same data. It thus sets the scene for the contributions to follow.

In order to demonstrate all the techniques with one example, we must make several simplifications. We forego a comprehensive presentation of the results which can be obtained when applying each technique and concentrate instead on an illustration of the main insights the different techniques offer. We also have to relax some of the usual requirements for some of the techniques, in particular concerning the scale level of the variables normally required. Thus, this chapter should be read as an illustration of the logic of these techniques rather than a primer for applying them. For more advanced techniques, we refer both to introductory literature and to additional empirical applications.

A variety of approaches are covered, from the most basic to more advanced and sophisticated approaches. This distinction does not fully overlap with a categorization into exploratory and confirmatory techniques. Exploratory techniques are found not only among the basic but also among the more advanced methods. The techniques introduced in the four following chapters belong to the most advanced. In this chapter we consider more advanced techniques alongside more basic ones. While more advanced techniques allow for a greater flexibility and thoroughness in the analysis, basic techniques such as considered here also have advantages. First, they are easier to implement—a considerable plus point for researchers who are usually more interested in substantive problems than in methodology. They may also allow for a more immediate grasp of potential pitfalls when comparing survey data across nations or cultures. Some of the more advanced techniques might involve considerably greater effort in order to arrive at the same general conclusions. The basic techniques are also useful to familiarize oneself with the data. This could be crucial in deciding whether surprising results resulting from (improper) application of more complicated techniques are in fact artifactual. Moreover, as we will show, a superficial application of the more advanced techniques, e.g., exclusively taking into account fit indices without also examining other relevant information (such as modification indices in confirmatory factor analysis), might lead to a failure to detect problems. Some of the statistically more advanced methods of analysis are no more difficult to apply than the basic methods. Also, as software becomes more advanced and user-friendly, some of the advanced techniques may become easier to use and consequently more popular.

Criteria for the selection of appropriate statistical techniques for cross-national and cross-cultural analyses might include, in addition to the usual criterion of scale level, the degree to which they permit a quick overview of problems, the degree to which they can handle a large number of countries or cultural groups, the degree to which they allow an investigator to identify individual countries/cultures that are analytically problematic, the degree to which they are able to provide summary measures of comparability for the entire set of countries/cultures, and the degree to which they address problems at the item or test level.

In Table 20.1, we identify some of the more common analytic approaches currently available for evaluating the comparability of survey measures across nations or cultures. In the remainder of this chapter, we briefly illustrate and evaluate these different techniques, outlining what each procedure can do to assist the researcher. For a more detailed treatment of these and other methods for the analysis of cross-national and cross-cultural survey data, see van de Vijver and Leung (1997), De Beuckelaer (2005), and Hox, de Leeuw, and Brinkhuis (Chapter 21, this volume).

20.1 DATA

To illustrate these techniques, we use the gender-role battery of the International Social Survey Program's (ISSP) Family and Changing Gender Roles module, which was fielded in 1994. Attitudes toward consequences of parental labor force participation for children are measured with four items (Table 20.2).

Items 1–3 were repeated from the first Family and Changing Gender Roles module of the ISSP, fielded in 1988. They have a pedigree inasmuch as partly identical items have been used since the early 1970s and 1980s in the U.S. General Social Survey and the German ALLBUS (Davis, Smith, & Marsden, 2007; Smith, Kim, Koch, & Park, 2005). The third item (*warm relation*) has been found to work differently in some countries, while the first and the second items (*child suffers* and *family suffers*) have been shown to function in a nearly identical manner across countries. Braun (2006) suggests that the likely reason for these differences is that the third item is understood differently in more traditional countries than it is in less traditional countries. In less traditional countries, all three items are interpreted in terms of negative consequences for the children arising from the

TABLE 20.1. Techniques for Evaluating Survey Data Comparability Across Nations and Cultures

Distributions across response categories, including nonresponse
Means
Correlations with benchmark items
Correlations with explanatory variables
Interaction plots
Exploratory factor analysis
Reliability
Multiple correspondence analysis (MCA)
Multidimensional scaling (MDS)
Multilevel modeling (see Chapter 21, this volume)
Item response theory (IRT) models (see Chapter 22, this volume)
Multigroup structural equation modeling (MGSEM) (see Chapters 21 and 22, this volume)
Multilevel structural equation modeling (MLSEM) (see Chapter 21, this volume)
Multilevel latent class analysis (see Chapter 21, this volume)
Multitrait-Multimethod (MTMM) analysis (see Chapter 23, this volume)

TABLE 20.2. Items Used in the ISSP 1994 Study[a]

Consequences of parental labor force participation for the children

1. *Child suffers*: A preschool child is likely to suffer if his or her mother works.

2. *Family suffers*: All in all, family life suffers when the woman has a full-time job.

3. *Warm relation*: A working mother can establish just as warm and secure a relationship with her children as a mother who does not work.

4. *Men work too much*: Family life often suffers because men concentrate too much on their work.

Benchmark item for gender ideology

5. *Housework women's job*: A man's job is to earn money; a woman's job is to look after the home and family.

[a] Response scale from 1 = strongly agree to 5 = strongly disagree; *warm relation* is reverse-coded.

mother's labor force participation. Since these differences in interpretation are thus related to the degree to which traditional attitudes are present in a society, we use former West Germany, the United States, and Canada for comparison here. In 1994, western Germany was still one of the most traditional regions with regard to the consequences of parental labor force participation (together with Austria, Bulgaria, Hungary, and Russia), while Canada was one of the least traditional countries, together with Norway, Sweden, and eastern Germany (cf. Braun, 2006; Haller, Höllinger, & Gomilschak, 1999).

When the ISSP 1994 questionnaire was originally drafted in the early 1990s, Item 4 (*men work too much*) was developed to address the male role in this context of work and family. Unfortunately, this item suffers from a conceptual weakness (Braun, 2006): While it can be expected that egalitarian respondents would be in favor of a more substantial role for fathers in childcare (which would make them endorse this item), they would at the same time be more supportive of parental labor force participation and, thus, should also be tolerant with regard to men's work role (which would make them tend to disagree with this item). This means that Item 4 has opposing implications for the consequences dimension and the more general gender-ideology dimension. Therefore, respondents will have considerable difficulty giving an answer and, as the results reported below show, the resulting data are fairly useless for substantive data analysis. Table 20.2 also lists a benchmark item representing general gender ideology, *housework women's job*, which according to McInnes (1998) is "... not only a classic statement of male breadwinner ideology, but captures one of the essentials of a patriarchal sexual division of labor: that men are naturally suited to public activity and women to private nurturance" (p. 243).

We use only the data of respondents who were at least 18 years old at the time of the survey. With the exception of the presentation of item nonresponse in Table 20.3, we exclude cases with missing values on any of the substantive items. This leaves 2,019 cases for western Germany, 1,355 for the United States, and 1,359 for Canada.

20.2 PRESENTATION OF RESULTS OBTAINED BY DIFFERENT TECHNIQUES

Some of the most basic methods for analyzing cross-national comparability include (1) examining the distributions of responses across response categories [including "don't know" (DK) responses], (2) comparison of means or other measures of central tendency of different items across countries, (3) comparison of correlations between the items under investigation and measures of the underlying dimension, with benchmark items that can be assumed to represent the dimension or a substantively related concept, and (4) comparison of correlations between the items under investigation with socio-demographic variables that are assumed to influence them.

20.2.1 Distributions (Including Item Nonresponse)

Table 20.3 shows the distribution of the answers for the four items in the three countries. Answers for the third item, *warm relation*, have been reverse-coded. While western Germans show much more traditional attitudes than Americans and Canadians when measured by the first two items, the Germans seem to be less traditional than the Americans and Canadians with respect to the third item. With the fourth item, differences between countries are very small. Although comparing distributions makes use of the full information in the data, this becomes rather confusing with an increasing number of countries (and response alternatives).

TABLE 20.3. Percentage Distribution of Answers Across Response Categories by Country (Including Item Nonresponse)[a]

Western Germany (n=2,019)	1 traditional	2	3	4	5 nontrad.	NA
Child suffers	26.6	41.9	10.4	14.1	3.4	3.7
Family suffers	23.5	36.0	14.0	17.3	5.9	3.4
Warm relation	4.0	15.8	4.0	35.6	39.9	3.7
Men work too much	11.7	48.8	14.8	14.2	2.8	7.8
United States (n=1,355)	1 traditional	2	3	4	5 nontrad.	NA
Child suffers	9.0	31.7	12.1	31.7	13.5	2.1
Family suffers	9.2	25.4	13.6	34.8	15.6	1.5
Warm relation	4.8	18.9	5.0	40.9	28.9	1.5
Men work too much	8.4	48.0	19.8	16.8	2.8	4.2
Canada (n=1,359)	1 traditional	2	3	4	5 nontrad.	NA
Child suffers	4.9	25.7	16.0	35.2	15.9	2.2
Family suffers	5.1	20.1	13.0	36.8	23.1	1.8
Warm relation	3.5	16.0	6.9	41.2	30.8	1.5
Men work too much	9.9	47.6	21.0	16.3	3.1	2.0

[a] For *child suffers*, *family suffers*, and *men work too much* the coding is:
1 = strongly agree, 2 = agree, 3 = neither agree nor disagree, 4 = disagree, 5 = strongly disagree.
For *warm relation*, the coding is reversed.
NA = not answered.

Traditional item analysis also considers item nonresponse as an indicator of respondents having problems answering single items. Table 20.3 thus also shows the percentage of missing values for the four items in the three countries. The results are not very revealing for our purpose. Only the fourth item, *men work too much*, shows higher item nonresponse, and only in western Germany and the United States. However, when item nonresponse and the use of the middle category, which might include no-opinions, are evaluated together, there are hardly any country differences. Western Germans show the highest item nonresponse, and Canadians the smallest. For the use of the middle category, the opposite holds. Americans are intermediate in both cases. For all of the following analyses, we exclude all cases that have a missing value on any of the four items.

20.2.2 Means

Using means for ordinal variables, as in this example, is not fully justified, although such usage abounds in substantive research. For ease of presentation, this fact is ignored both here and when discussing some of the advanced techniques.

As is well known, comparison of means between different countries is permitted only if certain stringent conditions are fulfilled, namely, scalar equivalence (Meredith, 1993; Steenkamp & Baumgartner, 1998). This has yet to be established in our example. However, this restriction relates to substantive conclusions about the relative position of countries. This is not what we intend to do here. What we are proposing is to look at differences in the ordering of countries when different items are used. A comparison of means of *different* items across countries is an easy first step to detect problems. For instance, using different items to rank countries should not result in different orderings. Such a finding could be taken as evidence for invariance across countries.

Table 20.4 presents the means of the four items in the three countries. For *child suffers* and *family suffers*, western Germany appears clearly to be the most traditional, and Canada the least traditional. The United States is closer to Canada, but still more traditional than its neighbor (with the sample sizes in the three countries, differences in the means of at least .2 are statistically significant at the .05 level). For *warm relation* and *men work too much*, however, there are essentially no country differences. Given that the four items were meant to measure the same dimension, these results are very suspicious. It is clear that a comparison between western Germany, on the one hand, and the United States and Canada on the other would yield different results—and lead to different conclusions—depending on which items are used. Statistical analysis alone cannot tell us which items are more appropriate for comparative purposes. Actually, the empirical patterns might reveal problems with the underlying theory in general and the assumption of unidimensionality of the substantive domain in particular. However, since we assume, that the last two items are problematic, the question we need to address is not whether all three countries are on the same level of traditionality—which we posit they are not—but which procedures are best able to detect the weaknesses of the last two items.

20.2.3 Correlations with a Benchmark Item for a Related Concept

Table 20.5 shows Pearson correlations of the four items with the benchmark item for gender ideology. The relationship with *child suffers* and *family suffers* is comparable and strong in all three countries. *Warm relation* shows a similarly strong relationship with gender ideology only in Canada, while the correlation in the United States is somewhat weaker, and in western Germany, the correlation is considerably weaker than for the first two items. Finally, *men work too much* has only a weak relationship to gender ideology in all three countries; in western Germany the correlation is close to being insignificant.

It should be noted that the relationships between the items under consideration with benchmark items representing the underlying dimension or socio-demographic variables can differ between countries for substantive reasons. However, in this case, they should differ in a homogenous way for all of the items, if the items truly have an identical meaning in all three countries.

TABLE 20.4. Item Means (Standard Deviations), by Country

	Western Germany	United States	Canada
Child suffers	2.2 (1.1)	3.1 (1.2)	3.3 (1.2)
Family suffers	2.4 (1.2)	3.2 (1.3)	3.5 (1.2)
Warm relation	3.9 (1.2)	3.7 (1.2)	3.8 (1.1)
Men work too much	2.4 (1.0)	2.6 (1.0)	2.5 (1.0)

TABLE 20.5. Correlations with the Benchmark Item for Gender Ideology (*housework is women's job*), by Country

	Western Germany	United States	Canada
Child suffers	.40[a]	.46[a]	.43[b]
Family suffers	.43[a]	.50[a]	.46[b]
Warm relation	.25[a]	.33[a]	.41[b]
Men work too much	.05[c]	.17[a]	.08[b]

[a] $p<.001$. [b] $p<.01$. [c] $p<.05$.

20.2.4 Correlations with Age of Respondent

A similar picture emerges when age (which in this context stands for a cohort and not a life-cycle effect) is used as an external variable, although all correlations are on a lower level (Table 20.6). For *child suffers* and *family suffers*, correlations are almost identical in all three countries. For *warm relation*, correlations are lower everywhere, but particularly so in western Germany, where the correlation reaches little more than one third of those of the other two countries. For the fourth item, correlations are still somewhat lower in the United States and Canada.

20.2.5 Interaction Plots

Interaction plots combine the analysis of means and correlations. Figure 20.1 shows the means of the four items across age groups. In general, these reveal:

- *Child suffers* and *family suffers* are very similar in their functioning and show the strongest relationship with age in all countries.
- *Warm relation* shows the highest level of nontraditionality everywhere, although the distance to the two benchmark items is most pronounced in western Germany, where the relationship with age is weakest.
- *Men work too much* produces values that hardly differ at all by age group and country.

TABLE 20.6. Correlations with Age of Respondent, by Country

	Western Germany	**United States**	**Canada**
Child suffers	$-.22^a$	$-.22^a$	$-.20^a$
Family suffers	$-.24^a$	$-.25^a$	$-.24^a$
Warm relation	$-.06^b$	$-.16^a$	$-.16^a$
Men work too much	$-.06^b$	$-.13^a$	$-.13^a$

$^a p<.001.$ $^b p<.01.$ $^c p<.05.$

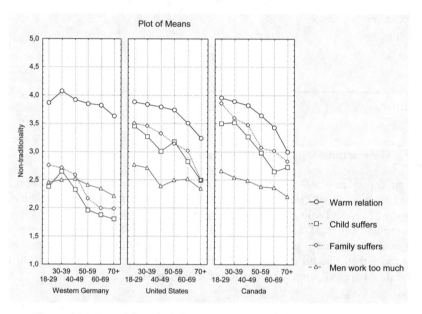

Figure 20.1. Interaction Plot of Item Means by Age Cohorts, by Country

20.2.6 Exploratory Factor Analysis

The four items were also submitted to an exploratory factor analysis with varimax rotation (van de Vijver & Leung, 1997). There is just one factor with an eigenvalue larger than 1.0 in all three countries. Table 20.7 shows the variance accounted for by this factor and the factor loadings for the items. These results provide further evidence that, for all three countries, *men work too much* is not as strongly associated with the parental labor force participation dimension as are the other measures. Country differences with regard to *warm relation* are less visible, although the loading of this item tends to be smallest in western Germany and biggest in Canada.

TABLE 20.7. Variance Accounted for by First Factor and Factor Loadings, by Country

	Western Germany	**United States**	**Canada**
Variance accounted for by first factor	46.6%	54.8%	55.6%
Child suffers	.60	.58	.58
Family suffers	.61	.58	.58
Warm relation	.46	.49	.53
Men work too much	.22	.29	.20

20.2.7 Comparative Reliability Assessment

Table 20.8 shows the reliability of the scales consisting of, first, all four items, then excluding *men work too much* and, finally, also excluding *warm relation*. A scale consisting of all four items has a distinctly lower reliability, as measured by Cronbach's alpha, in western Germany than in the United States and Canada. Excluding *men work too much* improves the reliability considerably everywhere, while excluding also *warm relation* leads to an improvement only in western Germany. Incidentally, western Germany has a significantly lower alpha coefficient across each of the three scales than the United States and Canada, using a test recommended by van de Vijver and Leung (1997). This result could be interpreted in the following way: western Germans find it more difficult to answer even the first two items, a conclusion which would also be supported by the generally higher level of item nonresponse in these data (cf. Table 20.3).

20.2.8 Multiple Correspondence Analysis (MCA)

Multiple correspondence analysis (MCA; Clausen, 1998; Greenacre & Blasius, 1994, 2006) makes no assumptions regarding scale level. Therefore nominal data can also be handled with this procedure. Actually, the procedure can be regarded as a principal components analysis for categorical data. A necessary condition for a high quality item is that the original order of the response categories is retained in low dimensional space (Blasius & Thiessen, 2006). MCA detects violations of ordinality in ordered data. It can also point to other methodological artifacts. Methodological applications of MCA include Blasius and Thiessen (2001) on political efficacy and trust and Blasius and Thiessen (2006) on gender roles.

Figure 20.2 presents the MCA representations for all four items. The numbers refer to the response categories (1 = traditional, 5 = nontraditional). While in Canada, the categories *child suffers, family suffers*, and *warm relation* are closely together, in western Germany *warm relation* is trailing the other two. The picture for the United States is similar to that for Canada, but not as pronounced. The item *men work too much* shows a similarly peculiar behavior in all three countries, with the middle categories clustered closely together near the origin.

TABLE 20.8. Cronbach's Alpha and Item-Total Correlations, by Country

	Western Germany	United States	Canada
Cronbach's alpha			
All 4 items	.59[a]	.72[b]	.71[b]
Without item 4	.67[a]	.78[b]	.81[b]
Without items 3 and 4	.73[a]	.80[b]	.80[b]
Item-total correlations			
Child suffers	.77	.83	.84
Family suffers	.78	.84	.83
Warm relation	.65	.73	.76
Men work too much	.45	.50	.45

Note: Values within the same row with different superscripts differ significantly ($p < .05$) between nations.

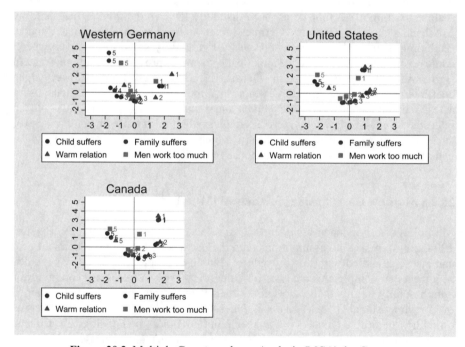

Figure 20.2. Multiple Correspondence Analysis (MCA), by Country

Disadvantages of correspondence analysis are low readability when many variables with many categories are involved, as each category of each variable is included separately in the graphical representation. However, single items deviating from the general pattern can be identified, if one can read them in the crowded graph. An alternative might be using the numerical output but this would mean foregoing one main advantage of this procedure. An advantage of MCA is, however, that categorical explanatory variables, such as age groups, can be added.

Methodological applications of MCA include Blasius and Thiessen (2001) on political efficacy and trust and Blasius and Thiessen (2006) on gender roles.

20.2.9 Multidimensional Scaling (MDS)

Multidimensional scaling (MDS; Borg & Groenen, 2005; Fontaine, 2003) graphically displays the intercorrelations between items. Data must at least be ordinal. A criterion for comparability is whether the items are located in corresponding regions in different countries. Dividing lines between regions have to be derived from theory and are not a result of the statistical procedure. MDS is mainly a graphical procedure. To compare different countries, the MDS representations for these countries have to be visually inspected. Lack of functional equivalence is demonstrated when items are located in different regions.

With regard to the four items introduced above, no suspicious differences between the three countries emerge (not presented here). Four items, however, are too few for MDS to produce meaningful results (which one cannot obtain even by visually inspecting the correlation coefficients). Even if we add other gender-role items in an additional model, MDS shows the lack of invariance only by the differences of the distances among the points that represent the items and, thus, is of limited use in this situation.

Some advantages of MDS include its ability to handle situations in which there are many survey items that potentially belong to different dimensions. MDS can also be implemented very rapidly and provides an instructive visual image of the data. However, no summary measures of comparability are provided by MDS. Instead, displays for different countries must be visually inspected. Results of MDS might differ for technical reasons (e.g., the choice of the starting configuration, lack of convergence, local minima solutions).

Methodological applications of MDS include those by Borg and Braun (1996) on work orientations and Schwartz and Sagiv (1995) on general values. Braun and Baumgartner (2006) present a substantive application on work orientations.

20.2.10 Multigroup Confirmatory Factor Analysis

Confirmatory factor analysis (CFA) produces a wealth of results, which can be used to gauge the quality of survey items within nations or cultural groups (Brown, 2006; Vandenberg & Lance, 2000). We restrict ourselves here to global fit indices. The question is: What conclusions can be drawn when these fit indices are taken into account? Table 20.9 lists both the Comparative Fit Index (CFI) and the Root Mean Square Error of Approximation (RMSEA) for separate CFAs in the three countries. Both measures point to a good fit of the four-item device in all countries, suggesting that the four items provide a good measurement of the underlying dimension. The problems demonstrated above go undetected.

TABLE 20.9. Fit Indices for Confirmatory Factor Analysis of All Four Items, by Country

	CFI	RMSEA		
		Estimate	Lower 90% CI	Upper 90% CI
Western Germany	.995	.039	.014	.068
United States	.999	.022	.000	.063
Canada	.996	.045	.013	.082

CFI=Comparative Fit Index.
CI=Confidence Interval.
RMSEA=Root Mean Square Error of Approximation.

Results from CFA, however, are more trustworthy when we choose a comparative perspective and examine multigroup, or stacked, CFAs for different numbers of items. Multigroup CFA answers the important question of whether a measurement instrument, which previously has been demonstrated to work reasonably well in one country, works equally well in other countries (Watkins, 1989). Different kinds of comparability are usually distinguished (see Brown, 2006; Vandenberg & Lance, 2000). Configural invariance is obtained, if all items belong to the same latent dimension in the different countries. Metric invariance requires equal loadings of the items on the factor everywhere. Finally, scalar invariance also stipulates that intercepts are equal. Here we impose the restriction of equal measurement weights and intercepts. Table 20.10 shows that multigroup CFA detects that neither the four- nor the three-item device is comparable across the three countries, but the two-item measurement, consisting only of *child suffers* and *family suffers* is comparable when equality constraints are in place.

An advantage of multigroup confirmatory factor analysis is that it gives a quantitative measure of the comparability across all countries involved. Bias in individual items can only be identified if additional measures beyond fit indices are checked, e.g., modification indices, which diagnose specific sources of misfit. However, if individual deviant cases need to be identified then comparisons between two countries have to be performed. For more detail see Chapter 21, this volume.

Examples of the application of multigroup CFA to evaluations of measurement invariance across cultures include work by Cepeda-Benito, Henry, Gleaves and Fernandez (2004) on smoking, Davidov (2008) and Davidov, Schmidt, and Schwartz (2008) on general values, Devins, Beiser, Dion, Pelletier, and Edwards (1997) on psychological well-being, and Griffin, Babin, and Christensen (2004) on materialism.

20.2.11 Hierarchical Linear Modeling / Multilevel Modeling

Hierarchical linear modeling (HLM) or multilevel modeling is a very versatile and flexible tool. It is commonly used when data are on two or more levels, for instance,

TABLE 20.10. Fit Indices for Multigroup Confirmatory Factor Analyses for Different Numbers of Items (Equal Measurement Weights and Intercepts)

	CFI	RMSEA		
		Estimate	Lower 90% CI	Upper 90% CI
All four items	.872	.132	.122	.142
Without item 4	.870	.195	.180	.210
Without items 3 and 4	.998	.052	.016	.100

CFI=Comparative Fit Index.
CI=Confidence Interval.
RMSEA=Root Mean Square Error of Approximation.

when individual respondents are nested within countries (Goldstein, 2003; Hox, 2002; Rabe-Hesketh & Skrondal, 2008; Raudenbush & Bryk, 2002; Skrondal & Rabe-Hesketh, 2004; Snijders & Bosker, 1999). There are models also for nominal data (described in the references listed above), but the presentation given here uses procedures for metric data only. Usually, a series of models is estimated beginning with a so-called "empty" or variance-component model. This model shows how much of the variance in the dependent variable that is located within countries (that is, individual variance around the country means) and how much is located between countries (that is, the variance of the country means around the grand mean of all countries). A second model typically includes the individual-level variables, as in normal regression. A comparison of the variance components of both models shows how much variance these individual-level variables are able to explain on both levels. The explanatory power of the individual-level variables with regard to the country-level variance is related to a composition effect, which is the part of the differences between countries that can be explained by taking into account the individual-level variables alone. A third model usually estimated is one in which, in addition to the individual-level variables, one or more additional variables at the country level are included. The aim is to further reduce and explain country-level variance, that is, the variance of the country means around the grand mean. All of these models are referred to as random-intercept models, as only the level of the dependent variable is allowed to vary across countries but not the effects of individual-level variables on the dependent variable.

To check for noncomparability, the response, irrespective to which item, is regarded as the dependent variable. In this case, the items constitute the first level, respondents the second, and countries the third. Item difficulties and item bias can then be analyzed. One extension is the random-slope model in which the relationship between individual-level variables and the dependent variable is allowed to vary across countries. Summary measures of variations in item difficulty and item bias are expressed by random coefficients. Significant random coefficients point to country-level differences. To explain these, variables on the country level can be included in the equation. However, this will only work when a specific pattern is shared by a number of countries that are similar with regard to these variables. If deviations only occur in one country (e.g., as the result of a translation error), they might go undetected, since the variance components might not become significant. Therefore, outlier detection could be useful here, for example, by using a graphical representation of the intercepts and slopes for the different countries. This, however, would not fit into the intrinsic logic of multilevel modeling.

For the sake of simplicity, these random-coefficient models are not considered here. Instead, cross-level interactions are immediately included in a random-intercept model, in which the effects of the independent variables on the dependent variable are not allowed to vary across countries. Only level differences between countries are possible. We also treat the first two items, *child suffers* and *family suffers*, as equal in their functioning. The remaining two items, *warm relation* and *men work too much*, are allowed to behave in distinct ways. Both individual- and country-level effects are included. The order in which the effects are presented does not follow the usual distinction between individual- and country-level effects. Instead, we make a division into substantive effects on the one hand and methodological effects on the other. As a multilevel model is inappropriate for small sample sizes at the country level (see Chapter 21, this volume), we include survey responses from a larger pool of 24 nations.

The first three effects are substantive in nature (Table 20.11). The first is an individual-level fixed effect that concerns the effect of age on traditionality with regard to parental labor-force participation (that is, a traditional or nontraditional response to any of the four items). The second is a country-level fixed effect. Living

TABLE 20.11. Multilevel Model for Consequences of Parental Labor Force Participation[a]

	Coefficient	Z-value
Substantive effects		
Age	$-.01^b$	-33.4
LS (Liberal/social-democratic state)	$.45^b$	4.1
Age X LS	$-.01^b$	-9.4
Method effects		
Item3	$.98^b$	119.2
Item4	$-.18$	-21.1
Age X Item3	$.01^b$	16.5
Age X Item4	$.01^b$	17.3
LS X Item3	$-.59^b$	-40.3
LS X Item4	$-.44^b$	-29.2
Age X LS X Item3	$.00$	0.3
Age X LS X Item4	$.00^b$	4.5
Constant	2.6^b	41.4

[a] Number of countries is 24; number of respondents is 31,023; number of observations is 123,161; [b] $p<.001$.

in a liberal or social-democratic welfare state (in the sense of Esping-Andersen, 1990) is pitted against living in a country that has a different type of regime. The third is a cross-level interaction between age and welfare regime. Z-values are reported to make assessment of the size of the effects easier. Attitudes are more traditional for older respondents than for those who are younger, and living in a liberal or social-democratic welfare state reduces traditionality. In addition, the interaction shows that the effect of age is more pronounced in liberal or social-democratic welfare states. These effects are all substantive in nature.

The lower panel shows the method effects. The first of these effects, labeled Item 3, shows that the third item, pertaining to *warm relation*, has a different item difficulty than do Items 1 and 2 (*child suffers* and *family suffers*, respectively). Respondents appear, on average, to be one scale point (on a 5-point scale) less traditional with this item than with the two baseline items. This is not problematic, as all items must not have the same difficulty. The second effect, labeled Item 4, shows that *men work too much* has a higher item difficulty than the benchmark items; that is, respondents with the same value on the latent variable are less inclined to disagree with this item.

The next effects show whether the item bias of the last two items differs by age of respondent and regime type. The interaction Age X Item3 shows that the relationship between *warm relation* and age is smaller than the relationship between the benchmark items and age. This is problematic and will result in attenuation of the correlation between an index which includes *warm relation* and age as an explanatory variable. The interaction Age X Item4 can be interpreted in the same way and is also of nearly equal strength.

The negative interaction LS X Item3 means that in liberal or social-democratic welfare states the difference in item difficulty between *warm relation* on the one hand and the two benchmark items on the other is much smaller than in the other countries, even though such a difference is also present in the latter. This suggests that there is nonuniform bias in the *warm relation* item. The negative interaction LS X Item4 can be interpreted in a similar way, although it is somewhat weaker.

A three-fold interaction, Age X LS X Item3, would refer to nonuniform bias (related to external validity), showing that the difference between the correlations of *warm relation* and age, on the one hand, and the benchmark items and age, on the other, differs across countries. This effect is not significant. However, the three-fold interaction Age X LS X Item4 is found to be significant, showing that the difference between the correlations of *men work too much* and age, on the one hand, and the benchmark items and age, on the other, vary across countries.

The main advantage of multilevel modeling is its usefulness when we analyze a large number of countries. In fact, a large number of countries is a requirement of this procedure. In addition, country-level variables can be included in the model to explain or even control comparability problems. This is not possible with multigroup CFA (see Chapter 21, this volume, where multigroup and multilevel structural equation modeling are compared). However, in the framework of multilevel analysis, identification of individual deviant cases is not possible, since outlier detection is not part of the method.

Methodological applications of multilevel modeling include Johnson, Cho, Holbrook, O'Rourke, Warnecke, and Chávez (2006) on the comprehension of questions and Johnson, Kulesa, Cho, and Shavitt (2005) on response styles. A substantive application on attitudes toward inequality is Hadler (2005). Numerous additional examples can be found in van de Vijver, van Hemert, and Poortinga (2008b).

20.2.12 Item Response Theory (IRT) Models

Item response theory (IRT) examines how individual items relate to the underlying dimension (Embretson & Reise, 2000; Hulin, 1987; van de Vijver & Leung, 1997). The origin of this approach is in developing and evaluating items for psychological tests. Thus, it comes as no surprise that this approach originally focused on the analysis of dichotomous items, such as correct versus incorrect answers. However, other scale types can also be handled within this framework (see Chapter 22, this volume, for an example with polythomous items). In our example, however, we used the traditional procedure for binary responses. We dichotomize the responses to the items, coding respondents who gave any of the two least traditional responses as 1 and the others as 0. It should be noted that, typically a larger number of items are used to represent the dimension of interest than are employed in this example. This is especially true in our case, as two of the four items can be suspected of having item bias. Nevertheless, here we use IRT for purely illustrative purposes. We use the program GLLAMM (Rabe-Hesketh, Skrondal, & Pickles, 2004) and the procedure outlined for conducting an IRT analysis with this program in Zheng and Rabe-Hesketh (2007).

Figure 20.3 presents the 1-parameter IRT model. In this model, difficulty (in the sense of the likelihood to give a nontraditional response to a specific item, given a specific value of the respondent on the latent dimension) can vary between items, but not the ability to discriminate between traditional and nontraditional respondents. The x-axis represents the underlying dimension, ranging from extremely traditional on the left to extremely nontraditional on the right, and the y-axis gives the probability for a positive response on each of the items. Although in all countries, *warm relation* is the least difficult item, this is most evident in western Germany.

In western Germany, the probability of giving nontraditional responses to this item requires a much lower position on the underlying dimension than on the other two items. In the United States and Canada, all three items behave in a more similar way. *Men work too much* is much more difficult than the two benchmark items in the United States and Canada, whereas in western Germany its curve coincides even with that of *family suffers*. This was to be expected, given the marginal distribution of these items (see Table 20.3) and the dichotomization chosen for the IRT model.

Figure 20.4 presents the 2-parameter IRT model. In this model, in addition to an item difficulty parameter, a discrimination parameter is also estimated. The steepness of the slope for the different items indicates how well an item discriminates between traditional and nontraditional respondents. A steep slope indicates a good discriminatory power. While for all countries, *warm relation* does not work as well as the two baseline items, this is particularly pronounced in western Germany. In stark contrast to the 1-parameter IRT model, *men work too much* demonstrates a degenerate pattern in all three countries.

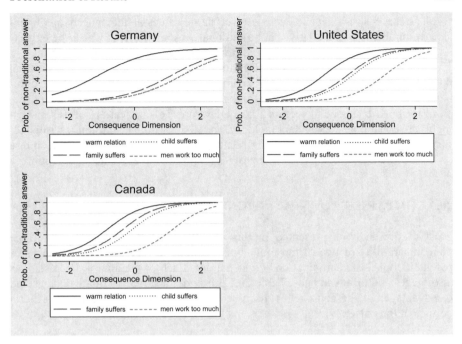

Figure 20.3. One-Parameter Item Response Theory (IRT) Model, by Country

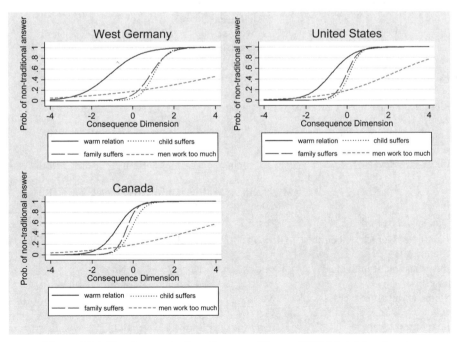

Figure 20.4. Two-Parameter Item Response Theory (IRT) Model, by Country

The curve is very flat and even with very large values on the underlying dimension, respondents do not all give "nontraditional" responses. This was to be expected, because it is unclear what a nontraditional response might be for this item.

IRT models are useful for demonstrating bias on the item level and indicating the form this bias assumes. IRT may become unwieldy, however, when a large number of countries are being compared. In this case, a multilevel variant has to be used, along the lines discussed in Section 20.2.11. Examples of applications of IRT modeling to cross-national survey data include Candell and Hulin (1986) on job satisfaction; De Jong, Steenkamp, Fox, and Baumgartner (2008) on extreme response style; and Ewing, Salzberger, and Sinkovics (2005) on advertisement perception.

20.3 COMPARISON OF STATISTICAL PROCEDURES

Most of the analyses presented in this chapter provide some evidence of the noncomparability of two of the four items intended to represent consequences of parental labor force participation across nations. Each of the approaches used has advantages and disadvantages. Table 20.12 provides a summary of advantages and disadvantages of the approaches discussed in the chapter in terms of the criteria defined at the outset.

TABLE 20.12. Some Advantages and Disadvantages of the Different Procedures

	Distributions, Means, Correlations	Exploratory Factor Analysis	Reliability
Quick overview	Yes	Yes	Yes
Handling of large number of countries	Difficult	Difficult	Easy
Identification of individual cases	Yes	Yes	Yes
Summary measures across countries	No	No	No
Item or test level	Item	Both	Both

	MCA and MDS	Multigroup CFA	Multilevel modeling	IRT models
Quick overview	Yes	No	No	No
Handling of large number of countries	Difficult	Difficult	Easy	Difficult
Identification of individual cases	Yes	No	No	Yes
Summary measures across countries	No	Yes	Yes	No
Item or test level	Item	Both	Both	Item

20.4 REMAINING CHAPTERS IN PART VII

The remaining chapters in Part VII present original applications of more advanced techniques to the analysis of cross-national survey data. Chapter 21 by Hox, de Leeuw, and Brinkhuis introduces multilevel and structural equation modeling and points to their relative strengths and weaknesses. In Woehr and Meriac (Chapter 22), polytomous item response theory (IRT) models are applied to substantive problems of the sociology of work and compared to CFA models. The chapter by Oberski, Saris, and Hagenaars (Chapter 23) introduces multitrait-multimethod (MTMM) designs. In the final chapter of this section (Chapter 24), van de Vijver and Chasiotis describe applications of mixed designs that integrate qualitative and quantitative strategies into the analysis of cross-cultural survey data, and thus go beyond the statistical techniques presented in the other chapters.

21

Analysis Models for Comparative Surveys

Joop J. Hox, Edith D. de Leeuw, and Matthieu J.S. Brinkhuis

21.1 INTRODUCTION

The aim of cross-national and cross-cultural surveys is to compare results across countries or cultural groups. Since the early international surveys in the 1970s (cf. Harkness, Mohler, & van de Vijver, 2003a) the number of comparative survey programs has grown and this trend is likely to continue (Lynn, Japec, & Lyberg, 2006). Over the years the methodological knowledge base on comparative studies has been growing as well and especially methods for questionnaire design (e.g., Harkness, van de Vijver & Johnson, 2003; Smith, 2003), harmonization (e.g., Braun & Mohler, 2003), and questionnaire translation (e.g., Harkness, 2009) have been developed. Although cross-national surveys aim to use common data collection methods, there generally remain a number of differences in the data collection process in the participating countries, in addition to the inevitable language differences. For instance, differences in the data collection methods (Kalgraff Skjak & Harkness, 2003) or in the implementation of mixed-mode strategies (de Leeuw, 2005), in details of the fieldwork of the data collection agencies (De Heer, 1999), and in survey climate, economic condition, and culture (e.g., de Leeuw & De Heer, 2002) may all affect the comparability of the data. Also differences in the achieved response rates can affect the comparability (Couper & de Leeuw, 2003).

There are three main statistical issues in comparative research. Firstly, there is the issue of measurement equivalence. Can we assume that the instruments measure the same constructs in the same way, how can we assess whether we have measurement equivalence, and if not how can we correct measures in order to achieve valid comparisons? Secondly, if measurement equivalence is achieved, the analysis must deal with the issues of analyzing relationships within and between countries (or other contexts). That is, relationships can be established at the individual level within each country, but in comparative research the central issue is often the question of

[1] Survey Methods in Multinational, Multiregional, and Multicultural Contexts, edited by Harkness et al.
Copyright © 2010 John Wiley & Sons, Inc.

whether such relationships are different between countries. Finally, the question is whether there are stable relationships between characteristics at the country level.

The classic statistical approach to deal with these questions is structural equation modeling (SEM) using a multigroup analysis. This analysis method makes it possible to test equivalence of measurement models and equivalence of structural (substantive) models. The SEM measurement model is a confirmatory factor model, where an explicit model specifies which survey questions are assumed to indicate which latent factor or construct. Modern SEM software can model categorical data, which implies that measurement models like Item Response Theory (IRT) models can also be subsumed under SEM.

However, when the number of groups or countries becomes larger, multi-group SEM becomes unwieldy. The software setups become complicated, especially if subtle differences in measurement properties must be included. The statistical model also becomes complicated. Multigroup SEM is a fixed effects model, which means that it takes each group or country as given and the set of countries as the complete universe to generalize to. Unless a great many equality constraints are imposed, SEM estimates a unique set of parameter values for each country, which results in a large model. A random effects model, such as multilevel modeling, treats the countries as a sample from a larger population. Instead of estimating a different parameter value for each country, it assumes a (normal) distribution of parameter values and estimates its mean and variance (and covariances). This makes multilevel SEM (MSEM) much more parsimonious than SEM when the number of countries becomes large. A second advantage is that differences between countries can in turn be modeled using country-level characteristics. Simulations show that multilevel modeling can be used with second-level (group-level) samples as small as 20 (Maas & Hox, 2005), which means that the larger collaborative comparative surveys involve enough countries to consider employing multilevel modeling methods.

Recently, latent class modeling (LCM) has come into use. LCM does not model differences between countries as random effects, but attempts to identify latent classes of similar respondents. This approach combines some advantages of SEM and MSEM: Differences between countries are modeled as differences between groups, but these groups are not the countries but latent classes of respondents, and the number of latent classes is assumed to be much smaller than the number of countries.

This chapter consists of three major sections. First, it compares the three statistical approaches outlined above (SEM, MSEM, LCM). The underlying statistical models are explained at a general, nontechnical level. The emphasis is on a comparison of the major characteristics: What is the structure of the model, what kind of questions can be answered using this approach and what are the important statistical assumptions underlying the model? The second section contains a small simulation study that compares the three approaches in a situation typical for comparative research: a relatively small number of groups (countries) but within groups relatively large sample sizes. This simulation addresses the question how accurate the estimates are with a small number of countries, and if there is sufficient power to detect anomalies in the measurement model. The third section applies and compares the three approaches on a real data set from a large scale comparative survey. A final section summarizes the findings and gives recommendations for model use and further methodological research.

21.2 STATISTICAL CONSIDERATIONS IN COMPARING MULTIGROUP SEM, MSEM, AND LCM

21.2.1 Multigroup Structural Equation Modeling (SEM)

The classic method for dealing with data from large cross-national surveys is to use multigroup SEM. This approach derives from the seminal work of Jöreskog (e.g., Jöreskog, 1971a). SEM provides a very general and convenient framework for statistical analysis that includes several traditional multivariate procedures as special cases, for example factor analysis, regression analysis, discriminant analysis, and canonical correlation. Structural equation models are often visualized by a graphical *path diagram* (see, for example, Figure 21.1). In the path diagram, observed variables are represented by a square and latent variables by a circle. The statistical model is usually represented in a set of matrix equations. Commonly, a distinction is made between the measurement part of the model and the structural part of the model. Figure 21.1 is a graphical presentation of the full structural equation model in the notation used by Bollen (1989). The diagram shows a measurement model for the latent factor ξ (ksi) and its associated observed indicators x and for the latent factor η_2 and its associated observed indicators y. The relationships between the latent variables ξ and η_1 and between η_1 and η_2 constitute the structural model. The latent variable ξ is denoted as exogenous, because there are only paths from it to other variables, and the latent variables η are denoted as endogenous because there are paths leading towards them. This distinction is important, because for the endogenous variables multivariate normality is assumed, but not for the exogenous variables (Bollen, 1989). Note that in this example the latent variable η_1 has no empirical indicators; it is solely defined by its role in the structural model. Such latent variables have their use in

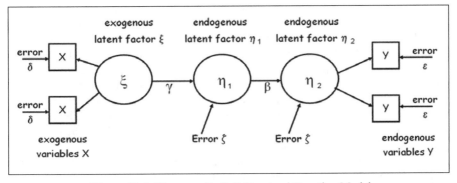

Figure 21.1. Diagram of a Full Structural Equation Model

specialized models, but it is more common to have empirical indicators for all latent variables in the model. It is also possible to have a path model with only observed variables. In that particular case there is no explicit measurement model. Implicitly, it is then assumed that all variables are measured identically in all countries or groups, but it is not possible to test this important assumption.

In SEM, it is usually assumed that the endogenous variables follow a multivariate normal distribution, which implies that the vector of means and the covariance matrix contain all the relevant information. The method most widely used for estimation is Maximum Likelihood (ML) estimation, assuming multivariate normal data and a reasonable sample size (e.g., about 200 observations). There are a variety of estimation procedures that can be used for nonnormal continuous data. With nonnormal data, including the types of ordinal categorical data typically collected in social surveys, the means and the covariance matrix do not represent all the information, and therefore alternative estimation methods require raw data.

Comparative research generally employs large samples for each country. Statistical tests for model fit have the general property that their power varies with the sample size. As a result, with large samples, we will almost always reject our model, even if the model actually describes the data very well. Conversely, with a very small sample, the model will always be accepted, even if it fits rather badly.

In comparative survey research, samples are typically large (e.g., above 1,000 respondents in each country). As a consequence, even small discrepancies between the model and the data will lead to a significant test result and a formal rejection of the model. Given the sensitivity of the chi-square statistic to sample size, researchers have proposed a variety of alternative fit indices to assess model fit. Most of these fit indices not only consider the fit of the model, but also its simplicity. A saturated model that specifies all possible paths between all variables, will always fit the data perfectly, but it is just as complex as the observed data itself. If two models have the same degree of fit, the principle of parsimony indicates that we should prefer the simpler model.

Modern SEM software computes a bewildering array of goodness-of-fit indices. For an overview and evaluation of a large number of fit indices, including those mentioned here, we refer to Gerbing and Anderson (1993). All fit indices are functions of the chi-square statistic, but some include a second function that penalizes complex models. For instance, Akaike's Information Criterion (AIC) is equal to the chi-square statistic plus twice the number of parameters in the model. Often used fit indices are the TLI (Tucker-Lewis Index) and the CFI (Comparative Fit Index), with values > 0.90 indicating a good fit, and 1.0 indicating a perfect fit (Bentler, 1990). A different approach to model fit is to accept that models are only approximations, and that perfect fit may be too much to ask for. Instead, the problem is to assess how well a given model approximates the true model. This view led to the development of an index called RMSEA (Root Mean Square Error of Approximation). If the approximation is good, the RMSEA should be small, with values < .05 indicating a good fit and values < .08 indicating an acceptable approximation (Browne & Cudeck, 1993).

If the fit of an SEM model is not adequate, it has become common practice to modify the model, by deleting parameters that are not significant and by adding parameters that improve the fit. To assist in this process, most SEM software can compute *modification indices* for each fixed parameter. The value of a given modification index is the minimum amount that the chi-square statistic is expected to decrease if the corresponding parameter is freed. Researchers often use this information to conduct a sequence of model modifications. At each step a parameter is freed that produces the largest improvement in fit, and this process is continued until an adequate fit is reached. For example, if in a confirmative factor model a loading that is fixed to zero shows a large modification index, we may free this parameter and estimate its value. This will improve the fit of the model at the cost of one degree of freedom.

The statistical model is usually described by separate equations for the measurement and the structural model. Thus, the equations for the measurement model in Figure 21.1 are in matrix format:

$$x = \Lambda_x \xi + \delta \tag{21.1}$$
$$y = \Lambda_y \eta + \varepsilon \tag{21.2}$$

and for the structural model it is:

$$\eta = B\eta + \Gamma\xi + \zeta \tag{21.3}$$

In this notation, Λ_x (lambda-x) is the factor matrix for the exogenous variables, and Λ_y (lambda-y) is the factor matrix for the endogenous variables. The notation used here distinguishes between independent variables x and ξ (ksi) and dependent variables y and η (eta). The paths from exogenous variables to endogenous variables are denoted γ (gamma, see Figure 21.1), collected in the matrix Gamma (Γ), while the path coefficients among endogenous variables are denoted β (beta, see Figure 21.1), collected in the matrix Beta (B). The residual errors are collected in matrix Zeta (ζ). The distinction between measurement model and structural model is conceptual rather than statistical, and most SEM software hides these complications from the user. However, in comparative research this distinction between measurement and structural model is of particular interest, because equivalence of the measurement model across different groups is of central importance in order to achieve valid comparisons (cf. Bechger, van den Wittenboer, Hox, & de Glopper, 1999). This means that before we can compare the structural (substantive) models for different countries, we must make sure that the factor matrices in the measurement model are in fact equal across countries.

The question of whether measurement invariance may be assumed is generally investigated using multigroup SEM. Multigroup SEM makes it possible to test hypotheses concerning equivalence between groups, such as the hypothesis of measurement equivalence (cf. Vandenburg & Lance, 2000). The weakest form of measurement equivalence is *functional equivalence*, sometimes also denoted as factorial equivalence, where the assumption holds that the different countries share a measurement model that has the same factor structure. This is a very weak form of measurement equivalence, because it only allows us to conclude that we are probably studying the same construct in each country, but there is no way to statistically compare the countries or to examine any differences. To analyze variation across countries, we need to prove first that the items that comprise a specific measuring instrument operate equivalently across the different populations or countries (Jöreskog, 1971a; Meredith, 1964, 1993). In our SEM notation, different groups are denoted by superscripts. Thus, the hypothesis that there is measurement equivalence across two groups for the latent variable η would be written as:

$$\Lambda_y^{(1)} = \Lambda_y^{(2)} \tag{21.4}$$

If the factor loadings are invariant across all countries, we have a form of equivalence that is referred to as *metric equivalence* (Vandenburg & Lance, 2000). Metric equivalence means that the measurement scale is comparable across countries. Although the ideal is achieving complete measurement invariance across all countries, in practice a small amount of variation is often judged acceptable, which leads to partial measurement invariance (Byrne, Shavelson, & Muthén, 1989; Steenkamp & Baumgartner, 1998).

When (partial) metric equivalence is achieved, it is possible to analyze differences between countries statistically. This includes the question whether paths in a specified causal structure are invariant across populations and the wider question of whether the same structural model holds in all countries. When comparisons of means of latent constructs are involved across countries, additional invariance restrictions are needed for the intercepts of the observed variables. If these intercepts can be considered invariant across countries, we have a form of equivalence that is referred to as *scalar equivalence* (Vanderburg & Lance, 2000). When scalar equivalence holds, the actual scores can be compared across countries. Again, the ideal is achieving complete invariance for all intercepts across all countries, but in practice a small amount of variation is judged acceptable, which leads again to partial measurement invariance. Regarding the minimal requirements for partial invariance, both Byrne et al. (1989) and Steenkamp and Baumgartner (1998) state that for each construct in addition to the marker item that defines the scale (marker item loading fixed at 1 and intercept fixed at 0) at least one more indicator must have invariant loadings and intercepts.

If (partial) metric equivalence has been established for the measurement model, we can use theoretical reasoning to specify substantive (structural) models for the relationships among constructs in different countries, and we can assess whether these relationships are the same across all countries. If (partial) scalar equivalence has been established, we can also test if the countries differ on the means of the constructs. However, a major shortcoming of multigroup SEM in the context of comparative survey research is that there are no provisions to specify models that include country-level variables to explain differences in these means.

It should be noted that the terminology used for the various forms of equivalence is not well standardized (Chapter 2, this volume). For instance, the term scalar equivalence is sometimes used for metric equivalence (cf. van de Vijver & Leung, 1997). Fortunately, there is consensus on the constraints that are needed to make specific comparisons valid. To compare relationships (regression, correlations) between countries we need equivalence constraints on the loadings, and to actually compare scores between different countries we need additional equivalence constraints on the intercepts. Typically, such constraints are not imposed stepwise, but they are imposed in one step across all countries in the analysis. Subsequently, chi-square tests and modification indices are used to investigate the adequacy of the constraints.

21.2.2 Multilevel Structural Equation Modeling (MSEM)

Multilevel models are specifically developed for the statistical analysis of data that have a hierarchical or clustered structure. Such data arise routinely in various fields, for instance in educational research where pupils are nested within schools or in family studies with children nested within families. Clustered data may also arise as a result of a specific research design. An example is longitudinal designs; one way of viewing longitudinal data is as a series of repeated measurements nested within individual subjects. Comparative surveys also lead to a multilevel structure with respondents nested within countries. In comparative research there are, in addition to respondent level variables, also variables measured at the country level. In contrast to multigroup SEM, MSEM can include these country-level variables as explanatory variables in the model.

The most used multilevel model is the multilevel regression model (Goldstein, 2003; Hox, 2010; Raudenbush & Bryk, 2002). It assumes hierarchical data, with one response variable measured at the lowest level (e.g., respondents) and explanatory variables at all existing levels (e.g., respondent and country). Conceptually, the model is often viewed as a hierarchical system of regression equations. For example, assume we have data in J groups, and a different number of individuals N_j in each group. On the individual (lowest) level we have the dependent variable Y_{ij} and the explanatory variable X_{ij}, and on the group (higher) level we have the explanatory variable Z_j. Thus, we have a separate regression equation in each group:

$$Y_{ij} = \beta_{0j} + \beta_{1j} X_{ij} + e_{ij} \tag{21.5}$$

The β_j are *random coefficients*, assumed to vary across groups. They are modeled by explanatory variables at the group level:

$$\beta_{0j} = \gamma_{00} + \gamma_{01} Z_j + u_{0j} \tag{21.6}$$

$$\beta_{1j} = \gamma_{10} + \gamma_{11} Z_j + u_{1j} \tag{21.7}$$

Substitution of (21.6) and (21.7) in (21.5) gives:

$$Y_{ij} = \gamma_{00} + \gamma_{10} X_{ij} + \gamma_{01} Z_j + \gamma_{11} X_{ij} Z_j + u_{1j} X_{ij} + u_{0j} + e_{ij} \tag{21.8}$$

In general, there will be more than one explanatory variable at each level. What these equations make clear is that there is a distinction between individual- (respondent-) level explanatory variables and group- (country-) level explanatory variables. Thus, in comparative surveys, we have respondent variables on the level of the respondents and country variables on the level of the countries. The outcome variable is always on the individual (respondent) level. For individual-level predictors only, we can hypothesize that the regression coefficients of these variables differ between countries. If they do, we can attempt to explain the variation between countries using country-level variables. As Equation 21.8 makes clear, this is done by adding a cross-level interaction ($X_{ij}Z_j$) to the model.

The multilevel regression model is a univariate model, although it can be used to analyze multivariate outcome data by introducing an additional lowest level for the outcome variables. Multilevel structural equation modeling (MSEM) is more flexible. In MSEM, we assume sampling at two levels, with both between-group (group level) and within-group (individual-level) covariation. This approach includes a measurement and structural model at each level, with random slopes and intercepts (Mehta & Neale, 2005). Muthén and Muthén (2007) and Skrondal and Rabe Hesketh (2004) have suggested extensions of the conventional graphic path diagrams to represent multiple levels and random slopes. We use the notation proposed by Muthén and Muthén here, since it is close to the usual SEM path diagram (see Figure 21.2).

The two-level path diagram in Figure 21.2 is based on the separate equations representation of the model, as provided by Equations 21.5-21.7. The within part of the model in the lower area specifies that Y is regressed on X. The between part of the model in the upper area specifies the existence of a group level variable Z. There are two latent variables represented by circles. The group level latent variable Y represents the group level variance of the intercept for Y. The group level latent variable XY slope represents the group level variance of the slope for X and Y, which is on the group level regressed on Z. The black circle in the within part is a new symbol, used to specify that this path coefficient is assumed to have random variation at the group level. This variation is modeled at the group level using group level variables.

MSEM provides a completely new approach to testing the equivalence of the measurement models across groups. Equivalent measurement models in this formulation mean that the same factor model must fit in all groups, with no factor loading having a coefficient that varies across groups. In other words, testing whether

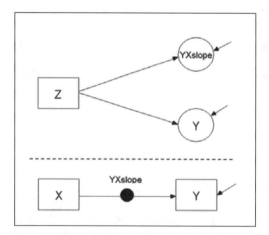

Figure 21.2. Path Diagram for a Two-Level Regression Model

factor loadings for the measurement model have significant variation across groups is a test on the metric equivalence of the measurement (c.f. Section 21.2.1).

Figure 21.3 shows the measurement model on the individual and the country level. There is no methodological need to constrain the loadings to be equal at the individual and the country level, but it is useful to investigate this possibility, because if there is metric equivalence across both levels, this suggests that the same process may be at work at both levels (cf. van de Vijver & Poortinga, 2002). This model handles unequal intercepts differently than the multigroup SEM. In multigroup SEM, unequal intercepts are basically prohibited. In MSEM, country-level variation in intercepts is modeled by allowing country-level residual measurement errors on the latent variables η. As a result, countries are allowed to have different intercepts, but these are modeled separately from the effect of the common latent variable η_B. Complete scalar invariance can be imposed by restricting the residual measurement errors on the between level to zero, but there is no need to do so.

In addition to testing measurement equivalence, using multilevel modeling to analyze substantive models and the potential differences between countries is an exciting approach. In MSEM, differences between countries can be modeled by explicitly including country-level explanatory variables in the analysis. In this chapter, we focus on assessing measurement equivalence, but we return to the more general issue of modeling country differences in the discussion.

21.2.3 Latent Class Modeling (LCM)

LCM was first described by Lazersfeld (Lazersfeld, 1950, Lazersfeld & Henry, 1968) who introduced the concept of latent structure analysis to describe the use of mathematical models for the association between latent variables (McCutcheon, 1987). Classic SEM is a form of latent structure analysis, with linear relations between continuous latent factors estimated on the basis of multivariate normal observed indicators. Latent class modeling assumes a discrete latent variable that represents latent classes, interpreted as subtypes of related cases in the population. The classical application of latent class analysis is to identify subtypes or segments in the population on the basis of observed categorical variables. In this application,

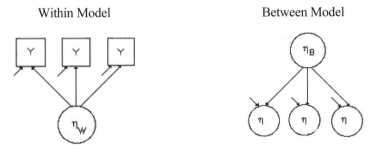

Figure 21.3. Measurement Model for Individuals (Within) and Countries (Between)

latent classes are defined by the condition of local independence, which stipulates that conditional on the class membership, the observed variables are independent. In other words, all of the association between the observed variables is explained by the categorical latent class variable.

A typical procedure in classical latent class analysis is to determine the minimum number of latent classes needed to achieve local independence, and to interpret the latent classes in terms of the response probabilities on the observed categorical variables. Although the classes are latent, and class membership is therefore unobserved, cases can be classified into their most likely latent class using recruitment probabilities. An introduction to the classic latent class model is given by McCutcheon (1987). An application in the survey field is the study by Biemer et al., who use latent classes to represent different types of misclassification in the U.S. census (Biemer, Woltmann, Raglin, & Hill, 2001).

The classic latent class model has been extended to allow continuous variables, and to allow class membership to be predicted by observed covariates. In addition, latent class models have been formulated that relax the condition of local independence. Instead, a specific model is assumed to hold that produces the associations between the responses, which can be either continuous or categorical variables. Usually it is assumed that the same model holds in the distinct latent classes, while different latent classes are characterized by having different values for the model parameters. The extended latent class model is also referred to as a mixture model, because it assumes that the data are generated by a mixture of different distributions, which in turn are generated by the corresponding models. For the purpose of investigating measurement invariance in comparative research, we will assume that the common model is a confirmatory factor model, and investigate whether there are different latent classes with different loadings. An introduction to these extended latent class models is given by Rost and Langeheine (1997) and Magidson and Vermunt (2004, 2005).

Thus, we assume that the observed variables are generated by a common confirmative factor model. Following this assumption, the latent class analysis follows the same reasoning as the reasoning in multiple group SEM. The relevant null hypothesis for measurement invariance across latent classes is

$$\Lambda_y^{(1)} = \Lambda_y^{(2)} \tag{21.13}$$

with the important difference that the superscripts in this case refer not to observed groups but to latent classes.

Although the methodological reasoning is the same, the actual testing procedure in latent class SEM is different than in multigroup SEM, due to differences in the underlying models. Searching for the correct number of latent classes is less straightforward than model testing in SEM, because there is no formal statistical test to test the 2-class against the 1-class model. In practice, decisions about the "correct" number of classes are based on information criteria like AIC or the Bayesian Information Criterion (BIC). In addition, LCM has no global test for the fit of the model, and there are no modification indices to suggest model improvements. In multigroup SEM we can pose functional equivalence, the weakest form of measurement invariance (the different countries share a measurement model that has the same factor structure, cf. Vandenburg & Lance, 2000), as a reasonable starting model. If that model is rejected, modification indices will inform us how to proceed. In latent class SEM, we can follow the same approach, but without the guidance of global tests and modification indices. Here we can specify the same measurement model for each latent class and search for the best number of classes. If there are two or more classes, we conclude that we have more than one subpopulation, and the latent classes are treated in much the same way as multiple groups in classical SEM. Thus, if the factor structure is the same in all classes we conclude that we have functional equivalence. If the factor loadings can be constrained to be equal across all latent classes, we have metric equivalence (Vandenburg & Lance, 2000). Just as in multigroup SEM, some amount of invariance is judged acceptable, which leads to partial measurement invariance. When (partial) metric equivalence is achieved, the model can be extended to allow for different structural models across groups. When comparisons of means of latent constructs are involved, we need additional invariance constraints across all latent classes on the intercepts of the observed variables to establish scalar equivalence (Vanderburg & Lance, 2000). Again, a small amount of variation is judged acceptable. The minimal requirements for partial invariance given by Byrne et al. (1989) and Steenkamp and Baumgartner (1998) (for each construct in addition to the marker item at least one more indicator with invariant loadings and intercepts) appear to be reasonable also in the context of latent class SEM modeling.

The next section presents and discusses a small simulation study, aimed at clarifying how well the three methods work with varying number of sampling units at the group level. The section following the simulation study reports the results of applying multigroup SEM, multilevel SEM, and LCM on a realistic data set.

21.3 A COMPARISON OF MULTIGROUP SEM, MSEM, AND LCM BY SIMULATION

Since measurement equivalence is of central importance, it is the topic of the simulation study. The simulation study uses two different data generating models, both simulating a simple measurement model. The data conform to the general structure of comparative studies, with a large sample within each country and a relatively small number of countries. We require that the latent variable is over-identified, which leads to four observed indicators for a single construct. For the model, there are two simulated conditions. In one condition, metric equivalence holds. The goal of simulating this condition is to investigate if the number of available countries permits accurate parameter estimates and standard errors. In the second condition, metric equivalence does not hold. The goal of simulating this condition is to investigate if the chosen analysis method has sufficient power to detect the violation of the equivalence of measurement.

21.3.1 Sample Sizes in SEM and Multilevel Modeling

Simulation research on single level SEM has shown that with a good model and multivariate normal data a reasonable sample size is about 200 cases (cf. Boomsma, 1982), although there are examples in the literature that use smaller samples. Simulation studies (e.g., Boomsma, 1982; Chou & Bentler, 1995) show that with nonnormal data, maximum likelihood estimation still produces good estimates in most cases, but that larger sample sizes are needed, typically at least 400 cases. Most surveys have sample sizes considerably larger than this, and carrying out a SEM analysis is feasible.

In comparative surveys, the sample size at the group or country level is often limited. Only a few simulation studies have investigated the sample size requirements for MSEM. These studies typically report that at all simulated sample sizes the individual-level coefficients and standard errors are estimated accurately. However, for the group level, it has been shown that MSEM often results in good parameter estimates and reasonable but not highly accurate standard errors. For instance, Hox and Maas (2001) found that a group level sample size of 100 is required for sufficient accuracy of the model test and confidence intervals for the parameters. With the group level sample size set to 50, the parameters are still estimated accurately, but the standard errors are too small. In a later simulation using a new and more sophisticated estimation method (full maximum likelihood instead of the approximate limited like-lihood), Hox, Maas, and Brinkhuis (2009) found accurate standard errors, even for group level factor loadings with only 50 groups, each of size 10. The coverage of the residual variances is, however, not as good (90%). In the context of multilevel regression, Hox and Maas (2005) find similar results for multilevel regression modeling. Typically, regression coefficients can be accurate with higher level sample sizes as small as 20; standard errors require somewhat larger sample sizes, but accurate estimation and testing of variances requires group level sample sizes of at least 50.

The simulations reported above are based on the usual multilevel designs, with group sizes reflecting applications in educational and organizational research. Comparative surveys typically use very large sample sizes within each country, and the number of countries in large scale comparative surveys is reaching the 20–40 range. Given the large samples within the countries, this smaller number of countries (the group level sample size) is likely to be sufficient to make multilevel SEM a serious analysis option. To gain a better understanding of how well the different analysis approaches fare under sample size conditions typical for comparative surveys, we report here the results of a small simulation study that investigates these issues.

21.3.2 A Small Simulation Study

To represent the number of countries as found in large scale surveys, three different values have been chosen for the Number of Countries (NC) in the simulation (NC = 20, NC = 30, and NC = 40). Within each country, 1,500 respondents are simulated. The model assuming metric equivalence is presented in Figure 21.4. It should be noted that means are fixed at 0 and that all simulations are performed 1,000 times in each condition.

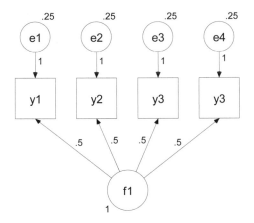

Figure 21.4. Path Diagram for a Factor Model

For the model where metric equivalence is violated, data are simulated where the fourth factor loading is different from the others for half of the countries, namely 0.3 instead of 0.5. In terms of Cohen's effect size conventions, the loading in these countries changes from *medium* to *small*. Thus, half of the countries have different weights than the other half. The same simulated data are used for the multigroup model, the multilevel model, and the latent class model.

21.3.3 Simulation Results

When the three approaches are used to analyze the data where metric equivalence holds, the results are straightforward. Multigroup SEM performs very well in all simulated conditions. Latent class SEM with only one class ignores the country level, but performs just as well. MSEM performs well with 30 or 40 countries; with 20 countries the standard errors are too large, resulting in an operating coverage for the estimated 95% confidence interval that is actually between 92% and 93%. In all cases, the chi-square was significant in approximately five percent of the simulations, and the fit indices TLI, CFI, and RMSEA all indicate a good fit.

When the three approaches are used to analyze the data that are simulated to violate metric equivalence by having one factor loading that varies across countries, the results are more complex. One striking result is that only the chi-square test in the multigroup model is able to detect this model violation. The latent class model totally lacks power to detect the violation. The chi-square model test does not reject the model even once. The multilevel model fares somewhat better. The chi-square test does reject the model in the majority of cases, and the model improves significantly if the loading that violates the assumption of measurement invariance is allowed to vary across countries. The general fit measures CFI, TLI, and RMSEA all perform very poorly. They look well in all simulated conditions, indicating that they lack power to detect the violation. Only the RMSEA in the multigroup model provides some indication of a nonperfect fit.

In sum, when the model is incorrect because metric equivalence is violated, the most striking result is a massive lack of power to detect this violation. Only in the classic multigroup analysis does the global chi-square test reject the model when the data that violate metric invariance are analyzed. But even in that case, the fit indices would indicate a very good fit, and most analysts would probably argue that given the large total sample size, the chi-square test is overly powerful and the model rejection therefore can be ignored. All the same, they would be ignoring a clear violation of measurement invariance. The conclusion is that a strong reliance on global fit indices is misleading. If most of the model is correctly specified, with only a few misspecifications in specific parts of the model, global fit tests and fit indices are overly optimistic. It is better to examine more specific indicators of lack of fit, such as modification indices and the corresponding estimated parameter change. In contrast to the chi-square test and associated fit indices, the modification indices are related to lack of fit for a specific parameter constraint. As such, when there is a specific fit problem in a model that fits well globally, the modification index has a much better power to indicate the source of this problem, and the estimated parameter change indicates how different the unconstrained parameter estimate is likely to be from the constrained estimate. This is the approach that is taken in the next section, where a realistic data set is analyzed.

21.4 A COMPARISON OF MULTIGROUP SEM, MSEM, AND LCM ON EXISTING DATA

The substantive analyses presented here illustrate the three statistical approaches on a realistic data set. We use data from the first round of the European Social Survey (ESS). The data collection took place in 22 countries between September 2002 and March 2003, the total number of subjects in these data is 41,207. For details on the data collection we refer to the ESS website (http://www .europeansocialsurvey.org/). We analyze a set of four items that measure "religious involvement." The items are as follows (the C/E item codes refer to the question identification code in the ESS questionnaire):

- Regardless of whether you belong to a particular religion, how religious would you say you are? (C13)
- Apart from special occasions such as weddings and funerals, about how often do you attend religious services nowadays? (C14)
- Apart from when you are at religious services, how often, if at all, do you pray? (C15)
- How important is religion in your life? (E18)

Items C13 and E18 were measured on an 11-point scale ranging from 0 = extremely unimportant to 10 = extremely important. Items C14 and C15 were measured on a 7-point scale ranging from 1 = every day to 7 = never. The scores are reversed for C14 and C15, and as a consequence, high scores are associated with high religiosity for all items.

These items have also been analyzed by Billiet and Welkenhuysen-Gybels (2004a) who concluded that a single factor underlies them, and that partial measurement invariance exists. Our example data are somewhat different because we include France (not available for the 2004 analysis), and treat the data as continuous. On the country level, we have added the religious diversity index by Alesina, Devleeschauwer, Easterly, Kurlat, & Wacziarg (2003). Since the data set is large and the amount of missing data on these four items is small, we use listwise deletion, which leaves 41,207 subjects. All analyses were carried out using Mplus 5 (Muthén & Muthén, 2007).

21.4.1 Analysis Results, Multigroup SEM

The typical approach in investigating measurement invariance using multigroup SEM is to set up a series of models that specify a common factor model for all groups, starting with a model with no constraints (the functional equivalence model) and then adding constraints on factor loadings (the metric equivalence model) and intercepts (the scalar equivalence model) in two successive steps. Since these models are nested, both formal chi-square tests and inspection of fit indices can be used to decide whether measurement invariance holds. This approach has the advantage that if the functional equivalence model is rejected the analysis process may stop, because statistical analysis of the differences between countries is not valid. However, when the number of countries is large, it can take many analysis steps to find out which equality constraints are allowed. When the number of groups (countries) is large, we suggest using the opposite strategy: start by fitting a model with all constraints needed for scalar invariance, and in addition for the purpose of comparison the functional equivalence model (no constraints). Again, if the functional equivalence model is rejected, the analysis should stop. If the functional equivalence model holds, but the scalar invariance model does not fit, the modification indices for the scalar invariance model and the differences between countries in their parameter estimates in the functional equivalence model both provide information about the model modifications that are needed to achieve some form of equivalence.

The scalar equivalence model is specified by constraining the latent variable mean to zero and its variance to one in the first country (Austria), and constraining all equivalent loadings and intercepts to be equal across all countries. A more common representation is to fix one intercept to zero and one loading to one for all countries, and allow the factor mean and variance of all countries to be estimated. The representation we have chosen here has the advantage that all intercepts and factor loadings are estimated, and therefore modification indices are available for all intercepts and factor loadings.

Since the sample sizes are very large, we do not use the chi-square statistic as the measure of model fit. Instead, we rely on the CFI and the Tucker-Lewis Index (TLI), with values > 0.90 indicating a good fit, and the RMSEA, with values < .05 indicating a good fit and values < .08 indicating an acceptable approximation (cf. Section 21.2.1). Given the simulation results discussed earlier, which show that a small violation of measurement invariance is easily masked when the remainder of the model fits well, we inspect all the modification indices that refer to the factor loadings and intercepts on a country-by-country basis, and follow up by comparing the constrained estimates to the unconstrained estimates in the functional equivalence model.

Table 21.1 presents the fit information for a sequence of models. Model 1, which states functional equivalence, shows a reasonable fit according to the CFI and TLI fit indices, but the RMSEA is not so good. The modification indices suggest correlated errors in some countries, mostly between C13 and E18 or between C14 and C15. Model 2, which states complete scalar equivalence, has a much worse fit. The three countries with the largest contribution to the overall chi-square are Ireland (1603.8), Poland (1255.0), and Israel (991.7), perhaps not by accident, countries where religion plays an important role.

Taking the scalar equivalence model as a starting point, modification indices were inspected to discover the source of the lack of fit. This showed a striking result: The intercept of item C14 (attendance of religious services) had a large modification index for almost all countries. When this intercept was set free the chi-square dropped by more than 3100. The CFI/TLI fit indices were acceptable for this solution (both 0.94) but the RMSEA was judged too large at 0.14. Following large values of the modification indices, intercepts and loadings were allowed to be freely estimated in specific countries, with the restriction that no country could have more than two intercepts or loadings free. This way, for all countries partial invariance is maintained. The result is the partial scalar invariance model in row 3 of Table 21.1. This model fits well. The way this model is specified, the first country (Austria) has a mean constrained to zero and a variance constrained to one. The means and variances in all other countries can be freely estimated. If the means are all constrained to be zero, the model is rejected again (chi-square = 8876.3, $df = 142$, CFI/TLI = 0.90, RMSEA = 0.18), and the difference between the two models is clearly significant (chi-square = 77585.9, $df = 142, p = < .001$). We conclude that the countries differ in the level of religious involvement. We present the country means and variances in Section 21.4.4, where we compare the results from the three analysis approaches.

21.4.2 Analysis Results, MSEM

In multilevel SEM a measurement model is specified at both the individual level and the country level. The first model is a model with no varying slopes, which

TABLE 21.1. Model Selection in Multigroup SEM

Model	Chi-square	df	CFI	TLI	RMSEA
1: Functional	1021.9	21	0.97	1.00	0.11
2: Scalar	8783.5	170	0.90	0.92	0.16
3: Partial	1290.4	121	0.99	0.99	0.07

implies metric invariance and in addition equal loadings across both levels. The latter implies metric invariance across the individual and the country levels. This is not a necessary condition, but if it holds, it has the advantage that it allows statistical comparisons of within and between factors. The within model was identified by constraining the individual-level factor mean at zero and the variance at one. All four loadings were estimated, and within country intercepts are zero by definition. Since the loadings are constrained to be equal across the two levels, the between model was identified by constraining the factor mean to zero, allowing all intercepts to be estimated. The variance of the between-level factor can be estimated freely.

The model fits reasonably, but there is one very large modification index that suggests a covariance between the residuals of C13 (religiosity) and E18 (importance of religion). When this covariance is added, the model fits very well: chi-square = 30.8 ($df = 6$, $p = 0.00$), CFI/TLI = 1.00, RMSEA = 0.01, and there are no large modification indices. In this model, the variance of the between factor is estimated at 0.25, which implies an intraclass correlation coefficient (ICC) of 0.20, meaning that 20% of the variance is at the country level. The ICCs for the observed variables are lower: They range from 0.12 to 0.17. Finding a larger ICC for the latent variables is typical, since measurement error in the observed variables ends up at the lowest level (Muthén, 1991).

Despite the good model fit, given the lack of power of the global tests in the simulations, four additional models were estimated that allowed each of the four loadings in turn to vary across countries. This shows a significant variance component for the loading of C14 (attendance of religious services). This variance component is small (0.094), but significant (*SE* is 0.03, $p = .001$). To interpret the size of the variance component, it is useful to refer the corresponding standard deviation to the normal distribution. The loadings of C14 have a normal distribution with a mean of 1.27 (the average loading) and a standard deviation of 0.31 ($= \sqrt{0.094}$) across countries. Since about 95% of a normal distribution is within two standard deviations from the mean, we can conclude that for 95% of all countries this loading is between 0.65 and 1.89. This is a sizeable difference, but we can also conclude that there are no countries where this loading is actually negative.

The conclusion is that in the multilevel measurement model partial metric invariance appears to hold. For further modeling, we can include explanatory variables at either the individual or the country level to explain variation in religious involvement, or even to explain the variation in the loadings for C14 in the measurement model. Since the loadings can be constrained equal across the individual and the country level, it is likely that the within-country and the between-country variation are caused by the same explanatory variables. Therefore, we may hypothesize that the difference between countries is not so much an effect of contextual differences in country-level variables, but in differences in the composition of the countries' populace. If this is indeed the case, country-level variation is likely to be explained by aggregated individual variables, rather than by country-level variables. Such analyses are not pursued here. To enable a comparison between the different statistical approaches, we estimate the factor scores for the between and the within factor. The mean factor scores across countries are presented in Section 21.4.4, where we compare the results from the three analysis approaches.

21.4.3 Analysis Results, Latent Class SEM

In latent class SEM, a number of latent classes is postulated that are each characterized by their own structural model. The approach is analogous to multigroup SEM, with the latent classes replacing the groups. The difference is that the number of latent classes is assumed to be much smaller than the number of countries. In our case, we investigate both metric and scalar invariance. Consequently, a confirmatory factor model was specified with equality constraints for the loadings and intercepts across all latent classes. Interestingly, when this model was specified for two classes, it did not converge. This could be solved by allowing the intercept of C14 (attendance of religious services) to be estimated freely in the two classes. The model for three classes did not converge, and a search for additional intercepts or loadings to be freed did not result in convergence. In the two-class model, C15 had a small negative residual variance in the first class; this was corrected by constraining it to zero. Table 21.2 presents the AIC and BIC measures of model fit for these models.

In the two-class model, the first class (about 44% of the respondents) constrains the latent religion variable mean to zero and the variance to one. The second class (about 56% of the respondents) estimates the mean as 8.9 and the variance as 7.7. The first class can be described as individuals for whom religious involvement is not part of their life. The second class can be described as individuals for whom religious involvement is important: they have a much higher mean but also display a larger variance in religious involvement.

Similar to the multilevel SEM approach, it is possible to add explanatory variables that explain variations in the latent variable religious involvement. For example, individuals for whom religious involvement is important have a much higher mean but also display a larger variance in religious involvement. In addition, it is possible to add explanatory variables that explain class membership. This will not be pursued here. For comparison purposes, the factor scores and latent class probabilities are estimated, and discussed in the next section.

TABLE 21.2. Fit Measures for Different Latent Class Structural Equation Models

Model	AIC	BIC
1 Class	675316.2	675428.4
2 Class	632512.8	632702.7
2 Class corrected	632564.7	632702.7
3 Class	no convergence	no convergence

21.4.4 Analysis Results, Comparison of Approaches

The results from the three different approaches tend to converge on similar conclusions. All three approaches result in a verdict of partial metric invariance. All three approaches conclude that item C14 is problematic. There is also a covariance between the residuals of items C13 and E18 that is necessary in all approaches to achieve a good fit.

Table 21.3 shows the factor loadings for the various measurement models. It is clear that the different approaches do not result in identical measurement models. The multigroup and the latent class model both analyze only within groups variation, and the variation between groups is modeled as differences in latent variable means and variances between observed or latent groups. In contrast, the multilevel approach estimates a common factor model for variation within and between countries. The multigroup and latent class approach both agree that C13 obtains the lowest loading and E18 the highest. As a result, the interpretation of the latent factor is subtly different in each approach. In each case, the interpretation of the latent factor is "religiosity," but there are some subtle differences in the meaning of this construct. To investigate the correspondence between the different approaches, we estimated factor scores for all latent variables in the final model in each of the three analysis approaches. The correlations between the multigroup factor score, the within factor score of the multilevel SEM, and the latent class factor scores vary from 0.87 to 0.95. Thus, although the factor scores are not exactly the same, they are all very similar. When the factor scores are aggregated to the country level, the correlations between the multigroup factor score, the between factor score from the multilevel SEM (no aggregation is needed here), and the latent class factor score vary from 0.96 to 0.99. Clearly, when the objective is to compare and analyze countries, all three approaches are effectively equivalent. Figure 21.5, which plots for all 22 countries their (Z-score transformed) multigroup means, country-level factor scores from the multilevel analysis, and aggregated latent class factor scores, highlights the similarities at the country level.

The latent class model provides additional information. The two class solution in the latent class approach suggests that the observed differences between the 22 countries can be the result of a different composition of their populations in terms of these two classes. To investigate this possibility, a crosstabulation was done for estimated class membership against country. The correlation between these vari-

TABLE 21.3. Measurement Models in Different Approaches (Raw Loadings)

Item	Multigroup	Multilevel	Latent Class
C13 (religious)	0.99	2.17	0.28
C14 (services)	2.35	1.04	0.45
C15 (pray)	1.79	1.83	0.45
E18 (importance)	2.80	2.52	0.54

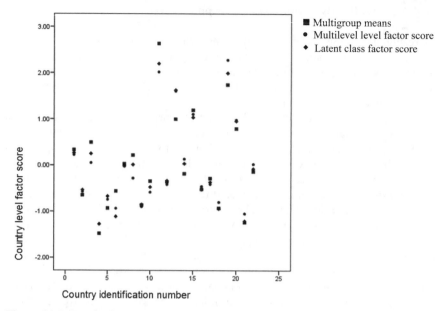

Figure 21.5. Standardized Factor Scores at the Country Level,
Three Analysis Approaches

ables expressed as the phi coefficient is 0.39. The class membership per country is given in Table 21.4, in ascending order of membership in the first class. Estimated class membership indeed shows a large variation across the countries. To investigate to which degree differences between countries can be attributed to differences in the composition of the populace, mean factor scores for the factors in the multigroup solution are calculated, both for the total population and separately for the two classes. The results are presented in Table 21.5. The last column presents the factor means in the total population adjusted for a class membership indicator.

Class membership accounts for approximately half of the variation in religious involvement between countries. The other half consists of variation in religious involvement within classes between countries. The effect of adjusting for differences in class membership are striking when we compare the total means with the covariate adjusted means. The total means vary from −0.62 to 1.17, while the adjusted means vary from −0.15 to 0.53.

To further analyze the differences between countries, we can use the country level religious diversity index (Alesina et al., 2003). In the multiple group approach, there is no formal way to include this variable in the model. We can correlate the countries' estimated means on the religiosity factor with the religious diversity index, which produces a correlation of −0.38 ($p = .08$). The same analysis can be done for the aggregated factor scores in the two-class latent class solution; this produces a correlation between the aggregated factor score and religious diversity of −0.40 ($p = .07$). In the multilevel model, we can directly predict the between-countries religiosity by religious diversity, which produces a standardized

TABLE 21.4. Estimated Class Membership (%) Per Country

Country	Class 1	Class 2	Country	Class 1	Class 2
Greece	8.4	91.6	Netherlands	55.1	44.9
Poland	10.1	89.9	Hungary	55.3	44.7
Ireland	15.4	84.6	United Kingdom	55.9	44.1
Portugal	28.0	72.0	Belgium	56.7	43.3
Italy	28.4	71.6	Luxembourg	57.1	42.9
Switzerland	40.0	60.0	Germany	60.5	39.5
Austria	40.1	59.9	France	62.5	37.5
Israel	44.2	55.8	Norway	69.0	31.0
Spain	45.6	54.4	Denmark	70.5	29.5
Slovenia	45.7	54.3	Czech Republic	73.2	26.8
Finland	50.2	49.8	Sweden	73.5	26.5
All countries	46.8	53.2	Sweden	73.5	26.5

regression coefficient of -0.18 ($p = .78$). In all approaches we conclude that there is a negative but nonsignificant relationship between religious diversity and religiosity. In the latent class model, we can predict class membership with religious diversity, this produces a raw regression coefficient of -1.29 ($p <. 001$), indicating that high religious diversity is related to less membership in Class 2. Although the analyses methods differ, the conclusions converge: high religious diversity in a country is (weakly) related to lower feelings of religiosity in general.

21.5 DISCUSSION

21.5.1 Multigroup SEM, MSEM, and LCM

In the context of comparative research, we have a multilevel data structure consisting of respondents within countries. If we investigate the measurement model, we need to model respondent-level variation. Both the simulations and the example indicate that the classic analysis approach using multigroup SEM is still attractive. It is feasible with more than 20 countries, and the most powerful approach when violations of measurement equivalence are to be detected. The approach taken in the example data, which contrasts the functional equivalence model (same form, no constraints) with the scalar equivalence model (all constraints on loadings and intercepts) has proven to be effective. Latent class SEM is much like multigroup SEM: it is a fixed effects model, but is expected to include far fewer classes than we have countries, which makes it more parsimonious than multigroup SEM.

If we investigate differences between countries, we need to model country-level variation. Multilevel modeling treats the observed countries as a sample from

TABLE 21.5. Differences Between Countries

Country	Percent Class 1	Total Mean	Class 1 Mean	Class 2 Mean	Adjusted Total Mean
Greece	8.4	1.17	−.47	1.32	.53
Poland	10.1	.73	−1.05	.93	.12
Ireland	15.4	.40	−1.03	.66	−.13
Portugal	28.0	.33	−.72	.73	.00
Italy	28.4	.52	−.54	.94	.20
Switzerland	40.0	.18	−.65	.73	.05
Austria	40.1	.13	−.74	.71	.01
Israel	44.2	−.06	−.94	.63	−.10
Spain	45.6	.03	−.78	.70	.02
Slovenia	45.7	−.04	−.85	.64	−.07
Finland	50.2	.10	−.57	.78	.13
Netherlands	55.1	−.10	−.80	.77	.05
Hungary	55.3	−.14	−.88	.78	.00
United Kingdom	55.9	−.12	−.82	.76	.04
Belgium	56.7	−.18	−.84	.68	−.02
Luxembourg	57.1	−.19	−.90	.76	−.01
Germany	60.5	−.34	−1.01	.68	−.10
France	62.5	−.33	−.91	.65	−.06
Norway	69.0	−.34	−.80	.68	.01
Denmark	70.5	−.20	−.57	.68	.18
Czech Republic	73.2	−.62	−1.13	.78	−.15
Sweden	73.5	−.47	−.87	.65	−.01
R-squared		0.17	0.07	0.09	

a larger population and can include country-level variables in the analysis. In our analyses, we have used multilevel SEM instead of the more usual but less powerful multilevel regression models. Multilevel SEM is more parsimonious than multigroup SEM, but does assume a reasonable sample size at the country level. Our simulations suggest 30 countries as a reasonable sample, but even at 20 countries, the estimates and standard errors appear accurate enough.

The simulation results make clear that reliance on global model fit tests or indices can be dangerous. Such tests and indices take the entire model into account, and do not have sufficient power regarding very specific model violations, such as one factor loading that differs across countries. The simulations indicate that the multigroup SEM approach, although laborious, provides the most detailed information on the amount and the sources of lack of measurement invariance.

A comparison of the three approaches on the example data shows that some features of the data are identified in each of the three approaches. All approaches identify item C14 as problematic, and the residual correlation between items C13 and E18 is also present in each final model. Thus, the results of the different analyses do converge. There are also major differences between the final models, which are based on the elementary fact that we are considering models that are qualitatively very different. When issues of measurement invariance are involved, the multigroup SEM is clearly the best approach, because it provides very specific information on sources of misfit, and allows also very specific modifications of the strict invariance model. When the number of countries is large, latent class SEM offers a way to make the model more parsimonious. However, when we want to use country-level variables to predict individual- or country-level outcomes, multilevel modeling is better suited, because the country-level variables can be included in the model. In multigroup SEM, and latent class SEM, we need to use a two-step approach where country-level variables and aggregated individual-level variables are combined and analyzed in a separate analysis step.

It appears that the different analysis approaches should be seen as complementing each other, rather than competing. When measurement invariance has been established using multigroup SEM, subsequent analyses including country-level variables can best be done using a multilevel model. Since the models that can be specified in the framework of multilevel regression are limited, because multilevel regression is essentially a univariate approach, this is likely to involve multilevel SEM. Given the generally limited sample sizes at the country level, such models must at the country level be kept parsimonious. Latent class SEM is an interesting addition to the statistical toolkit, but not a replacement for either multigroup or multilevel SEM. In our case, we have kept the latent class SEM example simple, by investigating the latent class structure only at the within country level. This assumes that differences between countries are the result of a different composition of their populace in terms of the latent classes. Latent class SEM can be expanded to include a multilevel structure, that is, latent classes both at the within-country (individual) and the between-country level. This possibility is not pursued here, because it is not well suited to investigate issues of measurement equivalence. After measurement equivalence has been established, multilevel latent class methods appear useful to explore differences at the country level with respect to the relationships between the country-level variables and the way these influence relationships at the individual level (within countries).

21.5.2 Software Issues

The advent of powerful and user-friendly software for multilevel modeling has had a large impact in research fields as diverse as education, organizational research, demography, epidemiology, and medicine. Multilevel modeling has had less impact on the analysis of international and cross-cultural survey data. This is not caused by a lack of interest in statistical techniques on the part of cross-cultural researchers; comparative researchers in intercultural and organizational research were among the first to use multigroup SEM to assess measurement comparability. The almost exclusive focus on multigroup SEM reflects specific challenges in comparative surveys. Until recently, few comparative surveys included a large enough number of countries to make multilevel modeling an attractive option. The sporadic use of latent class modeling in comparative survey research reflects mostly a lack of powerful software. To be useful in the context of comparative surveys, latent class modeling must be able to include multiple and distinct observed groups such as countries in the model. That is, the latent class models and accompanying software must include either multigroup or multilevel capacities. Until recently, such software was simply not available.

The most flexible software capable of multigroup SEM, multilevel analysis, and latent class analysis at the moment are M*plus* (Muthén & Muthén, 2007) and GLLAMM (Rabe-Hesketh, Skrondal, & Pickles, 2004). These programs can both estimate models with varying combinations of multiple groups, multilevel, and (latent class) structural equation models, that are far more complicated than the examples given above. They can also deal with incomplete data and nonnormal variables. A practical limitation for both programs is that when complex nonlinear models are estimated for nonnormal or incomplete data, the estimation methods that must be used are computationally very demanding, to the point where estimation is actually impossible.

Other, more limited, software packages are Latent GOLD, which can analyze multilevel latent class models and has limited capacities for structural equation modeling, and the SEM packages LISREL and EQS, which have limited capacities for multilevel modeling and cannot estimate latent class models. The multilevel regression package HLM has limited capacity for multilevel structural equation models, which can be estimated, provided they are recursive.

ACKNOWLEDGMENT

We gratefully acknowledge comments on earlier drafts from Michael Braun, Timothy Johnson, Remco Feskens, and Ana Villar.

22

Using Polytomous Item Response Theory to Examine Differential Item and Test Functioning: The Case of Work Ethic

David J. Woehr and John P. Meriac

22.1 INTRODUCTION

Establishing the measurement invariance of an instrument is a logical prerequisite to conducting any meaningful substantive cross-group comparisons. This is especially true for complex constructs. In their review and integration of the literature on measurement invariance in organizational research, Vandenberg and Lance (2000) state that "violations of measurement invariance assumptions are as threatening to substantive interpretations as is an inability to demonstrate reliability and validity" (p. 6). In evaluating measurement invariance across different cultures, we ask whether members of both groups interpret items in the same way, or whether there are group differences that preclude the comparison of responses across groups. In this chapter we will examine, and provide illustrative examples of, two approaches toward evaluating measurement invariance. Specifically, we directly compare confirmatory factor analytic (CFA) and item response theory (IRT) based approaches for the evaluation of measurement invariance across groups.

22.1.1 Work Ethic and the Measurement of Work Ethic Across Groups

The primary focus of this chapter is on the application of polytomous item response theory as a tool for the examination of cross-cultural measurement invariance in the assessment of complex work-related constructs. A secondary goal is the comparison of this approach to more typical CFA-based approaches. We base this application and comparison in the context of the measurement of

---

work ethic. Miller, Woehr, and Hudspeth (2002) presented a historical and conceptual review of the work ethic construct. Drawing on the large body of literature stemming from Weber's original work, they posit that work ethic is not a single, unitary construct but a constellation of attitudes and beliefs pertaining to work behavior. Further, they suggest that the work ethic construct: (a) is multidimensional; (b) pertains to work and work-related activity in general, not specific to any particular job (yet may generalize to domains other than work, such as school, hobbies, etc.); (c) is learned; (d) refers to attitudes and beliefs (not necessarily behavior); (e) is a motivational construct reflected in behavior; and (f) is secular and not necessarily tied to any one set of religious beliefs. Based on previous literature as well as original empirical research, Miller, Woehr, and Hudspeth (2002) highlight seven components or dimensions that they argue comprise the work ethic construct (that is, centrality of work, self-reliance, hard work, leisure, morality/ethics, delay of gratification, and wasted time). A full definition of each dimension and sample items is presented in Table 22.1.

TABLE 22.1. MWEP Dimensions, Dimension Definitions, and Sample Items

Dimension	Definition	Sample Items
Centrality of work	Belief in work for work's sake and the importance of work.	Even if I inherited a great deal of money, I would continue to work somewhere. It is very important for me to always be able to work.
Self-reliance	Striving for independence in one's daily work.	I strive to be self-reliant. Self-reliance is the key to being successful.
Hard work	Belief in the virtues of hard work.	If you work hard you will succeed. By simply working hard enough, one can achieve one's goals.
Leisure	Proleisure attitudes and beliefs in the importance of non-work activities.	People should have more leisure time to spend in relaxation. The job that provides the most leisure time is the job for me.
Morality/ ethics	Believing in a just and moral existence.	People should be fair in their dealings with others. It is never appropriate to take something that does not belong to you.
Delay of gratification	Orientation toward the future; the postponement of rewards.	The best things in life are those you have to wait for. If I want to buy something, I always wait until I can afford it.
Wasted Time	Attitudes and beliefs reflecting active and productive use of time.	I try to plan out my workday so as not to waste time. Time should not be wasted, it should be used efficiently.

Miller and colleagues (2002) developed and provided initial support for a multidimensional work ethic inventory—the Multidimensional Work Ethic Profile (MWEP). The MWEP is a 65-item self-report measure tapping 7 conceptually distinct (that is, divergent) dimensions. Each of the 7 dimensions is assessed with 10 items with the exception of Delay of Gratification (7 items) and Wasted Time (8 items). The MWEP has been translated and used in several different languages including Spanish, Korean, Romanian, and Dutch. Woehr, Arciniega, and Lim (2007) provided initial evidence for the measurement invariance of some of these various measures. Our goal in the present chapter is to further examine the measurement invariance of the MWEP using IRT. In addition, we review and compare the results of both CFA and the IRT analyses in the present study.

22.1.2 Traditional Approaches to Measurement Invariance

Measurement invariance can be assessed in several ways, including mean differences across groups, relationships with external variables, or internal covariance differences across item responses. Although a number of approaches have been used to evaluate measurement invariance (cf. Hui & Triandis, 1985; Vandenberg & Lance, 2000), there is general agreement that the multigroup confirmatory factor analytic (CFA) model (Jöreskog, 1971b) provides a versatile technique for testing cross-group measurement invariance. Based on their review, Vandenberg and Lance (2000) called for an increased application of measurement invariance techniques before substantive comparisons are considered. Multiple-group CFA is essentially a test-level approach for examining the invariance of tests across subgroups (although it may also be used to assess item-level differences), which has been used to examine test invariance across racial, gender, and cultural subgroups, to name a few applications.

CFA approaches to measurement invariance are based on Classical True score Theory (CTT), where variance in observed scores is viewed as a linear composite of true score variance and error variance. Configural invariance is established when the factor structure of a measure from one group is confirmed in another group (that is, equivalent factor structure across groups). Metric invariance imposes further constraints on this model and requires not only the same factor structure but also equal factor loadings across groups (that is, equivalent factor loadings across groups). Error invariance constrains the model still further by requiring not only factor loadings but error variances to be equal across groups (that is, equivalent error variance for the same items across groups).

Despite the popularity of CFA approaches to assessing measurement invariance, there are a number of potential problems. One limitation arises from the difficulties inherent in factor analyzing categorical item-level data (see Bernstein & Teng, 1989; Nunnally & Bernstein, 1994, for detailed discussion of these problems). Traditional CFA approaches based on item-level (as opposed to scale level) data tend to overestimate the number of factors underlying the data. One solution to this problem is the use of item parceling (e.g., Bagozzi & Heatherton, 1994; Gibbons & Hocevar, 1998). Parceling involves summing or averaging item scores from two or more items and using these parcel scores in place of the item scores as manifest indicators in a CFA. Here it should be noted that a key limitation of the item parceling approach is that it does not allow for a direct comparison of items across groups. However, the rationale for this approach is to avoid the difficulties associated with categorical item-level data described above and to achieve a higher level of reliability for each of the scores on which the confirmatory factor analyses were based than would be realized from responses on each of the individual items. So, for example, for a subscale with 12 individual items, 3 sets comprised of 4 randomly selected items could be identified. A composite score based on the 4 items would then be computed for each set. The 3 composites would then serve as manifest indicators of the latent variable (as opposed to each 1 of the 12 items separately) in the latent factor analysis.

The literature provides a good deal of support for this approach and suggests that the use of composite-level indicators leads to far more interpretable and meaningful results than an analytic approach based on large numbers of individual items (e.g., Bagozzi & Heatherton, 1994; Gibbons & Hocevar, 1998; Hall, Snell, & Foust, 1999; Landis, Beal, & Tesluk, 2000; Paik & Michael, 1999). However, there are a number of issues and limitations associated with the use of item parceling. One issue examined is the effect of the specific parceling strategy on subsequent analytic outcomes. Specifically, multiple strategies have been suggested for selecting the set of items to be included in each parcel. The most common approach is a random parceling in which scale items are randomly selected for each parcel. However, it has been demonstrated that this strategy is inappropriate if the construct assessed by the scale is multidimensional (Hall, Snell, & Foust, 1999). In these cases, more sophisticated parceling strategies are required. As noted above, however, one key limitation of the item parceling approach is that it does not permit a direct comparison of individual items, because results are restricted to the parcel level. This limitation is particularly relevant with CFA approaches to assessing measurement invariance in that it prevents the comparison of specific items across groups. Woehr, Arciniega, and Lim (2007) provide a detailed example of the use of the item-parceling approach with the 65-item MWEP.

22.1.3 Item Response Theory as an Alternative for Assessing Measurement Invariance

Item response theory (IRT) differs from traditional Classical True score Theory (CTT) approaches, such as confirmatory factor analysis (CFA), in several ways (Embretson & Reise, 2000; Hambleton, Swaminathan, & Rodgers, 1991). One primary distinction between these methods is that IRT evaluates psychometric properties at the item level, whereas CFA approaches primarily focus on the test as a whole. If an item-level examination is of interest, then IRT provides a clearer picture of test properties over CTT approaches. Also, CTT methods provide indices of item characteristics such as difficulty (p-values) and discrimination (item-total correlations) that are dependent on group differences in traits (Raju & Ellis, 2002). On the contrary, IRT parameter estimation techniques, control for individual differences in respondent ability, yielding values that are independent of the group to which the set of items was administered (Raju & Ellis, 2002).

Item response theory also takes a somewhat different approach toward the examination of measurement invariance than do CFA approaches. IRT approaches view invariance in terms of "differential functioning" of items and tests (Raju, van der Linden, & Fleer, 1995). IRT approaches to measurement invariance have an advantage over CFA in that item parameters are not dependent on group differences on the trait; in other words, they allow for the control of group differences in ability when estimating parameters (difficulty and discrimination) so they are not confounded with group differences (Raju & Ellis, 2002).

When items demonstrate invariance across two groups, the probability of endorsing a response option is the same across groups for a given trait level. However, when individuals possess the same level of the trait or ability and have different probabilities of endorsing a given response alternative on an item, the item displays differential item functioning, hereafter referred to as DIF (Hambleton, Swaminathan, & Rodgers, 1991; Hulin, Drasgow, & Parsons, 1983). More specifically, DIF refers to group differences in responses to test items when ability or trait level is held constant, and would be indicative of some form of group-specific response tendency or bias. Differential Test Functioning (DTF) is the extension of this same phenomenon to test-level differences in response tendencies (Raju & Ellis, 2002). Raju, van der Linden, and Fleer (1995) proposed a method for evaluating the differential functioning of items and tests (DFIT). The DFIT approach further offers two indices of differential item functioning (DIF): compensatory (CDIF) and noncompensatory (NCDIF). CDIF is calculated as the sum of the differences between the two subgroups' item-level scores and as such differential functioning in one item may compensate for differential functioning in another item. NCDIF represents the average squared difference between the two subgroups' item-level true scores, and thus does not allow for compensation across items. In addition, CDIF is used in the computation of a differential test functioning (DTF) index such that the sum of the CDIF values is the DTF value. CDIF is also helpful in that these values allow us to see which items can be deleted to result in an overall lack of invariance at the test level. For a full explanation of these indices readers are referred to Raju and Ellis (2002) and Raju, van der Linden, and Fleer (1995). However, it is important to note that a key aspect of compensatory test-level fit indices is that they allow for item differences to be compensatory. That is, if two items demonstrate reciprocal differences in each of two groups in the sense that one item is more "difficult" in one group than in another and a second item demonstrates the opposite pattern, then it is possible that there may be no test-level differences across groups despite pronounced item-level differences. In other words, items that possess DIF can have a "cancellation effect" on each other to nullify DTF. Although there are several options for evaluating DIF, only the Raju, van der Linden, and Fleer. (1995) DFIT procedure provides an index of DTF. This is analogous to CFA approaches, which index invariance at the test level. Table 22.2 provides a general overview comparison of both CFA and IRT approaches to assessing measurement invariance.

TABLE 22.2. A Brief Comparison of CFA and IRT Approaches for Assessing Measurement Invariance[a]

Issue	CFA	IRT
Level of analysis	Test or "factor"	Item
Model	Linear: $X = T + E$	Nonlinear: Normal ogive
Analytic input	Variances/covariance	Item responses
Focal outcomes	Factor structure Factor loadings Proportion of item error variance	Item parameters [b] b = threshold parameter, which indexes the likelihood of endorsing a particular response category over another adjacent category a = discrimination parameter, which indicates how well the item distinguishes among levels of the latent trait (θ) across the θ continuum
Test statistics	Model goodness of fit	CDIF [c] NCDIF DTF

[a] A more detailed comparison of CFA and IRT approaches to invariance is presented in Ellis & Mead (2002).
[b] The item parameter descriptions in the table above pertain directly to Samejima's (1969) graded response model, different IRT models provide different information.
[c] The DIF indices presented in the table correspond directly to the Raju et al. (1995) DFIT procedure; different DIF methods use different indices.

22.2 MEASUREMENT INVARIANCE OF THE MWEP ACROSS RESPONDENTS FROM THE UNITED STATES AND MEXICO

We examined the measurement invariance of the MWEP across respondents from both the United States and Mexico, taking both a CFA and an IRT approach. Miller, Woehr, and Hudspeth (2002) note that the MWEP was initially created in English. The Mexican sample of respondents completed a Spanish version of the MWEP. The U.S. respondents were on average 28 years old and 57% were males. The Mexican respondents were on average also 28 years old, and 46% were males. Responses were obtained from several different types of industrial settings and college students. All of the Mexican respondents were full-time employees, and 46% of the U.S. sample had a part-time job at least at the time of data collection. Woehr, Arciniega, and Lim (2007) provide more complete information on the development and evaluation of the Spanish versions.

22.2.1 Application of Multiple-Group CFA to Evaluate Measurement Invariance of the MWEP Hard Work Scale

Woehr, Arciniega, and Lim (2007) applied multiple-group CFA with the MWEP to examine its invariance across respondents from the United States ($n = 238$) and Mexico ($n = 208$). Analytically, this process involves constructing a common CFA model for both subgroups, and constraining parameters to equality as mentioned above to evaluate configural, metric, and error invariance. Hence, each successive step in this sequence is increasingly restrictive and examines whether (1) the pattern of items load onto the latent variables in the same configuration, (2) their weights are of equal size across both groups, and (3) the weights of the error terms are invariant across groups. Inferences regarding the level of measurement invariance are made by examining model-data fit indices.

In constructing the CFA model that would be examined in their study, Woehr and colleagues (2007) randomly parceled items, which served as the manifest indicators on each latent variable. Specifically, three parcels loaded onto each latent MWEP factor (e.g., dimension or subscale). Hence, instead of having 65 manifest indicators, their analyses only contained 21, thereby creating a more parsimonious model. Their results showed that the 7-factor model with the item composites fit the data well for each subgroup, indicating that the factor structure held prior to constraining any item parameters to equality. Next, the parameter-nested sequence comparisons between the U.S. and Mexican samples were conducted and showed that the English and Spanish versions of the MWEP demonstrated metric invariance, which permits one to proceed to make cross-cultural comparisons between groups. Making such decisions is based on an examination of the change in the various fit indices that are examined. Here, one examines whether the change in the magnitude of the fit values exceed several critical values suggested by Cheung and Rensvold (2002). These comparisons indicated that the versions of the MWEP were conceptually invariant (that is, metric invariance), which indicates that the composites relate to the subscales in the same way for both cultural groups. More specific information regarding the results of these analyses can be found in Woehr, Arciniega, and Lim (2007).

There is, however, a major shortcoming associated with this approach. That is, the indices for determining invariance are at the overall MWEP level (model-data fit). This indexes the extent to which all dimensions of the MWEP are invariant in general. In the absence of overall invariance, it is possible to examine the degree of invariance or "partial invariance" by successively freeing parameters. Still, given that the CFA analyses in this case are based on item parcels as opposed to individual items, there is no indication of which items are invariant or not. For diagnostic purposes, if one happened to find a lack of measurement invariance, the information regarding which items specifically vary across groups is limited. In this case, no invariance was found, and there was no reason to examine invariance at an item level. In situations in which invariance is found, however, item-level information may be beneficial for making adjustments to the questionnaire, specifically targeting items that may be more prone to differential interpretation. Consequently, we chose to reexamine the measurement invariance of the MWEP's subscales using individual items rather than item parcels. Given our intent to provide an illustrative example, we present results for only one subscale for purposes of brevity. Specifically, we examine the degree of measurement invariance for the "hard work" subscale at both the scale and item level. Our goal is to provide a more relevant comparison of the CFA approach to the IRT approach elaborated below.

In order to test the measurement invariance of the hard work subscale across the two samples, we used a multiple-group CFA application of LISREL 8.51 (Jöreskog & Sörbom, 2001) to test the three models representing different levels of measurement invariance. All three models were operationalized as a one-factor model corresponding to the hard work dimension. Model 1 posits an equivalent factor structure (that is, items relate to the same factor) across groups and thus represents a test of configural invariance. Model 2 is based on the same measurement model as Model 1, but with more constraints placed on the model parameters. That is, factor pattern coefficients for like items are constrained to be equal across groups. Thus Model 2 provides a test of metric invariance. Finally, Model 3 places even more constraints on the model parameters. Specifically, in addition to equivalent factor loadings, the like item error loadings were constrained to be equal across groups. Thus, Model 3 provides a test of error invariance.

Results of the CFA analyses are presented in Table 22.3. Given that the models represent a parameter-nested sequence, we initially used the χ^2 test statistic and a difference of χ^2 test to evaluate the fit of each model in the series. We also focused on four additional overall fit indices: Steiger's (1990) root mean square error of approximation (RMSEA), McDonald's (1989) noncentrality index (NCI), Steiger's (1990) gamma hat, and the comparative fit index (CFI; Bentler, 1990). The RMSEA provides an overall test of model fit that compensates for the effect of model complexity. Lower RMSEA values indicate better fit with values below .08 generally indicating good fit. In addition, both the NCI and gamma hat represent absolute fit indices providing an indication of overall model fit and the CFI is an incremental (comparative) measure of fit providing an indication of fit relative to a null model. These three latter indices range from 0 to 1 with larger values indicating better fit and values of .90 or larger are typically interpreted as indicating acceptable levels of fit. More importantly Cheung and Rensvold (2002) show that across a wide range of fit measures, the changes in fit from one group to another reflected in the CFI, gamma hat, and NCI indices provide the most robust statistics for testing the between-group invariance of CFA models. In a large-scale simulation, they demonstrated that, when testing across two groups, a change in the value of CFI smaller than or equal to .01 indicates that the null hypothesis of invariance should not be rejected (that is, the versions are equivalent). Similarly, critical values for the gamma hat and NCI were .001 and .02, respectively. On the basis of these values, the results presented in Table 22.3 indicate that Model 2 representing metric invariance provides the optimal representation of the data across groups. Thus, in general our results using item-level indicators are consistent with those of Woehr, Arciniega, and Lim (2007) and indicate that the two versions of the MWEP may be viewed as invariant.

Here it is important to note that the results presented above speak only to scale-level measurement invariance. Given our finding of subscale invariance, it is not necessary to examine item-level differences. However, had we found non-

TABLE 22.3. Results for the Sequence of Measurement Invariance Tests for the MWEP Measure[a]

Model	df	χ^2	$\Delta\chi^{2b}$	RMSEA	CFI	NCI	Δ NCI[b]	Gamma Hat	Δ Gamma Hat[b]
1. Factor invariance (same factor structure, unconstrained factor loadings across groups)	60	927.99		0.14	0.96	0.733		0.976	
2. Metric invariance (same factor structure, factor loadings constrained to equality across groups)	69	1009.75	81.8	0.14	0.96	0.714	0.019	0.968	0.008
3. Error invariance (same factor structure, factor loadings and error terms constrained to equality across groups)	79	2822.91	1813	0.22	0.92	0.385	0.358	0.89	0.088
Item 1	61	929.55	1.560	0.14	0.96	0.739	0.000	0.976	0.000
Item 2	61	929.25	1.260	0.14	0.96	0.739	0.000	0.976	0.000
Item 3	61	929.84	1.850	0.14	0.96	0.739	0.000	0.976	0.000
Item 4	61	929.45	1.460	0.14	0.96	0.739	0.000	0.976	0.000
Item 5 [c]	61	959.09	31.100[d]	0.14	0.96	0.732	0.008	0.975	0.001
Item 6 [c]	61	943.98	15.990[d]	0.14	0.96	0.735	0.004	0.975	0.001
Item 7	61	929.56	1.570	0.14	0.96	0.739	0.000	0.976	0.000
Item 8	61	928.15	0.160	0.14	0.96	0.740	0.000	0.976	0.000
Item 9 [c]	61	939.87	11.880[d]	0.14	0.96	0.737	0.003	0.975	0.001
Item 10 [c]	61	932.75	4.760[d]	0.14	0.96	0.738	0.001	0.976	0.001

[a] RMSEA = root mean squared error of approximation; CFI = comparative fit index; NCI = McDonald's noncentrality index.
[b] Items that contributed to a lack of invariance as identified by the CFA approach.
[c] Difference for each model is relative to Model 1.
[d] Statistically significant at $p < .05$ or less.

427

equivalence, the next step would be to examine differences across individual items. Specifically, the fit of a model in which any single given item is constrained to be equal across groups is compared to the fit of the model in which no items are constrained (Model 1). For purposes of illustration, results of this test for each item are also presented in Table 22.3. These results indicate a significant reduction in model fit when equality constraints are placed on items 5, 6, 9, and 10. Thus, these items are identified as those reflecting differences across groups. It must be noted, however, that these differences are based solely on the difference of χ^2 values. The additional fit indices suggest no item-level differences.

22.2.2 Application of IRT to Evaluate Measurement Invariance of the MWEP Hard Work Scale

We applied the IRT analyses described below to responses from the U.S. ($n = 1230$) and Mexican ($n = 208$) respondents on each of the MWEP subscales. However, for purposes of brevity, in our comparison we focus primarily on the 10-item hard work subscale. Items were evaluated in terms of differential functioning using the Raju, van der Linden, and Fleer (1995) DFIT procedure. Compared with the CFA approach, the level of analysis of IRT differs; specifically, IRT's focus is foremost at the item level, where CFA operates at the test level. In the present case, we first estimated item parameters and scores using Multilog 7.03 (Thissen, 2003). Next, parameter estimates and latent trait distributions were equated using the Equate 2.1 program (Baker, 1995). Finally, differential item and test functioning were evaluated using Raju's (1999) DFITPS6 program. Each of these steps is described in detail in the following sections.

A requirement for IRT model convergence is that there are a sufficient number of responses for each of the response categories for each of the items in each group. In the present example, several items had a limited number of responses for the lowest response option on the scale, "strongly disagree." Hence, it was necessary to collapse the MWEP into a four-point scale, where we essentially treated the lower two response options as the same. Although this is not preferable due to variance reduction, it is sometimes necessary as a low probability of endorsing response options can result in inadmissible solutions in IRT software programs. In other words, if there are few or no instances where a particular response is chosen for a given item, the IRT solution cannot converge.

The first step in conducting the series of analyses is a check for unidimensionality. Here, we evaluated whether one or more factors underlie the data. If the unidimensionality assumption cannot be reasonably met, then results from DIF analyses can be spurious and potentially result from additional latent factors, rather than actual DIF. This is typically conducted using a full-information exploratory factor analysis approach. Here, we examined the unidimensionality of each MWEP subscale for each population separately. To evaluate unidimensionality, we conducted an exploratory factor analysis, using principal axis factoring, and examined the scree plots and percentages of variance explained by each component. Reckase (1979) found that IRT parameter estimation methods are robust to minor departures from strict unidimensionality, and proposed that if at least 20% of the variance is explained by the first component, then a scale can be interpreted as effectively unidimensional. For the hard work subscale, the first component in the U.S. subgroup explained 62% (eigenvalue = 6.22) compared to a much smaller second component that explained 16% of the variance (eigenvalue = 1.58), with a ratio of 3.93. In the Mexican subgroup, the first component explained 42% of the variance (eigenvalue = 4.15) compared to a second component that explained 12% of the variance (eigenvalue = 1.20). Further, in comparison with the second component, it is evident that a dominant factor emerged in each of the analyses. Based on this information, we interpreted these results as indicating that MWEP hard work subscale reasonably met the unidimensionality assumption in both countries.

We proceeded to estimate item parameters using Multilog 7.03 (Thissen, 2003). In the current example we used Samejima's (1969) graded response model (GRM), since the MWEP utilizes a Likert-type rating scale with five options and the GRM has often been used in this context. The GRM is a polytomous IRT model that is an extension of the two-parameter logistic (2PL) model, for responses that fall into ordered categories (Ostini & Nering, 2006). The GRM provides a single discrimination parameter (a) for the item, as well as $k-1$ boundary response functions (b_j). The discrimination parameter, also known as the slope, indexes the extent to which items differentiate among respondents, and the boundary response functions model the probability of a positive response for any given pair of successive response alternatives. Table 22.3 displays the item parameters for each of the items within the hard work subscale. Here it is evident that the discrimination and boundary response functions differ across groups in each of the subscales. However, the extent to which these differ when the latent level of work ethic on this dimension is held constant is not evident. We next proceeded to evaluate the DIF of these items.

To allow a comparison of item parameters across groups by assessing DIF), it is necessary to rescale the item parameters so that both groups are on the same metric. This is commonly referred to as "equating" or "linking" the items. In the language of DIF, the two groups that are equated are referred to as the reference group and the focal group. Essentially, the responses of the reference group (in this case the United States) must be on the same metric of the focal group (in this case Mexico) to make comparisons such as whether items exhibit DIF. In this example, we employed the equating procedures developed by Stocking and Lord (1983) that are implemented in the Equate 2.1 program (Baker, 1995). Equating parameters is an iterative process that necessitates choosing items that do not exhibit DIF to serve as "linking" items.

This process begins by first using all items in the subscale as linking items, then evaluating whether any items are identified as exhibiting DIF (as detected in the DFIT discussion that will follow). After any items that exhibit DIF are identified, they are not used as linking items, and the equating process is repeated. Hence, the properties of these items are not used in linking the items in the other subgroup. This process continues until successive iterations identify the same items as exhibiting DIF. When the process converges on the same items that do not exhibit DIF, the process stops. With the hard work subscale, this process converged in three iterations.

Finally, to evaluate invariance properties of the items, we evaluated differential functioning using Raju's (1999) DFIT procedure. The DFIT procedure is conducted using a software program written by Raju (1999), which provides information regarding the differential functioning at the level of items and at test level. As mentioned above, there are two primary item-level indices of interest: CDIF and NCDIF. For example, in Table 22.3, DIF and DTF values are displayed for the 'hard work' subscale. To make decisions on whether items are judged to possess DIF, one examines whether DIF and DTF values exceed critical values and are statistically significant as detected by a χ^2 test. Critical values are empirically derived (Raju, 1999), and serve as a criterion for deciding whether items do or do not exhibit DIF. Critical values differ based on the number of response alternatives for an item. In the current analyses, the critical value of .054 was used for determining whether NCDIF was present since the MWEP items were rated on a four-point scale. This specific critical value is based on the recommendations by Raju (1999). The DTF critical value is determined by multiplying this cutoff score by the number of items, such that the critical value for DTF for the hard work subscale was .540 (since the subscale contained 10 items). Results of the DFIT analyses for the 'hard work' subscale are presented in Table 22.4.

Examination of these results indicated that 7 of the 10 items exhibited differential functioning across groups. Specifically, the NCDIF values for these items (2, 3, 4, 5, 6, 9, and 10) exceeded .054 and were statistically significant at $p < .001$. However, examination of the overall differential test function (DTF) index, which is based on the compensatory differential item function values (CDIF), indicates that the subscale as a whole did not exhibit a statistically significant DTF. This is because the DTF index must reach a certain critical value (in this case .540) and be statistically significant. Hence, the critical value was reached, yet statistical significance was not. These results signify that several of the items are not invariant across subgroups, yet the hard work subscale as a whole is invariant. Another feature of the DFIT approach is that it indicates which items can be removed to obtain a scale without DTF, e.g., by examining items based on CDIF. This was not necessary in the hard work subscale, but if the DTF value had happened to be statistically significant, then the removal of item 9 would have brought the DTF value below the critical value, resulting in an overall lack of DTF.

22.3 COMPARISON OF CFA AND IRT ANALYTIC RESULTS FOR THE HARD WORK SUBSCALE

Interestingly, the results of the CFA and IRT analyses were remarkably similar. That is, both approaches indicate that, at the scale level, the measure demonstrates adequate measurement invariance between the Mexican and the U.S. subgroups. The CFA approach supports both configural and metric invariance, since the measure demonstrates the same factor structure and loading across groups. The IRT approach also indicates relative invariance of the measure across groups. However, both approaches highlight potentially problematic items. Specifically, both approaches suggest that items 5, 6, 9, and 10 may not be strictly equivalent. The IRT approach further identifies items 2, 3, and 4 as potentially problematic. As noted

TABLE 22.4. Hard Work Equated Parameter Estimates, DIF and DTF Information

Subscale	Item	United States				Mexico					
		a	b_1	b_2	b_3	a	b_1	b_2	b_3	CDIF	NCDIF
Hard work	1	1.351	−2.079	−1.084	.484	1.552	−1.930	−0.883	0.559	.038	.016
	2[+]	2.272	−1.766	−1.128	.604	1.341	−2.754	−0.933	0.912	.075	.057
	3[+]	2.429	−1.792	−.987	.715	1.325	−2.804	−1.669	0.479	−.005	.115
	4[+]	1.182	−2.504	−.575	.525	1.038	−1.876	0.219	2.365	.198	.160
	5[+]	.533	−2.569	−.455	.459	3.029	−1.396	−0.570	0.939	−.102	.300
	6[+]	.535	−2.534		.255	2.333	−1.263	−0.405	1.016	−.049	.308
	7	2.596	−1.942	−1.052	.080	2.511	−1.768	−0.843	0.872	.111	.036
	8	2.405	−1.700	−1.253	.842	1.648	−2.339	−1.440	0.655	−.012	.040
	9[+]	1.759	−1.582	−1.467	.059	0.600	−9.105	−5.362	0.060	.457	1.565
	10[+]	1.151	−2.088	−.814	.897	2.445	−1.296	−0.558	1.013	.043	.084

Note. The results presented are for one of the MWEP subscales, and item parameters are presented for respondents from both the United States and Mexico. The *a* parameter for each group indexes item discrimination, which indicates how well the item distinguishes among levels of the latent trait (θ) across the θ continuum. There are $j-1$ category threshold parameters (b_j) for each item, which indexes the likelihood of endorsing a particular response category over another adjacent one. This is based on an individual's θ level (e.g., level of the hard work trait). After equating groups (e.g., putting the groups on the same metric, or linking parameters), the item-level true scores are compared using the DFIT procedure, which provides the compensatory DIF (CDIF), noncompensatory DIF (NCDIF), and DTF values. The DIF values index the differences in the item-level true scores of the response item response functions, which are the nonlinear functions indexing the relationship between the latent trait and the tendency for choosing a particular response option. The CDIF value is this difference when the effects of the items are considered on one another. The NCDIF value indexes DIF without the effects of other items. The DTF value is presented at the bottom of the column containing the CDIF values, where DTF is the sum of CDIF values.

by Billiet and Welkenhuysen-Gybels (2004b), the CFA approach may be less powerful in its ability to detect deviations from invariance. Thus, the DFIT approach may be more powerful in its detection of invariance at both the item and scale level.

Here it is tempting to search for substantive explanations for why these items may be operating differently across groups, and it is not inappropriate to do so. A list of the items from the hard work subscale is presented in Table 22.5. For example, the item "Working hard is the key to being successful" has a much larger a parameter for the U.S. respondents than Mexican respondents. This could be partially attributable to a more individualistic approach (versus a collective orientation) by U.S. workers, where it is more reflective of the hard work construct to U.S. respondents than to Mexican respondents. This may stem from different interpretations of what exactly constitutes success. Alternatively, "working hard" itself may have different meaning to each respondent subgroup — specifically, it could mean working efficiently or may refer to physically demanding work. This is pure speculation, but could point to a meaningful line of inquiry. Yet an important caveat should be noted. In the absence of scale-level nonequivalence, the examination of item differences is not necessary. That is, if the measure as a whole demonstrates invariance, it is appropriate to make comparisons across groups with respect to the target construct.

How is it that several items can be identified as nonequivalent, while the scale as a whole is equivalent? In the present study, this is very likely an artifact of the metrics of invariance examined. Specifically, for the item-level evaluations we focused primarily on χ^2 tests of statistical significance. Here, statistical significance was viewed as evidence for noninvariance. Yet it is widely known that such tests are overly sensitive to sample size. Given a large enough sample even trivial differences may be significant. This issue has been highlighted by several authors

TABLE 22.5. Hard Work Items

Item 1. Nothing is impossible if you work hard enough.

Item 2. Working hard is the key to being successful.[a]

Item 3. If one works hard enough, one is likely to make a good life for oneself.[a]

Item 4. Hard work makes one a better person.[a]

Item 5. By working hard a person can overcome every obstacle that life presents.[a]

Item 6. Any problem can be overcome with hard work.[a]

Item 7. If you work hard you will succeed.

Item 8. Anyone who is able and willing to work hard has a good chance of succeeding.

Item 9. A person should always do the best job possible.[a]

Item 10. By simply working hard enough, one can achieve one's goals.[a]

---

[a] NCDIF = Noncompensatory Differential Item Functioning Index

and alternative fit indices proposed (e.g., Cheung & Rensvold, 2002). Using these indices, there is no inconsistency between the scale- and item-level analyses.

22.4 SUMMARY AND CONCLUSIONS

We utilized two approaches (CFA and IRT) to the assessment of measurement invariance. In addition, we demonstrated that the results of both approaches are quite similar. Both indicate that cross-cultural comparisons based on English and Spanish versions of the MWEP "hard work" subscale are appropriate. Which approach is preferable? The answer is likely it "depends." There are pros and cons associated with each. One of the decisions one must make in deciding is what level of evaluation (test or item) is most important. As noted by Drasgow and Hulin (1990), an important consideration is to choose metrics relevant to the level at which decisions are made. With the MWEP, subscale or dimension-level (e.g., test-level) scores are most relevant, as this is how information is used. Hence, the CFA procedure or DTF information provides more meaningful information at this level, and these types of analyses are the most important in this instance.

However, in other situations where larger item pools exist and different configurations may yield scores (e.g., tailored or computer adaptive testing), the choice of items that do not demonstrate DIF is crucial for the study of cross-cultural differences. Also, information at the item level can be helpful in test development or refinement situations. Specifically, the ability to evaluate item properties at the item level can guide test developers toward eliminating items that exhibit DIF, have low discrimination parameters, or have inappropriate (e.g., too high or too low) difficulty parameters. However, if researchers propose removing items in order to obtain DTF-free scores, then potentially some substantive content may be lost.

At this point it is important to note that DIF indices have only been developed for some IRT models, and DFIT specifically is not available for most IRT models. DIF can, however, be detected using several other non-IRT approaches. Based on the structure of one's data and the number of response options, other approaches may be more appropriate in other situations. For some response formats such as multiple choice tests one cannot readily implement the DFIT procedure if one is interested in all response options, but it can be used in dichotomously scored items such as correct or incorrect. Also, approaches toward evaluating DIF in multidimensional models are not yet well developed. For an introduction to popular IRT models, readers could consult Embretson and Reise (2000), and for a review on various approaches toward computing DIF, Raju and Ellis (2002).

In summary, there are numerous options one can pursue in evaluating measurement invariance, two of which have been discussed in the present chapter. Interestingly, the approaches we took here, multiple-group CFA and DFIT, showed a reasonable degree of convergence in item- and test-level results. Based on these results, the hard work subscale of the MWEP can be used to make meaningful comparisons across subgroups. However, as several of the items show a lack of invariance, it may be worthwhile to reexamine the individual items for content that may contribute to this lack of invariance.

23

Categorization Errors and Differences in the Quality of Questions in Comparative Surveys

Daniel Oberski, Willem E. Saris, and Jacques A. Hagenaars

23.1 INTRODUCTION

Differentials in measurement error and in question quality can bias findings in comparative survey research. This chapter examines methods to identify and to remedy such bias. The first section will discuss models used here to estimate the measurement error coefficients of survey questions, starting with a basic response model. We then present questions and data from the European Social Survey (ESS) used in our chapter. A short discussion of our previous findings follows. We present estimates from our prior research and briefly review possible explanations for large differences found in these estimates across countries. The next section presents the model focused upon here to account for categorization errors. We illustrate what we mean by such errors and how we compare the results obtained from categorical versus continuous models; next we introduce our method of estimation and then discuss our results. Finally, we present a meta-analysis of the many findings and discuss general conclusions on the basis of this.

23.1.1 Differences in Measurement Errors

Measurement errors can invalidate findings in comparative survey research if their magnitudes differ across countries. Thus attention should not only be given to absolute levels of errors, but also to the differences between them. Different strategies have been developed to deal with the problem, for example, within the context of invariance testing in the social sciences (Jöreskog, 1971a), differential item functioning in psychology (Muthén & Lehman, 1985), and differential measurement error models in epidemiology and biostatistics (Carroll, Ruppert, & Sefanski, 1995).

[1] Survey Methods in Multinational, Multiregional, and Multicultural Contexts, edited by Harkness et al.
Copyright © 2010 John Wiley & Sons, Inc.

The quality of questions is related to this. Despite efforts to minimize errors in the ESS (Harkness, 2002, 2007; Häder & Lynn, 2007), considerable variation in question quality can be observed across countries. It is important to study these because they can cause differences in relationships between variables in different countries which in fact have no substantive meaning but are a consequence of differences in measurement quality (Saris & Gallhofer, 2007b). In order to avoid such differences, we need to study the reasons behind them.

In an earlier study, we investigated differences in translations, design of experiments, and complexity of questions as possible reasons for the range of question quality across countries (Oberski, Saris, & Hagenaars, 2007). However, we found that those factors did not explain much of the difference. In this chapter we turn to consider differences in (response) categorization errors as a potential source of differences between countries.

23.1.2 Categorization Errors

Categorization errors are part of the discrepancy between an unobserved continuous variable and a discrete observed variable that measures the unobserved continuous variable. Specifically, categorization errors are the differences between the score on the latent variable and the observed category that are due solely to the categorization process. For example, suppose a person's age is known only to belong in one out of three categories. These are assigned the scores one, two, and three. Suppose also that there are never any mistakes in this categorization. In spite of this absence of mistakes in categorization, there is still a discrepancy between the age of the person and the category she is assigned to, first, because people of different ages have been lumped together. Second, the distance between the categories in terms of average age may not be equal to the distances of unity between the numbers one, two, and three assigned to the categories. This means that if one treats the observed variable as an interval level measure, the result of calculations such as correlations will also differ from what would have been obtained if the original age variable had been used.

In general, one can say that categorization errors arise when a continuous latent response variable is split into discrete categories. This leads to two types of errors: grouping errors and transformation errors (Johnson & Creech, 1983). Grouping errors occur when different opinions are grouped together in one category. Transformation errors occur when the differences between the numerical values of adjacent categories do not correspond to equal distances between the means of the latent response variables in those categories. If, for instance, the distances between categories are not the same in two different countries, this can lead to larger categorization errors in one country than another, resulting in turn in lower question quality. This is why distance between response categories is one possible explanation for differences in question quality across countries.

23.2 STATISTICAL MODEL

Figure 23.1 shows the basic response model (Saris & Gallhofer, 2007b) used here as our starting point. The difference between the observed response (y) and the variable of interest (f) or concept by intuition (cf. Saris & Gallhofer 2007a, pp. 15–62) is both random measurement error (e) and systematic error due to the respondent's reaction to the method (M).

The coefficient q represents the quality coefficient and we call q^2 the total quality. It equals $\mathrm{Var}(f)/\mathrm{Var}(y)$ and can be interpreted as the proportion of variation in the observed variable due to the unobserved trait of interest. The correlation between the unobserved variables of interest is denoted by $\rho(f_1, f_2)$.

Note that the correlation $\rho(y_{ij}, y_{kj})$ between two observed variables measured with the same method is a composite of:

$$\rho(y_{ij}, y_{kj}) = \underbrace{\rho(f_i, f_k)}_{\text{Correlation of interest}} \cdot \underbrace{q_{ij} \cdot q_{kj}}_{\text{Attenuation factor}} + \underbrace{m_{ij} \cdot m_{kj}}_{\text{Correlation due to method}} \qquad (23.1)$$

where $i \neq k$ indicates the concepts by intuition and j, a method.

This implies that the correlation between the observed variables is typically smaller than the correlation between the unobserved variables of interest, but can be larger if the method effects are considerable. In addition, one cannot compare correlations across countries if their quality coefficients are very different: this follows directly from Equation (23.1). Moreover, one cannot estimate the quality from this simple design with two observed variables. It is impossible to estimate these five parameters from just one observed correlation.

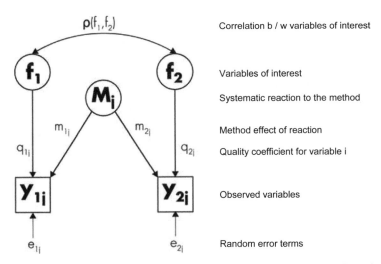

$\rho(f_1, f_2)$	Correlation b / w variables of interest
f_1 f_2	Variables of interest
M_i	Systematic reaction to the method
m_{1j} m_{2i}	Method effect of reaction
q_{1i} q_{2i}	Quality coefficient for variable i
y_{1i} y_{2i}	Observed variables
e_{1i} e_{2i}	Random error terms

Figure 23.1. The Continuous Response Model Used in the MTMM Experiments

There are two different approaches to estimate these coefficients. One approach is to use multiple traits and multiple methods (MTMM) to evaluate the quality of measurement instruments (Campbell & Fiske, 1959). The classical MTMM approach recommends the use of at least three traits measured with three different methods, leading to nine different observed variables. Table 23.1 provides an example of such a design. Given the responses on all the variables, the coefficients described above can be estimated. The second approach is the use of the Survey Quality Predictor program (SQP; http://www.sqp.nl/). SQP predicts the quality coefficient and method effect of a single question from many of its characteristics such as the topic, the number of categories, and so forth. The SQP prediction is currently based on a meta-analysis of 87 MTMM experiments and 1028 different questions; many more experiments are soon to be added (Oberski, Kuipers, & Saris, 2007). In the work reported here, we use the MTMM approach. See Saris and Gallhofer (2007a) for more detail on MTMM and SQP.

23.3 DATA FROM THE EUROPEAN SOCIAL SURVEY

The ESS is fielded in more than 20 countries. In each round of the ESS, MTMM experiments are conducted to evaluate the quality of a limited number of questions, providing an exceptional opportunity to observe differences in question

TABLE 23.1. The Quality of 18 Questions in the Main Questionnaire Included in the Experiments

Country	Mean	Median	Minimum	Maximum
Portugal	0.79	0.81	0.63	0.91
Switzerland	0.79	0.84	0.56	0.90
Greece	0.78	0.79	0.64	0.90
Estonia	0.78	0.85	0.58	0.90
Poland	0.73	0.85	0.51	0.90
Luxembourg	0.72	0.73	0.53	0.88
United Kingdom	0.70	0.71	0.56	0.82
Denmark	0.70	0.70	0.52	0.80
Belgium	0.70	0.73	0.46	0.90
Germany	0.69	0.70	0.53	0.83
Spain	0.69	0.64	0.54	0.90
Austria	0.68	0.68	0.51	0.85
Czech Republic	0.65	0.60	0.52	0.87
Slovenia	0.63	0.60	0.46	0.82
Norway	0.59	0.59	0.35	0.83
Sweden	0.58	0.58	0.43	0.68
Finland	0.57	0.54	0.42	0.78

quality over a large number of countries. The findings discussed here are from MTMM experiments conducted in the second round of the ESS in 2004 (http://www. europeansocialsurvey.org/).

The questions examined in these experiments pertained to four domains: (1) the social distance between doctor and patients; (2) opinions about job; (3) the role of men and women in society; and (4) political efficacy.

Three questions, presented in three different formats/formulations, were asked for each domain. The first format/formulation, presented to all respondents, was used in the main questionnaire. The two alternative forms were presented in a supplementary questionnaire, completed after the main questionnaire. All respondents were presented with only one of the two alternative forms, but different groups received different versions of the same questions (Saris et al., 2004). Details of the specific questions in the six experiments are available on the ESS website and the ESS archive website (http://ess.nsd.uib.no/) in the form of documentation on all the questionnaires in source and target language versions.

Each experiment manipulated a different aspect of how the questions were presented to respondents. The "social distance" experiment examined the effect of choosing arbitrary scale positions as a starting point for agreement-disagreement with a statement. The "job" experiment compared a 4-point true-false scale with direct questions using 4- and 11-point scales. In the "role of women" experiment, agree-disagree scales were reversed; in addition, there was one negative item and a "don't know" category was omitted in one of the treatments. Finally, the political efficacy experiment compared agree-disagree scales with direct questions.

A special ESS working group ensured that the samples in the different countries were probability samples and as comparable as possible (Häder & Lynn, 2007). In addition, the questions asked in the different countries were carefully translated from the English source questionnaire (see Chapter 26, this volume; Harkness, 2002, 2007). Despite these efforts to make the data as comparable as possible, large differences in measurement quality were found across the different countries. Table 23.1 shows the mean and median standardized quality of the questions in the main questionnaire across the experiments for the different countries.

A remarkable feature of this table is that the Scandinavian countries have the lowest quality, while the highest quality is obtained in Portugal, Switzerland, Greece, and Estonia. The other countries are in between these two groups. The differences are considerable and statistically significant across countries ($F = 3.19$, $df = 16$, $p < 0.001$) and experiments ($F = 92.65$, $df = 5$, $p < 0.0001$). The highest mean quality is 0.79 in Portugal, while the lowest is 0.57 in Finland. If the correlation between the constructs of interest is 0.60 in both countries and the measures for these variables have the above quality, then the observed correlation in Portugal would be 0.47, and the observed correlation in Finland would be 0.34. Many researchers would say that this large difference in correlations requires a substantive explanation. However, this difference has no substantive meaning; it can be expected because of differences in data quality. Not all of these differences are necessarily due to categorization error, however. We discuss other possible explanations for some of the differences below.

23.4 EXPLANATIONS FOR CROSS-COUNTRY DIFFERENCES IN QUESTION QUALITY

The previous section showed that in some cases, large differences were found in question quality across the ESS countries. In a previous study, we examined a few possible explanations of these discrepancies (Oberski, Saris, & Hagenaars, 2007).

In some cases, we found artificial differences in quality that were likely to be due to an erroneous translation or nonstandardized implementation of the experimental design. However, these cases were not so numerous that they could explain the large overall variations in question quality found in the ESS. We then explored the possibility that the distance between the categories in the categorical questions differs across countries. Before discussing the influence of categorization errors on the quality in different countries and experiments, we present a more detailed explanation of the model used to estimate the distances between the categories.

23.4.1 The Categorical Response Model

The response model discussed so far makes no mention of the fact that many of the measures used are in fact ordinal. Two types of measurement models have been proposed for such situations. The first assumes that there is an unobserved discrete variable and that errors arise because the probability of choosing a category on the observed variable given a score on the unobserved variable is not equal to one. Such models are often referred to as latent class models (Lazarsfeld & Henry, 1968; Hagenaars & McCutcheon, 2002).

The second approach deals with the case in which a continuous scale or "latent response variable" (LRV) is thought to underlie the observed categorical item. Such models are sometimes called latent trait models. Several extensions are possible, but we focus here on a special case described by Muthén (1984) used in our analysis of the data (Figure 23.2). It can be shown that analyzing polychoric correlations is a special case of the model we use, equivalent to the multi-dimensional two-parameter graded response model in item response theory (Muthén & Asparouhov, 2002).

Errors may arise at two stages. The first is the connection between the latent response variable (LRV_{ij} in Figure 23.2) and its latent trait (f_i). This part of the error model is completely analogous to factor analysis or MTMM models for continuous data: the scale is modeled as a linear combination of a latent trait (f_i), a reaction to the particular method used to measure the trait (M_j), and a random error (e_{ij}), and interest focuses on the connection between the trait and the scale (q_{ij}), which we again term the "quality coefficient" (see also Figures 23.1 and 23.2).

The second stage at which errors arise differs from the continuous case. This is the connection between the variables LRV_{ij} and y_{ij} in Figure 23.3. Here, the continuous latent response variable is split into different categories, such that each category of the observed variable corresponds to a certain range on the unobserved continuous scale. The size of each range is determined by threshold parameters. In Figure 23.3, this step function is represented by a black triangle. Examples of step functions are illustrated in Figure 23.3.

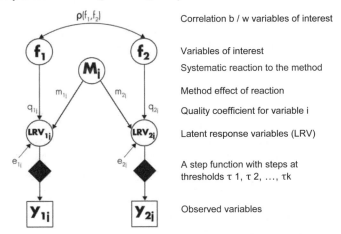

Figure 23.2. The Categorical Response Model Used in the MTMM Experiments

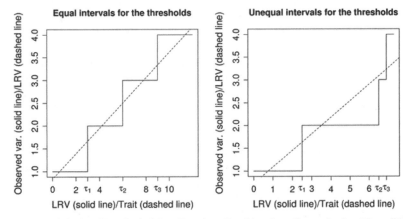

Figure 23.3. Two Hypothetical Step Functions Resulting from Categorization. The solid lines plot the observed categorical variable as a function of the latent response variable (LRV). The diagonal dotted lines plot the expectation of the LRV as a function of the latent trait on the same scale. The thresholds used for categorization are denoted by the symbols τ_1, τ_2, and τ_3.

In Figure 23.3, the steps (solid line) show the relationship between the LRV and the observed variable, while the straight (dotted) line plots the expectation of the LRV given the latent trait. In the step function on the left-hand side, the LRV has been categorized using equal intervals. The error added by the categorization is the vertical distance between the dotted line and the step, that is, the distance between the dotted line and the horizontal segments of the solid line. It can be seen that the error is zero when the straight line crosses the steps, and that at each step, the error is the same (at 3, 6, and 9). The expectations within the categories have the same interval as the thresholds of unity; so if the values 1, 2, 3, and 4 are assigned to the categories, no transformation occurs. Errors still occur, because the values along the dotted line have been grouped into the four categories formed by the solid line. Relationships of the observed categorical variable with other variables will therefore be attenuated.

Conversely, the right-hand side shows a latent response variable that has been categorized with unequal steps. The figure shows that the distances between the thresholds τ_1, τ_2, and τ_3 are very different from each other. The consequence is that at the second step, i.e., in between τ_2 and τ_3, there is almost no extra error, while at the first and third steps the errors are much larger. Here a transformation occurs. Suppose, as is often the case, that the categories are given the numerical values 1, 2, 3, and 4. The distances between the expectations of the LRV in each of the categories then do not equal unity, which is the distance between the values chosen for the categories.

To sum up, two types of errors can be distinguished at this stage (Johnson & Creech, 1983): grouping errors and transformation errors.

Grouping errors occur because the infinite possible values of the latent response variable are collapsed into a fixed number of categories (the vertical distances between the diagonal line and the steps in Figure 23.3). These errors will be higher when there are fewer categories.

Transformation errors occur when the distances between the numerical scores assigned to each category are not the same as the distances between the means of the latent response variable in those categories. This happens when the thresholds are not equally spaced, or when the available categories do not cover the unobserved opinions adequately.

We have described the categorization process here. It is important to note, however, that normally this process is not observed; usually one only observes a discrete variable that is assumed to be the result of the categorization process.

Thus categorization can be expected to be another source of measurement error, in addition to random errors and method variance. If these errors differ across countries, so will the overall measurement quality, and differences in means, correlations, regression coefficients, and cross tabulations across countries will result which are due purely to differences in measurement errors.

Thus, the model we have used allows us to some extent to distinguish between errors due to the categorization, errors due to the reaction to the method, and random errors. In the next section, we take advantage of this to compare the amount of error due to categorization introduced across countries.

23.4.2 Categorization Errors in Survey Questions

Using the MTMM design, it is possible to obtain a measure (q^2) of the total quality of a question. If a continuous variable model (hereafter referred to as a CV model) is used, this quality is influenced by errors in both stages of the categorical response model: not only random errors and method effects are included, but also errors due to the categorization. For this reason, Coenders (1996) argued that the linear MTMM model assuming continuous variables does not ignore categorization errors, but absorbs them in the estimates of the random error and method correlations. How the absorption functions will depend on the model used and this has not been extensively studied. The extent to which it holds in general is thus a matter of ongoing debate.

However, since the quality coefficient is estimated from the covariance matrix of the measures, it can be both reduced and increased by categorization errors. In general, all correlations between measures increase after correction for categorization, but they need not all increase equally. If categorization errors are higher using one of the methods, for instance, the correlations between the latent response variables will increase more, relative to the observed correlations, than will the correlations of each variable with its repetition using a different method. In this case, the amount of variance in the response variable due to method will be larger in the categorical model than in the CV model, and the estimated quality of the measure in the categorical response model can become lower than the estimated quality in the continuous MTMM model. This is because there are method effects (correlated errors) on the level of the continuous latent response variables that do not manifest themselves in the observed (Pearson) correlations between the categorical variables. In some instances, categorization can inflate estimates of the quality of categorical observed variables, even though it can simultaneously cause errors that reduce the quality. Thus two processes are at work which have opposite effects on the estimates of the quality.

As noted before, the quality of a variable is defined as the ratio of the true trait variance to the observed variance (see also Figure 23.1):

$$q^2 = \frac{\text{Var}(f)}{\text{Var}(y)}. \tag{23.2}$$

However, we have now seen that y is itself a categorization of an unobserved continuous variable (c), and therefore the above Equation (23.2) can be "decomposed" into:

$$q^2 = \frac{\text{Var}(f)}{\text{Var}(\text{LRV})} \cdot \frac{\text{Var}(\text{LRV})}{\text{Var}(y)}. \tag{23.3}$$

The scale of LRV, the latent response variable, is arbitrary, except that it may vary across countries due to relative differences in variance (Muthén & Asparouhov, 2002). However, the ratio $\text{Var}(\text{LRV})/\text{Var}(y)$ can easily be calculated once q^2_{con}, the quality from the continuous analysis, and $\text{Var}(f)/\text{Var}(\text{LRV})$, the quality from the categorical MTMM analysis (q^2_{cat}), have been obtained. So Equation (23.3) shows that $q^2_{\text{con}} = q^2_{\text{cat.}} \cdot c$ and $c = q^2_{\text{con}}/q^2_{\text{cat.}}$, where c is the categorization effect, or $\ln(q^2_{\text{con}}) = \ln(q^2_{\text{cat.}}) + \ln(c)$. This correction factor is a useful index of the relative differences between the quality estimates of the continuous and categorical models. In the present study, we estimate this "categorization factor" for different countries and experiments, and examine the extent to which it can explain differences in quality across countries.

23.5 METHODS

As mentioned earlier, respondents in almost every ESS country completed a supplementary questionnaire containing repetitions used in the experiments. Not all respondents completed the same questionnaire; a random half answered the first and second form of the questions, the other half the first and third form.

This split-ballot MTMM approach lightens the response burden for the experiment by presenting respondents with fewer questions and fewer repetitions. Saris, Satorra, and Coenders (2004) showed that the different parameters of the MTMM model can still be estimated using this design. Since we can identify the necessary covariances in the categorical model, it is identified as well (Millsap & Yun-Tein, 2004).

Two different models were estimated. The continuous analysis used the covariance matrices as input and the maximum likelihood estimator in LISREL 8. The results presented in the tables below were standardized after the estimation.

In principle, the categorical model can also be estimated using maximum likelihood. However, in order to deal with the planned missing data (split-ballot) a procedure such as full-information maximum likelihood would be necessary. This requires numerical integration in the software we used (Mplus 4), making the procedure prohibitively slow and imprecise. We therefore used an alternative two-step approach, whereby in the first step the covariance matrices of the latent response variables are estimated, and in the second step, the MTMM model was fitted to the estimated matrices. The estimation in the first step was made using the weighted least squares approach described by Flora and Curran (2004), and the second step again employed the maximum likelihood estimator.

This approach has the advantage that consistent and numerically precise estimates can be obtained quickly (Muthén & Asparouhov, 2002). The disadvantages are that the standard errors of the estimates in the model are incorrect and that the chi-square statistic and modification indices may be inflated. Although these problems can be remedied by using the asymptotic covariance matrix of the covariances as weights in the estimation (Jöreskog, 1990), in the present study we compare only the consistent point estimates of this model.

We use threshold parameters to model categorization errors. These thresholds are the theoretical cutting points where the continuous latent response variable (LRV) has been discretized into the observed categories. If the thresholds are different across countries, the questions are not directly comparable, since differences in the frequency distribution are partly due to differences in the way the LRV was discretized. If the thresholds are the same across countries, the questions may still not be comparable because of differences in linear transformations (loadings) and random errors. In that case, it is not categorization error that causes incomparability. If loadings, random errors, and thresholds are the same across countries, the frequency distributions can be directly compared.

In this chapter, we will present only a basic invariance test on thresholds. If the thresholds are equal, categorization error is not a likely cause of differences in quality. However, we do not continue with tests for invariance on loadings and error variance; instead we compare the results of the two different models.

The two models are the same with respect to the covariance structure of the response variables, the "MTMM part" of the model. However, they differ in their basic assumptions about the "observation part": the CV model assumes that the response variables are directly observed, while the categorical model assumes a threshold connection between the response variables and the observed ones.

Both models assume normality of the response variables, but the differences in basic assumptions cause the categorical model to be more sensitive to departures from the norm. While in the CV model, under quite general conditions, violation of normality will not affect the consistency of the estimates (Satorra, 1990), this is not so in the categorical model. There, the threshold estimates are derived directly from quantiles of the normal distribution that the latent response variable is assumed to follow. Therefore, if the LRVs are not normally distributed, threshold estimates will be biased. The MTMM estimates depend on the thresholds and may also change, although the precise conditions under which such estimates would change significantly have, to our knowledge, not been investigated analytically. Several simulation studies found empirically that bias occurs especially when latent response variables are skewed in opposite directions (Coenders, 1996).

Thus, while the categorical model may be more realistic in modeling the observed variables as ordinal rather than interval level measures, the CV model may be more realistic, in that it is robust to violations of normality. Whether one or the other model provides a more adequate estimate of the quality of the questions in any given analysis depends therefore on the degree to which these assumptions are violated, as well as on the impact such violations have on the estimates. This should be kept in mind when interpreting results.

We estimated the quality of the measures based on the CV model and based on the categorical model for four experiments focusing on answer scales of five categories or less in the main questionnaire. For each experiment, the countries with the highest and the lowest qualities in the CV model were analyzed. For each of the questions we took the ratio, called "categorization factor," of the two different quality measures as an index of the effect that categorization has on the continuous quality estimates. The next section presents the results.

23.6 RESULTS

23.6.1 Results of the Experiments

The first experiment concerned opinions on the role of women in society (see Table 23.2). We first turn to the hypothesis that all thresholds are equal across different countries. If this hypothesis cannot be rejected, there is little reason to think that the categorization is causing differences in the quality coefficients.

We selected Portugal and Greece, the countries with the highest quality coefficients, and Slovenia, with the lowest. In this experiment, the wording of the

TABLE 23.2. "Role of Women": Questions and Threshold Estimates (in z-scores)

	Agree strongly		Agree		Neither disagree nor agree		Disagree		Disagree strongly
	1	τ_1	2	τ_2	3	τ_3	4	τ_4	5

"A woman should be prepared to cut down on her paid work for the sake of her family."

Slovenia		−1.4		−0.1		0.6		1.8	
Greece		−1.1		−0.2		0.5		1.4	

"A woman should not have to cut down on her paid work for the sake of her family."

Slovenia		−1.5		−0.0		0.6		2.0	
Greece		−1.5		−0.3		0.4		1.5	

"Men should take as much responsibility as women for the home and children."

Slovenia		−0.5		1.3		1.9		2.6	
Greece		−0.6		0.7		1.6		2.3	

"Women should take more responsibility for the home and children than men."

Slovenia		−1.7		−0.7		−0.2		1.2	
Greece		−1.6		−0.5		0.0		1.4	

"When jobs are scarce, men should have more right to a job than women."

Slovenia		−1.8		−0.8		−0.3		0.9	
Greece		−0.9		0.1		0.6		1.4	

"When jobs are scarce, women should have the same right to a job as men."

Slovenia		−0.8		0.7		1.1		1.9	
Greece		−1.1		−0.1		0.7		2.0	

question was changed in the second method. For example, the statement *When jobs are scarce, men should have more right to a job than women* from the main questionnaire was changed to *When jobs are scarce, women should have the same right to a job as men* in the supplementary questionnaire. To separately study misspecifications in the categorization part of the model, we imposed no restrictions on the covariance matrix of the latent response variables at this stage.

In the first analysis, all thresholds were constrained to be equal across the five countries. This yielded a likelihood ratio statistic of 507 on 48 degrees of freedom. The country with the highest (128) contribution to this chi-square statistic was Portugal. Examining the expected parameter changes, we found that for Portugal these standardized values were very large, with some values close to 0.9, while in other countries the highest obtained and exceptional value was 0.6. For some reason, the equality constraint on the Portuguese thresholds appears to have been a particularly gross misspecification.

As it turns out, this particular misspecification was probably due to a translation error. The intention of the experiment was to alter the wording of the question in the second method. However, in Portugal the changed wording provided above was not used, and in the supplementary version the same version was presented as in the main questionnaire. To prevent incomparability when the MTMM model was estimated, we omitted Portugal from further analyses and continued with two countries.

The model in which all thresholds were constrained to be equal yielded a likelihood ratio of 351 and 36 degrees of freedom ($p < 0:00001$). This model was rejected, as the thresholds were significantly different across countries.

We used the procedure of Saris, Satorra, and van der Veld (2008) to determine whether misspecifications were present in the model. For this test, we needed the Expected Parameter Change (EPC), Modification Index (MI), and the power of the test. The EPC gives direct estimates of the size of the misspecification for all fixed parameters, while the MI provides a significance test for the estimated misspecification (Saris, Satorra, & Sörbom, 1987).

However, these two indices were not sufficient for determining misspecifications because the MI depends on other characteristics of the model. For this reason, the power of the MI test must be known in order to determine whether a restriction is misspecified. We used these quantities to incrementally free parameters that were indicated to be misspecified. Using the modification indices and power as guides, we formulated a new model in which some thresholds were constrained to be equal, while others were allowed to vary.

The resulting model had an approximate likelihood ratio of 2.8 on 2 degrees of freedom ($p = 0.24$). The resulting estimates of the threshold parameters are presented in Table 23.2. These estimates have been expressed as z-scores.

Table 23.2 presents three different traits, each asked in two different forms. The first form of each trait is the version asked in the main questionnaire, while the second form was asked in the supplementary questionnaire (the third form has been omitted for brevity).

The thresholds in this model represent how extreme the "agreement" has to be before the next category is preferred to the previous one. This strength is expressed in standard deviations from the mean. Take, for instance, the third statement in the table: *Men should take as much responsibility as women for the home and children.* Slovenians need to have an LRV score 2.6 times the standard deviation above the country mean before they will respond "disagree strongly."

Note that the threshold part of the relationship between LRV and observed response is deterministic. However, not all Slovenians with an *opinion* on the indicator that is 2.6 standard deviations or more away from the mean will necessarily answer "disagree completely." This is so because the latent response variable is also affected by random measurement error. Since random error plays an important role in this relationship, not only the thresholds should be discussed here, but also the quality coefficients.

For the first question, the distances between the thresholds are unequal for these two countries and different from one. In addition, the endpoints are somewhat distant, especially in Slovenia: there the category "disagree strongly" is 1.8 standard deviations or more away from the mean, reducing the number of scale points that are available for some respondents. The second form of the same question is similar to the first in this respect, except that here both of the endpoints are rather distant in each country, again reducing the number of scale points. As noted above, a reduction in scale points is expected to increase grouping errors.

The second trait, "responsibility," presents a radically different picture. In both countries, the "disagree" and "disagree strongly" categories are quite far away from the mean. This again reduces the number scale points, while, at the same time, the scale is cut off in this manner from only one side. Large transformation errors can be expected. Moreover, in Slovenia this effect is much worse than in Greece: the category "neither disagree nor agree" is already 1.3 standard deviations or more away from the mean, reducing the amount of information provided by this variable for Slovenia even further. The second phrasing of this question seemed to provide better coverage of the prevailing opinions on women and men's responsibility for the home and children.

For the third and last trait, "the right to a job," the most striking feature of the thresholds was that in Slovenia, the first three categories represented opinions below the mean, while in Greece only the first category did. Beyond this, it is difficult to say which scale would have produced fewer categorization errors. However, the supplementary questionnaire form of the same question seemed to produce much more comparable scales with respect to the thresholds than the main questionnaire form in this case.

It is also clear from the table that the two forms of phrasing were not exactly opposite in the way they were understood and/or answered. This was especially true for the "right to a job" item. However, the design choice for one phrasing or the other seems arbitrary. This particular way of phrasing a question is therefore inadvisable, because a decision that seems arbitrary is not arbitrary in its consequences. The key problem in this case may be the complex sentence structure in which men are compared to women, given an attribute (right to a job) under a certain condition (when jobs are scarce), and then a "degree of agreement" with a norm ("should have") is asked. Asking respondents directly about the rights men and women should have could provide a more accurate measurement that would be less sensitive to arbitrary shifts in response behavior.

The thresholds provided some insight into the nature of differences in categorization. However, the quality of the measure in the continuous model depended also on parameters of the categorical response model, such as the method effects and the error variances, and on the latent response variable distribution.

The correlations between the LRVs were also estimated. Estimates of the quality and method effects of the measures corrected for categorization were then estimated for all questions on the basis of these correlations. The quality and method effects of the CV model were also estimated. Based on these results, the categorization effect was derived from the ratio of the two coefficients. These results are presented in Table 23.3.

The top two rows of Table 23.3 show that, using the CV model, the quality in Greece was higher than in Slovenia; this is, indeed, the reason we chose these particular countries to compare. The quality in Slovenia is lower for the first question, dramatically lower for the second question, and very similar for the third question. In principle, these findings are in line with the expectations of categorization errors. However, Table 23.3 also shows that such interpretations of the possible influence of the thresholds are not as straightforward as they might seem. We fitted the MTMM model to the estimated covariance matrix of the latent response variables,

TABLE 23.3. Quality (Q^2) and Method Effects (M) According to the Continuous and Categorical Models, with Categorization Factors for the Experiment on Opinions About the Role of Men and Women in Society

		Cut Down	"Women" Respnsib.	Men Right
Continuous analysis				
q^2	Greece	0.71	0.66	0.71
	Slovenia	0.54	0.25	0.68
m	Greece	0.15	0.15	0.15
	Slovenia	0.17	0.24	0.15
Categorical analysis				
q^2	Greece	0.51	0.35	0.48
	Slovenia	0.69	0.29	0.65
m	Greece	0.49	0.14	0.32
	Slovenia	0.33	0.75	0.19
Categorization factor				
	Greece	1.4	1.9	1.5
	Slovenia	0.8	0.9	1.0

and obtained a model which seemed to fit reasonably well ($\chi^2 = 20$, $df = 10$, $p = 0.02$). While for the first and second questions the low qualities are indeed corrected upwards somewhat after the categorization has been taken into account, the opposite happens in Greece. In that country, all of the quality coefficients are lower using the categorical analysis than they are in the continuous analysis.

One consequence of this is that, using the CV model, a higher quality is obtained in Greece than in Slovenia, while the reverse is true in the categorical model for the first and last items. This is rather striking, given that over all items in the main questionnaire Greece had a substantially higher quality estimate than did Slovenia (see Table 23.1 earlier).

The analyses of the other three experiments showed that sometimes no large differences between the countries were present, while in others the thresholds were rather different. In particular, we found several cases in which the same question did not cover the distribution of the opinion in one country, but provided more information in another. We also found examples of cases where differences in the quality do not go together with differences in the thresholds, and examples of cases where they do.

The question remains whether there is a connection between the categorization factor and the quality of the question. The next section presents the results of a meta-analysis we conducted on the categorization factors.

23.6.2 A Meta-Analysis of the Results

The question remains whether the categorization factor affects the quality or not. Using the results presented in the previous sections, we constructed a data set consisting of the categorization factor for all questions—including those from the supplementary questionnaire not shown above—in the four different experiments for which this index was available. This yielded 72 cases in total.

As shown before, the categorization factor equals $c = q^2_{con}/q^2_{con}$ and so $q^2_{cat}(c)$ = q^2_{con}. If there were no effect of the categorization, then there would be no relationship between c and q^2_{con}, since q^2_{cat} would be higher or lower by a constant factor. If c and q^2_{con} are plotted against one another, one would then expect to find the points randomly distributed along a horizontal line. Figure 23.4 shows the scatterplot of these two quantities. Estimates from different experiments have been indicated with different symbols.

The clear relationship evident in the figure indicates that high quality coefficients from the continuous model tended to be lower in the categorical model, and vice versa. Figure 23.4 shows that categorization factors above unity were mostly found for questions with a high quality. We estimated the relationship between the quality from the continuous model for each experiment by the transformation $\ln(q^2_{con}) = \alpha_k + \beta_k \ln(c)$. Here k indexes the four different experiments. Note that the base level of q^2_{con} is $\exp(\alpha_k)$. We then fit a linear regression to the transformed variables. The resulting predictions for each experiment are shown in Figure 23.5 on the original scales.

Figure 23.5 shows that both the intercepts and slopes for the "efficacy" and "job" experiments are rather similar, while the coefficients for the "role of women" and "social distance" experiments are completely different. The effect of the categorization factor was strongest in the "social distance" experiment, where also some

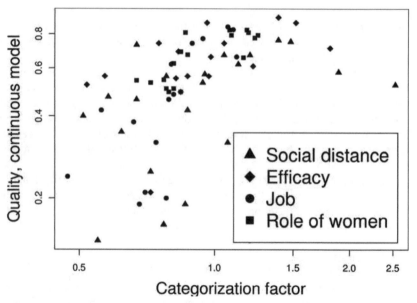

Figure 23.4. Scatterplot of the Categorization Factor (c) and the Total Quality of a Measure (q^2_{con}) across the Experiments. Note the log-log scales.

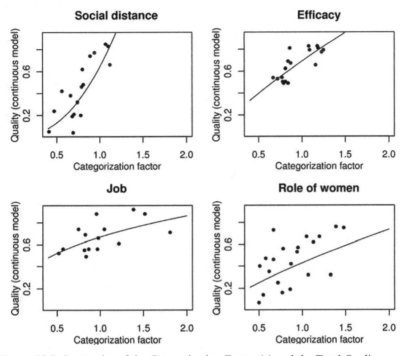

Figure 23.5. Scatterplot of the Categorization Factor (c) and the Total Quality of a Measure (q^2_{con}) by Experiment. The prediction line of the model as estimated for each experiment separately, is also given.

large differences between the threshold distances were found. The experiment with the smaller number of categories, "job," did not have a high coefficient.

We now turn to the question of whether these factors were different between the countries with "high" and "low" quality coefficients. Splitting the sample by "high" and "low" quality, we found that the means of the categorization factors of the two groups are 1.25 and 0.85, respectively, for the questions in the main questionnaire ($t = 3{:}7$, $df \approx 18$, $p = 0{:}002$. For the questions in the supplementary questionnaire, the difference is in the opposite direction, but not statistically significant ($t = -1{:}70$, $df = 28$, $p = 0{:}10$). This suggests there is a considerable effect of the categorization, at least in the main questionnaire.

One possible explanation for the interaction effect found here is that the method factors were often constrained to zero for the main questionnaire. The questions in the main questionnaire were selected because they were expected to have the highest quality and lowest method effects. After the initial continuous analysis, the model indicated that the questions in the main questionnaire indeed had zero method correlation. Given that the categorical model tends to increase the correlations, if the monomethod correlations for the main questionnaire increase more than the other correlations, a method factor is found where none was found before. This lowers the quality estimates.

A test was done of the hypothesis that questions for which the method effect was constrained to zero in the continuous model have the same categorization factor as other questions, controlling for country effects. A hierarchical linear model was fit using the computer program $, version 2.6.1, with fixed effects of country and restricting the method to zero or not (0/1), and a random intercept across topics to account for the dependency among the observations. The hypothesis that the method restrictions coefficient equals zero was rejected ($p = 0.02$). The explanation that near-zero method factor variances cause the interaction found therefore seems plausible.

23.7 DISCUSSION AND CONCLUSION

We found large differences between countries in the quality of survey questions. Because such differences can have important implications for cross-national research and survey design, we set out to discover whether these differences could be attributable to errors due to the use of a small number of categories.

Overall, we found that categorization errors do occur along with random errors and method effects. These errors have two types of effects on the quality of the questions, which can work in opposing directions. The first is that the quality is lower when there is more categorization error. The second is that categorization attenuates the relationships between different variables in the model differently, affecting not only the quality, but also the method effects and other parameters of the model. This in turn means that the quality parameter under the CV model is not always smaller than the quality under the categorical model, as evidenced by the many "categorization factors" above unity which we found.

A caveat should be added to the interpretation of this result, because a violation of the assumptions of the models (no categorization error versus bivariate normality) can have different consequences for the estimates. A categorization factor above unity does not necessarily indicate overestimation of the quality in the CV model. Several studies of the robustness of factor analysis models to categorization errors exist (see, for example, Olsson, 1979). However, their results do not necessarily apply in the MTMM model, which also includes method factors. Given the ubiquity of correlated errors in survey questions, it would be useful to study more closely the robustness of this particular type of measurement error model to categorization error more closely.

In a meta-analysis, we gathered the results from our four different experiments and analyzed the relationship between the categorization factor and the quality in the continuous model. Effects were found for all four experiments.

If the categorization factors were equal for countries with the highest and lowest quality coefficients, they could not explain the differences in quality that we found earlier. The meta-analysis suggested that there is a considerable difference in the categorization factor between countries where the highest and the lowest quality coefficients were found, depending on whether the question was part of the main or supplementary questionnaire.

The methods in the main questionnaire were chosen beforehand as the ones least likely to cause method effects. For example, direct questions rather than batteries were used. After re-examining the experiments on which the meta-analysis was based, it appears this is closely related to the interaction effect found there.

The main reason for the interaction effect in the meta-analysis appears to be that the method variance for the main questionnaire method was often close to zero. The general rise in correlations that results from correction for categorization seems to have "pushed" the monomethod correlations of the main questionnaire variable to the point where the method variance could no longer be constrained to zero. And, as the method variance rises, quality decreases.

In other words, the correction for categorization has a negative influence on the quality. When the method factors were constrained to zero in the first instance, the effect was that the quality was in general lower in the categorical model than in the continuous model. This is contrary to what one might expect, considering that all the polychoric correlations are higher than their Pearson counterparts.

We have shown in this chapter that it is possible to split the measurement error model into three parts: (1) a part due to random errors; (2) a part due to systematic errors; and (3) a part due to splitting the variable into just a few categories: "categorization error."

This chapter has been largely descriptive of the effects of categorization error. Given our findings, it seems important to judge the relative merits of the continuous and categorical models better, as well as the effects that different question characteristics have, not only on quality and method effects, but also on the categorization errors.

Our study also has some limitations due to assumptions made to attain the above separation. These are: normality of the latent response variables, linearity of the relationship between the latent traits and latent response variables, and interval measurement of the latent traits. Future studies will examine ways to relax these assumptions. Future research might also focus on finding other explanations for differences in quality across countries.

24

Making Methods Meet: Mixed Designs in Cross-Cultural Research

Fons J. R. van de Vijver and Athanasios Chasiotis

24.1 MAKING METHODS MEET

Qualitative and quantitative research methods have usually been treated as independent or even mutually exclusive. We refer here to qualitative data collection as dealing with methods that produce reports, transcripts, or other non-numerical output, whereas quantitative data collection yields numerical output (Axinn & Pearce, 2006). Qualitative data analysis involves the transformation of data (usually of a qualitative nature) to a more condensed and communicable form that addresses research questions, infers meaning to observations, tests hypotheses, or serves some other study purpose; quantitative data analysis involves similar transformation aimed at similar purposes, but now with numbers as input.

The toolkits of qualitative and quantitative data collection methods are very different. The same goes for data analyses. The nonoverlap of both types of data collection and analyses in both traditions conveys the impression of incompatibility, if not incommensurability. A cursory reading of the literature is sufficient to demonstrate that both fields have indeed shown their own independent developments. The debate between the proponents of qualitative methods and of quantitative methods often has ideological undertones which hamper their successful integration. Thus, the quantitative paradigm is often associated with positivism, which argues that there is an objective truth that can be known. Theories and methods are more adequate when they give a better representation of this truth. On the other hand, the qualitative paradigm is based on the idea that multiple realities or truths exist and that these realities are constructed. In their extreme forms, these two paradigms are incompatible. However, this incompatibility can be easily overrated in the everyday practice of cross-national research. There is ample experience in quantitative research which shows that multiple realities, as experienced by individuals in different cultures, are open to empirical scrutiny. The study of meaning is also crucial for quantitative cross-cultural research. There is an extensive research tradition that uses statistical techniques to address the question of overlap of meaning of a construct in different cultures (van de Vijver & Leung, 1997). It is therefore not surprising to see a growing recognition that the two methods can be fruitfully combined (Tashakkori & Teddlie, 2003). It is counterproductive in our view to associate research questions with particular paradigms. Successful applications of combinations of qualitative and quantitative methods were often not set up as deliberate attempts to integrate the two approaches but as studies that tried to solve substantive questions that could not be addressed by the reliance on a single method. We contend that many differences between qualitative and quantitative methods can be overcome if we replace the discussion of methodological purity and rigor by a discussion of which method is more adequate in which conditions (Tashakkori & Teddlie, 2003).

The domain of application of the mixed methods described in the present chapter is cross-cultural survey research. The main motivation for this choice is the great potential of mixed methods for culture-comparative research. Studies of cross-cultural similarities and differences in attitudes and behaviors often require different kinds of methods and evidence. For example, we may observe differences in political preferences across countries in a quantitative study, while an exploration of the source of these preferences may require an historical analysis of the political systems of the countries involved. The latter analysis would be qualitative. Similarly, a qualitative method, such as semi-structured interviews, could be used to explore the attitudes and behaviors that individuals living in a specific country associate with well-being, while a comparison of test scores based on a quantitative analysis could be used to determine whether countries have equally high scores on their common indicators of well-being.

Section 24.2 describes the terminology and the metatheoretical framework of the present chapter. Drawing on this framework, Section 24.3 describes strengths and weaknesses of qualitative and quantitative methods. Section 24.4 deals with mixed method designs in cross-cultural research. The fourth section describes two examples of mixed method studies in cross-cultural psychology. Conclusions are drawn in Section 24.5.

24.2 TERMINOLOGY AND CONCEPTUAL FRAMEWORK

The common denominator of studies using mixed methods is their implementation of both qualitative and quantitative methods. More specifically, a mixed methods study "involves the collection or analysis of both quantitative and/or qualitative data in a single study in which the data are collected concurrently or sequentially, are given a priority, and involve the integration of the data at one or more stages in the process of research" (Creswell, Plano Clark, Gutmann, & Hanson, 2003, p. 212). The most important feature of mixed methods is their collection of both qualitative and quantitative evidence to strengthen the quality of the study. Leech and Onwuegbuzie (2008) have proposed a taxonomy of mixed methods that is based on three underlying dichotomies (presented in Figure 24.1; see also Morse, 2003). These are level of mixing (partially mixed versus fully mixed), time orientation (concurrent versus sequential), and emphasis of approaches (equal status versus dominant status).

The first level of mixing involves the question of whether the qualitative and quantitative methods are used side by side or are fully integrated. An example of a study that uses the methods side by side would be a cross-cultural study of depression in which interviews are held with participants from different cultures asking for their personal experiences and in which also a standardized question-naire is administered to assess depression. The methods would be fully mixed if only the standardized questionnaire would be administered and participants would be asked to explain their answers (and the latter would be asked as open-ended questions).

The second dimension refers to the timing of the different methods: Are the two methods used concurrently or sequentially? Both previous examples exemplify concurrent designs, because the instruments are administered more or less at the same time. An example of a sequential design would be a study in which a qualitative method in each culture would be used to identify valid markers of depression in that culture and a quantitative instrument would then be administered that builds on findings of the qualitative study.

The third dimension refers to the relative dominance of the qualitative and quantitative methods employed in a study. The qualitative and quantitative evidence

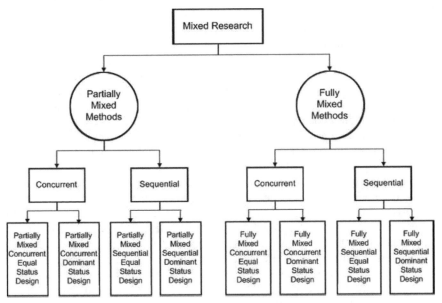

Figure 24.1. A Typology of Mixed Methods Research Designs (adapted from Leech & Onwuegbuzie, 2008)

is more or less equally important in the four previous examples. However, many mixed studies use methods in which one kind of evidence is more important than the other. A common example can be found in cross-cultural psychological studies in which unexpected findings are explained post hoc. These explanations are often based on qualitative evidence.

Qualitative and quantitative methods have their own criteria for deciding when inferences are correct. However, there are important commonalities in these criteria. In our view, correctness of inferences can be seen as a mapping issue: reality is represented in a set of statements, such as models or hypotheses, which are correct if they provide an adequate rendering of one or multiple realities. We refer here to a representation of these realities in models, concepts, and other tools used by science. A commonly applied concept in quantitative methods is *validity* (Messick, 1989). Applying this reasoning to psychological tests, the AERA/-APA/NCME Standards document (1985) stresses that validity refers to the degree to which evidence supports the inferences that are made from the scores (American Educational Research Association, American Psychological Association, and National Council on Measurement in Education, 1985). The terminology in qualitative methods to describe this mapping issue is different although the underlying problem is identical. Lincoln and Guba (1985) coined the term *trustworthiness* (or *credibility*) to indicate to what extent findings are persuasive and worth taking into account. More recently, the concept of *inference quality* has been proposed (Teddlie & Tashakkori, 2003). The current chapter does not aim at describing the subtle differences between the concepts within their different paradigmatic origins but at building bridges between different qualitative and quantitative approaches. Therefore, we treat the terms here as exchangeable. After all, each of these concepts can be used to address the correspondence between a statement, model, or theory and the depicted reality. This position is in line with the general tenet of the present chapter that a rapprochement between different methodologies can only be successful if we capitalize on communalities and complementarities and refrain from capitalizing on their differences.

In our view, concepts like validity, trustworthiness, and inference quality can be seen as essential elements in quality management. Assuring the quality of a study (and of its inferences) can be seen as the net result of a chain of decisions. These involve decisions about the choice of a theoretical framework of the study, the questionnaire, the sampling frame, the field testing, the statistical analysis, and many other aspects. Couched in management terminology, this view implies that ensuring a high quality of a study requires integral management or chain management. Cross-national studies often use mixed methods even if this is not mentioned explicitly. For example, the identification of culturally appropriate items to measure political involvement may be based on qualitative methods, such as focus groups. The evaluation of the adequacy of the items may be based on psychometric criteria such as internal consistency. There is a common distinction in the literature according to which qualitative methods are better suitable for the "context of discovery" while quantitative methods are better suited in the "context of justification" (Reichenbach, 1938). Although we think that the boundaries between the two methods are often more blurred than the example suggests, an important principle is illustrated: It is often counterproductive to be wedded to a single method.

An attractive feature of mixed methods is their focus on research questions as the main force behind the choice of data collection methods and analyses. It is fairly common in social sciences that methods determine not only how a study is conducted but also, remarkably, which questions are examined. Particularly when new methods have been developed, it is easy to see why methods can suggest research questions. For example, recent statistical advances in multilevel modeling have made it possible to study individual behavior (e.g., quality of life) as a joint function of individual characteristics, such as personality and education, and of societal characteristics, such as political freedom and affluence (Lucas & Diener, 2008). Various studies have been published in the last years that employ multilevel models to address research questions about cross-level interactions with a previously unattainable statistical rigor (van de Vijver, Van Hemert, & Poortinga, 2008a). It is fairly common in cross-cultural research that we deal with questions in which methodological aspects play an important role as tools to increase the validity of our studies. On the other hand, generating research ideas or research questions at the beginning of a study is always a qualitative endeavor. Thus, an appropriate choice of methods can help us in making our studies as compelling as possible, but methods cannot dictate our research agenda. (We return to this issue in Section 24.3.1.)

A crucial concept in the literature on the combination of qualitative and quantitative methods is *triangulation* (Denzin, 1978). Within the context of mixed methods, triangulation often refers to the combination of qualitative and quantitative evidence. If both types of evidence point in the same direction (e.g., both support the specific hypothesis), the convergence is seen as strengthening the conclusions of the research. The concept of triangulation refers to the integration of different kinds of evidence, which can be qualitative, quantitative, or a combination. A good example of triangulation is the comparison of quantitative evidence from surveys with field observations. Combining qualitative or quantitative evidence from interviews done in several countries with (either quantitative or qualitative) archival data from these countries is another example. An empirical example can be found in a study by Howe and McKay (2007) who were interested in poverty in Rwanda. They compared quantitative household survey data of poverty (mainly income based) with a qualitative approach, called Participatory Poverty Assessment, in which they tried to identify indicators of poverty. Triangulation was used to compare the results of the qualitative and quantitative evidence. Much convergent information was found. However, there were also some interesting complementarities. For example, the qualitative procedure identified extremely poor groups (without a regular shelter) that were not included in the quantitative survey; this undercoverage was due to the incomplete sampling frame used in the survey.

24.2.1 Some Myths Regarding the Relation Between Qualitative and Quantitative Methods

The often strict boundaries that are kept between qualitative and quantitative methods have created various dichotomies that obstruct their integration and impede progress in the field. We discuss here four of them. Firstly, *structured data collection methods such as observations, psychological tests, questionnaires, and (structured) interviews can only go together with quantitative methods, while unstructured data collection methods such as archival analysis and semistructured or unstructured interviews can only go together with qualitative methods.* The distinction between data collection methods and the underlying measurement paradigm, qualitative or quantitative, is not strict. Many applications of these techniques can be found in both measurement paradigms. Illustrative examples can be found in observation studies. Studies in developmental psychology that use observations tend to work in a quantitative paradigm, emphasizing interrater consistency as one of the criteria of study quality, whereas participant observation is used in ethnography as a qualitative tool. There seems to be an implicit rule, which says that once you start a study using a method, you should stay within the paradigm. For example, if you start with grounded theory and work in the qualitative paradigm, you should stick to qualitative methods in all stages of the study. It is our view that the link between data collection methods and methods of data analysis is often weak. Attempts to increase the global quality of a study may require one to sidestep the dichotomy. A good example where these pitfalls were avoided elegantly can be found in the extensive cross-cultural research program conducted by Keller and her collaborators, where they utilized "method triangulation" (cf. Keller, 2006). Interviews and verbal material of observed interactions are analyzed with qualitative methods and a quantitative methodology was employed in the analysis of questionnaires and videotaped or in-situ spot observations of behavior. While the goal of the qualitative codings, namely to gather instances for further examination, is more pragmatic, the applied quantitative methodology additionally allows the analytical testing of hypotheses generated by qualitative means (for more examples of cross-cultural developmental studies with mixed designs, see van de Vijver, Hofer, & Chasiotis, 2009). Another good example is the use of projective tests which are often associated with a qualitative paradigm, but have recently been employed successfully in cross-cultural research using quantitative methodology (Hofer & Chasiotis, 2003, 2004; Hofer, Chasiotis, Friedlmeier, Busch, & Campos, 2005).

A second myth refers to *the link between the study of cultural specifics and qualitative methods on the one hand and the study of cross-cultural universals and quantitative methods on the other hand.* The idea behind this myth is that qualitative methods are needed to identify culture specifics, while quantitative methods are needed to identify universals. The simple dichotomy between culture specifics and universals is counterproductive (Berry, Poortinga, Segall, & Dasen, 2002). Many constructs that are studied in cross-cultural survey research have both universal and culture-specific aspects. Thus, the question of whether depression is universal has shifted to the question of which aspects of depression are universal and which aspects are culture-specific. Methodological flexibility is needed to deal with this new position. Both culture-specific studies that are largely qualitative and culture-comparative studies that use standard surveys in a quantitative framework are needed to delineate common and specific aspects. A final example that demonstrates the false dichotomy can be found in equivalence analyses, which are typically conducted to assess cross-cultural similarity in meaning from a quantitative perspective. If properly conducted, the analysis of structural equivalence that employs a quantitative framework can identify both universal and culture-specific aspects of a measurement device such as questions on a standardized survey.

A third myth refers to *the value expected from mixed methods*. These methods have been proposed as the "third movement" (Tashakkori & Teddlie, 2003). The main problem with this position is the inherent danger that like both qualitative and quantitative methods, mixed methods become "reified" and are seen as an independent paradigm that provides the solution to the methodological problems in social and behavioral research. Mixed methods should not become an independent paradigm that is developed only by persons committed to the third movement. It would be counterproductive for their development if mixed methods would become a paradigm that is only or mainly focusing on the further development of their own paradigm. The existing qualitative and quantitative paradigms will be invincible Goliaths for mixed methods unless the latter can prove their incremental value. Moreover, integration of research methods can never be an end in itself. The value of a study does not intrinsically increase by employing mixed methods.

Fourthly, *all qualitative and quantitative methods can be combined*. This myth is based on the idea that all qualitative and quantitative methods are compatible. This compatibility does not even hold within a single paradigm. Some quantitative methods cannot be combined. Thus, statistical analyses can be based on assumptions that cannot be combined with one another. Models that are based on ordinal and interval-level data may yield different and even incompatible outcomes, that is, correlations between ranks (based on ordinal values) may not be significant whereas product moment correlations (based on interval-level data) of the same data may be significant. Obviously, this incompatibility cannot be blamed on the underlying models. Decisions about the appropriateness of the underlying model rest with the researcher. This incompatibility also holds for some combinations of qualitative and quantitative techniques. There are research methods in the qualitative paradigm that do not accept validity, credibility, and inference quality as ultimate criteria for determining scientific value. It is mainly in the so-called postmodern tradition that the criterion of validity is abandoned and is replaced by the extent to which other researchers are convinced by the arguments proposed by an author. We see this crucial role for rhetorical skills as incompatible with more conventional approaches. Only those qualitative methods are considered here that can be adequately combined with quantitative methods that use validity and related concepts as the ultimate goal.

24.3 MIXED-METHOD DESIGNS IN CROSS-CULTURAL RESEARCH

Mixed methods build on qualitative and quantitative procedures. These methods try to generate incremental value by standing on the shoulders of their parental methods. Creating incremental value and trying to get the best of both worlds should start from an analysis of the strengths and weaknesses of qualitative and quantitative methods.

24.3.1 Strengths and Weaknesses of Qualitative and Quantitative Methods

Qualitative methods tend to employ procedures that are less structured than quantitative methods. Working with less structured instruments holds the promise of finding more novel information (Denzin & Lincoln, 2000). The fewer restrictions we impose on our instruments, the more diverse the information that can be collected with these instruments. When we study novel research topics or work with cultural groups that are not well researched, relatively unstructured instruments and an open-minded attitude by the researcher help to gain much information in a short time period. Qualitative methods display their main strength in the context of discovery (Denzin & Lincoln, 2000; Lincoln & Guba, 1985). They are helpful to get information about various cultural characteristics of an ethnic group we are dealing with for the first time, to build models, and to generate hypotheses. The main weakness of qualitative methods is constituted by their infrequent usage of available testing procedures.

The strengths and weaknesses of quantitative methods mirror those of qualitative methods. Structured instruments such as questionnaires with fixed response formats such as Likert scales are useful to test specific theories of cross-cultural differences, but less suitable for more exploratory approaches in which there is no theory that guides the choice of an instrument.

The mirroring of strengths and weaknesses can make mixed designs valuable. Adato, Lund, and Mhlongo (2006) conducted a qualitative study (using focus group discussions and key informant interviews) that built on a longitudinal poverty survey in KwaZulu-Natal, South Africa. The authors were interested in various aspects of poverty that were not covered by the survey, such as how poverty is experienced by the participants and identifying the mechanisms by which members of communities cope with adverse conditions. Other interesting examples can be found in systematic anomalous case analyses. Axinn and Pearce (2006, Chapter 4) studied the role of religion and family size preferences in Nepal. Religious denomination was used as predictor of family size preference (after controlling for various background characteristics, such as age, education, and gender). Deviant cases were defined as participants with observed scores that differed more than two standard deviations from their predicted values. Interviews were held with a random sample of these deviant cases to explore the reasons for the discrepancies.

24.3.2 Methodological Promises and Challenges of Mixed Designs

Mixed designs "inherit" both the advantages and disadvantages of their parental disciplines. For example, guidelines for sampling frames as developed for quantitative survey research apply to mixed methods inasmuch as representative samples are required. Similarly, guidelines for sampling designs as developed in qualitative methodology apply to mixed methods inasmuch as the study is conducted among specific groups of informants that can provide key information about the construct being studied. The question can be asked to what extent there are specific guidelines for sampling in mixed-methods studies that do not apply to either quantitative or qualitative studies. More generally, are there specific methodological challenges for mixed methods that do not apply to either quantitative or qualitative designs? Very few methodological features that are unique to mixed designs have emerged in the literature. The incremental value that can be gained by mixing data collection methods can be achieved at no extra cost. Dealing with the methodological issues in mixed-methods research amounts to dealing with the challenges of the qualitative parts and quantitative parts of the study separately.

Suppose that we want to conduct a cross-cultural study of health perception in different countries. In the first stage of the project, the qualitative stage, interviews could be held in each country with experts, such as health care professionals, clinical psychologists, and patients. These groups could provide valuable information about which features of health are particularly salient in that culture. A quantitative instrument could then be developed that contains items that are partly common to all cultures and partly unique to specific cultures. This instrument can then be applied to a probability sample in each country. The challenges in the sampling schemes of both stages are presumably the adequacy of the expertise of the groups in the qualitative stage and the representativeness of the sample in the quantitative stage. These problems are well known from the qualitative and quantitative literature. No problems are expected that would be unique to a mixed design. Problems of sampling have been described in great detail in the literature. A classic text is Kish (1965). Issues such as adequacy of information provided by local experts and the need to increase sample size until saturation is reached and no new information is gained are well known from the qualitative literature (Green & Thorogood, 2004; Sandelowski, 1995). Similarly, problems of nonprobability sampling and of selection bias due to the nonrandom-ness of refusals are well documented in the quantitative literature (Groves & Couper, 1998). Mixed methods can build on much valuable experience of the parental methods. A good knowledge of both methods makes it possible to maximize their complementary strengths while trying to deal with the unavoidable nonoverlapping weaknesses (Brewer & Hunter, 1989; Johnson & Turner, 2003).

As argued before, one of the main strengths of qualitative methods is their flexibility, which makes them highly suitable for formulating new models, while quantitative methods are more suitable for testing these models. We expect that in mixed designs, hypothesis generation will remain the stronghold of qualitative methods and hypothesis testing the stronghold of quantitative methods. Still, this difference in functions can be easily overrated. Qualitative studies can be hypothesis testing and quantitative studies can have exploratory purposes. One of the main challenges of mixed research is indeed a better integration of qualitative and quantitative methods in all stages of a study.

It has been argued that one of the shortcomings of mixed-method studies is the absence of specific criteria to appraise their quality. Sale and Brazil (2004) conducted a meta-analysis of mixed-method studies and concluded that they could not find any specific criterion to evaluate such studies. All they could find were criteria that applied either to qualitative or quantitative studies. Do we need criteria beyond the usual criteria of qualitative and quantitative studies to evaluate the adequacy of mixed-method studies? Lincoln and Guba (1986) argue that the ultimate goals of both qualitative and quantitative studies are trustworthiness (referring to the veridicality of study inferences) and rigor (referring to the procedure used to obtain the information), which come close to the criteria we mentioned in the introduction. These authors point to important similarities in how these two goals are achieved in qualitative and quantitative studies. More specifically, the two ultimate goals amount to four more proximal goals, which apply to both kinds of studies: (1) truth value of study inferences (internal validity for quantitative methods versus credibility for qualitative methods); (2) applicability involving the specification of the context in which the information was obtained and new contexts in which the same information would hold (external validity for quantitative methods versus transferability or fittingness for qualitative methods); (3) consistency involving the question to what extent other researchers or procedures would yield similar outcomes (reliability for quantitative methods versus dependability for qualitative methods); (4) neutrality involving the influence of the researcher and his or her ideas on study outcomes (objectivity for quantitative methods versus confirmability for qualitative methods). Lincoln and Guba's argument can be taken to imply that no specific criteria for mixed-method studies are needed and that conventional criteria, as developed in the qualitative and quantitative literature, can be applied to evaluate these studies. We concur with this view.

A final challenge of mixed-method studies involves the appropriate use of triangulation. Within the context of mixed designs, triangulation virtually always refers to combining qualitative and quantitative information. Yet, the concept of triangulation is broader and could also refer to combining qualitative information (or quantitative information, for that matter).

Converting qualitative data to quantitative data may be particularly attractive given the advanced techniques for triangulating data that are available in the quantitative paradigm. Data from different quantitative sources that address the same issue can be combined in so-called multimethod matrices (Campbell & Fiske, 1959; see also Bechger & Maris, 2004; Schmitt & Stults, 1986; Wothke & Browne, 1990). These matrices are a special case of multitrait-multimethod matrices.

An example can be found in a study by Arends-Tóth and van de Vijver (2007). These authors were interested in the question of whether different methods of assessing acculturation orientations proposed in the literature would yield identical results. One-, two-, and four-statement methods were employed to measure acculturation attitudes of Turkish first- and second-generation immigrants in the Netherlands. Items of the one-statement measurement method referred to the preference for any of the two cultures on a five-point scale. For example, one item asked for preferred ethnicity of friends, ranging from exclusively Dutch friends to exclusively Turkish friends (five response options). The items covered 15 life domains (partly the private domain, such as child rearing; partly the public domain, such as education). Items of the two-statement method assessed preferences in the same life domains with independent measures of attitudes toward the Dutch culture and the Turkish culture. For example, separate items dealt with the importance of having Dutch friends and Turkish friends. The third acculturation measure, the four-statement measurement method, had four scales with independent measures of assimilation, integration, separation, and marginalization. These four orientations are based on a widely employed model proposed by Berry (1997). Each statement in this measurement method refers to a comparison of the Dutch and the Turkish cultures. Integration items express a positive attitude toward both cultures (e.g., "It is important to me to speak Dutch well and it is also important to me to speak Turkish well"). The three other acculturation orientations were assessed in a similar manner. Assimilation items refer to a negative evaluation of the Turkish culture and a positive evaluation of the Dutch culture. Separation refers to the opposite orientation. Finally, marginalization refers to negative attitudes toward both cultures. It has been argued that the four-statement method has the disadvantage of asking double-barreled questions, which are discouraged both by cross-cultural survey researchers (Rudmin & Ahmadzadeh, 2001) and monocultural survey researchers (Sudman & Bradburn, 1974). However, the use of the double-barreled items is based on a substantive argument: Berry and Sam (1997) have argued that double-barreled items should not be avoided, because acculturation orientations are about making choices between cultures. One of the purposes of our study was to examine the adequacy of these double-barreled questions.

The Turkish culture was more valued than the Dutch culture in the private domain, while both cultures were about equally favored in the public domain. As can be seen in Figure 24.2, a confirmatory factor analysis of the multitrait-multimethod data yielded a general method factor on which all three measurement methods loaded, and a general acculturation attitude factor with positive loadings for two indicators (some items dealt with the private domain such as the importance of having friends from one's own ethnic group and other items dealt with the public domain such as items dealing with importance of schooling). Finally, the four-statement method (using the double-barreled questions) showed the largest method effects, but its factor loadings were in line with the other methods.

The linear decomposition of trait and method effects as used in this study is widely employed in the literature, notably in combination with hierarchically nested models in confirmatory factor analysis (Widaman, 1985). Nonlinear techniques have also been applied to multitrait-multimethod data. Wothke and Browne (1990) have proposed a multiplicative relation between trait and method effects and showed how this model can be reparametrized so that it becomes amenable to linear techniques; Levin, Montag, and Comrey (1983) used multidimensional scaling procedures for multitrait-multimethod data. So, various statistical techniques are available for examining these data. In the mixed methods literature the focus is often on the use of multiple methods to assess a single construct, which are known in the literature as monotrait-multimethod matrices. The statistical techniques described can help to examine the convergent validity of multiple methods (Cole, 1987). More specifically, confirmatory factor analysis can address the question to what extent different methods constitute a single underlying factor, thereby providing support for the convergent validity of the triangulated measures.

Procedures to combine qualitative evidence are less formalized. Still, the same basic question of assessing the extent of convergence of different sources of

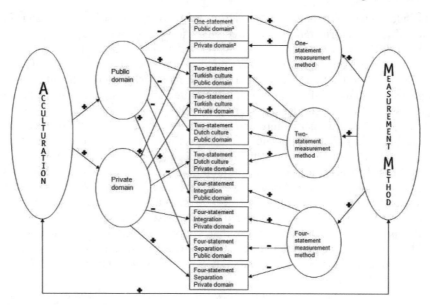

[a]Higher score refers to more adaptation.
[†]Fixed at 1 in nonstandardized solution. *ns* = not significant.
p < .05. **p* < .01.

Figure 24.2. Confirmatory Factor Analysis of Multitrait-Multimethod Data Assessing Acculturation Orientations Using Three Methods of Assessment (adapted from Arends-Tóth & van de Vijver, 2007, p. 1480)

evidence has to be addressed. Triangulation of qualitative evidence is a fairly common procedure in ethnographic research. For example, accounts from different sources about past events may need to be combined. Accounts of the same events provided by various informants or sources can have three relations with each other. Firstly, the information may be converging, which is usually interpreted as strengthening the quality of the observations and inferences based on them. Secondly, the information may be complementary when different sources address different aspects of past events. Such evidence does not support the convergence, but it does not contradict it either. So, from a methodological point of view, complementary information broadens our inference but may not make it more valid. Thirdly, the information may be incompatible. Various ways have been proposed how this incompatibility can be dealt with. The most obvious conclusion would be that the incompatibility points to the inadequacy of at least one of the sources and makes it impossible to draw any firm conclusions. However, this conclusion may not be the only or best option. For example, the quality of accounts depends on the way human memory is organized (Freeman, Romney, & Freeman, 1987; Romney, Batchelder, & Weller, 1987). Informant quality may depend on the involvement, his or her role in past events, and the time elapsed since the event. As another example, a distinction can be made between disagreement on major issues and disagreement on minor issues. More weight can be assigned to the former than to the latter.

There are no formalized procedures to compare qualitative and quantitative evidence. Yet, at least two different kinds of procedures can be employed after suitable modification of the evidence base so as to find a shared standard of comparison. Practically speaking, the procedures amount to transforming all evidence to a single domain (by transforming quantitative evidence into qualitative evidence or vice versa). The conversion of qualitative evidence in quantitative evidence deserves some further attention (Boyatzis, 1998). More specifically, a caveat is required because sample sizes are often relatively small in qualitative research. Little can be gained from converting qualitative to quantitative data if the sample size is very small and the quantitative data offer little scope for adequate statistical analysis. In general, conversions of data to another domain should be in line with recommended practices in the new domain.

There is a rich literature on content analysis which is useful for bringing qualitative data within the realm of quantitative analysis (Holsti, 1969; Krippendorff, 1980; Mohler & Züll, 2001, 2000; Neuendorf, 2001; Weber, 1985). An essential element in content analysis is the preparation of a coding scheme, which is the basis for converting qualitative information into numerical scores. A qualitative data analysis in which the main features of the data source are extracted may not be difficult to transform into a coding scheme.

After the qualitative data have been quantified, the previously described techniques such as confirmatory factor analysis can address the convergence of the measures. In practice, though, it is common to address the agreement between the quantitative and originally qualitative data as an interrater reliability issue. A commonly used measure is the percentage of agreement of raters. The statistic has a straightforward interpretation, but it also has some drawbacks. A simple percentage may not correct for chance agreement (e.g., if raters have to choose among a limited number of alternative codes, there is a risk that some agreement occurs due to chance) and a percentage of agreement may give an inflated number when the number of codes from which can be chosen is very large while only a few categories are actually chosen. Another widely employed statistic is Cohen's (1960) kappa. The original version is sometimes used although there is an improved version that corrects for chance (Brennan & Prediger, 1981). The statistic compares observed frequencies of (dis)agreement of two raters with frequencies that would be expected on the basis of chance. Despite its wide usage, Cohen's kappa has some problems. One of these is that the index was originally developed for nominal data where any difference in rating by two observers points to an inconsistency. However, the statistic has often been applied to ordinal- and interval-level data (Maclure & Willett, 1987). Differences in codings that are further apart on the scale should decrease the kappa value more than differences in codings that are closer to one another on the scale. Weighted kappa coefficients have been developed to deal with this problem (e.g., Cicchetti & Heavens, 1981). Another problem of Cohen's kappa involves the difficulty in interpreting the value of the statistic. Although there is a rule of thumb, which says that values of Cohen's kappa of at least .60 are adequate, the value of the statistic does not have a simple interpretation.

The third kind of interrater agreement index is the family of intraclass correlation coefficients (DeVellis, 2005; McGraw & Wong, 1996; Shrout & Fleiss, 1979). In their seminal paper, Shrout and Fleiss (1979) showed how interrater reliability can be assessed by computing intraclass correlations in analysis of variance designs. The versatility of this approach makes it highly suitable to assess the agreement in a wide variety of study designs. However, the same versatility forces the researcher to make informed choices about what kind of interrater reliability is desired. For example, how should we deal with the situation in which one observer gives systematically higher scores than another observer? If this differential leniency reflects acceptable differences in response styles, we can compute consistency coefficients, such as Cronbach's alpha. However, if these differences between coders should be treated as response inconsistencies, an absolute agreement index should be computed.

It can be concluded that various procedures for triangulation have been developed. Quantitative procedures are particularly strong in all kinds of tests of triangulation. These procedures go far beyond the simple question of whether findings and different methods are convergent or divergent and allow for a fine-grained analysis of the level of convergence and divergence.

24.4 EXAMPLES OF MIXED-METHOD STUDIES

24.4.1 A Comparison of Formal and Informal Theories of Acculturation

Arends-Tóth and van de Vijver (2004) were interested in the question of to what extent theoretical models of acculturation that have been proposed in cross-cultural psychology are similar to the implicit theories immigrants have about acculturation. Informal (or everyday or folk) theories of acculturation express views of immigrants or mainstream groups on important aspects and components of acculturation (e.g., what are the main problems faced by immigrants? Should immigrants maintain or give up their ethnic culture?).

Theoretical models ask essentially the same question. Current theories of acculturation orientations are based on any of three models (Berry & Sam, 1997; Sam & Berry, 2006; Ward, Bochner, & Furnham, 2001). Firstly, the unidimensional model assumes that acculturation always implies a shift from the culture of the country of origin to the culture of the country of settlement, and that immigrants eventually adopt the culture of the country of settlement (Gordon, 1964). Secondly, the bidimensional model argues that immigrants do not need to choose between the two cultures and that the adoption of the mainstream culture is not necessarily accompanied by a loss of the ethnic culture (LaFromboise, Coleman, & Gerton, 1993). Thirdly, in the fusion model, immigrants create a new culture that combines features of the cultures of the mainstream and their ethnic group (Arends-Tóth & van de Vijver, 2006). The bidimensional model is now most popular in psychological research. The authors were also interested in the role of domain specificity in acculturation. Again, the question was whether the important role of domain specificity as proposed in cross-cultural psychology was also present in immigrants' implicit theories.

Interviews were held with 147 Turkish-Dutch adults (77 women and 70 men; 71 first generation and 76 second generation; mean age of 30.44 years, SD = 8.91). Part of the interview was comprised of open-ended questions. Participants were asked to indicate which aspects of the Dutch and the Turkish cultures and people they evaluated positively and negatively. The second topic, how to integrate the two cultures, was addressed with the question of how they combined the two cultures. Perceived cultural differences were addressed by asking participants to describe areas of similarities and differences of the two cultures.

Another part of the interview comprised of quantitative instruments. The four acculturation strategies (integration, assimilation, separation, and marginalization; Berry, 1997) were presented one by one as short statements and the participants had to indicate their level of agreement. As an example, the integration item was: "Turkish people in the Netherlands should adapt to the Dutch culture and they also should maintain their Turkish culture." Furthermore, the rank order of perceived importance of eight domains (education, language, news, child-rearing, religion, social contacts, celebrations, and food) was also assessed for the Turkish and Dutch cultures separately. Acculturation preferences were measured with eight items, largely covering the same life domains (language, news, child-rearing, social contacts, cultural habits, neighborhood, celebrations, and food). Scores ranged from 1 (nearly only Turkish) to 5 (nearly only Dutch).

This study can be classified in terms of the dimensions distinguished by Leech and Onwuegbuzie (2008). The qualitative and quantitative aspects of the study are both essential for the argumentation. Therefore, the study exemplifies a design in which both aspects have an equal status. Furthermore, both methods are fully mixed from the beginning of the study. Finally, the comparison of informal theories, based on the semistructured interview, with the responses to the questionnaire illustrates a concurrent design, although as shown below, part of the data analysis was sequential.

Each interview was transcribed. A detailed categorization scheme was first constructed using responses of Turkish-Dutch to each open-ended question, resulting in more than 150 labels. Because this number was still too large for drawing adequate conclusions, a new category system was constructed after lengthy discussions among the researchers which reduced the 150 labels to 17 categories (see Table 24.1). These categories were used as a coding system to quantify the data. (This part of the study illustrates a successive mixed design.) Three aspects of each utterance were scored: category (to which of the 17 categories did the utterance refer?), culture involved (did the utterance involve the Turkish or Dutch culture?), and valuation (did the utterance describe a positive or negative aspect?). The interrater reliability was determined on the basis of five arbitrarily chosen interviews. The positive and negative aspects of both cultures were independently scored for each of the 17 categories by two researchers. The average percentage of agreement (defined as the average of the cells agreement divided by the maximum agreement, which was 68, and multiplied by 100) was 95%.

The patterning of the likes and dislikes for both cultures were examined by computing phi values (correlation coefficients between dichotomous variables) between culture and valuation. The phi values were positive and significant for social-emotional, private domains (e.g., family and child-rearing practices,

TABLE 24.1. Proportion of Participants Mentioning Each of the 17 Categories, Their Positive and Negative Evaluation, and the Relationship between Culture and Evaluation

| Categories | Turkish | | Dutch | | Phi^a |
	Valued positively	Valued negatively	Valued positively	Valued negatively	
Family, child-rearing	.68	.19	.15	.36	.48[b]
Religion	.54	.00	.00	.00	—
Amount/way of social contacts	.46	.04	.08	.37	.74[b]
Language	.11	.00	.03	.01	.36
Cultural habits, pride	.32	.05	.01	.05	.57[b]
Marriage, sexuality	.17	.10	.03	.18	.52[b]
Celebrations, food	.16	.01	.03	.01	.39[d]
Leisure activities	.05	.00	.03	.03	.58[d]
Clothes	.03	.05	.01	.08	.32
Decency	.04	.01	.06	.25	.50[b]
Society, social security	.00	.08	.41	.01	-.91[b]
Education	.01	.05	.09	.00	-.82[b]
Open-mindedness, mentality	.02	.18	.18	.03	-.77[b]
Freedom, independence	.00	.32	.29	.14	-.69[b]
Communication style	.03	.14	.50	.07	-.65[b]
Gender-role differences	.00	.20	.09	.01	-.95[b]
Prejudice, discrimination	.00	.00	.00	.52	—

Source: From Arends-Tóth & van de Vijver, 2004, p. 26).

[a]Phi is the correlation between culture (Turkish-Dutch) and evaluation (liking-disliking). Proportions do not add up to a fixed sum per row or column, because scores are derived from free responses and participants were not forced to mention each category or to choose any of the four cells of a row.
[b]$p < .001$
[c]$p < .01$
[d]$p < .05$

intensity of social contacts, cultural habits and pride, marriage and sexuality, celebrations and food, leisure activities, and decency), as can be seen in Table 24.1. The Turkish culture was more positively valued than the Dutch culture in these domains. In addition, religion was mentioned as a highly important, positively valued domain of the Turkish culture. The values of phi were negative and significant for domains that were related to functional, utilitarian, and public aspects of both cultures (e.g., society and social security, education, open-mindedness and mentality, freedom and independence, communication style, and gender-role differences).

As for the quantitative data, the acculturation preferences in eight domains (asked in a closed-format part in the questionnaire) were first factor analyzed. Two factors were extracted. The first represents the private domain (including child-rearing, cultural habits, celebrations, and food). The second factor is defined by the more public and utilitarian domains (language, news, contacts, and neighborhood). The mean scale score of the public domain was much higher than the mean scale score of the private domain, which means that the Turkish culture was more preferred in the private domains. Like the qualitative data, these quantitative results support the domain-specificity model of acculturation.

Support was found for the unidimensional model (Turkish aspects on one side and Dutch aspects on the other, meaning that the Turkish and Dutch aspects are negatively related), but the bidimensional model (Turkish and Dutch culture, which are positively related) and the fusion model (creating a new culture) were also present in the implicit theories. However, the popularity of the bidimensional acculturation models in cross-cultural psychology is not matched in the implicit theories. Even when explaining how they combine the cultures, many participants indicated that, depending on the life domain and on whether they are in a more public or private context, they focus more on either culture. It seems that implicit theories of Turkish-Dutch are more in line with a unidimensional, domain-specific model of acculturation than with a bidimensional model.

24.4.2 Implicit Measures

Other mixed-method approaches can be found in cross-cultural applications of measures of implicit motives on life satisfaction (Hofer, Chasiotis, & Campos, 2008), generativity (Hofer, Busch, Chasiotis, Kärtner, & Campos, 2008), and parenthood (Chasiotis, Hofer, & Campos, 2006). Implicit motives reflect dispositions defined as recurrent preferences for particular qualities of affective experience. Because implicit motives develop in the pre-linguistic period, they are introspectively less accessible, but express themselves in individuals' fantasies, and are therefore measured by fantasy-based, narrative methods. Picture-story exercises have been routinely used to assess implicit motives (McCelland & Pilon, 1983). The three basic implicit motives are affiliation, achievement, and power (cf. McClelland, 1987). The affiliation or intimacy motive represents a concern for warm, close relationships and for establishing, maintaining, or restoring a positive affective relationship with a person or a group (McAdams, 1982).

As an example of a mixed-method design, Hofer, Chasiotis, and Campos (2006) replicated earlier findings in monocultural studies with German (Brunstein, Schultheiss, & Grässmann, 1998) and Zambian adolescents (Hofer & Chasiotis, 2003) in a cross-cultural study among Germans, Costa Ricans, and Cameroonians using bias-free implicit (qualitative) and explicit (quantitative) measures of affiliation as predictors of life satisfaction. The methodological approach which realized an integrated examination of construct, method, and item/picture bias (van de Vijver & Poortinga,1997) is described by Hofer et al. (2005), using the Thematic Apperception Test as an example. Log-linear analysis was used as a technique to identify bias. An item shows nonuniform bias when a model including score level and culture does not fit the data. Furthermore, a fitting model that includes only score level indicates the absence of uniform bias, that is, participants with the same overall score on average have the same score with respect to a given picture/item irrespective of the culture they pertain to (Kok et al., 1985). As an explicit measure, the Benevolence Scale of the Schwartz Value Survey was used. As an implicit measure, a bias-free picture-story test measuring the need for affiliation-intimacy was administered. The test is derived from the Thematic Apperception Test, which uses a series of pictures and the participant is asked to tell a story related to the picture (e.g., What events led to the picture? What is happening in the picture?). Results revealed that an alignment of implicit motives and explicit, self-attributed social values was associated with an enhanced life satisfaction across cultures.

Chasiotis et al. (2006) assessed explicit and implicit motivation for parenthood combined with a cross-cultural developmental perspective. They assumed that childhood context is important for the emergence of care-giving motivation. A model was tested across cultures in which being exposed to interactive experiences with younger siblings in childhood elicits nurturant implicit affiliative motivations which, in turn, lead to more conscious feelings of love toward children in adulthood, which are linked to parenthood. The path model describing this developmental pathway was valid in male and female participants and in all cultures under examination. This developmental pathway supports the view that childhood context variables such as birth order might exert similar influences on psychological, somatic, and reproductive trajectories across different cultures (cf. Chasiotis, Keller, & Scheffer, 2003). Importantly, this pathway would not have been identified without the combined use of qualitative and quantitative methods, since the qualitative, implicit measure mediates the relationship between having younger siblings and the quantitative, explicit self-reported fondness of children (see also van de Vijver et al., 2010).

24.5 CONCLUSION

There is an increasing interest in mixed methods. It is easy to appreciate the value of mixed methods for cross-cultural research. A good cross-cultural study starts with an examination of its cultural context. This part of the study is often exploratory and qualitative. In many cross-cultural studies this qualitative stage is followed by a quantitative component. These parts are complementary here. The results of the qualitative study will have ramifications for the quantitative study. Large-scale cross-national projects that are aimed at quantitative comparisons often spend much effort on the development of stimulus materials. This developmental stage often relies mainly or exclusively on qualitative methods. So, qualitative methods are vital in the early, exploratory stages. If more in-depth studies are made of relatively few cultures, qualitative and quantitative methods could be integrated in more stages of the project. Various examples have been described in the literature in which closed and open-ended questions were both used to get information about cross-cultural similarities and differences in target constructs. These studies were typically conducted to study target constructs in a nonwestern context (Cheung et al., 1996: personality; Hayati, Karami & Slee, 2006: poverty; Miller et al., 2006, and Patel, 1995: mental health; Nastasi et al., 2007: stress reactions).

Mixed methods are in their adolescence (Teddlie & Tashakkori, 2003). The methodological tools are largely available to conduct studies using mixed methods. There seem to be two main reasons for their less than enthusiastic reception in the social and behavioral sciences. The first is the paradigm clash between qualitative and quantitative studies (Teddlie & Tashakkori, 2003). These paradigms have acted too much as closed fortresses that block emerging developments such as mixed methods. The second reason is related to the first. Mixed methods are not widely known among editorial boards and reviewers. Neither authors nor editors have templates as to how mixed-method studies should be reported and evaluated. Editors may find it relatively difficult to evaluate mixed-method manuscripts and to identify reviewers who can evaluate the appropriateness of mixed methods. Not many reviewers will be able to appraise both the qualitative and quantitative parts of a study. However, evaluations that focus on either aspect may not show the incremental value of mixed methods.

Both reasons may obstruct the current development of mixed methods. However, it is unlikely that either factor can block this development in the longer run. The number of researchers who do not know or do not believe in mixed methods may steadily decrease. There is a scarcity of mixed-method studies in cross-national research. Yet, we expect a further increase of studies using this approach. It will be important to demonstrate the incremental value of using these methods and to illustrate how their usage can generate insights that cannot be obtained with conventional methods. The future will tell whether mixed methods can live up to their important potential.

PART VIII

GLOBAL SURVEY PROGRAMS

25

The Globalization of Survey Research

Tom W. Smith

25.1 INTRODUCTION

The world may be shrinking, but survey research is expanding. More and more countries are routinely and freely conducting surveys and the volume of cross-national survey research is increasing. This heralds an era of unprecedented opportunity for the scientific and comparative study of human society, but also underscores the great challenges that face survey research in general and comparative survey research in particular. As the "total survey error paradigm" indicates, conducting valid and reliable survey research is a complex and daunting activity with many sources of error distorting results and invalidating findings. When it comes to comparative analysis, the challenges multiply. Research needs not only to be valid and reliable in each society measured, but the measurements need to achieve functional equivalence across surveys (see Chapter 2, this volume).

This chapter will examine: (1) the development of cross-national survey research in general, focusing on the evolution of the field of comparative survey research and the major programs that it encompasses, (2) the contemporary situation including conditions in the academic, governmental, and commercial sectors, data access and archives, international academic, professional, and trade associations, and the development of international standards, and (3) future prospects for further growth and methodological improvement.

25.2 HISTORICAL DEVELOPMENT

Cross-national survey research has gone through three distinct periods of development. In the first phase lasting until about 1972, comparative survey research was ad hoc, consisting of a fairly limited number of studies that covered a small number of societies and were conducted on a one-time, topic-specific basis.

---

[1] Survey Methods in Multinational, Multiregional, and Multicultural Contexts, edited by Harkness et al.
Copyright © 2010 John Wiley & Sons, Inc.

Soon after the start of national representative surveys in the United States in the mid-1930s (Converse, 1987), survey research took root in other countries. In 1937 Gallup established a counterpart to its American Institute for Public Opinion (AIPO) in the United Kingdom, the British Institute for Public Opinion (BIPO), and at least as early as 1942, AIPO and BIPO were asking parallel questions on their surveys. Another early example of cross-national survey research was the Strategic Bombing Surveys conducted by the U.S. government in Germany and Japan at the end of World War II (MacIsaac, 1976). Early collaborations by social scientists included the How Nations See Each Other Study in nine countries in 1948–49 by William Buchanan and Hadley Cantril (1953), the Civic Culture Study in five nations in 1959–60 by Gabriel Almond and Sidney Verba (1963), the Attitudes towards Europe Study in five countries in 1962 as part of the European Communities (http://www.gesis.org/dienstleistungen/daten/umfragedaten/eurobaro meter-data-service/standard-special-e b /study-overview/attitudes-towards-europe-1962-za-0078-feb-mar-1962/), and the Political Participation and Equality Study in seven nations in 1966–1971 by Verba, Nie, and Kim (1978).

In the second phase from 1973 to 2002, comparative survey research expanded in scope and became sustained and collaborative. Rather than being only one-time, intermittent enterprises directed by a small research team often representing only a few of the covered countries, cross-national research was established on an ongoing basis with research teams drawn from most, if not all, of the participating societies or with a study formally representing an association of countries such as the European Community (EC).

Other major examples are collaborative research programs of social scientists:

1. The inter-connected European and World Value Surveys (EVS/WVS) that started in 1981 and which across four rounds have grown from 20 to 71 countries (http://www.worldvaluessurvey.org/)
2. The International Social Survey Program (ISSP) which has conducted 24 annual studies from 1985 through 2008 while expanding from 4 to 44 countries (Smith, 2007a; http://www.issp.org/)[2]
3. The Comparative Study of Electoral Systems (CSES) which is now engaged in its third round since its start in 1996 (http://www.cses.org/)
4. The Comparative National Elections Project (CNEP) which started in the late 1980s and now has 24 participating countries (http://www.cnep.ics. ul.pt/)
5. The various loosely related Globalbarometers (Lagos, 2008) (http://www. globalbarometer.net) consisting of the New Democracies/New European Barometers (1991+) (http://www.abdn.ac.uk/cspp/nebo.shtml/), the Latino-barometers (1995+) (http://www.latinobarometro.org/), the Afrobarome-

---

[2] The ISSP started as a collaboration between existing social indicators programs in the United States [the National Opinion Research Center's General Social Survey (GSS)], Germany (the Zentrum fuer Umfragen und Methoden's ALLBUS), the United Kingdom (the Social Community Planning Research's British Social Attitudes Study), and Australia (the Australia National University's National Social Science Survey) and extended bilateral studies carried out as part of the GSS and ALLBUS in 1982–1984.

ters (1999+) (http://www.afrobarometer.org/), the Asianbarometers (2001+) (http://www.asianbarometer.org/), and the Arab Barometers (2005+) (http: // www.arabbarometer.org/)

Of course the ad hoc studies that characterized the first period also continued during this phase, with the World Fertility Study carried out in 61 countries (including 41 developing nations) in 1974–1982 being the most prominent example (Cleland & Scott, 1987; Cornelius, 1985).

In the third phase starting in 2002, cross-national survey research has become a coordinated part of the social science infrastructure. This development is marked by the establishment of the European Social Survey (ESS) in 2002 which carries out surveys biennially (Jowell, Roberts, Fitzgerald, & Eva, 2007b) (http://www. european socialsurvey.org/). Like the WVS, ISSP, and CSES, the ESS is a collaboration of social scientists, but unlike these earlier consortia, it has centralized funding for the design, direction, and methodological monitoring of the national surveys. Although the surveys themselves are funded nationally, the centralized resources and coordination of the ESS distinguishes it from the earlier collaborations.

25.3 THE CONTEMPORARY SITUATION

At present cross-national survey research can be classified into several major types. First, there are the global, general social science collaborations discussed above (e.g., the CNEP, CSES, Globalbarometers, ISSP, and WVS). These are large, on-going, and expanding collaborations that seek comprehensive coverage of societies across the globe.

Second, there are global, general-population studies on specialized topics such as the International Mental Health Stigma Survey (http://www.indiana.edu/~sgcmhs/index.htm/), International Adult Literacy Survey (IALS)/Adult Literacy and Life Skills Survey (ALL; http://nces.ed.gov/surveys/all/), Demographic and Health Surveys (DHS; http://www.measuredhs.com/), Multinational Time Use Study (MTUS; http://www.timeuse.org/mtus/), World Health Survey (WHS; http://www.who.int/healthinfo/survey/en/index.html/), and World Internet Project (WIP; http://www.worldinternetproject.net/).

Third, there are global special-population studies such as student surveys like the Programme for International Student Assessment (PISA; http://www.pisa.oecd.org/), the Relevance of Science Education (ROSE; http://www.ils.uio.no/english/rose/), Progress in International Reading Literacy Study (PIRLS; http://nces.ed.gov/surveys/pirls/), and Trends in International Mathematical and Science Study (TIMSS; http://nces.ed.gov/timss/).

Fourth, there are regional general-topic surveys such as the ESS (http://www.europeansocialsurvey.org/), East Asian Social Survey (EASS; http://www.eass.info/), and various regional barometers (Lagos, 2008). Like the global general-topic surveys, these are directed by social scientists and operate on a continuing basis.

Fifth, there are regional special-population surveys like the Survey of Health, Ageing, and Retirement in Europe (SHARE; http://www.share-project.org/), the European Working Conditions Survey (EWCS; http://www.eurofound.europa.eu/

ewco/surveys/), and the European Quality of Life Survey (EQLS; http://www. eurofound.europa.eu/areas/qualityoflife/eqls/). These are especially common in European Union (EU) countries.

Sixth, there are global polls conducted by large commercial companies such as the Gallup Organization (http://www.gallup.com/), Gfk NOP (http://www. gfknop.com/), Harris Interactive (http://www.harrisinteractive.com/), IPSOS (http://www.ipsos.com/), Synovate/Aegis Group (http://www.synovate.com/), and Taylor Nelson Sofres (TNS; http://www.tnsglobal.com/). In recent years, there have been a series of mergers creating larger and more international commercial firms. Many now routinely conduct cross-national research, such as the Gallup World Poll (GWP), and all offer comparative research as a standard product.

Seventh, there are consortia and allied associations of commercial firms. Some represent long-term general collaborations, such as the Gallup International Association[3] (http://www.gallup-international.com/), and Globescan (http://www. globescan.com/); others are more project-specific, such as the Pew Global Attitudes project in 2001–2005 (http://pewglobal.org/).

Finally, there are harmonization projects that adjust and make more comparable studies not originally designed for comparative purposes, such as the Luxembourg Income Study (LIS; http://www.lisproject.org/), and many efforts by the United Nations (http://unstats.un.org/) and Eurostat (http://epp.eurostat.ec. europa.eu/).

Data from most of the cross-national surveys carried out by social scientists and governments and some of those conducted by commercial firms are stored in and accessible from major international survey archives such as the following:

1. Interuniversity Consortium for Political and Social Research, University of Michigan (ICPSR)—http://www.icpsr.umich.edu/
2. Roper Center for Public Opinion Research, University of Connecticut— http://www.ropercenter.uconn.edu/
3. Norsk samfunnsvitenskapelig datatjeneste (NSD; Norwegian Social Science Data Services), University of Bergen—http://www.nsd.uib.no/
4. UK Data Archive, University of Essex (UKDA)—http://www.data-archive.ac.uk/
5. Zentralarchiv fuer Empirische Sozialforschung (ZA; Central Archive for Empirical Social Research, University of Cologne—http://www.gesis.org /en/za/index.htm/

While all of these have extensive international and cross-national holdings, none focuses on comparative survey-research data.

In addition, many cross-national programs make documentation and data available from their project websites. These include the CSES, ESS, ISSP, and WVS. Also, some commercial projects make reports and occasionally data

---

[3]The Gallup Organization is the company founded by George Gallup Sr. and is headquartered in the United States. The Gallup International Association (GIA) is not affiliated with the Gallup Organization and is headquartered in London. Some GIA affiliates did have ties to George Gallup in the past. After recent mergers many GIA affiliates are now part of TNS.

available at corporate sites. Full access may be limited to subscribers or otherwise restricted, however.

A third important component of the comparative survey-research community are the academic, professional, and trade associations. These include (1) general academic and professional associations such as the International Political Science Association (IPSA; http://www.ipsa.org/), the International Sociological Association (ISA; http://www.isa-sociology.org/), the International Statistical Institute (ISI; http://isi.cbs.nl/) and its affiliate the International Association of Survey Statisticians (IASS; http://isi.cbs.nl/iass/), (2) survey-research-specific professional and academic organizations like the market-research-oriented ESOMAR (formerly the European Society for Opinion and Market Research; http://www.esomar.org), the World Association for Public Opinion Research (WAPOR; http://www.unl.edu/wapor/), and the European Survey Research Association (ESRA; http://www.surveymethodology.eu/home/), (3) archival groups like the International Association for Social Science Information, Service, and Technology (IASSIST; http://www.iassistdata.org), the Council of European Social Science Data Archives (CESSDA; http://www.cessda.org/), and the International Federation of Data Organizations for the Social Sciences (IFDO; http://www.ifdo.org/), (4) survey-research-methodology collaborations such as the Comparative Survey Design and Implementation Workshop (CSDI; http://www.csdiworkshop.org/)[4] and the series of International Workshops on Household Survey Nonresponse (http://www.nonresponse.org/), and (5) other social science organizations like the International Social Science Council (http://www.unesco.org/ngo/issc/) and the newly formed (2007) International Data Forum (http://www.internationaldataforum.org/).

Finally, an important development has taken place in the creation of formal international standards for survey research (Smith, 2008b). Most comprehensive are the Standards for Market, Opinion, and Social Research which were issued by the International Organization for Standards in 2006 (http://www.iso.org/). Other examples are *Standard Definitions: Final Dispositions of Case Codes and Outcome Rates for Surveys* first promulgated by AAPOR in 1998 and later adopted by WAPOR, the ISSP, and other groups (http://www.aapor.org/response ratesanoverview/), the International Guidelines for Opinion Surveys of the Organisation for Economic Co-operation and Development (http://www.oecd.org/dataoecd/3/20/37358090.pdf/), and the Cross-Cultural Survey Guidelines of the Comparative Survey Design and Implementation Guidelines Initiative (http://projects.isr.umich.edu/csdi/).

25.4 EXAMPLES OF MAJOR CROSS-NATIONAL STUDIES

The six studies described in this section are illustrative of the current range of major cross-national survey-research collaborations mentioned above. Two, the ESS and SHARE, are regional and four, CSES, GWP, ISSP, and TIMSS/PIRLS, are global. Two sample sub-populations, TIMSS/PIRLS covering students and

---

[4] The CSDI, of course, was the germinator for the International Conference on Survey Methods in Multicultural, Multinational, and Multiregional Contexts (3MC; http://www.3mc2008.de/).

SHARE older adults, and four, CSES, ESS, GWP, and ISSP, are of adults in general. Two focus on particular topics, TIMSS/PIRLS on student achievement and SHARE on health, aging, and retirement and four, CSES, ESS, GWP, and ISSP, are general and variable in scope.

The descriptions of the sextet of studies in this section reveals many similarities. Either via convergence or explicit adoption, most of the studies have formulated similar solutions to many of the common challenges of cross-national collaboration. Administratively, with the exception of GWP, all have formed coordinating bodies to direct and monitor collaboration and have delegated national representatives to be responsible on data collections in individual countries. Similarly, most have adopted a questionnaire reporting form such as the Study Monitoring Questionnaire of the ISSP or the Survey Activities Report of TIMSS/PIRLS.

Methodologically, they all of course face the task of maximizing functional equivalence in general and trying to achieve comparability in the various survey components such as sample, mode, response rate, question development, translation, data collection, data processing and cleaning, and file documentation and distribution. In many instances, the studies have adopted similar procedures and standards. For example, all conduct pretests, have minimum sample sizes, create a merged system file, and prepare documentation.

But in other cases protocols and emphasis differ. Several examples will illustrate the across-study variation. First, the ESS, SHARE, and TIMSS/PIRLS, for example, have tried to eliminate mode effects by allowing only a single mode—in-person interviews for the ESS, computer-assisted personal interviews for SHARE, and classroom self-administration for TIMSS/PIRLS, while multiple modes are allowed by CSES (in-person, telephone, postal, and mixed), GWP (telephone and in-person), and the ISSP (postal and in-person). Second, the ESS has tried to minimize variation in response rates by setting minimum targets and has tried to reduce and adjust for nonresponse bias by collecting and utilizing information on contacts. Likewise, SHARE and the ISSP have examined the levels and impact of nonresponse on data quality and comparability, while CSES and GWP have focused less on this element. Third, augmenting the case-level survey data with metadata has been a key aspect of CSES with its political and electoral variables, the ESS with its event-reporting measure, the planned linkage of information from other databases by SHARE, and the school-level variables of TIMSS/PIRLS, while GWP and ISSP have included little metadata. Finally, all studies must draw samples based on the differing national data resources that exist and are accessible to researchers (e.g., national censuses, population registers), the target populations of interest (e.g., students, older adults, all adults), and the mode of data collection utilized. But beyond this variation there are important differences with the ESS, for example, shunning random-route sampling and the use of case-level substitution and the GWP using both approaches.

In addition, each of the six studies has special features. For example, CSES is a component of larger national election studies in many countries. ESS's event reporting and its collection and dissemination of call records are unique. GWP covers more developing countries than any of the other collaborations and has given special attention to issues of doing surveys in areas disrupted by natural disasters and civil strife and in countries with more limited survey-research infrastructure. It is also distinctive in using dichotomies for virtually all of its questions. ISSP has pioneered in improving translation techniques and experimentally testing for mode effects. SHARE is the only study based primarily on a panel design and is also distinctive in making extensive use of retrospective life-events batteries and in its plans to add biomarkers to the questionnaire data. TIMSS/PIRLS have routinely used item-response-theory procedures in both the pretesting of items and in the analysis of the final content.

25.5 FUTURE DEVELOPMENTS

Survey research has greatly expanded over the past seven decades since its inception in the United States in the mid-1930s. Survey research now spans countries covering over 90% of the world's population (Geer, 2004). On the one hand, only a few nations such as North Korea and Myanmar have no survey research. On the other hand, for less than 30% of the world's population is survey research both essentially unrestricted and common. In many countries survey research is limited in scope and range, not being able to cover politically sensitive topics and/or restricted to large urban areas and more developed regions. The situation in China illustrates both the growth and limitations that still exist in much of the world. While rare just a generation ago, survey research is now commonplace in China. Surveys routinely explore many topics like consumer preferences with little restriction. But other topics, like evaluations of the Communist Party, are strictly forbidden. Between these two poles, there is a large gray area. In other societies the limitation are not political, but financial. In many developing countries surveys are rare because there are not the resources to conduct them. Moreover, when conducted, they are often restricted to urban areas and less remote regions. Finally, civil disturbances and natural disasters hinder or prevent survey at certain times in other areas. Despite these remaining barriers, over time political and economic barriers to survey research have fallen and it is likely that coverage will continue to expand in the future.

Likewise, cross-national survey research has expanded. Comparative studies have become more common, covered more countries, and gained methodological rigor. Comparative surveys of one kind or the other are continually underway both regionally and globally. The CSES, Globalbarometers, ISSP, and WVS each cover scores of countries and are continuing to expand. The ESS, with its extensive program of methodological research, coordinated, data-collection standards, and study monitoring, represents the best current practice in comparative survey research and serves as a model for other collaborations to emulate.

Besides the expansions and methodological improvements within each of the major cross-national programs, there is also greater collaboration across projects with the emerging possibility of formal coordination of methodological and substantive research amongst the major global and regional programs. As illustrated by the six programs featured in this section, each program has distinctive features and few features are universally applied across all studies. Further collaboration across programs would lead to improvements in each of the individual efforts and the advancement of cross-national survey research collectively.

The 3MC conference demonstrates the great challenges that stand in the way to achieving valid, reliable, and comparable measurements across surveys. Minimizing total survey error in a single survey is difficult, doing so in two or more surveys conducted in one monolingual society or culture is still more difficult (Smith, 2007a), and doing this in multiple surveys across languages, societies, and cultures is the most difficult of all. Multiple surveys are, of course, more complicated and error prone simply because there are more components that have to be planned, executed, and verified. But doing cross-national/cross-culture surveys is especially challenging because both measurement and the target population interact with one another and are confounded, making methodological and substantive explanations for differences equally possible. For example, typically a cross-national survey conducted in two countries will be in two languages, the data collection done by two organizations, and with differences in many other aspects such as the sampling frame used, interviewer training, and data-cleaning protocols. To meaningfully study the true cross-national differences across these societies, one needs to establish that despite all of these inherent variations in measurement, functional equivalence is achieved. Rigorous methodological work, including experimental designs, are needed to improve data quality in general and functional equivalence in particular.

But achieving that laudable goal is notably hampered by two main factors. First, despite all of the progress that has been achieved in improving survey methodology, there is still much to learn about the sources of measurement error in general and how to maximize measurement comparability in particular. The many papers at the 3MC conference demonstrate both the fine work that is underway to improve data quality and comparability, but also how much additional work is needed.

Second, comparative surveys are often unable to utilize the best existing methods and therefore do not achieve the best results that the current state of the art would allow. Sometimes this may be because of lack of expertise on the part of the principal researchers, but more often it is simply a lack of resources. That is, the know-how and even the desire to do better comparisons are present, but the funds to carry out those optimal comparisons are simply unavailable.

In brief, cross-national survey research has achieved notable growth and progress over recent decades and reached a level of development unimaginable just a few decades ago. But it has enormous work yet to do before its full potential can be reached. Both innovative and applied methodological research and experimentation are needed to advance our knowledge of total survey error and to maximize data quality. Then, cross-national collaborations must find the will and means to incorporate rigorous methodologies into their study designs and ensure their implementation. Cross-national studies built on solid methodological foundations designed to achieve functional equivalence will establish a social science infrastructure that will yield important and generalizable findings about global society and the human condition.

26

Measurement Equivalence in Comparative Surveys: The European Social Survey (ESS) — From Design to Implementation and Beyond

Rory Fitzgerald and Roger Jowell

26.1 INTRODUCTION

The European Social Survey (ESS) was launched in 2001 after about five years of meticulous design under the aegis of the European Science Foundation and representatives of the main European academic funding councils. The aim was not only to set up a new high quality time series on changing social values in Europe but also to improve methods of cross-national attitude measurement.

It is no coincidence that such a development took place in Europe. Increased multinational governance and the interdependence it promotes requires accurate cross-national data. This interdependence is by no means confined to government but extends to other sectors, including business and academia. Eurostat has been meeting the need for harmonized behavioral and socio-demographic data, but a shortage of high quality comparative attitudinal data has long been evident.

Although it is an academic endeavor, the ESS required backing not just from academic funders in participating countries, but also from the European Commission for its central design and coordination. Thus a pan-European infrastructure was envisaged even during the ESS's planning stages in an attempt to build a robust vehicle for social measurement and analysis.

To achieve these objectives, the ESS has introduced a cluster of new approaches to measurement equivalence. We outline here the case for these approaches and describe how they are achieved via an unusual combination of methodological and organizational measures. We also try to identify the ESS's successes and failures and strategies for improvement.

---

[1] Survey Methods in Multinational, Multiregional, and Multicultural Contexts, edited by Harkness et al.

26.2 PROFILE OF THE ESS

The ESS is a biennial survey. One half of its questionnaire is repeated at each round and comprises three broad domains:

- People's values and ideological orientations (their worldviews, including their religiosity, their socio-political values and their moral standpoints)
- People's cultural/national orientations (their sense of nation and cultural attachment and their — related — feelings toward outgroups and cross-national governance)
- The underlying social structure of society (people's social positions, including class, education, degree of social exclusion, plus standard background socio-demographic variables, and media usage)

As Table 26.1 shows, these domains are reflected in a series of sub-modules. The rotating half of the ESS questionnaire consists of two or more modules per round, the topics and authors of which are determined via a round-by-round competition across Europe (Table 26.2). The duration of the whole interview is about one hour.

The hallmark of the ESS is its exacting methodology, including meticulous (and equivalent) probability samples, detailed question-testing procedures, closely specified translation, field work and response enhancement protocols, event-recording and impressive strides in documentation and data access. All protocols and data are freely and immediately available on the Web.

TABLE 26.1. Core Topics

Trust in institutions	National, religious, ethnic identities
Political engagement	Well-being and security
Socio-political values	Demographic composition
Social capital, social trust	Education and occupation
Moral and social values	Financial circumstances
Social exclusion	Household circumstances

TABLE 26.2. Rotating Module Topics to Date

Round 1	Round 3
Immigration and asylum	*Indicators of quality of life*
Citizen engagement and democracy	*Perceptions of the life course*
Round 2	Round 4
Family, work, and well-being	*Attitudes to welfare*
Economic morality	*Ageism*
Health and care-seeking	

Europe is fortunate in possessing a number of time series on socio-political change, but there has been a shortage of academically driven trend data on changes in public attitudes, perceptions, and social values. The European Community's (EC) Eurobarometers have provided valuable insights into social change over the years, as have the European and World Values Surveys and the International Social Surveys Programme. But each time series has its own well-established focus and methodology (O'Shea et al., 2001), which—whether for one reason or another—make them resistant to change. So the ESS was to fill a gap (CCT, 2006) and provide robust social indicators to complement many economic indicators (Jowell & Eva, 2009). Table 26.3 shows the breadth of ESS participation.

ESS is built on the "principle of equivalence" (Jowell 1998; Jowell et al., 2007a, p. 6) in respect of questionnaire design, sampling, translation, and so on. All cross-national studies need to overcome cultural, organizational, financial, and methodological barriers, and the ESS was no exception. Apart from its ambitious substantive aims, the ESS's methodological imperative was to prevent Europe from "sleepwalking" toward lower standards of social measurement.

26.3 ORGANIZATION

26.3.1 Organizational Structure and Coordination

Effective project management is a pre-condition of successful survey research and is especially important in a cross-national study where overall management is

TABLE 26.3. ESS Participating Countries to Date

Country	R1	R2	R3	Country	R1	R2	R3
Austria	☐	☐	☐	Latvia			☐
Belgium	☐	☐	☐	Luxembourg	☐	☐	
Bulgaria			☐	Netherlands	☐	☐	☐
Cyprus			☐	Norway	☐	☐	☐
Czech Republic	☐	☐		Poland	☐	☐	☐
Denmark	☐	☐	☐	Portugal	☐	☐	☐
Estonia		☐	☐	Romania			☐
Finland	☐	☐	☐	Russia			☐
France	☐	☐	☐	Slovakia		☐	☐
Germany	☐	☐	☐	Slovenia	☐	☐	☐
Greece	☐	☐		Spain	☐	☐	☐
Hungary	☐	☐	☐	Sweden	☐	☐	☐
Iceland		☐		Switzerland	☐	☐	☐
Ireland	☐	☐	☐	Turkey		☐	
Israel	☐			Ukraine		☐	☐
Italy	☐	☐		UK	☐	☐	☐

Note: Number of countries in Round 1: 22; Round 2: 26; Round 3: 25.

remote and compliance problematical. Much of the ESS governance structure derives from the original blueprint for the survey (European Science Foundation, 1999) but was revised somewhat in the course of its implementation (see Jowell et al, 2007a, p. 13).

The benefit of adequately staffed central coordination in the running of a large cross-national study cannot be over-estimated. Many academic multinational surveys have to rely essentially on the unfunded commitment of people in different locations and institutions. That they cope in these circumstances is admirable, but it takes its toll (Park and Jowell, 1997b). A defining feature of the ESS has been its Central Coordinating Team (CCT) funded by the European Commission. Supported and reinforced by specialist committees and individuals, it produces the detailed rules of engagement and guidance on how to achieve the required standards. Like all ESS documents it appears on the ESS website (www.europeansocialsurvey.org/). All participating countries agree through their funders to comply with the specifications, which have been determined in consultation with methodologists and practitioners across Europe.

26.3.2 Country Coverage

As noted, 32 countries took part in at least one of the ESS's first three rounds. More than half participated in all three, while others have missed one or more for financial reasons. But no country has so far withdrawn and new countries join at every round, the latest having been Croatia and Lithuania. From Round 4 ESS countries will include the whole EC except Malta, in addition to several non-EC countries.

Welcome as it is, such wide coverage poses challenges. True, it offers impressive analytical potential from a wide range of national contexts. But the more diverse the range of countries, the more difficult it is for the study to achieve equivalence. For instance, the entrance of Turkey into the ESS in Round 2 as its first Muslim country raised immediate issues about the Judaic-Christian assumptions behind the existing questions on religion. Similarly, questions on "democracy" cue in different issues among the "new" democracies of Eastern Europe from those in Western Europe. And questions on pensions pose problems because of large differences in provision between countries. We annotate the source questionnaire to convey to translators which connotation we refer to, but we cannot be certain that a comparable form of words is available in all languages.

26.3.3 National Arrangements

Each country appoints and funds a National Coordinator (NC) for each round and finances its own field work. NCs have a variety of roles, such as:

- Contributing expertise and "local" knowledge for question design
- Designing and drawing a representative probability sample

- Overseeing questionnaire translation according to ESS procedures
- Commissioning and overseeing fieldwork and pursuing high response
- Collecting and codifying event data
- Ensuring data and documentation are deposited in the required format

ESS practice is to select survey houses in each country (usually after competitive bids) according to their suitability for fulfilling the central specifications. This helps to ensure the most appropriate survey house in each country (sometimes the national statistical institute) rather than relying on a single multinational firm.

26.3.4 Advisory Groups

In addition to expert groups on sampling and translation, which are a core element of the CCT, the project draws on advice from many external experts. The Scientific Advisory Board (SAB) comprises senior social scientists nominated by each national funder, ensuring both national voices in the project and access to considerable collective wisdom and experience. The SAB's role is advisory, but its single decision-making role is its responsibility at each round, following a Europe-wide competition, for selecting the teams to design the rotating modules (see Table 26.2).

These Question Module Design teams (QDTs) represent a critical bottom-up element of the project, enhancing its responsiveness to priorities and issues identified by the academic community. The QDTs receive no funding for contributing to questionnaire design. Their sole reward is the inclusion of questions on their selected topic administered to national samples in over 25 countries, amounting to around 50,000 interviews per round. As they work closely with the CCT, the rewards gained by the ESS team are also considerable.

A specialist Methods Group and a Funders' Forum complete the advisory framework on which the ESS depends (Jowell et al., 2007a, pp. 13–14).

26.4 METHODOLOGICAL FEATURES

The ESS is based on a model of input harmonization, recognizing nonetheless that equivalence across countries does not necessarily depend on identical inputs, but on adherence to the same detailed design principles. Countries differ in their opportunities and circumstances, such as in the availability or otherwise of accessible and reliable sampling sources. Even so, random sampling is *de rigueur* if the highest academic standards are to be achieved. Thus, no quota controls or substitution are permitted at any stage of the ESS sampling process because they would compromise these standards. Countries also differ in respect of the way in which they boost response rates, but all countries are expected to make strenuous efforts to optimize response rates and minimize response bias.

ESS procedures are at their least flexible when it comes to transparency and documentation. Potential users of a publicly available comparative dataset are entitled to information about the scope and nature of any national input variations.

Since not all users of the ESS wish to analyze data from every country on every topic, they may well select countries to analyze on the basis of methodological as well as substantive considerations. Good science dictates that the data on which to make such choices are available.

26.4.1 Sampling

The samples are of all resident adults aged 15+ (with no upper age cut-off), regardless of their citizenship.[2] In some countries, the first point for selection is a population register, in others a postcode address file, and so on. A few countries select a simple random sample of individuals, but most use a multistage design, the last stage of which is the random selection of individuals within households or addresses (Häder & Lynn, 2007). Random route procedures are adopted only where no alternative is available, and in these cases careful controls are implemented. All national sample designs are "signed off" centrally by the Sampling Panel before adoption.

Each national sample is designed to be of the same "effective" size—that is of a size equivalent to a *simple* random sample of 1,500 people aged 15+ nationwide.[3] The likely design effects due to clustering and anticipated response rates are calculated before settling on the starting sample size in each country, and the achieved sample sizes are thus planned to differ as a result of these factors. Apart from countries such as Finland, which opt for a simple random sample, actual national sample sizes therefore exceed 1,500—sometimes considerably so.

The ESS's rigorous sampling is one reason for its high reputation, and in some countries the ESS experience has reinstated random sampling in major studies for which it had begun to be more or less a thing of the past.

26.4.2 Questionnaire Design

Developed in British English and subsequently translated into other languages (see 26.4.3 below), almost all questions are closed and are administered in the same format in all countries with the same answer categories. But some concepts—such as occupation and education—require country-specific questions that are later coded to a standard classification. All concepts and dimensions in the survey are ultimately represented in the integrated dataset in an identical format for all countries, facilitating easy comparison.

In developing the core questionnaire, attempts were made to draw on validated questions from other cross-national studies (CCT, 2002). This proved more difficult than anticipated and it was often necessary to adapt existing items or develop new ones. A range of techniques was used to develop and test the core items, and the same methods are now used each round for the rotating modules. They include:

---

[2] People residing in institutions are, however, omitted.

[3] Small countries (with a population of less than 2 million) are allowed to have a smaller effective sample size of 800 (usually amounting to an *actual* sample size of over 1,000).

- Expert papers by substantive specialists
- Multidisciplinary specialist review
- Consultation with NCs
- Using the Survey Quality Predictor program (Saris & Gallhofer, 2007a) to estimate the likely reliability and validity of new items
- Large-scale two-nation quantitative pilots followed by extensive analysis of item nonresponse, scalability, factor structure and expected correlations
- Split ballot multitrait multimethod (MTMM) experiments (Saris & Gallhofer, 2007a)

Data corrections based on MTMM experiments will also soon be available aiding data analysis and helping to assess the performance of key concepts (Saris and Gallhofer, 2007b).

Like every other time series of this kind, the ESS faces tough choices at each round between the conflicting needs of continuity and change. But the ESS is reaching a stage at which difficult decisions will also need to be made about whether or when to discard certain questions that have possibly served their purpose, and when to repeat certain rotating modules to measure change.

26.4.3 Translation

As noted, questions are framed in English and are subsequently translated into all languages spoken as a first language by at least 5% of the population of each country. For budgetary reasons a large-scale pilot, whose purpose is to test items, scales, and the coherence of concepts, takes place in just two countries, one of which is English speaking (the United Kingdom or Ireland). But small pre-tests later take place in all countries to identify problems in the translated questionnaires.

No "live" translations by interviewers are permitted. The protocol for translations is based on a method involving translation, review, adjudication, pre-testing, and documentation (TRAPD)—see Harkness (2007). It involves a team-based approach that includes adjudication and marks a major departure from the longstanding reliance on back translations.

Unlike ESS sampling procedures, there is no centralized procedure for "signing off" each language translation as—with over 20 languages involved—it would be too resource intensive. So a few preventable errors in national translations have inevitably occurred, and in some cases they were identified only once the data had been collected and released.

26.4.4 Field Work

The ESS has a devolved system of field work execution and management. As noted, the NCs and research councils are responsible for selecting and supervising their national field work agencies. The CCT's role is to provide detailed specifications and issue a field work checklist which is then "signed off" prior to the start of field work. The specification outlines numerous requirements including the permitted period of field work, maximum interviewer assignment size, and quality control checks.

Our records show that compliance is high on most aspects of field work, with the notable exception of timetable adherence. Delays in countries tend to arise mainly as a result of the timing of different funding decisions, but this problem may well be mitigated under a longer-term funding regime.

Reflecting their respective survey infrastructures, some countries use paper and pencil data collection (PAPI) while others use computer-assisted interviewing (CAPI). But, as noted, all data collection to date has been face-to-face. Although other national surveys within several ESS countries have changed over to telephone or Web-based modes, these switches were often made without first ensuring that they generated the same or better quality data. As a cross-national time series, the ESS has to be particularly cautious about such changes if it is to maintain quality and comparability. Any decisions taken will be evidence-based, based on a series of mode experiments now being undertaken. Meanwhile, pressure is mounting to allow more flexibility in choice of mode at the national level.

26.4.5 Response Rates and Nonresponse Bias

The ESS sets an ambitious minimum response rate target of 70% and a target non-contact rate of 3%, using standard methods for their definition and calculation. To ensure the required effort, at least four contact attempts to each sampling unit are required, at least one of which must be in the evening and another at the weekend. Although the actual number of contact attempts sometimes falls short, as does (more frequently) the actual response rate, these targets were set in the knowledge that not all countries would, in the event, achieve them. Even so, the targets seem to have had the virtuous effect of promoting above average response rates in several countries, and—perhaps more important—they have raised response rate expectations more generally and correspondingly boosted efforts to achieve them.

Although response rates are, of course, no guarantee of survey quality and are imperfectly associated with response bias, they remain important in an academically led cross-national study conducted in an era of declining response rates. Moreover, there appears to be no single operationally viable alternative strategy for minimizing likely nonresponse bias in a cross-national survey where national sampling frames and paradata differ so greatly. In any case, the use of such data to minimize nonresponse bias during field work would be limited to countries using CAPI, which applies to only around half of the ESS participating countries. Furthermore using demographic data as an indication of data quality for attitudinal surveys is also sub-optimal.

The overall response rate figures for Rounds 1–3 suggest that countries that achieve disappointing response rates in their first round seem to have been spurred on to produce significant improvements subsequently. Unfortunately, however, there seems to be an equivalent regression to the mean among countries which achieved impressively high response rates in their first round. The full response rate tables for each round are available on the data website (http://ess.nsd.uib.no/).

Despite Couper and de Leeuw's call (2003, p. 157) for more information about how the response process works across different countries, few cross-national studies publish such data (see Chapter 18, this volume). This may be because some of those studies use a heterodox range of sampling approaches, making comparable response rates difficult to compute. The ESS set out to rectify this by producing round by round, country by country data not only on aggregate response outcomes, but also on the individual components of nonresponse. Uniform "contact forms" allow such an assessment of the response rate process for each sample unit (Stoop et al., 2003). Every attempt to contact a potential ESS respondent is recorded in the field, resulting in equivalent records, irrespective of different sample designs. From 2006 onwards, these data have been archived.

A series of papers based on these data are produced on the quality of field work at each round (Billiet et al., 2007b), which also assess potential field work strategies for minimizing nonresponse bias in future rounds.

26.4.6 Event Reporting

An innovative element of the ESS is its event reporting—a procedure to ensure the recording and archiving of major news events in each country that might influence responses to ESS questions at each round (Stoop, 2007). The premise is not only that certain national incidents are likely to have a particular effect on responses, but also that certain *international* events (such as the Iraq war) may also play differently in different countries. Analysts who use ESS data soon after field work are likely to be aware of such differences, but users in the distant future may not have that advantage. All NCs thus keep a log of events which in their judgment might produce "blips" in the time series and which occur shortly before and during field work. These records are archived alongside each country's substantive data. A program of development work is in progress with a view to replacing the current approach with a system of standard coding, thus making the data more comparable cross-nationally.

26.4.7 Data Processing, Documentation, and Transparency

From the outset ESS policy has made its data immediately, freely, and fully available to all with no privileged access (Kolsrud et al., 2007, p. 139). The data are released in a fully documented form, enabling both immediate online analysis and immediate downloads for more detailed analyses.

ESS process data and documentation are also made available so that all elements that might influence the findings are visible to data users. Thus ESS datasets are accompanied by an array of metadata, paradata, and complementary datasets. The ESS data website (http://ess.nsd.uib.no/) contains a host of materials, including questionnaires in all languages, methodological protocols, and an online searchable list of publications. Interviewer and contact data files are also available. Meanwhile the ESS main website (www.europeansocialsurvey.org) contains all key documents and protocols, including—from Round 4 onwards—the dialogues and exchanges leading to the form and content of all questions in the rotating modules.

The ESS's principle of transparency also extends to compliance deviations—whether deliberate or through negligence—which are reported in each end-of-round report. This may be an unusual practice, but is in our view essential in order to inform potential data users in advance about flawed measures in certain countries as a result of, say, incomplete sample coverage or mistranslations of key questions. Otherwise such details would be hidden from those who need them most—future data analysts—many of whom may well have a choice as to which countries to include in their analyses.

As a further protection for data users, we omit from the ESS combined datasets any national dataset for which sampling details have not been deposited, and any variable whose structure has been altered (say by adding or omitting answer categories or by a gross translation error), with a note saying why. These "offending" pieces of data remain available only in national data files, but not in the combined one.

In the four years since they were first released, ESS datasets have attracted over 19,000 registered users and have generated an ever-growing number of publications—including nine books, many journal articles, and countless conference papers. Since first becoming an EU-financed "infrastructure" in 2006, the ESS has also produced a series of online (EduNet[4]) and in-person (ESSTrain[5]) training courses, covering methods of analysis, research design, and comparative data collection methods. Other courses are planned and it augurs well that students form the largest single group of ESS data users, thus promising a European social science community of the future that will be both more quantitative and comparative than hitherto.

26.5 LOOKING BACK AND MOVING AHEAD

The ESS started with three primary aims: first, to produce rigorous data about long-term changes in socio-political attitudes within and between European nations; second, to rectify longstanding measurement deficits in the equivalence of comparative quantitative data, particularly in respect of public attitudes; and third, to gain legitimacy for selected data on social attitudes and public perceptions to be more prominent as indicators of national progress.

The story to date has been highly encouraging (Mohler, 2007; Groves et al., 2008). Not only was the ESS successfully launched with Europe-wide support, but it has so far negotiated four rounds of unusually high quality data collection, and with further rounds assured. With over 30 independent funding decisions every round to contend with, and with many of these decisions contingent on the others, it is surprising indeed that the ESS has progressed so smoothly. It has recently

---

[4] See http://essedunet.nsd.uib.no.
[5] See www.europeansocialsurvey.org.

been selected as one of only three social science projects eligible to become a long-term European infrastructure (ESFRI, 2006), and en route it was awarded the coveted Descartes Prize, Europe's top science award, "for excellence in collaborative scientific research"—the first social science project even to be shortlisted. Even more recently, an independent scientific review of the project commissioned by its funders gives extravagant praise for the project alongside valuable suggestions for improvement and expansion (Groves et al., 2008).

Probably the project's proudest achievement, however, is its creation of three multinational datasets to date, all released publicly and on time, which have already acquired multiple users worldwide. Books, articles, and papers based on ESS data—both substantive and methodological—are already flowing, and with the availability of more robust change data we may now anticipate even greater levels of output.

Even so, the ESS still falls short of its goal of spreading high standards of social measurement throughout Europe and beyond. Uncertain long-term funding at both national and European levels make this goal elusive, as does inertia in a system in which funders and researchers have too often made a Faustian bargain to do what they consider to be "good enough research" without recognizing the corrosive effect this tends to have on good science. If and when the uncertainty surrounding future rounds of the ESS is dispersed, the focus can once more return to this central aim.

Meanwhile certain key aspects of ESS methodology still demand attention. Among them are:

- Learning more about the performance of questions cross-nationally
- Tailoring translation tools to manage and maximize equivalence
- Developing a strategy for balancing consistency and change in the questionnaire
- Reassessing the ESS's adherence to exclusive face-to-face interviewing
- Considering more nuanced approaches to boosting national response rates
- Updating the ESS's cyber infrastructure to accommodate ever-increasing data flows

Accelerated plans to derive workable social indicators from the ESS data are under way. A reliable long-term time series of this sort should surely provide a unique source of data on societal change both in respect of citizens' overall contentment with their lives and their cognitive evaluations of how well or otherwise their societies are functioning. The ESS offers just such an opportunity to supplement rather than supplant existing measures (Jowell & Eva, 2009).

27

The International Social Survey Programme: Annual Cross-National Social Surveys Since 1985

Knut Kalgraff Skjåk

27.1 ORGANIZATION

The International Social Survey Programme (ISSP) was established in 1984 by four survey research institutions[2] in Australia, Germany, Great Britain, and the United States, and the first module, "Role of Government 1985," was fielded in six countries. Since then the ISSP modules have been conducted annually in a continuously growing number of countries; the "Citizenship 2004" module for example was fielded in 37 countries. The ISSP organization currently has 45 member countries in Asia, Oceania, Africa, Europe, and Latin and North America (see http://www.issp.org/).

The ISSP data are widely disseminated and used. The ISSP bibliography contains more than 3,226 titles, of which 1,058 are journal articles or book chapters (Smith, 2008a). Since 2006 all ISSP data files have been available online from the ISSP Archive's Web pages,[3] and 3,337 data downloads were registered at the website during 2008 (GESIS, 2009).

The ISSP modules address topics that are of central importance for social science research internationally (Table 27.1). By combining a cross-national perspective with a cross-time perspective through regular replications of the modules, the ISSP provides a powerful research design for studies of societal variation and changes over space and time. ISSP questions are asked as a single

---

[1] *Survey Methods in Multinational, Multiregional, and Multicultural Contexts*, edited by Harkness et al. Copyright © 2010 John Wiley & Sons, Inc.
[2] Research School of Social Sciences at the Australian National University (RSSS); Zentrum für Umfragen, Methoden und Analysen (ZUMA), Mannheim (now Gesellschaft Sozialwissenschaftlicher Infrastruktureinrichtungen GESIS, Abteilung GESIS-ZUMA); Social and Community Planning Research (SCPR), London (now National Centre for Social Research, NATCEN); National Opinion Research Center at the University of Chicago (NORC).
[3] http://zacat.gesis.org/webview/index.jsp/

TABLE 27.1. ISSP Modules 1985–2012

Role of Government	1985, 1990, 1996, 2006
Social Networks and Support Systems	1986
Social Inequality	1987, 1992, 1999, 2009
Family and Changing Gender Roles	1988, 1994, 2002, 2012
Work Orientations	1989, 1997, 2005
Religion	1991, 1998, 2008
Environment	1993, 2000, 2010
National Identity	1995, 2003
Social Networks	2001
Citizenship	2004
Leisure Time and Sports	2007
Health	2011

block in identical order in each country. The samples are nationally representative random samples of the adult population and are designed to achieve a norm of 1,400 cases. All decisions, including the detailed content and design of the annual topics and methodology, are made by majority votes at the annual General Meetings.

There are no central funds in the ISSP. Each member funds its own costs and must field the designated module and attend the annual plenary meeting at least every second year. The member acting as Secretariat covers the additional Secretariat costs, while the ISSP archives cover all costs of the merging and dissemination of the integrated data files. From the beginning, GESIS Data Archive and Data Analysis in Cologne has carried out this work, aided since 1997 by the Spanish ISSP partner Análisis Sociológicos, Económicos y Políticos (ASEP) in Madrid.

In 1989 when the number of members had increased to 11, two of the founding fathers described the growth in these terms: "the ISSP has grown and developed somewhat haphazardly, and this pattern shows every sign of continuing as long as it seems to work" (Davis & Jowell, 1989). The number of members has since quadrupled, and in many respects the organization works better than ever before. One important reason for this is that, as early as 1993, the ISSP established a methodological framework in order to come to grips with the methodological and organizational challenges resulting from the increasing diversity and complexity of the organization. A Methodological Committee, which is assisted by Methods Working Groups, covers ongoing challenges such as nonresponse, background variables, questionnaire design, translation, and mode of data collection.

This chapter looks at the ISSP's ability to produce cross-national comparable data and discusses some major methodological challenges that we expect the organization will face in the coming years. In general, the focus will be on what we consider to be the most fundamental issue of equivalence, namely, questionnaire development and design.

27.2 DEVELOPMENT AND EQUIVALENCE OF ISSP MEASUREMENT INSTRUMENTS

A central goal of cross-national surveys is to provide the research communities with data that enable them to explore and explain cross-cultural differences and similarities. In order to be able to compare structures, relationships, and indicators across countries and cultures, it must be shown that the instruments are comparable at measurement levels adequate for the research design of each particular study.

In the review of equivalence of ISSP measurement instruments that follows below, equivalence is, wherever possible, conceptualized as the measurement level at which comparable scores or analyses can be obtained, also called "invariance of measurement instruments" or "procedural equivalence" (Coromina, Saris, & Oberski, 2007; Johnson, 1998; van de Vijver, 2003a; van de Vijver & Leung, 1997). We distinguish three hierarchal forms of measurement equivalence: (a) if the instrument ensures that the same construct is measured across the cultures involved, construct equivalence has been achieved; (b) the next level is metric (or measurement unit) equivalence, which assumes construct equivalence. Metric equivalence is established if the measurement unit is identical; (c) the third is scalar equivalence, i.e., in addition to having metric equivalence, no groups give systematically higher or lower responses on the response scale (same intercept across cultures). While various methods of exploring or confirming structures in different groups can be applied when construct equivalence is established (which might be sufficient for most studies in comparative social science using country or culture as context only), only metric and scalar equivalence allow for direct comparison and testing between different groups, for example, of mean scores (i.e., country as object of analysis).

The questionnaire development phase and source questionnaire translation are crucial steps in establishing equivalence in cross-national social surveys, but may be also the weakest links (Harkness, 2003; Harkness, van de Vijver, & Johnson, 2003; Smith, 2003). In addition to language differences and incompatibilities, structural differences can mean that distinctions and wordings used in one country do not exist in another. Even small nuances in the translation of question and answer scale labels can have considerable impact on data. We return to questionnaire development and translation later, and just point out here that a minimum requirement for the development of cross-cultural questionnaires should be to ensure that the questions are suitable for all cultural groups to be studied with respect to meaning as well as language.

27.2.1 Questionnaire Development in ISSP

The development goals for ISSP questionnaires are to design questions that are relevant in all countries and which retain construct equivalence in translation. The questionnaire development is regarded as a cornerstone in ISSP collaboration. The modules are developed over at least three years, starting with a theoretically grounded proposal arguing for a new module or (part) replication of an existing one advanced by one or more members at the General Meeting. If the proposal is adopted, the General Meeting elects a sub-committee (a drafting group), normally with members from six countries representing different regions and cultures of the world. In the second year, on the basis of material provided by the drafting group, the General Meeting establishes its priorities regarding the dimensions and concepts to be emphasized in the final draft. During the last year of module preparation, (part of) the new questionnaire is piloted in several countries before the Drafting Group presents the final draft questionnaire in English. This is then discussed in great detail at that year's General Meeting; majority votes are made on each item in the module. The source questionnaire is drafted in English and then translated into other languages by the research teams in each country.

As just described, ISSP questionnaires, whether on new or replicated topics, are discussed extensively at several consecutive annual General Meetings. All members have the possibility to contribute to the development, relevance, and equivalence of the questionnaire, based on their knowledge of the topics in the context of their culture and other cultural/linguistic factors.

27.2.2 Results

To what degree has the ISSP succeeded in achieving measurement equivalency? Using different ISSP modules, the few publications we refer to below have addressed the issue of equivalence in varying degrees of detail. Far more could be mentioned, but these should provide representative examples of the state of the art. Examining data from Australia, Austria, Great Britain, the United States, and West Germany from one of the first ISSP modules, "Social Inequality 1987," Kelley and Evans (1995) studied people's subjective images of class and class conflict and how these images and their political consequences reflect materialist (objective class) as well as reference-group forces. Data from Switzerland were also available as Switzerland, although not a member, had fielded the module. The unrestricted overall measurement model fitted well for Australia, Great Britain, and the United States, supporting construct equivalence. There was, however, a significant lack of fit in the three Central European countries. Discrepancies between model and data were unsystematic and the authors found that the dangers of improving fit outweighed the potential benefits. The class conflict measurement (an additive function of scores on two items) was found to be metrically equivalent between all countries.

Zucha (2005) analyzed 12 items on attitudes regarding social inequality and income distribution in the "Social Inequality 1999" module. Her proposed measurement model has 4 latent variables: group-level egalitarianism; individual-level egalitarianism; group-level individualism; and individual-level individualism. The analysis included Austria, the Czech Republic, and Germany, the last divided into former East and West Germany. When comparing all 4 regions/countries, only construct equivalence could be confirmed. However, excluding the Czech Republic from the confirmatory factor analysis, higher levels of equivalence were obtained, especially when only former West Germany and Austria were included in the model.

Coenders and Scheepers (2003) used data from another widely used ISSP module, "National Identity 1995," in an analysis of the effect of education on nationalism and ethnic exclusion. They analyzed data from the 22 industrialized countries that participated in the survey (Australia, Canada, Japan, and the United States, plus 18 countries from most regions of Europe). The measurement model they tested included 2 latent variables of nationalism (chauvinism and patriotism) and 3 of ethnic exclusion (exclusion of immigrants, of political refugees, and in-group membership). The invariance of the unrestricted model was rather high, that is, construct equivalence was well established. In addition, an acceptable loss of fit when the model was restricted to invariant factor loadings justified using simultaneous regression analysis.

Also using the 1995 module data for Norway and Sweden, Knudsen (1997) analyzed the relationships between the latent variables national chauvinism (partly based on different items from those used by Coenders & Scheepers, 2003), regime legitimacy and xenophobia. His analyses confirmed the robustness of (parts of) this module.

The ISSP modules on "Work Orientations" deal with working conditions, work experience, and attitudes toward work. In "Work Orientations 1997," affective commitment and continuance commitment were introduced as constructs in the questionnaire. In testing these for Germany (eastern and western), Great Britain, Hungary, Japan, Slovenia, and the United States, Andolsek and Stebe (2004) found that a two-factor solution was superior to a one-factor solution in all countries and that even if reliabilities were a bit under the generally acceptable level, the latent variables could be said to represent construct equivalence.

The publications just mentioned all applied structural equation modeling in testing equivalence. Blasius and Thiessen (2006), in contrast, used multiple correspondence analysis to test data quality and equivalence using data from the "Family and Changing Gender Roles 1994" module. Multiple correspondence analysis makes no assumptions about the measurement level of items. It was applied to explore the quality of data and the underlying structure of 11 items measuring support for single- and/or dual-earner families. Twenty-four countries/regions were included in the analysis. Based on different scenarios on the dependence between support for single-earner and dual-earner family structures, the authors were able to group the countries into 3 clusters of different, but internally consistent, structures, while another 2 clusters proved more or less inconsistent. Blasius and Thiessen (2006) concluded that the countries in the first 3 clusters (the member countries in Oceania, North America, Northwestern Europe, and Italy and the Czech Republic) could be relatively easily compared, at least with a carefully selected set of items (construct equivalence at minimum), but that it would be "hazardous" to include the others (countries in East Asia, Eastern Europe, and Israel and Spain) in comparative analysis.

27.3 THE PICTURE SO FAR

The ISSP literature reviewed here, as well as the large number of publications employing more widely used procedures of data quality tests such as exploratory factor analysis and validity and reliability testing (see, for example, Bean & Papadakis, 1998; Kelley & Evans, 1993; Knudsen & Wærness, 1999) are good examples of why ISSP data have a good reputation. The ISSP produces questionnaires containing highly relevant topics, concepts, and questions and data with equivalence on construct level or higher. On the other hand, the outcomes of the ISSP surveys indicate that data users, as always, should take care in testing the quality of data and selecting items and countries. It is also important to notice that the analyses presented above used data from a relatively homogenous group of countries compared to what the ISSP is today.

Until the end of the 1990s only two or three countries outside the Continental European and Celtic-Anglo spheres were included in the integrated data files. Since then data from several countries in East Asia, Latin America, and one African country have been included. Given the time it takes from when data are available until in-depth studies of equivalence can be published, we can expect very soon to get broad empirical evidence on how ISSP data works cross-nationally within the "new" regions as well as compared to the long-term members.

The 10 "rules of thumb" that Jowell (1998) offers for comparative survey research in general (see Chapter 2, this volume) are of continuing high relevance for existing and forthcoming ISSP data. At the very least, social scientists should observe his 2 first recommendations: (i) not to interpret data relating to countries about which the scientist knows little or nothing and (ii) not to compare too many countries. "When that happens explanations and interpretation soon give away to league tables" (Jowell, 1998, p. 174).

So, what are the main issues for continued improvement of cross-cultural equivalence in the ISSP? We turn to these in the following sections.

27.4 MAKING QUESTIONNAIRES IN A GLOBAL SOCIAL SURVEY

27.4.1 Question Wording

As suggested earlier, question design and translation are probably the weakest links in cross-cultural surveys. Concepts and wordings that at first glance seem to be the same can differ even between closely related languages. For example, "equality/égalité" is understood differently in the United States, English-speaking Canada, and French-speaking Canada, while the meaning of "educación" for Spanish-speaking immigrants in the United States includes social skills of proper behavior that are essentially missing from the meaning of "education" in English (Smith, 2004a). Furthermore, nuances and ambiguity (intended or not) in wordings of the source language may easily cross the borders to some countries and languages but may produce questions with different meanings, measuring other social realities in others.

However, to find and follow the tracks of language crossings can be difficult. One interesting example is found in Braun and Scott (1998), where the translations in Italian, German, and Hungarian were evaluated using multidimensional scaling findings. Here we focus only on the Italian case.[4] The authors analyzed the four following items from "Family and Changing Gender Roles 1988":

A. All in all, family life suffers when the woman has a full-time job.
   Tutto considerato la vita familiare risente negativamente se la madre lavora a tempo pieno.
B. A woman and her family will all be happier if she goes out to work.
   Una donna e i suoi familiari sono più sereni se la donna ha un lavoro.
C. A job is all right, but what most women really want is a home and children.
   Un lavoro è una buona cosa ma quello che realmente vuole la maggioranza delle donne è una casa e dei bambini.
D. Being a housewife is just as fulfilling as working for pay.
   Essere una casalinga è altrettanto soddisfacente quanto avere un lavoro retribuito.

The means of items A and C indicated more traditional attitudes in Italy compared to the United States and Germany, *and* more traditional attitudes in southern Italy than in northern Italy. On the other hand, items B and D indicated less traditional attitudes in Italy than in the United States and Germany, and the means were more or less equal for southern and northern Italy. How could this come about?

As Braun and Scott point out, item B. as formulated in English, is likely to tap several aspects and the ambiguity could raise problems if people in different countries (and subgroups in countries) understood the question in a different way. Furthermore, the aspect of employment may have been more salient to Italians than the issue of being away from home because the Italian question asked about having a job (*ha un lavoro*) rather than going out to work. Thus, respondents may have been more inclined to agree with the Italian wording, especially in southern Italy. The same argument applies to item D. According to Braun and Scott, economic consequences might come more easily to mind in the Italian version where the expression "having a paid job" (*avere un lavoro retribuito*) is used rather than "working for pay."

However, this is not the complete picture. If we take a closer look at item A, we see that the Italians were asked about "the mother" (*la madre*) instead of "the woman." This could in fact explain why the mean of this item (out of the four) expressed the most traditional attitudes, within Italy as well as compared to the other countries. More subtle, but still possible, is that using the terms *la maggioranza delle donne* ("the majority of women") and *realmente* ("really," with reference to a objective reality) in item C might have resulted in more traditional attitudes than if *la maggior parte delle donne* ("the bigger part of women/most women") and *davvero* ("really," without reference to something objective) had been used.

So which of the four items produced the somewhat surprising results? Finding the right answer seems impossible, but the example above is an excellent illustration of what a difficult and intricate journey it is to take survey questions safely across borders (and back).

---

[4] Many thanks to the PI of ISSP in Italy, Cinzia Meraviglia, the University of Eastern Piedmont "Amedeo Avogadro," for useful help in understanding the Italian wording.

27.4.2 Response Scales

In addition to the body of the item where the substance and the stimulus are presented, a survey question typically consists of a response scale, and the equivalence of response scales across countries must be established to the same extent as question concepts and substance. In many respects response scales represent greater challenges than the substance of questions. Verbal scales very often consist of sentence segments with adverbs and adjectives—small nuances can have great impacts on response behavior. Furthermore, countries and field work organizations with established survey traditions tend to stick to what is already tested and considered as best in their own context. Shishido, Iwai, and Yasuda (2006) point out the fact that not only do existing cross-national surveys use various types of scales, but also that the translations differ between organizations in the same country. In an experiment organized by the Japanese General Social Survey (JGSS) they found that different translations of "agree" in Japan (where strong adverbs seem far more difficult to use than in other East Asian countries) resulted in different distributions. The balanced five-point verbal scale from "strongly agree" to "strongly disagree" is one of the most widely used response scales in the ISSP, and the problems in Japan are not unique. For example, the Eurobarometer in France, the ISSP France, and the ISSP French Canada apply different versions of this response scale (Harkness, 2003).

27.4.3 Concepts

The ISSP is, like most large-scale, long-term cross-national surveys, of Western origin. Regardless of how thorough and academically rigorous the development of methodologies, theories, concepts, and instruments are, there is always a danger that the focus on relatively homogenous countries in the beginning years of the ISSP could result in cultural constraints and ethnocentric bias when the studies expand to new cultural areas. As already mentioned, it is only during the last few years that ISSP data from a broader range of countries in Eastern Asia and Latin America have been available. We can, however, already illustrate some issues that the ISSP as a *global* social survey program has started to address, and which are likely to demand more attention in the coming years.

Religion. The first ISSP module on religion was conducted in 1991 and partly replicated in 1998. The combination of topic *and* participating countries each time probably makes this the most challenging ISSP module with respect to equivalence. Japan was not a member in 1991 and in 1998 was still the only country in ISSP where the large majority of the population could not be classified as monotheistic. Thus, the questionnaires did not include items that would tap fundamental concepts of religiosity and religious behavior in Japan and other East Asian countries, such as polytheism and syncretism (Jagodzinski & Manabe, 2002). The ISSP "Religion 2008" module aims to capture more aspects of Asian religious cultures by including two emic ("culture-specific") items, and by adding non-Western religious concepts as etic ("culturally neutral") items.

Working life. There are great variations in the organization of work, relationships between employees and employers, and work values around the world, and a challenge in future "Work Orientations" module will be to embrace the essence of these variations. Andolsek and Stebe (2004) note that "Work Orientation 1997" included measures of affective and continuance commitment, whereas work values like respect and obedience, conforming and fitting in, working for collective benefits, loyalty and security, etc. (associated with "normative commitment") were absent from the questionnaire. The salience of such work values varies greatly across countries now represented in the ISSP. They will therefore have to be addressed in the next module of this topic.

27.4.4 Mode of Administration

Mode of administration is the methodological aspect in social surveys that is most closely related to how substance and stimulus are presented to respondents in questions and response scales, and will be the last issue raised in this chapter. When the first ISSP modules were planned, the intention was that the ISSP add-ons primarily should be administered as self-completion questionnaires. But already in the first round of fielding, two out of six surveys were conducted as face-to-face interviews, mainly because of limitations in the survey infrastructure in those countries, but also because of illiteracy. Even if the issue of mode has been systematically addressed by the ISSP, it has turned out to be difficult to harmonize the mode of administration across countries.

Mode is considered to be a major source of survey error (de Leeuw, 1992; Krosnick, 1991; Ross & Mirowsky, 1984) in both mono-cultural and cross-cultural surveys. Based on the literature it is fair to say that in mono-cultural societies in the Western world mode of administration causes different response patterns. On the other hand, it is hard to find consistent patterns cross-culturally, apart from social desirability. A mixed-mode experiment conducted in six ISSP countries in 1996 (Kalgraff Skjåk & Harkness, 2003) indicated that the meaning and effects of modes varied between cultures and topics—mode differences found in one country within one survey instrument would not necessarily replicate in another country or in another instrument.

27.5 POSSIBLE ISSUES TO BE ADDRESSED

A continuous challenge for projects like the ISSP is to improve questionnaire equivalence and relevance in a global context. Regional cross-national social surveys have been undertaken for some time now in Africa, Asia, and Latin America, and experiences from these—and also regional surveys in Europe and North America—could provide amendments to and improvements of measurement of values and attitudes in existing global social surveys. The use of emic questions, as measurements of dimensions, could be expanded. Another important topic to address is the design of response scales. Are nonverbal numerical scales better understood than verbal? Are construct-specific response options better than agree/disagree options (Saris, 2008; Saris, Krosnick, & Shaeffer, 2005)? And, are there efficient methods for calibrating response scales across countries (Smith, 2003)?

This chapter suggests how an increased focus on translation may improve instrument equivalence. Not only translation procedures and protocols should be addressed, but also the extent to which translation should be an integrated part of the questionnaire development process (see Chapters 3 and 6, this volume).

Mode of administration would be another important issue to address, not only the current practice of different collection methods, but also to what extent acceptance and tailoring of several data collection methods could counteract the falling rates of survey participation across continents. In this respect ISSP's experience in designing modules for different modes could turn out to be very helpful.

Another challenge is how best to integrate metadata closer with the data files, permitting data users to undertake more extensive assessments. This could also lead to useful feedback from external users to the ISSP and point out priorities for methodological improvements. An underlying principle in the ISSP is that each member has the responsibility for the overall quality of the ISSP surveys in its own country. The ISSP Study Monitoring Report (Park & Jowell, 1997a) is an indispensable tool in making the ISSP surveys as transparent as possible at country level. These reports are available at the ISSP website from the 1995 module on. ISSP is also in the process of providing detailed information on the measurement and post-harmonization of background variables in each country.

Improving survey methodology must be based on empirical evidence from routinely conducted methodological experiments and in-depth knowledge of the landscapes that are mapped. With its organizational structure, scientific network and accumulated worldwide survey experiences, the ISSP is perfectly tailored for continued mutually cooperative methodological research and survey development.

ACKNOWLEDGMENT

Many thanks to Kirstine Kolsrud, NSD, for useful comments.

28

Longitudinal Data Collection in Continental Europe: Experiences from the Survey of Health, Ageing, and Retirement in Europe (SHARE)

Axel Börsch-Supan, Karsten Hank, Hendrik Jürges, and Mathis Schröder

28.1 POPULATION AGING IS A SOCIAL (SCIENCE) CHALLENGE—THE NEED FOR A LONGITUDINAL SURVEY OF HEALTH, AGEING, AND RETIREMENT IN EUROPE

To cope with the challenges of Europe's rapid population aging, it is important to improve our understanding of the complex linkages between economic, health, and social factors determining the quality of life of the older population. These interactions take place at the individual level in the first place, they are dynamic—as aging is a process, not a state in time—and they must be related to a country's welfare regime. So far, however, cross-nationally comparable, longitudinal micro-data on the economic, social, and health situation of older people in Europe were missing.

The "Survey of Health, Ageing, and Retirement in Europe" (SHARE) is closing this gap. So far, SHARE collected data on the health, social, and economic situation of more than 30,000 individuals aged 50 or older. In 2004, a baseline wave of data collection was conducted in 11 countries, ranging from Scandinavia (Denmark and Sweden) through Central Europe (Austria, France, Germany, Switzerland, Belgium, and the Netherlands) to the Mediterranean (Spain, Italy, and Greece). In 2005–06, further SHARE data were collected in Israel. For the second wave of data collection, which was conducted in 2006–07, two "new" European Union (EU) member states—the Czech Republic and Poland—as well as Ireland joined SHARE. The survey's third wave, scheduled for 2008–09, focuses on the collection of detailed life-histories of respondents who participated in previous waves. Further waves are being planned to take place on a biennial basis.

---

[1] Survey Methods in Multinational, Multiregional, and Multicultural Contexts, edited by Harkness et al.
Copyright © 2010 John Wiley & Sons, Inc.

Substantively, SHARE provides an infrastructure helping researchers to understand better the individual and population aging process: where we are, where we are heading, and how we can influence the quality of life as we age— both as individuals and as societies (cf. Börsch-Supan et al., 2005).

Methodologically, SHARE provides a unique opportunity to address a broad range of survey-research issues against the background of an ongoing large-scale cross-national study with a longitudinal perspective (cf. Börsch-Supan & Jürges, 2005). This chapter focuses on methodological issues of SHARE. It begins with a history of the SHARE baseline wave (Section 28.2), focusing on efforts made to ascertain cross-national comparability. We then describe the "longitudinal" experiences from the survey's second round (Section 28.3), followed by an overview of the preparations for collecting life-histories in wave 3 (Section 28.4). The final section concludes with an outlook on the future of SHARE.

28.2 GETTING STARTED—THE 2004 SHARE BASELINE WAVE

Based on the models of the U.S. Health and Retirement Study (cf. Juster & Suzman, 1995) and the English Longitudinal Study of Ageing (cf. Marmot et al., 2003), the SHARE development process started in January 2002 (see Börsch-Supan & Kemperman, 2005, for details). Draft versions of the questionnaire were tested in a series of pilot and pretest studies, which eventually resulted in the final SHARE baseline instrument in September 2004 (see Börsch-Supan & Jürges, 2005: Appendix B, for the main questionnaire). Already during this design stage, ascertaining cross-national comparability was a major concern for SHARE, which is particularly reflected in the project's efforts regarding (a) survey software, (b) translation, and (c) sampling design.

> *(a) Survey software* (see Das et al., 2005, for details): The SHARE data were collected using a centrally developed, Computer-Assisted Personal Inter- viewing (CAPI) program, which allowed each country involved to use exactly the same underlying structure of metadata and routing. The only difference across countries was the language. This mechanism, where ques- tion texts are separated from question routing, enforces the comparability of all country-specific translations with a generic questionnaire. The CAPI program was written in Blaise, a computer-assisted interviewing system and survey processing tool developed by Statistics Netherlands. The generic CAPI instrument was directly implemented in Blaise, and the generic texts (in English) were stored in an external database. The different countries translated their versions of the instrument using the Internet and a newly developed Language Management Utility (LMU). Another program was written converting all translated text from the LMU database into a country- specific survey instrument, based on the blueprint of the generic version. There were only few exceptions to the generic blueprint of the question- naire. Country-specific parts were introduced if institutions were fundamen- tally different or by skipping irrelevant answer categories (by adding new country-specific answer categories, respectively) in the LMU. These excep-

tions never led to a different sequence of questions for a specific country. Another new software development was a Sample Management System (SMS) to manage the coordination of the fieldwork. Only three countries used their own system: France, Switzerland, and The Netherlands. The SMS basically consists of a list of all households in the gross sample that should be approached by the interviewer. Contact notes and registrations, appointments with respondents, and area and case information could be entered in the system, and the system enforced common procedures for re-contacting respondents and how to handle nonresponse.

(b) *Translation* (see Harkness, 2005, for details): Although each country participating in the project organized its own translation effort, the SHARE coordinator initiated several activities to support the individual translation efforts and to ensure cross-national comparability. SHARE countries were provided with general guidelines for the translations process, similar to those used in the European Social Survey, for example. The guidelines advocated organizing a team to complete the translation and to review translations. The team would bring together the language and translation skills, survey questionnaire know-how, and substantive expertise needed to handle the SHARE questionnaire modules. Eventually, the coordinator commissioned a professional review of a sample of the first draft of SHARE translations. SHARE countries were provided with feedback from an external set of translators. The translators commented in detail on selected questions and submitted a brief general appraisal of the translation draft. This procedure was repeated for a later draft of the questionnaire and feedback again provided to SHARE participants. The pilot-and-pretest design of the SHARE study, coupled with the translation guidelines and appraisals, provided the SHARE project with a rare opportunity to refine and correct the source questionnaire and the translated versions.

(c) *Sampling design* (see Klevmarken et al., 2005, for details): In the participating SHARE countries the institutional conditions with respect to sampling are so different that a uniform sampling design for the entire project was infeasible. Good sampling frames for our target population of individuals 50+ and households with at least one 50+ individual did not exist or could not be used in all countries. In most countries there were registers of individuals that permitted stratification by age. In some countries these registers were administered at a regional level. Germany and the Netherlands are two examples. In these cases, we needed a two- or multistage design in which regions were sampled first and then individuals selected within regions. In the two Nordic countries Denmark and Sweden we could draw the samples from national population registers and thus use a relatively simple and efficient design. In France and Spain it became possible to get access to population registers through the co-operation with the national statistical office, while in other countries no cooperation was possible. In three countries, Austria, Greece, and Switzerland, we had to use telephone directories as sampling frames and pre-screening in the field of eligible sample participants.[2] As a result, the sampling designs used vary from simple random selection of households to rather complicated multistage designs. In the three countries that used telephone directories and in Denmark, the final sampling unit was a house-hold, while in all other countries the final unit of selection was an *individual*.

During the fieldwork period of the SHARE baseline study, which was mainly conducted from May through October 2004, field progress and quality of the incoming data were monitored thoroughly, contributing to ensuring cross-national comparability of the data also at this stage of the project (see de Luca & Lipps, 2005). After completion of the fieldwork period, considerable efforts were made to transform the SHARE raw data into a user-friendly database, resulting in a preliminary public Release 1 in May 2005 and a further Release 2 in June 2007 (see Table 28.1 for descriptive statistics).[3] Post-fieldwork activities included (i) extensive data cleaning, (ii) generation of user-friendly indicators (e.g., Jürges, 2005), (iii) computation of calibrated design weights (Klevmarken et al., 2005), (iv) nonresponse analysis (cf. de Luca & Peracchi, 2005; Kalwij & van Soest, 2005), and (v) imputation of missing income and asset information (cf. Brugiavini et al., 2005; Christelis et al., 2005).

TABLE 28.1. Description of 2004 SHARE Sample and Response Rates (Release 2)

Country	Total	Male	Female	Under 50	50 to 64	65 to 74	75+	Household Response Rate[a]	Individual Response Rate[a]
Austria	1,893	782	1,111	44	949	544	356	55.6%	87.5%
Belgium	3,827	1,739	2,088	178	1,991	986	672	39.7%	90.6%
Denmark	1,707	771	936	92	916	369	330	63.2%	93.0%
France	3,193	1,386	1,807	155	1,648	759	631	79.2%	92.3%
Germany	3,008	1,380	1,628	65	1,569	886	486	60.8%	86.4%
Greece	2,898	1,244	1,654	229	1,458	712	499	63.4%	92.4%
Israel	2,598	1,139	1,459	142	1,416	690	347	68.1%	83.9%
Italy	2,559	1,132	1,427	51	1,342	785	381	52.8%	79.1%
Netherlands	2,979	1,368	1,611	102	1,693	713	459	60.6%	88.0%
Spain	2,396	994	1,402	42	1,079	701	573	54.3%	73.7%
Sweden	3,053	1,414	1,639	56	1,589	816	592	47.3%	84.4%
Switzerland	1,004	462	542	42	505	251	204	38.8%	86.9%
Total	31,115	13,811	17,304	1,198	16,155	8,212	5,530	60.6%	85.0%

[a] Weighted average. *Source:* http://www.share-project.org.

[2] The proportion of the target population automatically excluded from the sampling frame because the household does have a telephone is relatively small; noncoverage resulting from unlisted numbers could be more serious. About 1.5% of all Swiss private households do not own a telephone; about 8% with telephones are unlisted. Cell-phone-only households were not (yet) felt to be an issue for SHARE with its relatively old target population.

[3] SHARE data, questionnaires, and documentation are available at http://www.share-project.org.

28.3 SHARE Goes Longitudinal—The Second Wave of Data Collection in 2006–07

When preparing and conducting the second wave of data collection, a major concern for the SHARE team was to maintain in the panel study the high level of cross-national comparability achieved in the baseline wave. The main fieldwork period of SHARE's second round lasted from October 2006 until September 2007. In some countries the fieldwork period was prolonged, as the specific sample requirements of SHARE—following respondents who had moved to their new residence (including nursing homes) and end-of-life interviews—required in some cases very time-consuming (administrative) efforts by survey agencies.

The survey software developed for the SHARE baseline wave was carefully adapted to serve the needs of a longitudinal survey. First, in some countries institutions had changed: new pension options had been introduced, particularly in the private market; some countries had health-care reforms; the set of available financial instruments had changed; transfer incomes had been reformed. We thus adapted the country-specific parts of the questionnaire in which these options and institutions were mentioned. Second, we adapted and improved the flow of the instrument by using preloaded information from the first wave. Such preloading, although it involves a lot of programming and testing effort, has several advantages in terms of data quality. For instance, it allows matching respondents easily across waves, to record changes in household composition, to monitor changes in labor market status, or to learn about the incidence of chronic conditions.

A longitudinal study requires a permanent-status update of all involved panel respondents. First, one wants to keep track of respondents who are moving. To this end, we maintain regular contact to panel members ("panel care"), for instance, by sending a Spring/Easter postcard each year with a response card attached that will be sent back in the case of a move with the new address, or by sending a brochure with new results from SHARE-based research that is of general interest.

Second, it is crucial to have a reliable account of what has happened to panel members who do not re-appear in the next wave, where one needs to distinguish between moving, temporary illness, and death, in particular when respondents live by themselves and in isolation from relatives, friends, and neighbors, as is often the case with the oldest. Interviewers have been advised and trained to verify the status of each sample person. In some countries, interviewers or fieldwork agencies had access to death certificates or registration records, being able to cross-reference the respondent database with register data. In other countries, such records are inaccessible or do not exist, requiring a coordinated approach of tracking panel members, for instance, by sending interviewers to addresses of respondents with unknown status and to ascertain the vital status of previous respondents.

Several methodological innovations have been introduced in wave 2, with cross-national comparability being a major concern. First, we added two new health measurements (respiratory peak flow and chair stand[4]) to our existing gait-speed

---

[4]The chair stand test measures strength and endurance in legs and lower body and speed and coordination. A stop watch is used to measure the time (in seconds) it takes a person to stand up from a sitting position and sit down again five times, while holding the arms crossed over the chest.

and hand-grip strength measurements (cf. Hank et al., 2008). Second, we included a set of anchoring vignettes (e.g., King & Wand, 2007) not only for a wide range of health domains but also for work disability; quality of life, employment, and health care; and satisfaction with political institutions. In a diverse continent like Europe, cross-national comparisons using surveys among households and individuals often from differences across countries and socio-economic groups in the way people answer survey questions, particularly self-evaluations of, for example, health or quality of work. Anchoring vignettes aim at solving this problem. Anchoring vignettes are short descriptions of, for example, the health or job characteristics of hypothetical persons. Respondents are asked to evaluate the hypothetical persons on the same scale on which they assess their own health or job. Respondents are thus providing an anchor, which fixes their own health assessment to a predetermined health status or job characteristic. These anchors can then be used to make subjective assessments comparable across countries and socio-economic groups. We have collected vignette ratings for a sub-group of about 600 respondents per country. The results are currently being used to construct improved cross-nationally comparable indicators of health, well-being, job satisfaction, and so on (see www.compare-project.org for detailed information).

Another innovation in wave 2 was the introduction of an "end-of-life" interview, also called exit interview. These data will give the analyst the rare opportunity to follow the lives of people right until the time of their death. In the exit interview, we have collected information on health, social well-being, and economic circumstances in the last year of life of all our first wave respondents that have died between the first two rounds of data collection. Overall we have so far conducted more than 500 end-of-life interviews (for 274 men and 247 women) with so-called proxy respondents, mostly with relatives, but also with neighbors, friends, or social workers. The average time between the decedent's death and the end-of-life interview was 14 months. We expect the exit interview data to be of good quality because our proxy respondents had very frequent contact with the decedent: 75.7% had daily contact with the deceased in the last year of his or her life, 13.3% had contact several times a week, and only 11% had less frequent contact. Frequency of contact clearly varies by proxy reporter type (i.e., relationship to the deceased). Quite naturally, immediate family had the most frequent contact with the decedent. However, even among other relatives and nonrelatives, more than 40% of the proxy reporters had daily contact.

28.4. SHARELIFE—PREPARING A RETROSPECTIVE SURVEY INSTRUMENT FOR SHARE'S WAVE 3

The third wave of SHARE—under the project name "SHARELIFE"—differs from the previous two conceptually, because here questions are asked about events that happened *throughout the respondents' lives* with the goal of constructing a detailed life-history. Although the study is still conducted as a panel to keep the longitudinal aspect of the survey, the questionnaire is completely new. SHARELIFE consists of five focus points, which correspond to the areas of interest from the regular SHARE questionnaire: Children, Partners, Accommodation, Work, and Health. For each of these different areas, the dates of certain events and the corresponding surrounding information will be collected. For example, not only the date of a residential move, but also information on region, ownership, and purchasing means of the specific residence is collected.

Similar to any survey, SHARELIFE relies on the respondent's ability to remember events in the past. Since the respondents have at least 50 years (and some much more than that) to look back upon, good techniques are needed to reduce the potential recall error. The method of questioning that is employed in SHARELIFE is based on a so-called life-history calendar (LHC; e.g., Belli, 1998). The respondent's life is basically represented graphically, with a grid that is filled through the course of the interview (see Figure 28.1). The idea of the LHC is to help the respondent remember by asking those life events first that are very likely to be remembered accurately. Thus, the interview usually starts with the names and birthdates of the respondent's children and is followed by the partner history. As soon as an event is entered in the LHC, it can be referred to by the interviewer to help; for example, when a respondent is not sure about the date of a job change, a useful probe may be: "Was that before or after your second child was born?" This principle extends to all other modules and is flexible as well: there is no need for the respondents to start with the children's module, if they feel that they better remember another part of their life-history.

The process of reaching the final SHARELIFE instrument can be described easily as a combination of those steps that were completed in wave one and two. As in the first wave, the questionnaire is developed with the use of generic English test versions followed by country-specific versions, which are tested in pilot and pretest studies. After each test, the questionnaire is evaluated using the results and improved accordingly. Similar to the development of the second wave, we will use the preload of previously obtained information and develop further the possibilities to follow our respondents, including the exit interviews that were already successfully used in wave two.

The SHARELIFE project started in the spring of 2007 with the first stages of questionnaire design, with fielding beginning in the fall of 2008.The completion date for the project is the end of 2009.

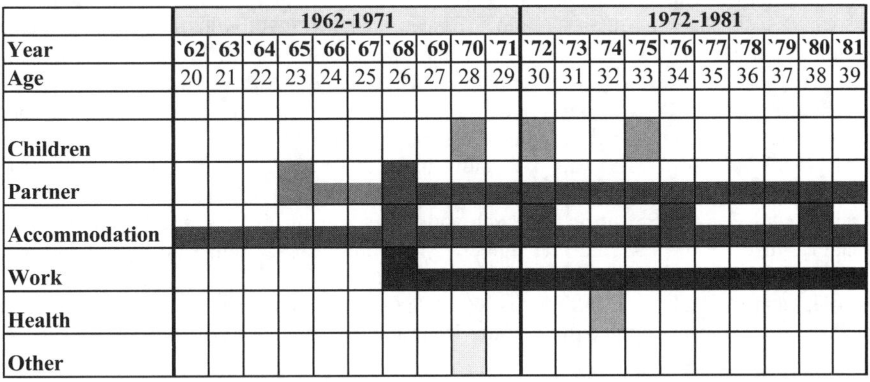

	1962-1971										1972-1981									
Year	'62	'63	'64	'65	'66	'67	'68	'69	'70	'71	'72	'73	'74	'75	'76	'77	'78	'79	'80	'81
Age	20	21	22	23	24	25	26	27	28	29	30	31	32	33	34	35	36	37	38	39
Children																				
Partner																				
Accommodation																				
Work																				
Health																				
Other																				

Figure 28.1. Part of a Completed Life-History Calendar

28.5 A LONG-TERM DATA INFRASTRUCTURE FOR RESEARCH ON AGING IN EUROPE AND BEYOND—PROSPECTS OF SHARE

In 2007, SHARE was selected to be included on the roadmap of the European Strategy Forum on Research Infrastructures as one of the 35 crucial pillars of the European Research Area. This allows a major upgrade of SHARE along two dimensions: *First*, it will prolong SHARE over the decade 2010–2020, generating a genuine eight-wave, biennial panel that follows individuals for up to 15 years as they age and react to the changes in the social and economic environment. From a research viewpoint, the time dimension is crucial since aging is a process that can only be understood if we observe the same individual at different points in time. *Second*, SHARE will expand to all 27 EU member states plus associated Switzerland and Israel.

Further methodological innovations are related to the envisaged inclusion of two fundamental sources of information which are currently not included in the instrument: social security numbers of respondents and so-called biomarkers. Social security numbers allow merging the SHARE data with economic data processed by various branches of the social security system. Biomarkers include physical measures such as body mass index, grip strength, lung volume, or blood pressure, as well as biochemical measures of saliva and blood. They significantly increase the precision of health measurement and allow important insights into the health history of the very old and the determinants of morbidity in old age.

The aim of a two-year "preparatory phase," which started in January 2008, is to bring the SHARE prototype to the level of financial, legal, governance, and technical maturity required to fill important knowledge gaps in individual and population aging. It will involve all stakeholders necessary for the major upgrade described above, among them research institutes and universities; national science ministries and foundations; two Directorates General of the European Commission; and the U.S. National Institute on Aging.

ACKNOWLEDGMENTS

The 2004 SHARE data collection was primarily funded by the European Commission through the 5th framework programme (project QLK6-CT-2001-00360 in the thematic program Quality of Life). Additional funding came from the U.S. National Institute on Aging (U01 AG09740-13S2, P01 AG005842, P01 AG08291, P30 AG12815, Y1-AG-4553-01 and OGHA 04-064). Data collection in Austria (through the Austrian Science Foundation, FWF), Belgium (through the Belgian Science Policy Office) and Switzerland (through BBW/OFES/UFES) was nationally funded. The SHARE data collection in Israel was funded by the U.S. National Institute on Aging (R21 AG025169), by the German-Israeli Foundation for Scientific Research and Development (G.I.F.), and by the National Insurance Institute of Israel. Further support by the European Commission for the 2006–07 SHARE wave through the 6th framework program (projects SHARE-I3, RII-CT-2006-062193, and COMPARE, CIT5-CT-2005-028857) is gratefully acknowledged. The SHARELIFE project is funded through the 6th framework program of the European Union.

29

Assessment Methods in IEA's TIMSS and PIRLS International Assessments of Mathematics, Science, and Reading

Ina V. S. Mullis and Michael O. Martin

29.1 DESCRIPTION OF TIMSS AND PIRLS

TIMSS (Trends in International Mathematics and Science Study) and PIRLS (Progress in International Reading Literacy Study) are projects of the International Association for the Evaluation of International Achievement (IEA). IEA is an independent international cooperative of national research institutions and government agencies that has been conducting studies of cross-national achievement in a wide range of subjects since 1959. The decision to participate in an IEA study is coordinated through the IEA Secretariat in Amsterdam and made solely by each country according to its own data needs and resources.

TIMSS and PIRLS achievement data inform countries about progress in student's learning in mathematics, science, and reading from three important perspectives—across time, in comparison to other countries, and in comparison to their own educational goals.

Conducted on a regular 4-year cycle, TIMSS assesses students' achievement in mathematics and science at the fourth and eighth grades. There have been TIMSS assessments in 1995, 1999, and 2003, with TIMSS 2007 conducted in more than 60 countries and planning underway for 2011. In addition, to providing trends in achievement at the fourth and eighth grades, TIMSS has a quasi-longitudinal component whereby students originally assessed as fourth graders move to the eighth grade. Thus, TIMSS also provides information about whether the relative performance of these students has changed in the intervening years. TIMSS Advanced 2008 provides information about changes since 1995 in achievement in advanced mathematics and physics for students in their final year of schooling (twelfth grade in most countries).

---

[1] Survey Methods in Multinational, Multiregional, and Multicultural Contexts, edited by Harkness et al. Copyright © 2010 John Wiley & Sons, Inc.

Inaugurated in 2001 and conducted every 5 years, PIRLS is IEA's assessment of students' reading achievement at the fourth grade. PIRLS 2001 was administered in 35 countries and PIRLS 2006 in 40 countries. According to the assessment cycle, PIRLS 2011 will be conducted together with TIMSS 2011, providing the opportunity for countries to collect reading, mathematics, and science achievement data at the fourth grade on the same students or schools according to their preference.

29.2 DESCRIBING EDUCATIONAL CONTEXTS

To provide comparative perspectives on trends in achievement in the context of different education systems, school organizational approaches, and instructional practices, TIMSS and PIRLS collect a rich array of background information. Countries contribute chapters to an "Encyclopedia" and complete questionnaires describing their educational systems, curricula, and resources devoted to education in mathematics, science, and reading.

To collect information about school and classroom contexts, each student's mathematics teacher and science teacher (for TIMSS) or reading teacher (for PIRLS) complete a questionnaire. The student's school principal also completes a questionnaire. To collect information about students' home and school environments and learning experiences, each student completes a questionnaire. This student questionnaire is the vehicle used by TIMSS and PIRLS for collecting information about students' attitudes toward mathematics and science or reading, respectively. For PIRLS, because of the particular importance of home support in developing literacy skills, the student's parents complete a questionnaire.

29.3 MANAGEMENT AND COORDINATION

TIMSS and PIRLS are major undertakings of the IEA, comprising the core of its regular cycles of studies. The TIMSS & PIRLS International Study Center at Boston College has responsibility for the overall direction and management of the projects, working closely with a team of experts around the world. The IEA Secretariat is responsible for verification of the instrument translations produced by the participating countries; the IEA Data Processing and Research Center in Hamburg is responsible for processing and verifying the data submitted by the participants; Statistics Canada in Ottawa is responsible for school and students sampling activities; and Educational Testing Service in Princeton, New Jersey, consults on psychometric methodology. Each project also has consultants and expert committees.

To work with the international team and coordinate within-country activities, each participating country designates an individual to be National Research Coordinator (NRC). The NRCs have the task of implementing the studies in their countries in accordance with the guidelines and procedures developed by the international team. The NRCs work with their colleagues in carrying out the very complex sampling, data collection, and scoring tasks involved in conducting TIMSS and PIRLS.

International projects of the magnitude of TIMSS and PIRLS require considerable financial support. IEA's major funding partners include the World Bank, the U.S. Department of Education through the National Center for Education Statistics, those countries that contribute by way of fees, and the United Nations Development Programme.

29.4 FRAMEWORK AND INSTRUMENT DEVELOPMENT

Because PIRLS and TIMSS are designed to measure trends, the assessment frameworks build on those from previous assessments, evolving from assessment to assessment based on reviews by the participating countries. The frameworks describe the content to be addressed in the achievement assessment, the contextual factors associated with students learning to be investigated in the background questionnaires, the assessment design, and guidelines for item development.

As described in the *PIRLS 2006 Assessment Framework and Specifications*, purposes for reading and processes of comprehension are the foundation of the PIRLS assessment of reading comprehension (Mullis, Kennedy, Martin, & Sainsbury, 2006). The two purposes for reading are (1) for literary experience and (2) to acquire and use information. The four processes of comprehension are: (1) focus on and retrieve explicitly stated information, (2) make straightforward inferences, (3) interpret and integrate ideas and information, and (4) examine and evaluate content, language, and textual elements. The four processes are assessed within each of the two major purposes for reading.

The mathematics and science assessments are framed by two organizing dimensions, a content dimension and a cognitive dimension, with each dimension having several domains (Mullis et al., 2005). The mathematics content domains are: Grade 4—number, geometric shapes and measures, and data display; Grade 8—number, algebra, geometry, and data and chance. The science content domains are: Grade 4—life science, physical science, and earth science; Grade 8—biology, chemistry, physics, and earth science. For each grade, each content domain contains topic areas, with each topic area elaborated with a list of assessment objectives. For both mathematics and science, and for both grades 4 and 8, the cognitive dimensions are: knowing, applying, and reasoning.

To monitor trends in achievement, about one-third to one-half of the test items and the bulk of the questionnaire items are carried forward from previous assessments and re-administered. However, each new assessment also involves creating new items and instruments. Developing the instruments for the TIMSS and PIRLS assessments is a cooperative venture, involving the NRCs from the participating countries throughout the entire process.

To develop the test items, the TIMSS & PIRLS International Study Center conducts an item writing workshop for the NRCs and their colleagues. Participants are assigned specific item writing tasks to meet the framework specifications. The items drafted at the workshop are reviewed extensively by content area and measurement specialists, and produced in booklets for the field test, with extensive translation and layout verification along the way. Participating countries field test the items with representative samples of students. The field test results, consisting of a variety of item statistics (primarily, percentages and biserial correlations), are widely reviewed by the international team and the countries. Finally, both the expert committees and the NRCs review the field test results and select the items for the assessment.

The background questionnaires do not undergo much change from assessment to assessment, because stability enables tracking changes in educational achievement in relation to contextual factors. However, at the beginning of each assessment cycle, just as there is a review of the content area specifications, the participating countries review the frameworks underlying the background questionnaires. Both TIMSS and PIRLS examine a range of factors, including curricular goals, educational resources, the teaching force and how it is educated, classroom activities, and home support for learning. Once they have reviewed and revised the framework at the beginning of the cycle, the international team works with the NRCs and the Questionnaire Development Committee to review the existing questionnaires and update them for field testing. After the field test, the results are reviewed and the committee and NRCs select the final questions to be included in the assessment.

International versions of the instruments are prepared in English prior to undergoing the translation and adaptation process. The translations of the newly developed items are checked twice, before the field test and before the assessment data collection, and trend items are carried forward from previous assessments using previously translated versions. Although most countries administer the assessment in just one language, there are a substantial number of countries that administer the assessments in two (or even more) languages. The translation effort includes: (1) explaining the explicit guidelines for translation and cultural adaptation, (2) the countries translating the instruments in accordance with the guidelines, (3) verification of translation quality using professional translators from an independent translation company, (4) corrections by the countries in accordance with the suggestions made, (5) verification that the corrections made and the layout of the instruments corresponds to the international version, and (6) a series of statistical checks after the testing to detect items that did not perform comparably across countries.

29.5 ASSESSMENT DESIGNS

Given the broad coverage required by the PIRLS and TIMSS frameworks and specifications, the accompanying test instruments require extensive testing time (about 6–8 hours to administer the assessment per student). Therefore, both studies use matrix-sampling together with a rotated-block design for the booklets.

PIRLS 2006 was based on 10 text passages, 5 for the literary purpose and 5 for the informational purpose. Four of the passages and accompanying items sets (2 literary and 2 informational) were carried forward from PIRLS 2001. The passages were "authentic" texts of approximately 800 words taken from the types of materials students are likely to experience in and outside of school, and included a special PIRLS reader printed in color. Half the items were in the multiple-choice format and half in the constructed-response format, totaling 126 items (167 score points, considering that some of the constructed-response items are worth 2 or 3 points and scored for partial or full credit). The 10 passages were distributed across 13 booklets, 2 per booklet, so that passages were paired together in as many different ways as possible. Thus, each student was administered one randomly chosen booklet consisting of two 40-minute blocks of passages and items. Booklet 13, the PIRLS reader, was presented in a colorized, magazine-type format to provide at least some passages in a more natural, authentic setting.

In total TIMSS 2007 included 777 mathematics and science items with at least half of the total number of score points coming from constructed-response items. The items were grouped into a series of item blocks, with approximately 12 items per block at fourth grade and 18 per block at eighth grade. Apart from the number of items per block and the amount of time expected to be spent on each block, the booklet design was the same for both grades. There were 28 blocks in total, 14 containing mathematics items and 14 containing science items. Half of the item blocks (7 in mathematics and 7 in science) were newly developed for TIMSS 2007 and half kept secure from 1999 or 2003 to be used in measuring trends in 2007. The 28 item blocks were distributed across 14 booklets, with 4 blocks of items in each booklet, 2 blocks of mathematics items and 2 blocks of science items. In half of the booklets, the 2 mathematics blocks came first, followed by the 2 science blocks, and in the other half, the order was reversed—science first, followed by mathematics. Additionally, 2 blocks in each booklet contained trend items and 2 contained newly developed items. Each student completed 1 booklet, assigned at random. Fourth-grade students were allowed 72 minutes to complete their booklet, and eighth-grade students, 90 minutes.

29.6 POPULATION SAMPLING

Both TIMSS and PIRLS have a target population of students enrolled in the fourth grade of formal schooling, counting from the first year of primary school as defined by the International Standard Classification for Education (UNESCO, 1999). According to the classification, Level 1 corresponds to primary education or the first stage of basic education, and the first year of Level 1 should mark the beginning of formal instruction in reading, writing, and mathematics. Accordingly, the fourth year of Level 1 is fourth grade in most countries. To avoid testing very young children, however, PIRLS has a policy that the average age of children in the grade tested should not be below 9.5 years old. Additionally, TIMSS has the second target population of students enrolled in the eighth grade of formal schooling, with the definition corresponding to that used for the fourth grade.

The assessments are administered to carefully drawn random samples of students from the target population in each country (Martin, Mullis, & Chrostowski, 2004a; Martin, Mullis, & Kennedy, 2007). The first stage is sampling schools with probability proportional to size, and the second stage is sampling intact classrooms with equal probabilities from the target grade in the sampled schools. Considering the desired population, within each country all schools with fourth- or eighth-grade students are eligible, including all school types (e.g., public, private, or vocational). However, countries can define a population that excludes a small percentage (less than 5%) of certain kinds of schools or students that would be very difficult or resource intensive to test (e.g., schools for students with special needs or schools that are very small or located in remote rural areas). The sampling approach has been established to achieve precision within plus or minus 5% for percentages, and plus or minus .1 SD for achievement means. Most countries sample 150 schools and 1 or 2 intact classrooms from each. This approach is designed to yield a representative sample of at least 4,000 students in each country.

Staff members from Statistics Canada and the IEA Data Processing and Research Center work with the NRCs on all phases of sampling, including training in how to use the within-school sampling software. Each step is thoroughly documented and the sampling documentation is used by the TIMSS & PIRLS International Study Center (in consultation with Statistics Canada and an independent sampling referee) to evaluate the quality of the samples. Upon acceptance of the sampling procedures, the data are weighted in accordance with the sampling design and the jackknife procedure is used to estimate the standard error associated with each statistic published in the international reports.

Most countries achieve the minimally acceptable participation rates—85% of both the schools and students, or a combined rate of 75% (the product of school and student participation). However, countries that meet the guidelines only after including replacement schools, and countries that fall just below the 75% (70–74%) are annotated. For further departures from the guidelines, data are segregated in the report (e.g., participation rates 50–70%) or not included at all (e.g., where random sampling approaches were not used).

29.7 DATA COLLECTION PROCEDURES

Each country is responsible for carrying out all aspects of its data collection, using standardized procedures developed for the assessments. Training manuals are created for school coordinators and test administrators that explain procedures for receipt and distribution of materials, as well as for the activities related to the testing sessions. These manuals covered procedures for test security, standardized scripts to regulate directions and timing, rules for answering students' questions, and steps to ensure that identification numbers on the test booklets and questionnaires corresponded to the information on the forms used to track students.

Each country is responsible for conducting quality control procedures and describing the effort in an online Survey Activities Report. In addition, the TIMSS & PIRLS International Study Center considers it essential to independently monitor compliance with standardized procedures. The IEA Secretariat obtains nominations from the countries for quality control monitors unconnected with the national testing centers to conduct interviews and observe the testing sessions in 15 schools in each country (typically, a 10% sample). The TIMSS & PIRLS International Study Center has developed manuals for the quality control monitors, and the monitors are briefed about their responsibilities in a two-day training session held at the IEA Secretariat. Based on the monitors' observations, there is evidence that the TIMSS and PIRLS data are collected in compliance with international procedures.

29.8 SCORING AND DATA ENTRY

TIMSS and PIRLS have elaborate procedures for reliably evaluating students' responses to constructed-response questions. The scoring guides, along with training packets containing extensive examples of student responses for practice in applying the guides, are used as a basis for intensive training in scoring the constructed-response items for both the field tests and the assessments. The training sessions are designed to train representatives from the participating countries, who then are responsible for training personnel in their own countries to apply the scoring guides reliably.

TIMSS and PIRLS gather and document three different types of empirical information about agreement among scorers, within-country, across countries, and across assessment cycles. Items with less than 70% agreement are not included in achievement scales, although this is rare (typically 1 or 2 items at most for an assessment). To collect data about within-country reliability, TIMSS and PIRLS arrange for randomly selected subsamples of at least 200 responses to each item to be scored independently by two readers. The percentage of exact agreement, on average, is typically higher than 90%. To monitor consistency across countries, TIMSS and PIRLS collect a sample of responses from each of the countries that administer the assessment in English, and devise a set of responses for 20–30 questions, that are digitally scanned and incorporated into custom-built presentation software. The responses are scored by at least 2 scorers from each country, and the percentage of exact agreement, averaged across the countries and the items, is generally higher than 85%. To ensure that responses to trend items are scored the same way from assessment to assessment, countries send a random sample of scored student booklets to the IEA Data Processing and Research Center where they are digitally scanned and incorporated into custom-built presentation software for rescoring in the next assessment cycle. Again, the agreement between assessments is generally high—90% exact agreement, on average, across countries.

29.9 CONSTRUCTING THE INTERNATIONAL DATABASE

To ensure accuracy and consistency across countries in the international database, the international team prepares manuals, variable codebooks, and data entry software for countries to use in creating and checking their data files before forwarding them to IEA's Data Processing and Research Center. At the data processing center, the data undergo an exhaustive quality-control process involving an iterative procedure of checking, editing, and rechecking. Participating countries are given multiple opportunities to review their data, and the international team reviews item statistics for the achievement item to identify poorly performing items. In general, the items exhibit very good psychometric properties. However, sometimes an item needs to be deleted from the achievement scaling in a particular country, usually because of a problem with translation.

29.10 IRT SCALING PROCEDURES

The primary approach to reporting achievement data is based on item response theory (IRT) scaling methods. Because IRT scaling is capable of estimating a student's score, even if that student has not responded to all of the items in the assessment pool, it is particularly appropriate for TIMSS and PIRLS where each student completes only one of many possible test booklets. The IRT scaling method produces a score by averaging the responses of each student to the items that he or she took in a way that takes into account the difficulty and discriminating power of each item. Achievement in reading, mathematics, or science, is summarized using 2- and 3-parameter IRT models for dichotomously scored items, and generalized partial credit models for items with two or more available score points.

Because individual students respond to relatively few items, particularly for the subsets of items assessing various content (e.g., algebra, literary reading) and process areas (e.g., knowing, reasoning), TIMSS and PIRLS use a process known as "conditioning" to improve the reliability of the group achievement measurement. The conditioning process combines students' responses to the items with information about the students' background characteristics. A distribution of achievement is constructed for each student conditional on the student's responses to the administered items and on the student's background characteristics. To provide student scores that may be used in analysis, TIMSS and PIRLS use this achievement distribution to predict or impute the achievement of each student.

These imputed scores, or "plausible values," are used as scale scores in analyses to create the exhibits in the reports, and are provided to countries and researchers throughout the world via the TIMSS and PIRLS databases accessible on the Web. To quantify any error in the imputation process, five plausible values are generated for each student and the TIMSS & PIRLS International Study Center conducts all analyses five times for any required statistic. The average of the results of the five analyses is taken as the best estimate of the statistic in question, and the differences between the five estimates of the statistic reflect the imputation error.

The IRT analysis provides a common scale on which performance can be compared across countries, and provides a basis for estimating mean achievement and percentiles as well as how students within countries vary. The achievement scales for reading, mathematics, and science were set in the initial assessments by weighting all participating countries equally, and setting the average across those countries at 500, and the standard deviation to 100. To preserve the metric from cycle to cycle, the scaling for the current assessment is based on students in all countries that participated in the current assessment and the preceding assessment. The re-administered items form the foundation for linking the two sets of assessment data, but all items are used in the scaling. After the item parameters have been reviewed, scores are computed for students who participated only in the current assessment. As well as the scales for achievement overall for the fourth grade in reading, and the fourth and eighth grades in mathematics and science, the TIMSS & PIRLS International Study Center produces scales for each of the content and process areas assessed at each grade (see previous description of frameworks).

29.11 REPORTING STUDENT ACHIEVEMENT AS INTERNATIONAL BENCHMARKS

To interpret trends in performance on the mathematics, science, and reading achievement scales in terms of students' knowledge, skills, and understandings, the TIMSS & PIRLS International Study Center uses four points on the achievement scales as international benchmarks. Selected to represent the range of performance shown by students internationally, the advanced benchmark is 625, the high benchmark is 550, the intermediate benchmark is 475, and the low benchmark is 400.

For comparability, the benchmark points are kept constant from assessment to assessment. However, after each assessment the TIMSS & PIRLS International Study Center conducts a scale anchoring analysis to update the description of achievement of students at those four points on the respective scales. The descriptions are in terms of the competencies demonstrated by students reaching each successively higher benchmark. The international reports contain the updated summary descriptions of performance at each of the benchmarks, accompanied by changes in the percentages of students in each country reaching the benchmarks (Martin, Mullis, Gonzalez, & Chrostowski, 2004b; Mullis, Martin, Gonzalez, & Chrostowski, 2004; Mullis, Martin, Kennedy, & Foy, 2007).

29.12 SUMMARY OF PROCEDURES FOR ENHANCING COMPARABILITY

TIMSS and PIRLS expend enormous effort to ensure the reliability and validity of the data. Test reliability coefficients (median KR-20 reliability across test booklets) are generally high, 0.9 or higher, on average. Assessment validity is addressed through widespread collaboration among participating countries in developing the assessment frameworks, and describing the assessment objectives in detail for each subject and grade. Also, items are reviewed by participating countries to ensure they are measuring objectives in the frameworks, and are appropriate for students in the countries.

TIMSS and PIRLS have the additional goal of ensuring that the data are internationally comparable. That is, TIMSS and PIRLS also must have comparative validity. The methods used to address comparative validity have been described throughout this chapter. For ease of reference, they are enumerated here.

Are the target populations comparable? TIMSS and PIRLS have detailed definitions of the target populations and comprehensive documentation about whether or not the definitions are met. There are consequences in data reporting for countries that do not meet the definitions.

Is the sampling conducted properly? Within the overall requirements of random sampling to achieve precision goals, TIMSS and PIRLS have expert sampling methodologists working with each country individually to ensure the sampling procedures are implemented correctly. There are consequences in data reporting for countries that do not implement sampling procedures correctly and do not achieve acceptable participation rates for schools and students.

Are translations comparable? The TIMSS & PIRLS International Study Center provides guidelines and instructions for translation of the instruments into the language(s) of instruction. The IEA Secretariat verifies each translation and refers issues to the countries for resolution. The TIMSS & PIRLS International Study Center verifies the final layout of the booklets prior to printing.

Are the tests administered appropriately? Countries are expected to follow the procedures thoroughly detailed in manuals developed by the TIMSS & PIRLS International Study Center. Adherence to the procedures is documented through a program of international quality control monitoring, and within-country quality control monitoring. The results from the international quality control monitoring conducted by the TIMSS & PIRLS International Study Center together with the IEA Secretariat are reported in the technical reports produced for each assessment.

Is the scoring of constructed-response scoring done correctly? A training set is prepared for each constructed-response item, including the scoring guide, explanations and example responses illustrating the scoring guide, and practice papers. Training is conducted for the field test and for the assessment data collection. Reliability data are collected for within-country scoring, across-country scoring, and scoring across assessment cycles. Items with below 70% exact agreement are not included in the scaling (usually at most 1 or 2 items for an entire assessment).

Are the data comparable? To standardize data preparation, countries are provided with manuals, variable codebooks, and data entry software. IEA's Data Processing and Research Center checks each country's data files for internal consistency and accuracy and interacts with countries to resolve data issues. The TIMSS & PIRLS International Study Center reviews achievement item statistics for every country and investigates any items with poor discrimination (typically a translation problem). Such items are not included in the scaling. This is relatively rare given broad field testing with representative samples in every participating country.

In summary, every effort is made to attend to the quality and comparability of the TIMSS and PIRLS data through careful planning and documentation, cooperation among participating countries, standardized procedures, and rigorous attention to quality control throughout.

30

Enhancing Quality and Comparability in the Comparative Study of Electoral Systems (CSES)

David A. Howell

30.1 INTRODUCTION

The Comparative Study of Electoral Systems (CSES, www.cses.org) was first conceived in the early 1990s, but given form in 1994 by a stimulus paper authored by American and European elections scholars John Curtice, Hans-Dieter Klingemann, Steven Rosenstone, and Jacques Thomassen (see Thomassen et al., 1994). Originally envisioned as an outgrowth of an organization named the International Committee for Research into Elections and Representative Democracy (ICORE), the CSES grew and eventually replaced ICORE. The project serves not only as an enabler of comparative, cross-national elections research, but as a network for scholars and practitioners interested in the interplay between institutional variation and citizen behavior, especially as applies to voting turnout and candidate/party preferences. While ICORE was comprised of directors of election surveys primarily from Europe, today the CSES project has grown to include collaborators from over 60 countries dispersed throughout the world. Geographical coverage of the CSES is indicated in Figure 30.1. The CSES is not designed to be a random sample of countries, but a selection that provides enough institutional variation to investigate the impact of different electoral arrangements on individual behavior. CSES data are a public good, distributed at no cost and without embargo, via the project website (http://www.cses.org/). The Planning Committee and collaborators receive no advance access—they must download the data from the website just like everyone else.

The CSES is not a centralized effort of a particular organization, country, region, or continent but a collective effort of the many scholars that participate in the project. A CSES Module 1 Planning Committee was convened in 1995 to develop and oversee administration of the first research module and associated materials, as well as recruit national election studies into the project. As of this

[1] Survey Methods in Multinational, Multiregional, and Multicultural Contexts, edited by Harkness et al.

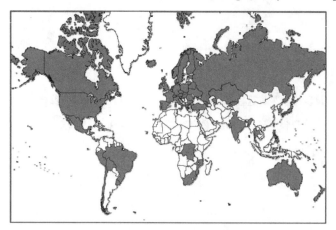

Figure 30.1. CSES Coverage (gray indicates countries and administrative regions in which CSES has collaborators, as of this writing)

writing, the CSES Module 3 Planning Committee is the active one, having grown from 12 scholars to 20, with broad representation from a variety of nations and cultures. The CSES has a Secretariat that is a joint effort of the University of Michigan and the GESIS Leibniz Institute for Social Science (GESIS), with financial support provided by the American National Science Foundation, GESIS, and the University of Michigan. Datasets for the first and second modules have been downloaded thousands of times from over 140 countries around the world. The CSES Bibliography currently contains over 400 citations. Hundreds of scholars have participated in the CSES collaborator network, and CSES has been used in some countries as a basis for launching new election studies. Planning is underway for a fourth module.

Ensuring quality and comparability is critical to the success of any comparative research project, and the CSES is no exception. Enhancing comparability across surveys and producing high quality data are the primary foci of the organizations that serve the CSES collective, and receive attention throughout every stage of the CSES project.

30.2 STUDY DESIGN AND PLANNING

The quality and comparability of any cross-national study begins in its theoretical and methodological design. The founders of CSES involved a broad set of collaborators from the very beginning at the first planning meeting in 1994 at the Wissenschaftszentrum Berlin für Sozialforschung, at which the basic principles and direction of the project were decided upon. This process benefitted from a careful construction of the membership and examination of efforts by comparative studies that preceded it. The initial planning committee included members with experience in comparative surveys such as the European Election Studies (EES), the International Social Survey Programme (ISSP), the Latinobarómetro, and the World Values Survey (WVS) and drew on their experiences to enhance comparability. Most members had previously managed a national election survey and were able to contribute to the development of data collection and quality guidelines. Data release policies were influenced by members from archives such as the Zentralarchiv für Empirische Sozialforschung (ZA; now part of GESIS) in Germany and the Inter-University Consortium for Political and Social Research (ICPSR) in the United States.

The procedures and guidelines set in place by the initial planning committee are nearly all still in use today, enhancing comparability not only within modules, but across them.

The project also puts significant effort into communicating with and learning from its peers. Just as the CSES is a network of election study scholars and practitioners, so is the CSES an involved member of an informal network of international comparative studies, maintaining connections with and staying informed about the best practices of other comparative studies worldwide.

30.3 SELECTION OF COLLABORATORS AND COUNTRIES

Collaborators participating in the CSES project must raise their own local funding and mount their own national data collection; centralized funds are not available for this purpose. Given this decentralization, an important step in ensuring data quality is the careful and conscientious selection of collaborators. No matter how much effort is put into developing quality materials, in the end it is in the hands of the project collaborators to collect high quality datasets that are appropriate for comparative analysis. For this reason, as much attention must be given to developing the collaborative network and making participation rewarding for election study teams as is given to the development of the scientific instruments themselves. Maintaining the investment and interest of collaborators and having them feel involved and responsible in the success of the project is critical. A collaborator knowing that they are receiving value in return pays dividends in adherence to standards, quality, and comparability.

CSES collaborators do not formally apply for membership, but are identified and recruited from among their country's foremost social scientists. They often have been recommended by other collaborators from the existing CSES network. To be selected, collaborators must have strong ties with, and interest in, academic research, and be capable of running a high quality national survey. As other comparative studies use similar criteria, collaborators on the CSES project are sometimes also collaborators on other comparative studies.

Also at issue in terms of data quality and comparability is the selection of countries or administrative regions to include in the CSES project. Elections sometimes occur in countries where the regime in place, and sometimes the elections themselves, are not free or fair. In terms of data quality, a concern is that the respondents in these regions do not have adequate freedom to answer the questions according to their actual opinion and that data quality might be negatively impacted as a result. The CSES Planning Committee has decided not to devote resources to specifically recruit collaborators from such regions, but will accept their data deposit if the studies meet all other criteria for inclusion. While the region may be at an early stage of democratic development now, it may later develop to be free and fair, at which time the early data could be valuable as a benchmark measurement against which to compare future studies from the region. As of Module 3, the CSES codebook contains a freedom status rating for each country that is included. This allows analysts to decide on a case-by-case basis whether each election study is appropriate for their particular analyses, rather than have the CSES project decide for them, as long as the election study meets the remaining criteria for inclusion.

30.4 QUESTIONNAIRE DESIGN

At the core of each CSES data collection is a substantive module, a common set of survey questions administered at least once each 5-year period to a national sample of respondents in all participating election studies. The set of questions that comprise a module are designed to require 10–15 minutes of interviewing time per respondent, depending on qualities of the electoral system such as the number of parties competing. Toward the end of each 5-year period, a new module is developed, with a new substantive theme. While initially each substantive theme was expected to be independent of those that preceded it, some fundamental questions became common to all modules, facilitating the use of data from different modules together, and allowing for the possibility of time series analysis.

One of the greatest benefits to comparability is the multiple years of effort that go into a CSES module before any data are collected. The first step in the preparation for a module is the election of a Planning Committee, selected from the full set of collaborators to be diverse geographically, intellectually, institutionally, and culturally. The Planning Committee spends considerable time soliciting and debating potential themes for the new module. Once some consensus has been reached, the theme is detailed in a stimulus paper and defended to an assembly of the collaborating election study teams. Upon approval of the collaborators, a questionnaire is drafted and undergoes multiple iterations of revision by the Planning Committee. The questions are typically borrowed from previous national studies, rather than developed anew. This ensures that wording of the questions and their ability to measure the appropriate concepts have already been validated, at least in a national context. The proposed questions are then reviewed by the Planning Committee and eventually the larger collaborator base, to evaluate their cross-cultural applicability. Some questions are rejected or revised at this stage because while they may have been successfully administered in some nations, they were found to be not adequately applicable cross-nationally and cross-culturally.

Once the draft questionnaire has passed this review process, the nearly finalized module is administered in a pretest using full, random national samples in at least three different nations. Data and reports from the pretests are made available to the Planning Committee, which revises the questionnaire and refines the questions accordingly. The questionnaire is then submitted to the collaborator base at an in-person plenary session for discussion and final approval.

Upon finalization of the questionnaire, collaborators are requested to ask the full set of questions exactly as they are specified, even if the collaborator has concerns as to their local applicability. This requirement ensures that the resulting variables will be available to analysts for all participating countries, and are collected in a comparable way. For a very few question batteries such as political ideology (left-right placement), collaborators are allowed an optional, additional question set to capture a comparable measure that is more relevant, or at least better understood, locally. This optional question set supplements, but does not replace, the original question set. Sometimes the additional question sets are valuable alternatives. For instance, in some countries in East and Southeast Asia, left-right placement is not always the most salient political spectrum.

When collaborators from national election studies deposit their survey data with the project, they include a number of demographic variables which have been allowed to be asked according to local standards. The project requests that, wherever possible, collaborators collect the variables in a way that allows later harmonization with a common, project-wide coding scheme that was established in advance. This common coding scheme was based in large part on standards already in place in other comparative surveys (see CSES Planning Committee, 1996). This flexibility on the part of CSES enables local election studies to ask demographic and background questions in different ways for legitimate reasons. For example, some participating studies have asked their demographic and voting questions in the same way for decades, and this policy allows their national time series to remain unbroken. The CSES plan for demographic variables allows local variations and cross-national comparability to coexist. The exception is when election studies ask demographic questions in a way that cannot be wedged into the CSES coding convention and thus are forced to be dropped.

If a national election study has too many questions missing (from both the core CSES module and the demographic section), or if the data have been collected in ways that do not allow for comparison cross-nationally, the election study is excluded from the final CSES data release. Absence of variables for particular countries reduces the overall utility of the dataset to analysts, and results in the dropping of those countries from analyses. The decision to exclude a national election study, for this reason or any other, is made on a case-by-case basis through coordination between the CSES Secretariat and Planning Committee, while in communication with the collaborators that collected the data.

30.5 DATA COLLECTION

Surveys appearing in CSES datasets are required to meet a set of "aspired to standards for data quality and comparability" as defined by the original Planning Committee (see CSES Planning Committee, 1996). The CSES Secretariat and Planning Committee members are available to collaborators both prior to and during their data collections to answer questions about these guidelines, and to provide advice as to study design and methodology. This is especially true for newly established national election studies. After the data collections are complete, collaborators fill out and deposit with the project a standardized report that details the methodology of their sample and data collection. This report is the basis of the project's assessment of the quality of each data collection effort.

The methodological report was designed to ask enough relevant, pointed questions to gain an understanding of the implementation of the included election studies, allow an accurate assessment of the quality of the data collection, and judge whether the data collection occurred in a way that allows comparison to other national studies. The project makes such a judgment about the quality of each study prior to its inclusion, and the original reports are also made publicly available on the CSES website so that individual analysts can draw their own conclusions.

Only post-election studies are accepted, without exception. In multiwave designs, it is allowable for some demographic variables to be collected prior to the election. But in order to be able to examine the relationship of behavior to actual vote (which can only be known after the election), questions from the core module must all be asked after the election.

Questions are requested to be asked in a single, uninterrupted block. In CSES Module 1, 24 of 34 election studies met this criterion (see Howell & Long Jusko, 2009).

The preferred mode of interview is face-to-face unless an alternate mode can be demonstrated to be superior. Face-to-face interviewing is considered to provide higher response quality than other modes such as telephone interviewing (see Holbrook, Green, & Krosnick, 2003). Face-to-face interviewing is also the preferred mode among CSES collaborators, with 70% of election studies being conducted face to face in CSES Modules 1 and 2. However, telephone, mail/self-completion, and mixed mode studies have also been accepted. While an Internet-based study has not yet appeared in a CSES dataset, the Planning Committee has determined it to be an acceptable mode of interviewing, if the study meets all other guidelines for inclusion.

Each study included in the CSES must be conducted as soon as possible after the election. In Module 1, 72% of data collections were completed within 3 months after the election. In Module 2, this number was 71%.

Among the most important methodological requirements of the CSES are that an election study's sample must be random at all stages of selection and have adequate coverage, so as to more likely be representative of the national population. Probability samples, including those using clustering or multistage random samples, are readily accepted. The use of nonprobability samples is disallowed in the CSES for the time being. Quota and replacement samples, despite having some basis in random selection, are discouraged but have been accepted on a limited basis, only when alternate sampling methods are unfeasible locally. Whatever the sampling method, the details are provided in full so that analysts can also make an independent judgment as to whether the election study is appropriate to include in their work.

Another requirement for inclusion in CSES is that the number of interviews must provide adequate statistical power to study the national population. A minimum of 1,000 age-eligible respondents per study is recommended. Module 1 and Module 2 had an average of 1,600 and 1,567 interviews, respectively, per election study.

CSES collaborators are also required to pay close attention to survey methods in general. They are requested to pretest their questionnaires. They are asked to use interviewers that are well-trained and familiar with the administration of the questionnaire, and to use modern field methods, including refusal conversion. Collaborators are asked to identify the number of contacts and methods of contact attempts, and to make every effort to achieve a high response rate.

Modern, sophisticated translation methods must be employed by each collaborator. The CSES encourages its collaborators to work together on translations and to share existing translations in common languages. While the project once relied heavily on back-translation to assess the quality of questionnaire translations, the project now focuses instead on assessment of the translation process itself, in accordance with modern questionnaire translation literature (see Harkness, Pennell, & Schoua-Glusberg, 2004). As of CSES Module 3, the translation section of the methodology report includes questions specifically drawn from the ISSP. Included in this section of the methodology report are questions as to who translated the questionnaire, whether it was checked or evaluated, whether the translation was pretested, and what problems were noticed when conducting the translation (see Scholz et al., 2008).

30.6 SURVEY DATA CLEANING

Collaborators are encouraged to clean their survey data to their national specifications prior to depositing the data with the CSES Secretariat. This cleaning according to national standards is a first defense against data quality problems. The CSES Secretariat, upon receiving the data, then cleans it anew. The Secretariat evaluates the data and documentation for completeness and makes note of missing information. It evaluates skip patterns both within the data deposit and against the data as originally collected, in order to identify errors and variations in the administration of the questionnaire. It recodes missing data—not ascertained, not answered, refused, don't know, etc.—to be consistent across studies. It compares the wordings in English back-translations against the original questionnaire. The project checks that identical code sets were used and that scales were administered in the correct direction. Outliers and unusual distributions are identified, noted, and reconciled. Weight variables are examined for unusual values and ranges, and used to evaluate the quality of the sample selection, as well as the impact of nonresponse. Weighted and unweighted distributions are compared to evaluate and document the extent to which the weights are correcting distributions of key dependent and independent variables. Demographic variable distributions are compared against national distributions to assess sample selection and nonresponse bias. Other variations, noncomparabilities, and concerns are also noted during this process.

30.7 MACRO DATA PREPARATION

A major strength of the CSES research design is that it allows analysts to examine individual behavior, as collected in respondent-level survey data, against macro characteristics of the electoral system and institutions within which individuals find themselves. The majority of macro data that appears in a CSES dataset is collected using a common macro report that each collaborator, or their expert designate, is responsible for completing. The report asks specific, detailed questions about the electoral system and the election itself, as of the time of the data collection. Other sections of the macro report are adapted to complement the specific themes of each module. The reporting tool is regularly revised to incorporate current best practices in the collection of macro-level data. One such example is the work of the International Institute for Democracy and Electoral Assistance (IDEA), which has developed a methodology for reporting turnout consistently across elections (see International IDEA, 2008). The components and definitions that IDEA has developed are now the basis for the turnout section of the CSES macro report.

The macro-level information, having been collected in a consistent way using the macro report tool, is then coded into the data by the Secretariat. Coding schemes are designed to group electoral systems similarly and maximize comparison across countries—keeping in mind that the most exact way to classify a system is not always that which maximizes the ability to compare it with other systems. This coding effort requires significant specialized knowledge about electoral systems, and so is performed by a political science graduate student or PhD with relevant substantive knowledge of electoral systems. The Secretariat independently verifies most of the information provided by the collaborator against publicly available sources such as official electoral commission websites and the Parline database on national parliaments (see Inter-Parliamentary Union, 2008). This serves as a quality control check and identifies alternate ways that the system might be classified. Where inconsistencies are found, and where multiple methods of classification are possible, the Secretariat works with the collaborator to decide on a classification that optimizes comparison with other countries in the dataset. After this quality control and review process, the macro report data are then supplemented with district-level results and additional macro-level data collected by the Secretariat from well-established, publicly available sources such as Adam Carr's Election Archive, Freedom House, and the World Bank. Once the macro data for an election study are finalized, the Secretariat merges them to the respondent survey data, enabling the micro-level to macro-level comparisons which make the CSES especially unique among its peers.

30.8 DATASET VETTING

After survey data cleaning and preparation of the district and macro data are complete, the Secretariat returns to the collaborator with a list of questions and clarifications, usually numbering 20 or more, which are required to be resolved. After the collaborator responds with advice, the Secretariat makes corrections as necessary to the individual election study. After all corrections are made, the data from the collaborator's own election study are returned to them for vetting. This process is a final review by the collaborator of the sum of the work of the Secretariat, and serves as a final check on data quality. When providing their final feedback to the Secretariat, collaborators also provide permission to include their data, as prepared and cleaned by the Secretariat, in an upcoming CSES data release.

30.9 CROSS-NATIONAL DATA CLEANING

Data cleaning and quality checks do not end with the vetting of the individual national files. After the national files are individually approved, they are merged together into a single file and undergo a new round of testing by the Secretariat, this time at the cross-national level. Question distributions are examined cross-nationally to identify outliers. For instance, a variable whose distribution is found to be in the opposite direction of comparable electoral systems could be either a real difference, or the result of scale endpoints being administered in the opposite direction. Macro variables are examined cross-nationally to ensure the systems have been categorized consistently across all election studies. Furthermore, when examining the macro variables cross-nationally, sometimes new classification or grouping possibilities become apparent that were not evident when coding the individual electoral systems originally. Last, the Secretariat uses the merged dataset to replicate well-established analytical models with known properties, to verify that the cross-national dataset performs similarly to these past, generally accepted outcomes. Should such models fail on the combined, cross-national dataset, the Secretariat investigates the potential causes. This sometimes uncovers issues in the dataset which can then be addressed, or at least well-documented, thus improving the quality and utility of the final data release.

30.10 DOCUMENTATION

In preparing its documentation, CSES subscribes to the same principles as the European Social Survey, which suggest that the imperfections of a study should not be hidden from, but highlighted for, the user community (see Jowell, Roberts, Fitzgerald, & Eva, 2007b). Within this philosophy, dataset users are empowered to consider such imperfections in their analyses and to adjust their interpretations and methods accordingly. CSES codebooks tend to be much longer than codebooks for other datasets with similar numbers of variables. This is because everything that the Secretariat notices throughout data processing which could have an impact on quality, comparability, or analytical outcomes is meticulously detailed in the codebook on an item-by-item basis. The project has received feedback that while the length of CSES codebooks are initially intimidating, ultimately analysts are appreciative of the detail, because nearly every unusual characteristic that they notice in the dataset has an associated note in the codebook. These notes document for each variable, election study by election study: variation in the administration or formulation of the question, coding issues (such as reversed scales, or scales of different formats), skip pattern problems, instances where local culture impacted question interpretation, and other miscellaneous issues. The macro data section of the codebook has notes on contextual data and justifications for the classification of the election studies in question.

In addition to a detailed codebook, the original methodology reports and macro reports deposited by the collaborator can be downloaded from the CSES website. The original language questionnaires, sometimes with a back-translation to English, are available for nearly every election study. The project's philosophy transparent and as complete as possible documentation is thought to ultimately enhance the credibility of the project, improve the quality of resulting analyses, and enhance the ability to make proper comparisons using the data.

30.11 POST-RELEASE

Improving data quality does not end with the release of a dataset. Although full releases of CSES modules are produced only every 5 years, on completion of data collection for module, advance releases are produced annually. These contain whatever studies have been deposited, processed, cleaned, and vetted prior to the advance release deadline. Advance releases provide data more quickly and serve to solicit feedback from the research community. Analysts regularly ask questions about what they find in advance releases, and their questions help us improve documentation, correct errors, and generally enhance the quality and comparability of subsequent data releases. Any errors discovered between data releases are posted as errata on the CSES website, allowing users to make corrections to their own datasets. The Secretariat also hosts conferences and reviews publications and presentations using CSES data. Keeping in close touch with scholarly uses of project data is an opportunity to gather further information that can improve the quality and comparability of future data releases and can results in refinement of methods used in future modules. Before the full release of a module, the election studies have been reviewed by the Secretariat individually and cross-nationally, as well as by a multitude of scholars in the course of their using of the data, at no incremental cost to the project.

30.12 IN CLOSING

Attention to quality and comparability is maintained at every stage of the CSES project: study design and planning, selection of collaborators and countries, questionnaire design, data collection, survey data cleaning, macro data preparation, dataset vetting, cross-national data cleaning, documentation, and post-release. This attention to quality and comparability is of critical importance so that analysts can draw proper inferences from the data and be positioned to achieve the ultimate goal of the project, which is to advance science.

31

The Gallup World Poll

Robert D. Tortora, Rajesh Srinivasan, and Neli Esipova

31.1 INTRODUCTION

Gallup began conducting its Gallup World Poll (GWP) in 2005 with the intention of regularly surveying "every" country around the world. The survey's main objectives are to (1) quantify the current state of well-being of those living in each country and (2) to collect additional data of importance in each of six regions around the world. The driving design principle is to conduct nationally representative surveys in each country.

Between mid-2005 and early 2007, the first round of data collection for the GWP was completed in 130 countries and areas, representing more than 95% of the world's population age 15 and older. The GWP is an ongoing survey funded entirely by Gallup. Only selected data from the Gallup World Poll are made available to the general public. Gallup does sell the GWP data and related consulting services to clients.

Not all of the approximately 200 countries in the world can be included in the survey. Countries are excluded from the GWP if (1) the country has a small population, (2) a national government does not allow Gallup to conduct a survey, or (3) if Gallup determines conditions in the country are too dangerous to safely conduct interviewing.

Gallup has offices in 29 countries. Thus, the company uses a combination of Gallup interviewers and Gallup-trained interviewers to collect GWP data. The latter are identified through local partners whom Gallup hires to conduct the survey. Gallup requires partners to meet worldwide specifications in regard to the sampling frame, sampling, translations, and quality control. In addition, Gallup physically participates in and monitors interviewer training in each country.

Two survey design methods are employed: a Random-Digit-Dial (RDD) telephone survey and a face-to-face personal interview survey. The former design is used in countries or regions where an adequate telephone frame exists, such as

[1] Survey Methods in Multinational, Multiregional, and Multicultural Contexts, edited by Harkness et al.
Copyright © 2010 John Wiley & Sons, Inc.

the United States and Canada, western Europe, Japan, Australia, and New Zealand. Gallup defines an adequate telephone frame as countries or regions in which at least 80% of the population has telephones. Each survey typically consists of 1,000 interviews, although in some cases Gallup uses an oversample for a main city. In some countries, an oversample is used in particular areas of interest. For example, in Russia, Gallup oversampled in Tatarstan and Dagestan, areas with predominantly Muslim populations. In India and China, urban populations were oversampled at two times the normal representation. Of course, because the survey's key goal is to have nationally representative samples, residents in urban and rural areas have a non-zero probability of being selected in each country's survey.

31.2 QUESTIONNAIRE CONTENT

The questionnaire generally consists of two parts, the core questions and a set of regional questions. In specific countries, Gallup also asks country-specific questions on topics such as consumerism in India and China.

While developing the questionnaires, Gallup wanted to minimize the difficulty of obtaining accurate translations, so it minimized the use of Likert scales. As a result, almost all GWP questions have dichotomous response categories such as Yes/No, Agree/Disagree, Satisfied/Dissatisfied, and Getting Better/Getting Worse. One exception to this approach is in the use of the Self-Anchoring Striving Scale (Cantril, 1965). These questions, which measure life satisfaction, ask respondents to think of a ladder with the steps numbered from 0 to 10, where "0" represents the worst possible life and "10" the best possible life. Respondents are asked to say which step of the ladder they presently stand on, with follow-up questions about where they stood five years ago and on which step they expect to stand in five years. Respondents are also asked the same three questions about their countries.

Gallup prefers to conduct data collection without interaction with national governments. However, in some countries government officials have to review the questionnaire. And in some of these countries, officials delete some questions. Countries where officials have deleted questions include China, Laos, Saudi Arabia, and Mauritania.

In some countries, we exclude some sensitive questions in order to avoid contacting the government and making the respondents and interviewers uncomfortable talking about topics that they are not accustomed to discussing. In Kazakhstan, Tajikistan, Pakistan, and Bangladesh, questions about homosexuals are not asked.

Only the core questions are asked in the telephone surveys. The core questions provide the basis for several indexes including Well-Being, Law and Order, Food and Shelter, Work, Economics, and Health (Gallup, 2007b). The core takes approximately 30 minutes to administer. The use of showcards is minimized because the core is also administered over the telephone. The only exception is for questions using the striving scale, where across a vast majority of personal interview countries in Asia, the GWP interviewer shows a picture of the ladder with the end points labeled. Showcards are used, when necessary, for face-to-face interviews for the regional questions.

For the sake of simplifying questionnaire content, the GWP is divided into several regions including the Americas, Europe, which includes the Balkans, the former Soviet Union, Latin America, predominantly Muslim countries, and sub-Saharan Africa. Surveys in countries where face-to-face data collection is used have tailored sets of questions that follow the core questions. These questions are about a variety of topics that are not mutually exclusive among regions. For example, topics asked about in the former Soviet Union include migration, religion, life before and after the Soviet Union, tolerance of people from different nationalities, and attitudes to a free-market economy. In the Balkans, topics include pre- and post-war comparisons on healthcare, standard of living, housing, transportation, and education. In Latin America, topics include migration, entrepreneurship and country friendliness toward business, natural resource use/misuse, gender roles, and corporate reputation. The regional questions asked in predominantly Muslim countries include obtaining opinions on attributes associated with Muslim societies, women's status and legal rights, Sharia, and perceptions of the West. In sub-Saharan Africa, topics include Millennium Development Goals, housing tenure, use of malaria-preventing bed nets, entrepreneurship, country friendliness toward business, natural resource use/misuse, education, HIV/AIDS, access to safe drinking water, and use of sanitation facilities. In the case of Asia, topics included general consumer attitudes, purchasing behavior, ownership of durables, entrepreneurship, and attitudes toward relationships with neighboring countries.

Once the questionnaire content is finalized, it is translated into the major languages. Gallup's ideal process includes two independent translations, independent back translations, and adjudication by a survey methodologist. In some cases, only one translation and an independent back translation are obtained.

The process of translation needs to be done carefully because some words and concepts are difficult to translate. For example, Gallup asks respondents questions about how much they experienced emotions such as enjoyment, depression, boredom, love, and worry the day before the survey. There are no exact translations for some of these emotions, and even if one exists, it may be a lengthy sentence substituting for the single word in English. In parts of Asia, the phrase "stress at work" is very commonly used to describe all kinds of situations ranging from hectic to hard work.

Cultural awareness must also be factored into the translation process. For example, in Senegal, one cannot ask how many children are in the household because many Senegalese believe that if they tell someone how many children they have, one of their children will die soon. So, in Senegal, rather than asking "how many children" one asks "how many little bits of God's wood" are in the household.

Different versions of Russian translations are used in the former Soviet Union (FSU) countries for some questions. For example, in order to convey the correct meaning of the word "identify," it had to be translated differently into Russian for Central Asian countries (Kazakhstan, Tajikistan, Kyrgyzstan, and Uzbekistan) from the translation used in the rest of FSU countries.

31.3 SAMPLING FRAMES

The GWP uses three different sampling frames: a frame of all telephone numbers, an area frame based on a country's census enumeration units, and an area frame using a non-census base. In the United States, Canada, western Europe, Japan, New Zealand, Hong Kong, Taiwan, South Korea and Australia, a sampling frame of telephone numbers is used for a Random-Digit-Dial (RDD) survey design. In the rest of the world, the GWP uses an area sampling frame. The area frame is almost always a frame from a country's central statistics office, typically, the frame from the latest population census. However, if the census is old and there have been many changes in the country, a new frame is constructed and used. Nigeria, Afghanistan, and Iraq are examples of countries where this type of area frame was used. In Iraq, for example, some census maps were available and these were combined with satellite imagery to construct the frame.

The survey design using the telephone frame is a typical RDD design using Cassady-Lepkowski list-assisted sampling. The area frame sample is selected as a stratified multistage random sample with proportional allocation of the sample to the following strata:

1. Cities with population of 1 million or more
2. Cities with population between 500,000 and 999,999
3. Cities with population between 100,000 and 499,999
4. Cities with population between 50,000 and 99,999
5. Towns with population between 10,000 and49,999
6. Towns/rural villages with population under 10,000

In smaller countries, one or more strata as defined above were non-existent. For example, in Kyrgyzstan and Rwanda, there are no cites with a population of 1 million or more. In larger countries such as India and China, we further expanded the stratification to include cities with 5 million or more.

Primary Sampling Units (PSU) are selected in each sampling stratum in proportion to the census population. Typically each PSU has a name that associates it with a geographic area in the country. To select the PSU within each stratum the PSUs are arranged in alpha order and a systematic sample of PSUs is selected. The sample consists of between 100 and 125 Primary Sampling Units. Depending on the number selected, in each PSU, 8 to 10 interviews are obtained for a total of 1,000 interviews per country. In some cases, an oversample is also selected to obtain more reliable estimates for a geographic area, say a country's capital city or urban area.

Two methods are used to select the adult (age 15 and older) to be interviewed. For telephone surveys, the GWP uses the latest birthday method and for face-to-face surveys, the latest birthday method or the Kish Grid is used for respondent selection.

31.4 COVERAGE

The GWP's driving design principle is to conduct nationally representative surveys in each country. In developing countries, an area frame is used, which generally provides complete coverage.

In countries with poor road and public transportation infrastructure, covering rural populations using face-to-face interviewing can be challenging and costly. The per-unit data collection costs in a geographically large country are typically higher for the rural areas. In sub-Saharan Africa, where more than 70% of the population lives in rural areas, the cost per interview in rural areas can be more than twice the cost of interviews in urban areas.

For various reasons, complete coverage is not always possible in some countries. For example, at the time of the GWP in 2006 and 2007, the Lord's Resistance Army was still active in northern Uganda and Gallup elected not to send any interviewers into this area. In Angola, areas that were known to still have landmines were not sampled. In a few countries, there are remote areas that are inaccessible to interviewers because they require a long trek by foot or a long journey by boat to interview there. In Indonesia, there was a massive earthquake in Java just before the fieldwork started and some parts of the province could not be easily accessed and had to be deleted from the sampling frame. In India, parts of the northeast and the islands of Andaman and Nicobar were not included in the frame. The loss of coverage from situations like these does not typically exceed 3–5% of the population.

A selected PSU that falls into either of these two categories (dangerous for interviewers or remote, difficult to reach and with very small populations) is replaced (by random selection) with another, more accessible, PSU. This situation occurred in 2006 in Madagascar, where two PSUs were replaced because the only way they could be reached was by boat. In the Philippines, PSUs were also replaced because of an ethnic conflict in the south, and in Vietnam because of resistance from local authorities.

31.5 FACE-TO-FACE DATA COLLECTION

The majority of the GWP data are collected through face-to-face interviews. For all face-to-face data collection, Gallup contracts with local vendors to provide local interviewers and collect the data according to Gallup specifications. This can be a challenging process that can potentially introduce slight differences into the data collection procedures. To minimize the chances of this happening, Gallup uses a set of standard specifications for the survey process that cover the sampling frame, stratification, field procedures, quality control, and data delivery. Five topics are included in the specification: the interviewer training method, household and respondent selection, the organization of the field process and interviewer teams, quality control, and external issues with obtaining permission to conduct interviews.

In most countries, a three- to four-day interviewer training session takes place. However, in the former Soviet Union one-day training sessions were conducted because the interviewers had considerable interviewing experience and were highly educated. In almost all Asian countries, Gallup dealt with existing research suppliers who had to be exposed to Gallup requirements and protocol.

Gallup participates in these interviewer training sessions. Depending on the size of the country, training/briefing is sometimes done in more than one location. The main topics of the interviewer training session, which is conducted in an interactive rather than lecture-style format, are household selection, respondent selection, and questionnaire content. Interviewers are provided with an interviewer brief that describes the survey and the processes. Most of the training focuses on household and respondent selection, as interviewers tend to pick up questionnaire content quickly.

Because there is no prelisting of households in selected PSUs, a random route procedure is used for household selection. The random route procedure works as follows: From a known starting point, interviewers, beginning with their back to the starting point, go to the right, and count three (or five in some countries) households; the third household counted is selected for an interview and called Main House No. 1. Unless refused, three attempts are made to obtain an interview. In remote rural areas, only two attempts are made. The second attempt is later the same day. Areas that are remote and have very poor accommodation facilities typically require interviewers to get in and out of a location in a day's time.

After a refusal or three unsuccessful attempts, the interviewer substitutes the first house on the right from the first main house and if unsuccessful there, substitutes the first house to the left of Main House No. 1. Left and right directions are always determined with the interviewer's back to the main door. In some countries, Gallup varies the number of households counted depending on whether the interview is in an urban or rural setting, higher in an urban setting and lower in a rural setting.

For each interview, the interviewer completes a Sample Management Sheet. Besides being designed to facilitate household selection, the sheet features a standard introduction that the interviewer has the option of using and the Kish Grid. The Sample Management Sheet assists with household selection because it is designed to show the interviewer whether he or she is attempting an interview at a Main House or a substitute house, the best time to attempt an interview if the selected respondent is not at home, and allows the interviewer to log why an attempted interview was not successful.

For interviewers, household selection is the most difficult part of the interview process to learn. An interviewer's inability to correctly execute household selection is almost always the reason why certain interviewers end up not being used. The second most common reason for not using interviewers is that they are unable to read at a high enough level. This typically happens in countries where there is little research going on. For example, in 2007 Gallup did not use some interviewers in the Central African Republic because they could not read Sangho well enough.

Once an interviewer arrives at a potential household, she or he must determine who the members of the household are. In sub-Saharan Africa, the people who usually eat from the same cooking pot define who is in the household. Polygamy, extended households, and heads of households that rotate among wives' housing units can complicate defining household memberships in certain cultures.

The most common error that interviewers make when filling out the Kish Grid is not listing household members in the proper order, from oldest to youngest. Part of this error may be attributed to large household sizes in the developing world where up to 20 adult household members can be listed.

Several exercises are used to help the interviewers learn the proper procedures for household and respondent selection to ensure that interviewers can implement these procedures.

About two-thirds of the training time is devoted to household and respondent selection. The remainder of the training is devoted to learning questionnaire content, doing mock interviews, and conducting pilot interviews. To learn questionnaire content, interviewers go around the room asking other interviewers questions. The "respondent" is not allowed to look at his or her questionnaire while being asked a question. Compared with a lecture-based approach, the interactive approach decreases learning time and makes it more difficult for interviewers to lose concentration. After going through the questionnaire in this fashion, interviewers are then paired up and they conduct two mock interviews. Again, the "respondent" is not allowed to look at his or her questionnaire. The training staff evaluates the mock interviewers and gives them one-on-one feedback. After successfully completing the household selection exercises and the mock interview, interviewers go to the "field" to conduct a pilot interview. Even though the training typically takes place in an urban location, the pilot interviews also include interviews in more rural settings around the city. The pilot interview also includes household selection. Each pilot interview is evaluated, and at the trainers' discretion, interviewers may be required to conduct a second pilot interview before starting data collection.

Once Gallup is satisfied with interviewer training, interviewer teams are organized and data collection begins. An interviewer team typically consists of one supervisor and 4 to 6 interviewers. The team uses public transportation to go to PSUs around the country. One interviewer works a PSU. Generally two days are allocated to complete data collection in a PSU.

31.6 QUALITY CONTROL PROGRAM

To ensure we collect high-quality data: (1) interviewers are accompanied by a supervisor for a minimum of 2 interviews; (2) the supervisor conducts spot checks on a minimum of 5% of the cases—he or she returns to an interviewed household to check whether the right adult was selected and the interview was completed; (3) a minimum of 25% of the cases are back checked, either by telephone or in person, where the supervisor again checks respondent selection, verifies that the interview has been completed, and verifies some answers; and (4) 100% of the questionnaires are checked for coherence and completeness.

In some of the former Soviet Union countries, two local vendors were used in each country: one conducted the data collection and another performed quality control. This way, the potential for co-workers to cover up each other's mistakes was minimized. In Tajikistan, 200 interviews could not pass this quality control procedure and the vendor had to redo all these interviews.

31.7 EXTERNAL ISSUES WITH OBTAINING PERMISSION TO INTERVIEW

External issues often need to be dealt with to obtain permission to conduct interviews. In some countries, interviewers are given letters that indicate they have been authorized to collect data. Interviewers often visit local police stations to tell the police they are in the area conducting interviews.

In rural areas of Asia, the village head is approached first to obtain informal permission to conduct the survey. In rural areas of sub-Saharan Africa, where tribal culture is strong, interviewers usually tell the chief about the interviews and obtain his permission to contact tribal members. The chief usually notifies tribe members to cooperate with the interviewers. The same procedure must be followed with some clans within tribes. For example, the headman of a Masai village must give his permission before other clan members can be interviewed.

Sometimes other situations dictate the need to obtain prior permission. For example, in Latin America, interviewers have had to obtain permission from the local drug dealer to conduct interviews in his area.

31.8 ITEM AND UNIT RESPONSE

In 2006, the face-to-face surveys in 26 countries in sub-Saharan Africa had a median response rate, using American Association of Public Opinion Research (AAPOR, 2008) definition 5, of 90% with a minimum of 59% and a maximum of 98%. The cooperation rate, using AAPOR definition 3, ranged from 80% to 98% and the unit refusal rate, AAPOR definition 3, ranged from 7% to 15%.

Only one question, the average monthly household income, had a serious item missing data (nonresponse) rate. Slightly more than 140,000 completed interviews were obtained in 2006 and the item nonresponse rate for average monthly household income was 31%. The distribution of item nonresponse varies considerably by country as does the distribution between Don't Know (DK) and Refused (R). For example, in sub-Saharan Africa, the highest item nonresponse rate for income was 30% in Ghana, composed of 20% DK and 10% R; the lowest was in Zimbabwe with an overall rate of 8%, composed of 7% DK and 1% R. In the FSU, the highest rate of 28% occurred in Lithuania which was composed of 7% DK and 20% R, and the lowest in Armenia and Georgia, both 5% overall, but all refusals in Georgia and with 2% DK and 3% R in Armenia. In Asia the highest item nonresponse rate for household income was 39% in Japan with 23% DK and 16% R, and the lowest was Thailand with the item nonresponse rate of 1%, all refusals. No unit or item imputation is currently used.

31.9 SUBSTITUTION

As noted previously, PSUs can be substituted if they fall into an area that is not enumerable for one of several reasons. These PSUs are replaced by another randomly selected PSU. Substitution can also occur at the household level within a selected PSU, but not at the person level within a household. If the selected person refuses to do the survey or is unavailable to do the survey, a new household is selected. Generally three attempts are made to complete the interview with the selected respondent. When an interview is not obtained, a household substitution is made. The interviewer stands with his or her back to this household and attempts to interview a randomly selected person in the household to the right. One attempt to obtain an interview is made at this household. If that attempt is not successful, the interviewer returns to the original household, stands with his or her back to that household and substitutes the household on the left. One attempt is made at this household to obtain the interview. If that attempt fails, the interviewer returns to the original household and moves three households to the right and starts the process all over again. This latter situation is an extremely rare event.

31.10 SUMMARY

The Gallup World Poll, funded entirely by Gallup, is the largest multinational survey ever undertaken. It is an ongoing survey and in its first year data was collected from more than 130 countries and areas representing 95% of the world population age 15 and older. In developed countries, the GWP uses an RDD design and in the developing world an area frame design with face-to-face interviewing. In 2007, the GWP expanded to include 142 countries.

The critical design features of the GWP are (1) standardized methodology to facilitate county and time comparisons, (2) representative sampling generally covering each country's population age 15 and older, and (3) the use of dichotomous response categories to reduce cultural and translation biases.

References

Abowd, J. and Stinson, M. H. (2007), Estimating measurement error in SIPP annual job earnings: A comparison of census survey and SSA administrative data. Paper presented at the 2007 Federal Committee on Statistical Methodology Research Conference, Arlington, VA (USA), November 5–7.

Abu-Lughod, L. (1999), *Veiled Sentiments: Honor and Poetry in a Bedouin Society*, Berkeley: University of California Press.

Adato, M., Lund, F., and Mhlongo, P. (2006), Methodological innovations in research on the dynamics of poverty: A longitudinal study in KwaZulu-Natal, South Africa, *World Development*, 35, 247–263.

Aday, L. A., Chiu, G. Y., and Andersen, R. (1980), Methodological issues in health care surveys of the Spanish heritage population, *American Journal of Public Health*, 70, 367–374.

Agresti, A. and Agresti, B. F. (1977), Statistical analysis of qualitative variation, *Sociological Methodology*, 1978(S), 204–237.

Alaminos, A. (2004*),* Improving survey methodology: The ESS Index and Measurement—The cultural borders of meaning. Paper presented at the Sixth International Conference on Social Science Methodology, Amsterdam.

Albaum, G., Roster, C., Yu, J. H., and Rogers, R. D. (2007), Simple rating scale formats. Exploring extreme response, *International Journal of Market Research*, 49, 633–650.

Alcser, K., Kirgis, N., and Guyer, H. (2008), Adapting U.S.-based general interviewing and study-specific training protocols for cross-national survey data collection. Presented at the International Conference on Survey Methods in Multinational, Multiregional, and Multicultural Contexts, Berlin.

Alesina, A., Devleeschauwer, A., Easterly, W., Kurlat, S., and Wacziarg, R. (2003), Fractionalization, *Journal of Economic Growth*, 8, 155–194. Retrieved March 2008 from http://www.stanford.edu/~wacziarg/papersum.html.

Allen, M. W., Ng, S. H., Ikeda, K. I., Jawan, J. A., Sufi, A. H., Wilson, M., et al. (2007), Two decades of change in cultural values and economic development in eight East Asian and Pacific Island nations, *Journal of Cross-Cultural Psychology*, 38, 247–269.

Almond, G. A. and Verba, S. (1963), *The Civic Culture: Political Attitudes and Democracy in Five Nations*, Princeton, NJ: Princeton University Press.

Alwin, D. F. (2007), *Margins of Error: A Study of Reliability in Survey Measurement*, Hoboken, NJ: John Wiley & Sons.

American Association for Public Opinion Research (AAPOR) (2000), *Standard Definitions—Final Dispositions of Case Codes and Outcome Rates for Surveys,* Ann Arbor, Michigan: AAPOR.

American Association for Public Opinion Research (AAPOR) (2006), *Standard Definitions—Final Dispositions of Case Codes and Outcome Rates for Surveys* (4th ed.), Lenexa, Kansas: AAPOR.

American Association for Public Opinion Research (AAPOR) (2008), *Standard Definitions: Standard Dispositions of Case Codes and Outcome Rates for Surveys,* Lenexa, KS: AAPOR. http://aapor.org/Content/NavigationMenu/Resourcesfor Researchers/ Standard_Definitions_07_08_Final.pdf.

American Educational Research Association, American Psychological Association, and National Council on Measurement in Education (1985), *Standards for Educational and Psychological Testing*, Washington, DC: American Psychological Association.

American Institutes for Research (AIR) (2002), USSS Survey Operations Manual, USSS Ref. No. 01-01-01, February 15.

Andolsek, D. M. and Stebe, J. (2004), Multinational perspectives on work values and commitment, *International Journal of Cross Cultural Management*, 4, 181–208.

Arends-Tóth, J. V. and van de Vijver, F. J. R. (2004), Dimensions and domains in models of acculturation: Implicit theories of Turkish-Dutch, *International Journal of Intercultural Relations*, 28, 19–35.

Arends-Tóth, J. V. and van de Vijver, F. J. R. (2006), Assessment of psychological acculturation: Choices in designing an instrument, in D. L. Sam and J. W. Berry (Eds.), *The Cambridge Handbook of Acculturation Psychology*, pp. 142–160, Cambridge, UK: Cambridge University Press.

Arends-Tóth, J. V. and van de Vijver, F. J. R. (2007), Acculturation attitudes: A comparison of measurement methods, *Journal of Applied Social Psychology*, 37, 1462–1488.

Armer, M. (1973), Methodological problems and possibilities in comparative research, in M. Armer and A. D. Grimshaw (Eds.), *Comparative Social Research: Methodological Problems and Strategies,* pp. 3–48, New York: John Wiley & Sons.

Axinn, W. G. and Pearce, L. D. (2006), *Mixed Method Data Collection Strategies*, Cambridge: Cambridge University Press.

Bachman, J. G. and O'Malley, P. M. (1984a), Black-white differences in self-esteem: Are they affected by response styles? *American Journal of Sociology*, 90, 624–639.

Bachman, J. G. and O'Malley, P. M. (1984b), Yea-saying, nay-saying, and going to extremes: Black-White differences in response style, *Public Opinion Quarterly*, 48, 491–509.

Bagli, M. and Sev'er, A. (2003), Female and male suicides in Batman, Turkey: Poverty, social change, patriarchal oppression and gender links, *Women's Health and Urban Life: An International and Interdisciplinary Journal*, 2, 60–84.

Bagozzi, R. P. and Heatherton, T. T. (1994), A general approach to representing multi-faceted personality constructs: Application to self-esteem, *Structural Equation Modeling*, 1, 35–67.

Baker, F. (1995), EQUATE 2.1, Computer program for equating two metrics in item response theory (Version 2.1), Madison: University of Wisconsin, Laboratory of Experimental Design.

Baker, M. and Saldanha, G. (Eds.) (2009), *Encyclopedia of Translation Studies* (2nd ed.), London: Routledge.

Barth, J., Gonzalez, E. J., and Neuschmidt, O. (2004), *TIMSS 2003 Survey Operations Procedures*, Boston, MA: Boston College.

Bastin, G. L. (2009), Adaptation, in M. Baker and G. Saldanha (Eds.), *Routledge Encyclopedia of Translation Studies* (2nd ed.), pp. 3–6, London: Routledge.

Baumeister, R. F. (1998), The self, in D. Gilbert, S. Fiske, and G. Lindzey (Eds.), *Handbook of Social Psychology* (4th ed.), Vol. 1, pp. 680–740, New York: McGraw-Hill.

Baumeister, R. F., Tice, D. M., and Hutton, D. G. (1989), Self-presentational motivations and personality differences in self-esteem, *Journal of Personality*, 57, 547–79.

Baumgartner, H. and Steenkamp, J.-B. E. M. (2001), Response styles in marketing research: A cross-national investigation, *Journal of Marketing Research*, 38, 143–156.

Baumgartner, H. and Steenkamp, J. E. M. (2006), Response biases in marketing research, in R. Grover and M. Vriens (Eds.), *The Handbook of Marketing Research: Uses, Misuses, and Future Advances,* pp. 95–109, Thousand Oaks, CA: Sage.

Bean, C. and Papadakis, E. (1998), A comparison of mass attitudes towards the welfare state in different institutional regimes, 1985–1990, *International Journal of Public Opinion Research*, 10, 211–236.

Bechger, T. M. and Maris, G. (2004), Structural equation modelling of multiple facet data: Extending models for multitrait-multimethod data, *Psicologica*, 25, 253–274.

Bechger, T. M., van den Wittenboer, G., Hox, J. J., and de Glopper, C. (1999), The validity of comparative educational studies, *Educational Measurement: Issues and Practice*, 18, 18–26.

Behr, D. (forthcoming), Translationswissenschaft und international vergleichende Umfrageforschung: Qualitätssicherung bei Fragebogenübersetzungen als Gegenstand einer Prozessanalyse, GESIS. [Translation: Research and cross-national survey research: Quality assurance in questionnaire translation from the perspective of translation process research.]

Belli, R. (1998), The structure of autobiographical memory and the Event History Calendar: Potential improvements in the quality of retrospective reports in surveys, *Memory* 6, 383–406.

Belson W. A. (1981), *The Design and Understanding of Survey Questions*. Aldershot, Hants., England: Gower.

Bennett, D. J. and Steel, D. (2000), An evaluation of a large-scale CATI household survey using random digit dialling, *Australian and New Zealand Journal of Statistics,* 42, 255–270.

Bentler, P. M. (1990), Comparative fit indexes in structural models, *Psychological Bulletin*, 107, 238–246.

Bentley, G. and Muttukrishna, S. (2007), Cross-cultural comparisons: Midlife, aging, and menopause, *Menopause*, 14(4), 668–679.

Bernstein, I. H. and Teng, G. (1989), Factoring items and factoring scales are different: Spurious evidence for multidimensionality due to item categorization, *Psychological Bulletin*, 105, 467–477.

Berry, J. W. (1969), On cross-cultural comparability, *International Journal of Psychology*, 4(2), 119–128.

Berry, J. W. (1990), Imposed etics, emics, and derived etics: Their conceptual and operational status in cross-cultural psychology, in T. Headland, K. L. Pike, and M. Harris (Eds.), *Emics and Etics: The Insider Outsider Debate*, pp. 84–99, Newbury Park, CA: Sage.

Berry, J. W. (1997), Immigration, acculturation, and adaptation, *Applied Psychology: An International Review*, 46, 5–68.

Berry, J. W. and Sam, D. L. (1997), Acculturation and adaptation, in J. W. Berry, M. H. Segall, and C. Kagitçibasi (Eds.), *Handbook of Cross-Cultural Psychology* (2nd ed.), Vol. 3., pp. 291–326, Boston, MA: Allyn & Bacon.

Berry, J. W., Poortinga, Y. H., Segall, M. H., and Dasen, P. R. (2002), *Cross-Cultural Psychology: Research and Applications* (2nd ed.), Cambridge, UK: Cambridge University Press.

Bethlehem, J. and Hundepool, A. (2004), TADEQ: A tool for the documentation and analysis of electronic questionnaires, *Journal of Official Statistics*, 20(2), 233–264.

Beullens, K., Billiet, J., and Loosveldt, G. (2008), The effect of elapsed time between the initial refusal and conversion contact on conversion success: Evidence from the second round of the European Social Survey. Working Paper of the Centre for Sociological Research CeSO/SM/2008-8, Leuven: Catholic University.

Beullens, K., Vandecasteele, L., and Billiet, J. (2007), Refusal conversion in the second round of the European Social Survey, Working Paper of the Centre for Sociological Research CeSO/SM/2007-5, Leuven: Catholic University.

Biemer, P. and Lyberg, L. E. (2003), *Introduction to Survey Quality*, Hoboken, NJ: John Wiley & Sons.

Biemer, P. and Stokes, S. L. (1989), The optimal design of quality control samples to detect interviewer cheating, *Journal of Official Statistics*, 5, 23–39.

Biemer, P., Groves, R. M., Lyberg, L. E., Mathiowetz, N. A., and Sudman, S. (Eds.) (2004), *Measurement Errors in Surveys,* Hoboken, NJ: John Wiley & Sons.

Biemer, P. P., Woltmann, H., Raglin, D., and Hill, J. (2001), Enumeration accuracy in a population census: An evaluation using latent class analysis, *Journal of Official Statistics*, 17, 129–148.

Billiet, J. B. and McClendon, M. J. (2000), Modeling acquiescence in measurement models for two balanced sets of items, *Structural Equation Modeling*, 7, 608–628.

Billiet, J. and Philippens, M. (2004), Data quality assessment in ESS Round 1: Between wishes and reality, London: Centre for Comparative Social Surveys, City University. Retrieved 23rd September 2008, from http://ess.nsd.uib.no/index.jsp?year=2003&module=documentation&country=.

Billiet, J. and Pleysier, S. (2007), Response based quality assessment in the ESS—Round 2: An update for 26 countries, London: Centre for Comparative Social Surveys, City University. Retrieved 23rd September 2008 from http://ess.nsd.uib.no/index.jsp?year=2005&module=documentation&country=.

Billiet, J. and Welkenhuysen-Gybels, J. (2004a), Assessing cross-national construct equivalence in the ESS: The case of religious involvement. Paper presented at the Sixth International Conference on Social Science Methodology, August 17–20, 2004 Amsterdam. Retrieved February 2008 on http://www.s3ri.soton.ac.uk/qmss/documents/BillietMainzpapermeasurementESSdefinite.pdf.

Billiet, J. and Welkenhuysen-Gybels, J. (2004b), Assessing cross-national construct equivalence in the ESS: The case of six immigration items, in C. Van Dijkum, J. Blasius, and C. Durand (Eds.), *Recent Developments and Applications in Social Research Methodology*, Barbara Budrich Pub. Proceedings of the Sixth International Conference on Social Science Methodology, Amsterdam.

Billiet, J., Koch, A., and Philippens, M. (2007), Understanding and improving response rates, in R. Jowell, C. Roberts, R. Fitzgerald, and G. Eva (Eds.), *Measuring Attitudes Cross-Nationally*, London: Sage.

Billiet, J., Philippens, M., Fitzgerald, R., and Stoop, I. (2007), Estimation of nonresponse bias in the European Social Survey: Using information from reluctant respondents, *Journal of Official Statistics*, 23,135–162.

Binstock, G. and Thornton, A. (2007), Knowledge and use of developmental thinking about societies and families among teenagers in Argentina, *Demografía* 50(5), 75–104.

Blair, J. and Lacy, M. G. (2000), Statistics of ordinal variation, *Sociological Methods & Research*, 28:S, 251–280.

Blair, J. and Piccinino, L. (2005), The development and testing of instruments for cross-cultural and multi-cultural surveys, in J. H. P. Hoffmeyer-Zlotnik and J. A. Harkness (Eds.), *Methodological Aspects in Cross-National Research*, pp. 13–30, Mannheim, Germany: ZUMA Nachrichten Spezial, Volume 11.

Blasius, J. and Thiessen, V. (2001), Methodological artifacts in measures of political efficacy and trust. An application of multiple correspondence analysis, *Political Analysis,* 9, 1–20.

Blasius, J. and Thiessen, V. (2006), Assessing data quality and construct comparability in cross-national surveys, *European Sociological Review*, 22, 229–242.

Bloch, A. (2007), Methodological challenges for national and multi-sited comparative survey research, *Journal of Refugee Studies,* 20(2), 230–247.

Blom, A. G. (2008), Measuring non-response cross-nationally. ISER Working Paper 2008-41, Colchester: University of Essex.

Blom, A. G., Jäckle, A., and Lynn, P. (2007), Explaining differences in contact rates across countries. Presented at the European Survey Research Association Conference, Prague.

Bodenhausen, G. V. and Wyer, R. S. (1987), Social cognition and social reality: Information acquisition and use in the laboratory and the real world, in H. J. Hippler, N. Schwarz, and S. Sudman (Eds.), *Social Information Processing and Survey Methodology*, pp. 6–41, New York: Springer.

Bogardus, E. S. (1925), Social distance and its origins, *Journal of Applied Sociology*, 9, 216–226.

Bollen, K. A. (1989), *Structural Equations with Latent Variables*, New York: John Wiley & Sons.

Bolton, P. and Tang, A. M. (2002), An alternative approach to cross-cultural function assessment, *Social Psychiatry and Psychiatric Epidemiology,* 37(11), 537–543.

Bond, M. H. and Yang, K.-S. (1982), Ethnic affirmation versus cross-cultural accommodation: The variable impact of questionnaire language on Chinese bilinguals from Hong Kong, *Journal of Cross-Cultural Psychology*, 13, 169–185.

Bond, R. and Smith, P. B. (1996), Culture and conformity: A meta-analysis of the Asch (1952b, 1956) line judgment task, *Psychological Bulletin*, 119, 111–137.

Boomsma, A. (1982), The robustness of LISREL against small sample sizes in factor analysis models, in K. G. Jöreskog and H. Wold (Eds.), *Systems Under Indirect Observation: Causality, Structure, Prediction*, Part I, pp. 149–173, Amsterdam: North-Holland.

Borg, I. and Braun, M. (1996), Work values in East and West Germany: Different weights, but identical structures, *Journal of Organizational Behavior*, 17, 541–555.

Borg, I. and Groenen, P. J. F. (2005), *Modern Multidimensional Scaling (*2nd ed.), New York: Springer.

Börsch-Supan, A. and Jürges, H. (Eds.) (2005), *The Survey of Health, Ageing and Retirement in Europe—Methodology*, Mannheim: MEA.

Börsch-Supan, A. and Kemperman, M.-L. (2005), The SHARE development process, in A. Börsch-Supan, and H. Jürges (Eds.), *The Survey of Health, Ageing and Retirement in Europe—Methodology*, pp. 7–11, Mannheim: MEA.

Börsch-Supan, A., Brugiavini, A., Jürges, H., Mackenbach, J., Siegrist, J., and Weber, G. (Eds.) (2005), *Health, Ageing and Retirement in Europe—First Results from the Survey of Health, Ageing and Retirement in Europe*, Mannheim: MEA.

Börsch-Supan, A., Jürges, H., and Lipps, O. (2003), *SHARE: Building a Panel Survey on Health, Ageing And Retirement in Europe*, Mannheim, Germany: Mannheim Research Institute for the Economics of Aging, University of Mannheim.

Bowers, A., Benson, G., Alcser, K., Clemens, J., and Orlowski, R. (2008), Ethical principles in cross-national research. Paper presented at The International Conference on Survey Methods in Multinational, Multiregional, and Multicultural Contexts, Berlin, Germany. http://www.csdiworkshop.org/pdf/3mc2008_proceedings/session_35/Bowersr.pdf.

Boyatzis, R. E. (1998), *Transforming Qualitative Information: Thematic Analysis and Code Development*, Thousand Oaks, CA: Sage.

Bradburn, N. M., Sudman, S.,and Wansink, B. (2004), *Asking Questions: The Definitive Guide to Questionnaire Design*, San Francisco: Jossey-Bass.

Brancato, G. (2006), Recommended practices for questionnaire development and testing in the European statistical system, presented at the European Conference on Quality in Survey Statistics, Cardiff, Wales.

Braun, M. (2003), Communication and social cognition, in J. Harkness, F. J. R. van de Vijver, and P. Mohler (Eds.), *Cross-Cultural Survey Methods*, pp. 57–67, Hoboken, NJ: John Wiley & Sons.

Braun, M. (2006), *Funktionale Äquivalenz in interkulturell vergleichenden Umfragen*, Mythos und Realität, Mannheim: ZUMA.

Braun, M. and Baumgärtner, M. K. (2006), The effects of work values and job characteristics on job satisfaction, in M. Braun and P. P. Mohler (Eds.), *Beyond the Horizon of Measurement. Festschrift in Honor of Ingwer Borg*. ZUMA-Nachrichten Spezial, 10.

Braun, M. and Harkness, J. A. (2005), Text and context: Challenges to comparability in survey questions, in J. H. P. Hoffmeyer-Zlotnik and J. A. Harkness (Eds.),

Methodological Aspects in Cross-National Research, pp. 95–107, Mannheim, Germany: ZUMA. Retrieved from http://www.gesis.org/fileadmin/upload/forschung/ publikationen/zeitschriften/zuma_nachrichten_spezial/znspezial11.pdf.

Braun, M. and Mohler, P. P. (2003), Background variables, in J. A. Harkness, F. J. R. van de Vijver, and P. P. Mohler (Eds.), *Cross-Cultural Survey Methods*, pp. 101–115, Hoboken, NJ: John Wiley & Sons.

Braun, M. and Scott, J. (1998), Multidimensional scaling and equivalence: Is *having a job the same as working?* in J. A. Harkness (Ed.), *ZUMA-Nachrichten Spezial No. 3. Cross-Cultural Survey Equivalence,* Mannnheim: ZUMA.

Braun, M. and Uher, R. (2003), The ISSP and its approach to background variables, in J. H. P. Hoffmeyer-Zlotnik and C. Wolf (Eds.), *Advances in Cross-National Comparison. A European Working Book for Demographic and Socio-Economic Variables*, pp. 33–47, New York: Kluwer Academic/Plenum Publishers.

Brennan, R. L. and Prediger, D. J. (1981), Coefficient kappa: Some uses, misuses, and alternatives, *Educational and Psychological Measurement*, 41, 687–699.

Brewer, J. and Hunter, A. (1989), *Multimethod Research: A Synthesis of Styles*, Newbury Park, CA: Sage.

Brislin, R. W. (1970), Back translation for cross-cultural research, *Journal of Cross-Cultural Psychology,* 1(3), 185–216.

Brislin, R. W. (1976), Comparative research methodology: Cross-cultural studies, *International Journal of Psychology*, 11(3), 215–229.

Brislin, R. W. (1986), The wording and translation of research instruments, in W. J. Lonner and J. W. Berry (Eds.), *Field Methods in Cross-Cultural Research*, Thousand Oaks, CA: Sage.

Brislin, R. W., Lonner, W. J., and Thorndike, R. M. (1973), Questionnaire wording and translation, in R. W. Brislin, W. J. Lonner, and R. M. Thorndike (Eds.), *Cross-Cultural Research Methods*, pp. 32–58, New York: John Wiley & Sons.

Brown, T. A. (2006), *Confirmatory Factor Analysis for Applied Research*, New York: Guilford Press.

Browne, M. W. and Cudeck, R. (1993), Alternative ways of assessing model fit, in K. A. Bollen and J. S. Long (Eds.), *Testing of Structural Equation Models*, pp. 136–162, Newbury Park, CA: Sage.

Brugiavini, A., Croda, E., Paccagnella, O., Rainato, R., and Weber, G. (2005), Generated income variables in SHARE Release 1, in A. Börsch-Supan and H. Jürges (Eds.), *The Survey of Health, Ageing and Retirement in Europe—Methodology*, pp. 105–113, Mannheim: MEA.

Brunstein, J. C., Schultheiss, O. C., and Grässmann, R. (1998), Personal goals and emotional well-being: The moderating role of motive dispositions, *Journal of Personality and Social Psychology*, 75, 494–508.

Bullinger, M., Schmidt, S., and Naber, D. (2007), Cross-cultural quality of life research in mental health, in M. S. Ritsner and A. G. Awad (Eds.), *Quality of Life Impairment in Schizophrenia, Mood and Anxiety Disorders. New Perspectives on Research and Treatment*, pp. 67–97, Dordrecht, The Netherlands: Springer.

Bulmer, M. (1998), The problem of exporting social survey research, *American Behavioral Scientist* 42(2), 153–167.

Bulmer, M. and Warwick, D. P. (Eds.) (1983/1993), *Social Research in Developing Countries: Surveys and Censuses in the Third World*, Chichester (West Sussex); New York: John Wiley & Sons.

Burrow, J. W. (1981), *Evolution and Society*, Cambridge: Cambridge University Press.

Byrne, B. M., Shavelson, R. J., and Muthén, B. O. (1989), Testing for the equivalence of factor and mean structures: The issue of partial measurement invariance, *Psychological Bulletin*, 105, 456–466.

Cameron, A. C. and Trivedi, P. K. (2005), *Micro-econometrics: Methods and Applications*, New York: Cambridge University Press.

Campanelli, P., Sturgis, P., and Purdon, S. (1997), *Can You Hear Me Knocking: An Investigation into the Impact of Interviewers on Survey Response Rates*, London: National Centre for Social Research.

Campbell, D. T. and Fiske, D. W. (1959), Convergent and discriminant validation by the multitrait-multimethod matrix, *Psychological Bulletin*, 56, 81–105.

Canberra Group [Expert Group on Household Income Statistics] (2001), *Final Report and Recommendations*, Ottawa.

Candell, G. L. and Hulin, C. L. (1986), Cross-language and cross-cultural comparisons in scale translations: Independent sources of information about item nonequivalence, *Journal of Cross-Cultural Psychology*, 17, 417–440.

Cannell, C., Miller, P., and Oksenberg, L. (1981), Research on interviewing techniques, in S. Leinhardt (Ed.), *Sociological Methodology*, pp. 389–437, San Francisco: Jossey-Bass.

Cantor, S. B., Byrd, T. L., Groff, J. Y., Reyes, Y., Tortolero-Luna, G., and Mullen, P. D. (2005), The language translation process in survey research: A cost analysis, *Journal of Behavioral Sciences,* 27(3), 364–370.

Cantril, H. (1965), *The Pattern of Human Concerns*, New Brunswick, NJ: Rutgers University Press.

Carey, S. (Ed.) (2000), Measuring adult literacy, the International Adult Literacy survey in the European Context, Office for National Statistics, UK.

Carlson, E. D. (2000), A case study in translation methodology using the Health-Promotion Lifestyle Profile II, *Public Health Nursing*, 17, 61–70.

Carr, L. G. (1969), The Srole items and acquiescence, *American Sociological Review*, 36, 287–293.

Carrasco, L. (2003), *The American Community Survey (ACS) en Español: Using cognitive interviews to test the functional equivalency of questionnaire translations*, Study Series, Survey Methodology #2003-17, Washington, DC: Statistical Research Division, U.S. Census Bureau.

Carroll, R. J., Ruppert, D., and Stefanski, L. A. (1995), Nonlinear measurement error models, in *Monographs on Statistics and Applied Probability*, Vol. 63, New York: Chapman and Hall.

Caspar, R. (2003), 2010 Census language program: Bilingual questionnaire research. Contract report submitted to the United States Census Bureau by RTI International, July 25, 2003.

Catford, J. C. (1965), *A Linguistic Theory of Translation: An Essay on Applied Linguistics*, London: Oxford University Press.

CCT (Central Coordinating Team) (2002), Questionnaire development report. .http://www. europeansocialsurvey.org.

CCT (Central Coordinating Team) (2006), ESS End of grant report. http://www. europeansocialsurvey.org

Cepeda-Benito, A., Henry, K., Gleaves, D. H., and Fernandez, M.C. (2004), Cross-cultural investigation of the questionnaire of smoking urges in American and Spanish smokers, *Assessment*, 11, 152–159.

Chasiotis, A., Hofer, J., and Campos, D. (2006), When does liking children lead to parenthood? Younger siblings, implicit prosocial power motivation, and explicit ove for children predict parenthood across cultures, *Journal of Cultural and Evolutionary Psychology*, 4, 95–123.

Chasiotis, A., Keller, H., and Scheffer, D. (2003), Birth order, age at menarche, and intergenerational context continuity: A comparison of female somatic development in West and East Germany, *North American Journal of Psychology*, 5, 153–170.

Chen, C., Lee, S., and Stevenson, H. W. (1995), Response style and cross-cultural comparisons of rating scales among East Asian and North American students, *Psychological Science*, 6, 170–175.

Cheung, F. M., Leung, K., Fan, R., Song, W. Z., Zhang, J. X., and Zhang, J. P. (1996), Development of the Chinese Personality Assessment Inventory (CPAI), *Journal of Cross-Cultural Psychology*, 27, 181–199.

Cheung, G. W. and Rensvold, R. B. (2000), Assessing extreme and acquiescence response sets in cross-cultural research using structural equations modeling, *Journal of Cross-Cultural Psychology*, 31, 187–212.

Cheung, G. W. and Rensvold, R. B. (2002), Evaluating goodness-of-fit indexes for testing measurement invariance, *Structural Equation Modeling*, 9, 233–255.

Chikwanha, A. B. (2005), Conducting surveys and quality control in Africa: Insights from the Afrobarometer, WAPOR/ISSC Conference, Ljubljana, Slovenia.

Chou, C.-P. and Bentler, P. M. (1995), Estimates and tests in structural equation modeling, in R. H. Hoyle (Ed.), *Structural Equation Modeling: Concepts, Issues and Applications*, Thousand Oaks, CA: Sage.

Christelis, D., Japelli, T., and Padula, M. (2005), Generated asset variables in SHARE Release 1, in A. Börsch-Supan and H. Jürges (Eds.), *The Survey of Health, Ageing and Retirement in Europe—Methodology*, pp. 114–127, Mannheim: MEA.

Christian, L. M., Dillman, D. A., and Smyth, J. D. (2008), Mode effects in the Canadian Community Health Survey: A comparison of CATI and CAPI, in J. M. Lepkowski, C. Tucker, J. M. Brick, E. de Leeuw, L. Japec, P. J. Lavrakas, et al. (Eds.), *Telephone Survey Methodology*, pp. 250–275, Hoboken, NJ: John Wiley & Sons.

Christopher, S., McCormick, A. K. H. G., Smith, A., and Christopher, J. C. (2005), Development of an interviewer training manual for a cervical health project on the Apsáalooke reservation, *Health Promotion Practice*, 6(4), 414–422.

Chun, K.-T., Campbell, J. B., and Yoo, J. A. (1974), Extreme response style in cross-cultural research: A reminder, *Journal of Cross-Cultural Psychology*, 5, 465–480.

Cicchetti, D. V. and Heavens, R. (1981), A computer program for determining the significance of the difference between pairs of independently derived values of kappa or weighted kappa, *Educational and Psychological Measurement*, 41, 189–193.

Clark, H. H. and Schober, M. F. (1992), Asking questions and influencing answers, in J. M. Tanur (Ed.), *Questions About Questions*, pp. 15–48, New York: Russell Sage.

Clarke, I., III (2000a), Extreme response style in cross-cultural research: An empirical investigation. *Journal of Social Behavior & Personality*, 15, 137–152.

Clarke, I., III (2000b), Global marketing research: Is extreme response style influencing your results? *Journal of International Consumer Marketing*, 12, 91–111.

Clarke, I., III (2001), Extreme response style in cross-cultural research, *International Marketing Review*, 18, 301–324.

Clausen, S. E. (1998), *Applied Correspondence Analysis. An Introduction*, Newbury Park CA: Sage.

Cleary, P. D., Mechanic, D., and Weiss, N. (1981), The effect of interviewer characteristics on responses to a mental health interview, *Journal of Health and Social Behavior*, 22(2), 183–193.

Cleland, J. (1996), Demographic data collection in less developed countries 1946–1996, *Population Studies*, 50(3), 433–50.

Cleland, J. and Scott, C, (Eds.) (1987), *The World Fertility Survey: An Assessment*, New York: Oxford University Press.

Clemenceau, A. and Museaux, J.-M. (2008), Harmonisation of household surveys from an official statistics perspective: Experience gained from HBS, ECHP, and EU-SILC. Paper presented at the 3MC Conference, Berlin, June.

Cochran, W. G. (1977), *Sampling Techniques*, Chichester: John Wiley & Sons.

Coenders, G. (1996), Structural equation modeling of ordinally measured survey data. Ph.D. thesis, Barcelona: Universitat Ramon Llull.

Coenders, M. and Scheepers, P. (2003), The effect of education on nationalism and ethnic exclusionism: An international comparison, *Political Psychology*, 24, 313–343.

Cohen, D. (1998), Culture, social organization, and patterns of violence, *Journal of Personality and Social Psychology*, 75, 408–419.

Cohen, D. and Nisbett, R. E. (1994), Self-protection and the culture of honor: Explaining southern violence, *Personality and Social Psychology Bulletin*, 20, 551–567.

Cohen, D. and Nisbett, R. E. (1997), Field experiments examining the culture of honor: The role of institutions in perpetuating norms about violence, *Personality and Social Psychology Bulletin*, 23, 1188–1199.

Cohen, D., Nisbett, R. E., Bowdle, B. F., and Schwarz, N. (1996), Insult, aggression, and the southern culture of honor: An experimental ethnography, *Journal of Personality and Social Psychology*, 70, 945–960.

Cohen, D., Vandello, J., and Rantilla, A. K. (1998), The sacred and the social: Cultures of honor and violence, in P. Gilbert and B. Andrews (Eds.), *Shame: Interpersonal Behavior, Psychopathology and Culture*, pp. 261–282, Oxford: Oxford University Press.

Cohen, J. (1960), A coefficient of agreement for nominal scales, *Educational and Psychological Measurement*, 20, 37–46

Cole, D. A. (1987), Utility of confirmatory factor analysis in test validation research, *Journal of Consulting and Clinical Psychology*, 55, 584–594.

Collins, S. D. (1946), The incidence of poliomyelitis and its crippling effects, as recorded in family surveys, *Public Health Reports*, 61(10), 327–355.

Converse, J. M. (1987), *Survey Research in the United States: Roots and Emergence 1890–1960*, Berkeley: University of California Press.

Converse, J. M. and Presser, S. (1986), *Survey Questions: Handcrafting the Standardized Questionnaire*, Thousand Oaks, CA: Sage.

Cornelius, R. M. (1985), The World Fertility Survey and its implications for future surveys, *Journal of Official Statistics*, 1, 427–433.

Coromina, L., Saris, W., and Oberski, D. (2007), The quality of the measurement of interest in the political issues in the media in the ESS. Paper presented 12th meeting of the ESS Scientific Advisory Board, 15–16 November 2007, Madrid.

Coronado, L. and Earle, D. (2002), Effectiveness of the American Community Survey of the U.S. Census in a borderlands colonia setting. Research report submitted to the U.S. Census Bureau, March 19, 2002.

Correll, S. J., Benard, S., and Paik, I. (2007), Getting a job: Is there a motherhood penalty? *American Journal of Sociology*, 112(5), 1297–1338.

Couch, A. and Keniston, K. (1960), Yeasayers and naysayers: Agreeing response set as a personality variable, *Journal of Abnormal & Social Psychology*, 60, 151–174.

Council for International Organizations of Medical Sciences (2002), *International Ethics Guidelines for Biomedical Research Involving Human Subjects*, Geneva, Switzerland: CIOMS.

Couper, M. (1998), Measuring survey quality in a CASIC environment. Paper presented at the Joint Statistical Meetings of the American Statistical Association, Dallas.

Couper, M. P. (2005), Technology trends in survey data collection, *Social Science Computer Review*, 23(4), 486–501.

Couper, M. and de Leeuw, E., D. (2003), Nonresponse in cross-cultural and cross-national surveys, in J. Harkness, F. J. R. van de Vijver, and P. Mohler (Eds.), *Cross-Cultural Survey Methods*, Hoboken, NJ: John Wiley & Sons.

Couper, M. and Lyberg, L. (2005), The use of paradata in survey research, *Proceedings of the 54th Session of the International Statistical Institute*, Sydney, Australia.

Couper, M. P., Holland, L., and Groves, R. M. (1992), Developing systematic procedures for monitoring in a centralized telephone facility, *Journal of Official Statistics*, 8(1), 63–76.

Creswell, J. W., Plano Clark, V. L., Gutmann, M. L., and Hanson, W. E. (2003), Advanced mixed methods research designs, in A. Tashakkori and C. Teddlie (Eds.), *Handbook on Mixed Methods in the Behavioral and Social Sciences*, pp. 209–240, Thousand Oaks, CA: Sage Publications.

Cronbach, L. J. (1946), Response sets and test validity, *Educational and Psychological Measurement*, 6, 475–494.

Cronbach, L. J. (1950), Further evidence on response sets and test design, *Educational and Psychological Measurement*, 10, 3–31.

Cross-Cultural Survey Guidelines (2008), *Harmonization of Survey and Statistical Data*, Chapter XII, http://ccsg.isr.umich.edu/harmonization.cfm.

CSDI (Comparative Survey Design and Implementation) Meeting (2007), Unpublished discussion related to creation of Best Practice Guidelines for Conducting Comparative Cognitive Tests. Meeting at National Center for Health Statistics, July, 2007.

CSES Planning Committee (1996), The comparative study of electoral systems: Final report of the 1995–1996 Planning Committee. Retrieved January 14, 2008, from http://www.cses.org/plancom/module1/95rpt2.htm.

Culpepper, R. A., Zhao, L., and Lowery, C. (2002), Survey response bias among Chinese managers, *Academy of Management Proceedings*, 2002 IM, J1–J6.

Darkovich, N., de Heer, W., Foy, P., Jones, S., Lyberg, L., Mohadjer, L., et al. (1999), The International Life Skills Survey Quality Assurance Specifications, Statistics Canada.

Das, M., Vis, C., and Weerman, B. (2005), Developing the survey instruments for SHARE, in A. Börsch-Supan and H. Jürges (Eds.), *The Survey of Health, Ageing and Retirement in Europe—Methodology*, pp. 12–23, Mannheim: MEA.

Dasborough, M. T., Sinclair, M., Russell-Bennett, R., and Tombs, A. (2008), Measuring emotion—Methodological issues and alternatives, in N. M. Ashkanasy and C. L. Cooper (Eds.), *Research Companion to Emotions in Organizations*, pp.197–208, Cheltenham, UK: Edwin Elgar Publishing (New Horizons in Management Series).

Data Documentation Initiative (DDI, 2003), Version 2.1: http://ddi.icpsr.umich.edu/specification/ddi2.1.

Data Documentation Initiative (DDI, 2008), Version 3.0: http://ddi.icpsr.umich.edu/specification/ddi3.0.

Davidov, E. (2008), A cross-country and cross-time comparison of the human values measurements with the second round of the European Social Survey, *Survey Research Methods*, 2, 33–46.

Davidov, E., Schmidt, P., and Schwartz, S. H. (2008), Bringing values back in: The adequacy of the European Social Survey to measure values in 20 countries, *Public Opinion Quarterly*, 72, 420–445.

Davis, D. W. (1997), Nonrandom measurement error and race of interviewer effects among African Americans, *Public Opinion Quarterly*, 61(1), 183–207.

Davis, J. A. and Jowell, R. (1989), Measuring national difference. An introduction to the International Social Survey Programme (ISSP), in R. Jowell, S. Witherspoon, and L. Brook (Eds.), 6th *British Social Attitudes. Special International Report*, Aldershot: Gower.

Davis, J. A., Smith, T. W., and Marsden, P. V. (2007), *General Social Surveys, 1972–2006: Cumulative Codebook*, Chicago: NORC.

Dawson, L. and Kass, N. E. (2005), Views of US researchers about informed consent in international collaborative research, S*ocial Science and Medicine*, 61(6), 1211–1222.

Dean, E., Caspar, R., McAvenchey, G., Reed, L., and Quiroz, R. (2005), Developing a low-cost technique for parallel cross-cultural instrument development: The Question Appraisal System (QAS-04), in J. H. P. Hoffmeyer-Zlotnik and J. A. Harkness (Eds.), *Methodological Aspects in Cross-National Research*, pp. 31–46, Mannheim, Germany: ZUMA Nachrichten Spezial, Volume 11.

Dean, E., Casper, R., McAvinchey, G., Reed, L., and Quiroz, R. (2007), Developing a low-cost technique for parallel cross-cultural instrument development: The question appraisal system (QAS-04), *International Journal of Social Research Methodology*, 10(3), 227–241.

De Beuckelaer, A. (2005), *Measurement Invariance Issues in International Management Research*, Diepenbeek: Limburgs Universitair Centrum.

de Heer, W. (1999), International response trends: Results of an international survey, *Journal of Official Statistics*, 15(2), 129–142.

De Jong, M. G., Steenkamp, J. B. E. M., Fox, J. P., and Baumgartner, H. (2008), Using item response theory to measure extreme response style in marketing research: A global investigation, *Journal of Marketing Research*, 45, 104–115.

de la Puente, M., Pan, Y., and Rose, D. (2003), An overview of proposed Census Bureau Guidelines for the translation of data collection instruments and supporting materials. Paper presented at the meeting of the Federal Committee on Statistical Methodology. Retrieved June 20, 2006 from the U.S. Census Bureau site: http://www.fcsm.gov/03papers/delaPuente_Final.pdf.

de Leeuw, E. D. (1992), *Data Quality in Mail, Telephone, and Face-to-Face Surveys*, Amsterdam: TT-Publikates.

de Leeuw, E. D. (2005), To mix or not to mix data collection modes in surveys, *Journal of Official Statistics*, 21(2), 233–255.

de Leeuw, E. D. (2008), Choosing the method of data collection, in E. D. de Leeuw, J. J. Hox, and D. A. Dillman (Eds.), *International Handbook of Survey Methodology*, pp. 113–135, New York: Lawrence Erlbaum.

de Leeuw, E. D. and de Heer, W. (2002), Trends in household survey nonresponse: A longitudinal and international comparison, in R. M. Groves, D. Dillman, J. Eltinge, and R. Little (Eds.), *Survey Nonresponse*, pp. 41–54, New York: John Wiley & Sons.

de Leeuw, E. D. and van der Zouwen, J. (1988), Data quality in telephone and face to face surveys: A comparative meta-analysis, in R. M. Groves, P. P. Biemer, L. E. Lyberg, J. T. Massey, W. L. Nicholls II, and J. Waksberg (Eds.), *Telephone Survey Methodology*, pp. 283–299, New York: John Wiley & Sons.

de Leeuw, E. D., Callegaro, M., Hox, J., Korendijk, E., and Lensvelt-Mulders, G. (2007), The influence of advance letters on response in telephone surveys: A meta-analysis, Second International Conference on Telephone Survey Methodology, Miami, FL.

de Leeuw, E. D., Hox, J. J., and Huisman, M. (2003), Prevention and treatment of item nonresponse, *Journal of Official Statistics,* 19, 153–176.

de Luca, G. and Lipps, O. (2005), Fieldwork and survey management in SHARE, in A. Börsch-Supan and H. Jürges (Eds.), *The Survey of Health, Ageing and Retirement in Europe—Methodology*, pp. 75–81, Mannheim, Germany: Mannheim Research Institute for the Economics of Aging.

de Luca, G. and Peracchi, F. (2005), Survey Participation in the First Wave of SHARE, in A. Börsch-Supan and H. Jürges (Eds.), *The Survey of Health, Ageing and Retirement in Europe—Methodology*, pp. 88–104, Mannheim: MEA.

DeMaio, T. D. (1984), Social desirability and survey measurement: A review, in C. E. Turner and E. Martin (Eds.), *Surveying Subjective Phenomena*, Vol. 2, pp. 257–282, New York: Russell Sage Foundation.

DeMaio, T. and Rothgeb, J. (1996), Cognitive interviewing techniques in the lab and in the field, in N. Schwarz and S. Sudman (Eds.), *Answering Questions: Methodology for Determining Cognitive and Communicative Processes in Survey Research*, pp. 177–195, San Francisco: Jossey-Bass.

Deming, W. E. (1944), On errors in surveys, *American Sociological Review*, 9(4), 359–369.

Denzin, N. K. (1978), *The Research Act: An Introduction to Sociological Methods*, New York: McGraw-Hill.

Denzin, N. K. and Lincoln, Y. S. (Eds.) (2000), *Handbook of Qualitative Research*, London: Sage Publications.

Dept, S., Ferrari, A., and Wäyrynen, L. (2008), Comparative overview of test adaptation and translation verification procedures in three international surveys. Paper presented at the International Conference on Survey Methods in Multinational, Multiregional, and Multicultural Contexts (3MC), Berlin, Germany. http://www.csdiworkshop.org/pdf/3mc2008_proceedings/session_09/Dept.pdf.

DeVellis, R. F. (2005), Inter-rater reliability, in K. Kempf-Leonard (Ed.), *Encyclopedia of Social Measurement*, vol. 2, pp. 317–322, Oxford, UK: Elsevier.

Devins, G. M., Beiser, M., Dion, R., Pelletier, L. G., and Edwards, R. G. (1997), Cross-cultural measurements of psychological well-being: The psychometric equivalence of Cantonese, Vietnamese, and Laotian translations of the affect balance scale, *American Journal of Public Health*, 87, 794–799.

Dillman, D. A. (2004), The conundrum of mixed-mode surveys in the 21st century. Paper presented at the Eighth Conference on Health Survey Research Methods, Hyattsville, MD.

Dillman, D. A. (2007), *Mail and Internet Surveys: The Tailored Design Method*, Hoboken, NJ: John Wiley & Sons.

Dillman, D. A., Phelps, G., Tortora, R., Swift, K., Kohrell, J., Berck, J., et al. (2009), Response rate and measurement differences in mixed-mode surveys using mail, telephone, interactive voice response (IVR) and the Internet, *Social Science Research*, 38, 1–18.

Dillman, D. A., Smyth, J. D., and Christian, L. M. (2008), *Internet, Mail, and Mixed-Mode Surveys: The Tailored Design Method*, Hoboken, NJ: John Wiley & Sons.

Dolnicar, S. and Grün, B. (2007), Cross-cultural differences in survey response patterns, *International Marketing Review*, 24, 127–143.

Dotinga, A., van den Eijnden, R. J. J. M., Bosveld, W., and Garretsen, H. F. L. (2005), The effect of data collection mode and ethnicity of interviewer on response rates and self-reported alcohol use among Turks and Moroccans in the Netherlands: An experimental study, *Alcohol and Alcoholism*, 40(3), 242–248.

Drasgow, F. and Hulin, C. L. (1990), in M. D. Dunnette and L. M. Hough (Eds.), *Handbook of Industrial and Organizational Psychology*, Vol. 1 (2nd ed.), pp. 577–636, Palo Alto, CA: Consulting Psychologists Press.

Duch, R. (1994), Documentation on the France 1994 Election Study. Retrieved at http://www.raymondduch.com/economicvoting/firststage/countrypages/codebooks/France1994.pdf.

Duijker, H. C. J. and Rokkan, S. (1954), Organizational aspects of cross-national social research, *Journal of Social Issues*, 10, 8–24.

Duncan, G., Kalton, G., Kasprzyk, D., and Singh, M. P. (1989), *Panel Surveys*, New York: John Wiley & Sons.

Eades, D. (2006), Lexical struggle in court: Aboriginal Australians versus the state, *Journal of Sociolinguistics*, 10(2), 153–180.

Edgar, J. and Downey, K. (2008), That didn't make any sense, but let's move on: Issues with scripted probes in cognitive interviewing. Paper presented at the American Association for Public Opinion Research conference, May 15–18, 2008, New Orleans, LA.

Edwards, B. (2009), Taking stock: Pretesting in comparative contexts. Paper presented at the International Workshop on Comparative Survey Design and Implementation, Ann Arbor, MI.

Edwards, B., Schneider, S., and Brick, P. D. (2008), Visual elements of questionnaires design: Experiments with a cati establishment survey, in J. M. Lepkowski, C. Tucker, J. M. Brick, E. de Leeuw, L. Japec, P. J. Lavrakas, et al. (Eds.), *Telephone Survey Methodology*, pp. 276–296, Hoboken, NJ: John Wiley & Sons.

Ehling, M. (2003), Harmonising data in official statistics: Development, procedures, and data quality, in J. H. P. Hoffmeyer-Zlotnik und C. Wolf (Eds.), *Advances in Cross-National Comparison. A European Working Book for Demographic and Socio-Economic Variables*, pp. 17–31, New York: Kluwer Academic/Plenum Publishers.

Elder, J. W. (1976), Comparative cross-national methodology, *Annual Review of Sociology*, 2, 209–230.

Ellis, B. B. and Mead, A. D. (2002), Item analysis: Theory and practice using classical and modern test theory, in S. Rogelberg (Ed.), *Handbook of Research Methods in Industrial and Organizational Psychology*, Malden, MA: Blackwell Publishers.

Embretson, S. E. and Reise, S. P. (2000), *Item Response Theory for Psychologists*, Mahwah NJ: Lawrence Erlbaum.

Endo, Y., Heine, S. J., and Lehman, D. R. (2000), Culture and positive illusions in relationships: How my relationships are better than yours, *Personality and Social Psychology Bulletin*, 26, 1571–1586.

Eng, J. L. V., Wolkon, A., Frolov, A. S., Terlouw, D. J., Eliades, M. J., Morgah, K., et al. (2007), Use of handheld computers with global positioning systems for probability sampling and data entry in household surveys, *American Journal of Tropical Medicine and Hygiene*, 77(2), 393–399.

Erkut, S., Alarcón, O., García Coll, C., Tropp, L. R., and Vázquez García, H. A. (1999), The dual-focus approach to creating bilingual measures, *Journal of Cross-Cultural Psychology*, 30, 206–218.

Ervin, S. and Bower, R. T. (1952–53), Translation problems in international surveys, *Public Opinion Quarterly*, 16, 595–604.

Erwin Tripp, S. M. (1964), An analysis of the interaction of language, topic and listener, *Supplement to American Anthropologist*, 66(6), 86–102.

ESFRI (European Strategy Forum on Research infrastructures) (2006), European roadmap for research infrastructures. Brussels: European Commission.

Esping-Andersen, G. (1990), *The Three Worlds of Welfare Capitalism*, Cambridge: Polity Press.

Esselink, B. (2000), *A Practical Guide to Localization*, Amsterdam: Benjamins.

European Science Foundation (1999), *Blueprint for a European Social Survey*. (http://www.esf.org/index.php?eID=tx_nawsecuredl&u=0&file=fileadmin/be_user/research_areas/social_sciences/documents/ESSreport.pdf&t=1248742612&hash=3cb4ddba2ba136c6d31263a052678770).

European Social Survey (2005), *European Social Survey, Round 3: Specification for Participating Countries*, London: European Social Survey.

European Social Survey (2006), *Appendix A1, Population Statistics*. http://ess.nsd.uib.no/index.jsp?year=2007&country=&module=documentation.

Eurostat (2003), Methodological Documents: Definition of quality in statistics, Working Group Assessment of Quality in Statistics2, meeting 2–3 October, Item 4.2C, Luxembourg: Eurostat.

Eurostat (2007), Task force on core social variables. Final report, Luxembourg: Office for Official Publications of the European Communities.

Ewing, M. T., Salzberger, T., and Sinkovics, R. R. (2005), An alternative approach to assessing cross-cultural measurement equivalence in advertising research, *Journal of Advertising*, 34, 17–36.

Eysenck, H. J. and Eysenck, S. B. G. (1964), *The Manual of the Eysenck Personality Inventory*, London: University of London Press.

Feld, K. and Mohsini, Z. (2008), Survey research in Afghanistan; Challenges and results. Paper presented at the International Conference on Survey Methods in Multinational, Multiregional, and Multicultural Contexts, Berlin. http://www.csdiworkshop.org/ pdf/3mc2008_proceedings/session_20/feld_aug08.pdf.

Ferber, R. (1966), Item non-response in a consumer survey, *Public Opinion Quarterly*, 30, 399–415.

Fetzer, J. S. (2000), *Public Attitudes toward Immigration in the United States, France, and Germany*, Cambridge, UK: Cambridge University Press.

Fischer, A. H., Manstead, A. S. R., and Mosquera, P. M. R. (1999), The role of honour-related vs. individualistic values in conceptualising pride, shame, and anger: Spanish and Dutch cultural prototypes, *Cognition & Emotion*, 13, 149–179.

Fischer, R. (2004), Standardization to account for cross-cultural response bias: A classification of score adjustment procedures and review of research in JCCP, *Journal of Cross-Cultural Psychology*, 35, 263–282.

Fisher, K., Gershuny, J., and Gauthier, A. H. (2009), *Multinational Time Use Study: User's Guide and Documentation*. Retrieved from http://www.timeuse.org/mtus/ documentation/docs/OnLineDocumentation.pdf.

Fiske, A. P., Kitayama, S., Markus, H. R., and Nisbett, R. E. (1998), The cultural matrix of social psychology, in D. Gilbert, S. Fiske, and G. Lindzey (Eds.), *Handbook of Social Psychology* (3rd ed.), Vol. 2, pp. 915–981, New York: Random House.

Fitzgerald, R. and Miller, K. (2009), Design and analysis of cognitive interviews for cross-national testing. Paper presented at the International Workshop on Comparative Survey Design and Implementation, Ann Arbor, MI.

Fitzgerald, R., Widdop, S., Gray, M., and Collins, D. (2008), Identifying sources of error in cross-national questionnaires: A new tool and scheme. Paper presented at the RC33 Conference, Naples, Italy.

Flora, D. B. and Curran, P. J. (2004), An empirical evaluation of alternative methods of estimation for confirmatory factor analysis with ordinal data, *Psychological Methods*, 9, 466–491.

Fontaine, J. (2003), Multidimensional scaling, in J. A. Harkness, F. J. R. van de Vijver, and P. P. Mohler (Eds.), *Cross-Cultural Survey Methods*, Hoboken NJ: John Wiley & Sons.

Forsman, G. and Schreiner, I. (1991), The design and analysis of reinterview: An overview, in P. P. Biemer, R. M. Groves, L. E. Lyberg, N. A. Mathiowetz, and S. Sudman (Eds.), *Measurement Errors in Surveys*, pp. 279–301, New York: John Wiley & Sons.

Forsyth, B. H., Stapleton Kudela, M., Levin, K., Lawrence, D., and Willis, G. B. (2007), Methods for translating an English-language survey questionnaire on tobacco use into Mandarin, Cantonese, Korean, and Vietnamese, *Field Methods*, 19, 264–283.

Fowler, F. J. (2004), The case for more split-sample experiments in developing survey instruments, in S. Presser, J. M. Rothgeb, M. P. Couper, J. T. Lessler, E. Martin, J. Martin, et al. (Eds.), *Methods for Testing and Evaluating Survey Questionnaire*, pp. 453–472, Hoboken, NJ: John Wiley & Sons.

Fowler, F. J. and Cannell, C. F. (1996), Using behavioral coding to identify problems with survey questions, in N. Schwartz and S. Sudman (Eds.), *Answering Questions: Methodology for Determining Cognitive and Communicative Processes in Survey Research*, pp. 15–36, San Francisco: Jossey-Bass.

Fowler, F. and Mangione, T. (1986), Reducing interviewer effects on health survey data, Washington DC: National Center for Health Services Research.

Fox-Rushby, J. A. and Parker, M. (1995), Culture and the measurement of health-related quality of life, *European Review of Applied Psychology*, 45, 257–263.

Freeman, L. C., Romney, A. K., and Freeman, S. C. (1987), Cognitive structure and informant accuracy, *American Anthropologist*, 89, 310–325.

Frey, F. W., Stephenson, P., and Smith, K. A. (1969), *Survey Research on Comparative Social Change: A Bibliography*, Cambridge, MA: MIT Press.

Frick, J. R. and Grabka, M. M. (2005), Item non-response on income questions in panel surveys: Incidence, imputation and the impact on inequality and mobility, *Allgemeines Statistisches Archiv*, 89, 49–61.

Frick, J. R. and Grabka, M. M. (2007), Item non-response and imputation of annual labor income in panel surveys from a cross-national perspective. IZA Discussion Paper No. 3043, September 2007, Bonn: IZA.

Frick, J. R., Goebel, J., Schechtman, E., Wagner, G. G., and Yitzhaki, S. (2006), Using analysis of Gini (ANoGi) for detecting whether two sub-samples represent the same universe: The German Socio-Economic Panel Study (SOEP) experience, *Sociological Methods & Research*, 34(4), 427–468.

Frick, J. R., Jenkins S. P., Lillard, D. R., Lipps, O., and Wooden, M. (2007), The Cross-National Equivalent File (CNEF) and its member country household panel studies, *Schmoller's Jahrbuch–Journal of Applied Social Science Studies*, 127(4), 627–654.

Fu, Y. and Chu, Y. (2008), Different survey modes and international comparisons, in W. Donsbach and M. W. Traugott (Eds.), *The Sage Handbook of Public Opinion Research*, pp. 284–293, Thousand Oaks, CA: Sage Publications.

Gabler, S., Häder, S., and Lahiri, P. (1999), A model based justification of Kish's formula for design effects for weighting and clustering, *Survey Methodology*, 25(1), 105–106.

Gallup, Inc. (2007a), *Gallup World Poll Research Design*, Princeton, NJ: The Gallup Organization.

Gallup, Inc. (2007b), *The State of Global Well-Being 2007*, New York: Gallup Press.

Gauthier, A. H. (2002), The promises of comparative research, *Schmollers Jahrbuch/Journal of Applied Social Science Studies*, 122, 5–20.

Geer, J. G. (Ed.) (2004), *Public Opinion and Polling around the World: A Historical Encyclopedia*, Santa Barbara: ABC Clio.

Georgas, J., Weiss, L. G., van de Vijver, F. J. R., and Saklofske, D. H. (Eds.). (2003), *Culture and Children's Intelligence: Cross-cultural Analysis of the WISC-III*, Amsterdam, Boston: Academic Press.

Gerber, E. R., Wellens, T., and Keeley, C. (1996), Who lives here?: The use of vignettes in household roster research, *Proceedings of the Section on Survey Research Methods, American Statistical Association*, 962–967.

Gerbing, D. W. and Anderson, J. C. (1993), Monte Carlo evaluations of goodness-of-fit indices for structural equation models, *Sociological Methods & Research*, 21, 132–161.

GESIS-ZA (2009), ISSP Data Distribution Statistics 2008, Report to the ISSP General Assembly, Vienna, April 26–29.

Gibbons, J. L. and Stiles, D. A. (2004), *The Thoughts of Youth: An International Perspective on Adolescents' Ideal Persons*, Greenwich, CT: Information Age.

Gibbons, J. L., Zellner, J. A., and Rudek, D. J. (1999), Effects of language and meaningfulness on the use of extreme response style by Spanish-English bilinguals. *Cross-Cultural Research*, 33(4), 369–381.

Gibbons, R. and Hocevar, D. (1998), Levels of aggregation in higher level confirmatory factor analysis: Application for academic self-concept, *Structural Equation Modeling*, 5, 377–390.

Gilmore, D. D. (1987), *Honor and Shame and the Unity of the Mediterranean,* Washington, DC: American Anthropological Association.

Glewwe, P. (2005), Overview of the implementation of household surveys in developing countries, *Household Sample Surveys in Developing and Transition Countries,* pp. 54–66, New York: United Nations.

Georgas, J., Weiss, L. G., van de Vijver, F. J. R., and Saklofske, D. H. (Eds.), (2003), *Culture and Children's Intelligence: Cross-Cultural Analysis of the WISC-III,* Amsterdam, Boston: Academic Press.

Goerman, P. (2006a), Adapting cognitive interview techniques for use in pretesting Spanish language survey instruments (Survey Methodology No. 2006-03), Washington, DC: U.S. Census Bureau. Retrieved from http://www.census.gov/srd/papers/pdf/rsm2006-03.pdf.

Goerman, P. L. (2006b), An examination of pretesting methods for multicultural, multilingual surveys: The use of cognitive interviews to test Spanish instruments, conducting cross-national and cross-cultural surveys, *ZUMA Nachrichten Spezial,* 12, 67–79.

Goerman, P. L. and Caspar, R. (2007), A new methodology for the cognitive testing of translated materials: Testing the source version as a basis for comparison. Paper presented at the American Association for Public Opinion Research conference, May 17–20, 2007, Anaheim, California and submitted to 2007 JSM Proceedings, Statistical Computing Section [CD-ROM], Alexandria, VA: American Statistical Association, 3949–3956.

Goerman, P. and Clifton, M. (2009), The use of vignettes in cross-cultural cognitive testing of survey instruments. Paper presented at the Sixth International Workshop on Comparative Survey Design and Implementation, Ann Arbor, MI.

Goldstein, H. (2003), *Multilevel Statistical Models* (3rd ed.), London: Arnold.

Gordon, M. M. (1964), *Assimilation in American Life,* New York: Oxford University Press.

Gordon, R. G., Jr. (2005), *Ethnologue: Languages of the World,* Dallas, TX: Summer Institute of Linguistics.

Goyder, J. (1985), Nonresponse on surveys: A Canada-United States comparison, *Canadian Journal of Sociology / Cahiers Canadiens de Sociologie,* 10, 231–251.

Green, J. and Thorogood, N. (2004), *Qualitative Methods for Health Research,* London: Sage.

Greenacre, M. J. and Blasius, J. (Eds.) (1994), *Correspondence Analysis in the Social Sciences,* San Diego: Academic Press.

Greenacre, M. J. and Blasius J. (Eds.) (2006), *Multiple Correspondence Analysis and Related Methods,* Boca Raton, FL: Chapman & Hall.

Greenfield, P. M. (1997), You can't take it with you: Why ability assessments don't cross cultures, *American Psychologist,* 52(10), 1115–1124.

Greenleaf, E. A. (1992a), Improving rating scale measures by detecting and correcting bias components in some response styles, *Journal of Marketing Research,* 29, 176–188.

Greenleaf, E. A. (1992b), Measuring extreme response style, *Public Opinion Quarterly,* 56, 328–351.

Gregg, G. S. (2005), *The Middle East: A Cultural Psychology,* New York: Oxford University Press.

Gregg, G. S. (2007), *Culture and Identity in a Muslim Society,* New York: Oxford University Press.

Grice, H. P. (1975), Logic and conversation, in P. Cole and J. L. Morgan (Eds.), *Syntax and Semantics,* Vol. 3: *Speech Acts,* pp. 41–58, New York: Academic Press.

Griffin, M., Babin, B. J., and Christensen, F. (2004), A cross-cultural investigation of the materialism construct assessing Richins and Dawson's materialism scale in Denmark, France and Russia, *Journal of Business Research,* 57, 893–900.

Grisay, A. (1999), Report on the development of the French version of the PISA test material (OECD/PISA Report), Melbourne: Australian Council for Educational Research.

Grisay, A. (2003), Translation procedures in OECD/PISA 2000 international assessment, *Language Testing*, 20(2), 225–240.

Grosh, M. E. and Muñoz, J. (1996), *A Manual for Planning and Implementing the Living Standards Measurement Study Survey*, Washington, DC: World Bank.

Groves, R. M. (1989), *Survey Errors and Survey Costs*, New York: John Wiley & Sons.

Groves, R. M. (2006), Nonresponse rates and nonresponse bias in household surveys, *Public Opinion Quarterly*, 70(5), 646–675.

Groves, R. M. and Couper, M. P. (1998), *Nonresponse in Household Interview Surveys*, New York: John Wiley & Sons.

Groves, R. M. and Heeringa, S. G. (2006), Responsive design for household surveys: Tools for actively controlling survey errors and costs, *Journal of the Royal Statistical Society: Series A (Statistics in Society)*, 169, 439–457.

Groves, R. M. and McGonagle, K. A. (2001), A theory-guided interviewer training protocol regarding survey participation, *Journal of Official Statistics*, 17(2), 249–265.

Groves, R., Bethlehem, J., Medrano, J. D., Gundelach, P., and Norris, P. (2008), Report of the Review Panel for the European Social Survey, European Science Foundation, April.

Groves, R. M., Couper, M. P., Lepkowski, J. M., Singer, E., and Tourangeau, R. (2004), *Survey Methodology*, Hoboken, NJ: John Wiley & Sons.

Groves, R. M., Dillman, D., Eltinge, J. L., and Little, R. J. A. (Eds.) (2002), *Survey Nonresponse*, New York: John Wiley & Sons.

Groves, R. M., Fowler, F.J., Couper, M. P., Lepkowski, J. M., Singer, E., and Tourangeau, R. (2004), *Survey Methodology*, New York: John Wiley & Sons.

Groves, R. M., Fowler, F. J., Couper, M. P., Lepkowski, J. M., Singer, E., and Tourangeau, R. (2009), *Survey Methodology*, 2nd ed., Hoboken, NJ: John Wiley & Sons.

Groves, R. M., Lepkowski, J., Van Hoewyk, J., and Schulz, P. (2009), Real-time propensity models for responsive survey design and post-survey adjustment through propensity models: Comparisons of model fit and model specification. Paper presented at the annual meeting of the American Association for Public Opinion Research, Fontainebleau Resort, Miami Beach, FL.

Groves, R. M., Singer, E., and Corning, A. (2000), Leverage-saliency theory of survey participation. Description and illustration, *Public Opinion Quarterly*, 64, 299–308.

Gudykunst, W. B. (2005), *Theorizing About Intercultural Communication*, Thousand Oaks: Sage.

Guizot, F. (1890), *The History of Civilization*, Vols. 1–3, translated by William Hazlitt, New York: D. Appleton & Co.

Gumperz, J. J. (1982), *Discourse Strategies*, Cambridge: Cambridge University Press.

Gutt, A. (1991), *Translation and Relevance*, Oxford: Blackwell.

Guyer, H., and Cheung, G-Q (2007), Michigan Questionnaire Documentation System (MQDS): A user's perspective. Paper presented at the 11th International Blaise Users Conference, September, Annapolis, MD.

Haberstroh, S., Oyserman, D., Schwarz, N., Kühnen, U., and Ji, L. (2002), Is the interdependent self more sensitive to question context than the independent self? Self-construal and the observation of conversational norms, *Journal of Experimental Social Psychology*, 38, 323–329.

Häder, S. and Gabler, S. (2003), Sampling and estimation, in J. A. Harkness, van de Vijver, F. J. R., and P. P. Mohler (Eds.), *Cross-Cultural Survey Methods*, pp. 117–134, Hoboken, NJ: John Wiley & Sons.

Häder, S. and Lynn, P. (2007), How representative can a multi-nation survey be? in R. Jowell, C. Roberts, R. Fitzgerald, and G. Eva, (Eds.), *Measuring Attitudes Cross-Nationally: Lessons from the European Social Survey*, London: Sage.

Hadler, M. (2005), Why do people accept different income ratios? A multi-level comparison of thirty countries, *Acta Sociologica*, 48, 131–154.

Hagenaars, J. A. and McCutcheon, A. L. (2002), *Applied Latent Class Analysis*, Cambridge University Press.

Hall, E. T. (1959), *The Silent Language*, Greenwich, CT: Fawcett.

Hall, E. T. (1976), *Beyond Culture*, New York: Anchor.

Hall, E. T. and Hall, M. R. (1987), *Hidden Differences: Doing Business with the Japanese*, New York: Anchor Books/Doubleday.

Hall, R. J., Snell, A. F., and Foust, M. S. (1999), Item parceling strategies in SEM: Investigating the subtle effects of unmodeled secondary constructs, *Organizational Research Methods*, 2, 233–256.

Haller, M., Höllinger, F., and Gomilschak,, M. (1999), Attitudes toward gender roles in international comparison. New findings from twenty countries, in R. Richter and S. Supper (Eds.), *New Qualities in the Lifecourse*, Würzburg: Ergon.

Hambleton, R. K. (1993), Translating achievement tests for use in cross-national studies, *European Journal of Psychological Assessment*, 19, 1.

Hambleton, R. K. (2002), Adapting achievement tests into multiple languages for international assessments, in A. C. Porter and A. Gamoran (Eds.), *Methodological Advances in Cross National Surveys of Educational Achievement*, pp. 58–79, Washington, DC: National Academy Press.

Hambleton, R. K., Merenda, P. F., and Spielberger, C. D. (Eds.) (2005), *Adapting Educational and Psychological Tests for Cross-Cultural Assessment*, Mahwah, N.J.: L. Erlbaum Associates.

Hambleton, R. K., Swaminathan, H., and Rodgers, H. J. (1991), *Fundamentals of Item Response Theory*, Newbury Park, CA: Sage Publications.

Han, J. J., Leichtman, M. D., and Wang, Q. (1998), Autobiographical memory in Korean, Chinese, and American children, *Developmental Psychology*, 34, 701–713.

Hank, K., Jürges, H., Schupp, J., and Wagner, G. G. (2009): Isometrische Greifkraft und sozialgerontologische Forschung: Ergebnisse und Analysepotentiale des SHARE und SOEP, *Zeitschrift für Gerontologie und Geriatrie*, 42(2), 117–126.

Hantrais, L. (2009), *International Comparative Research: Theory, Methods and Practice*, Basingstoke: Palgrave MacMillan.

Hantrais, L. and Mangen, S. P. (Eds.) (2007), *Cross-National Research Methodology and Practice*, London: Routledge.

Harkness, J. A. (1995), ISSP Methodology Translation work group report. Report presented to the general assembly of the International Social Survey Programme, Cologne, Germany.

Harkness, J. A. (1996), Thinking aloud about survey translation. Paper presented at the International Sociological Association Conference on Social Science Methodology, Colchester, August.

Harkness, J. (Ed.) (1998), *Cross-Cultural Equivalence*, Mannheim, Germany: ZUMA.

Harkness, J. (1999), In pursuit of quality: Issues for cross-national research, *International Journal of Social Research Methodology*, 2, 125–140.

Harkness, J. A. (2002), Round 1 ESS translation strategies and procedures, European Social Survey (mimeo), London: City University.

Harkness, J. A. (2003), Questionnaire translation, in J. A. Harkness, F. van de Vijver, and P. Ph. Mohler (Eds.), *Cross-Cultural Survey Methods*, pp. 35–56, Hoboken, NJ: John Wiley & Sons.

Harkness, J. A. (2004), Overview of problems in establishing conceptually equivalent health definitions across multiple cultural groups, in S. B. Cohen and J. M. Lepkowski (Eds.), Eighth Conference on Health Survey Research Methods, pp. 85–90, Hyattsville, MD: US Department of Health and Human Services.

Harkness, J. (2005), SHARE translation procedures and translation assessment, in A. Börsch-Supan and H. Jürges (Eds.), The Survey of Health, Ageing and Retirement in Europe—Methodology, pp. 24–27, MEA: Mannheim.

Harkness, J. A. (Ed.) (2006), ZUMA-Nachrichten Spezial Band 12, Conducting Cross-National and Cross-Cultural Surveys: Papers from the 2005 Meeting of the International Workshop on Comparative Survey Design and Implementation (CSDI), Mannheim, Germany: ZUMA. Retrieved from http://www.gesis.org/fileadmin/upload/forschung/publikationen/zeitschriften/zuma_nachrichten_spezial/znspezial12.pdf.

Harkness, J. A. (2007), Improving the comparability of translation, in R. Jowell et al. (Eds.) Measuring Attitudes Cross-Nationally, pp. 79–94, London: Sage.

Harkness, J. (2008a), Comparative survey research: Goal and challenges, in E. de Leeuw, J. Hox, and D. Dillman (Eds.), International Handbook of Survey Methodology, pp. 56–77, Mahwah, NJ: Lawrence Erlbaum Associates.

Harkness, J. (2008b), Round 4 ESS translation strategies and procedures. Retrieved November 18, 2008, from http://www.europeansocialsurvey.org/index.php?option=com_content&task=view&id=66&Itemid=112.

Harkness, J. (2009), Comparative survey research: Goal and challenges, in E. D. de Leeuw, J. J. Hox, and D. A. Dillman (Eds.), International Handbook of Survey Methodology, pp. 56–77, New York: Lawrence Erlbaum/Psychology Press, Taylor & Francis Group.

Harkness, J. and Behr, D. (2008a), How instrument quality affects translation: Insights from a film of a team translation review, International Sociological Association Research Committee 33, Naples, Italy, September.

Harkness, J. and Behr, D. (2008b), Assessing "versions" in multilingual surveys. Paper presented at the ESS Quality Assurance Meeting, Lisbon, November.

Harkness, J. A. and Mohler, P. P. (1997), Modifying expressions in the context of agreement-disagreement: Some findings from the MINTS project. Appendix to J. A. Harkness, P. Ph. Mohler, T. W. Smith, and J. A. Davis, Final report on the project "Research into the methodology of intercultural surveys" (MINTS), Mannheim: ZUMA.

Harkness, J. A. and Schoua-Glusberg, A. (1998), Questionnaires in translation, in J. A Harkness (Ed.), Cross-Cultural Survey Equivalence, pp. 87–126, Mannheim, Germany: ZUMA.

Harkness, J., Chin, T-Z., Yang, Y., Bilgen, I., Villar, A., Zeng, W., et al. (2007), Developing answer scales in multilingual studies. Paper presented at the Fifth International Workshop on Comparative Survey Design and Implementation Workshop, Chicago, March.

Harkness, J. A., Dinkelmann, K., Pennell, B.-E., and Mohler, P. Ph. (2007), Final Report and Translation Tool Specifications, ESSi Project Delivery No. 5, Mannheim: ZUMA.

Harkness, J., Mohler, P. Ph., and van de Vijver, F. (2003a), Comparative research, in J. A. Harkness, F. van de Vijver, and P. Ph. Mohler (Eds.), Cross-Cultural Survey Methods, pp. 3–16, Hoboken, NJ: John Wiley & Sons.

Harkness, J, Mohler, P.Ph., and van de Vijver, F. (2003b), Questionnaire design in comparative research, in J. A. Harkness, F. van de Vijver, and P. Ph. Mohler (Eds.), Cross-Cultural Survey Methods, pp. 35–56, Hoboken, NJ: John Wiley & Sons.

Harkness, J. A., Mohler, P. Ph., Smith, T. W., and Davis, J. A. (1997), Final Report on the Project: Research into the Methodology of Intercultural Surveys (MINTS), Mannheim: ZUMA.

Harkness, J., Pennell, B.-E., and Schoua-Glusberg, A. (2004), Questionnaire translation and assessment, in S. Presser, J. Rothgeb, M. Couper, J. Lessler, E. Martin, J. Martin, et al. (Eds.), *Methods for Testing and Evaluating Survey Questionnaires*, pp. 453–473, Hoboken, NJ: John Wiley & Sons.

Harkness, J. A., Schoebi, N., Joye, D., Mohler, P., Faass, T., and Behr, D. (2008), Oral translation in telephone surveys, in J. M. Lepkowski, C. Tucker, J. M. Brick, E. de Leeuw, L. Japec, P. J. Lavrakas, et al. (Eds.), *Advances In Telephone Survey Methodology*, pp. 231–249, Hoboken, NJ: John Wiley & Sons.

Harkness, J., van de Vijver, F. J. R., and Johnson, T. P. (2003), Questionnaire design in comparative research, in J. A. Harkness, F. J. R. van de Vijver, and P. Ph. Mohler (Eds.), *Cross-Cultural Survey Methods*, pp. 19–34, Hoboken, NJ: John Wiley & Sons.

Harkness, J., van de Vijver, F. J. R., and Mohler, P. (Eds.) (2003), *Cross-Cultural Survey Methods*, Hoboken, NJ: John Wiley & Sons.

Harkness, J. A., Villar, A., Kephart, K., Behr, D., and Schoua-Glusberg, A. (2009a), *Research on Translation Assessment Procedures: Back Translation and Expert Review*, Presented at the VI International Workshop for Comparative Survey Design and Implementation (CSDI); Ann Arbor, Michigan.

Harkness, J. A., Villar, A., Kephart, K., Schoua-Glusberg, A., and Behr, D. (2009b). Survey translation evaluation: Back translation versus expert review. Paper presented at the American Association for Public Opinion Research meeting (AAPOR), Hollywood, Florida.

Harkness, J. A., Villar, A., Kruse, Y., Branden, L., Edwards, B., Steele, C., and Yang, Y. (2009c), Interpreting telephone survey interviews. Paper presented at the VI International Workshop for Comparative Survey Design and Implementation (CSDI), Ann Arbor, MI.

Harkness, J. A., Villar, A., Kruse, Y., Branden, L., Edwards, B., Steele, C., and Yang, Y. (2009d). Using interpreters in telephone surveys. Paper presented at the American Association for Public Opinion Research meeting (AAPOR), Hollywood, FL.

Harris, M. (1968), *The Rise of Anthropological Theory*, New York: Thomas Y. Crowell Company.

Harris-Kojetin, B. and Tucker, C. (1999), Exploring the relation of economic and political conditions with refusal rates to a government survey, *Journal of Official Statistics*, 15, 167–184.

Harrison, D. E. and Krauss, S. I. (2002), Interviewer cheating: Implications for research on entrepreneurship in Africa, *Journal of Developmental Entrepreneurship*, 7, 319–330.

Harzing, A.-W. (2006), Response styles in cross-national survey research: A 26-country study, *International Journal of Cross-Cultural Management*, 6, 243–266.

Hayati, D., Karami, E., and Slee, B. (2006), Combining qualitative and quantitative methods in the measurement of rural poverty: The case of Iran, *Social Indicators Research*, 75, 361–394.

Headland, T. N., Pike, K. L., and Harris, M. (1990), *Emics and Etics: The Insider/Outsider Debate,* Newbury Park, CA: Sage.

Heath, A., Fisher, S., and Smith, S. (2005), The globalization of public opinion research, *Annual Review of Political Science*, 8, 297–333.

Heckman, J. (1979), Sample selection bias as a specification error, *Econometrica*, 47, 153–161.

Heeringa, S. G., Wells, J. E., Hubbard, F., Mneimneh, Z., Chiu, W. T., Sampson, N., and Berglund, P. (2008), Sample designs and sampling procedures for the WHO World Mental Health Initiative Surveys, in R. C. Kessler and B. Ustan (Eds.), *The World Mental Health Initiative*, Volume I, Cambridge, UK: Cambridge University Press.

Heine, S. J. (2007), Culture and motivation: What motivates people to act in the ways that they do? in S. Kitayama and D. Cohen (Eds.), *Handbook of Cultural Psychology*, pp. 714–733, New York: Guilford Press.

Heine, S. J., Lehman, D. R., Markus, H. R., and Kitayama, S. (1999), Is there a universal need for positive self-regard? *Psychological Review*, 106, 766–794.

Herdman, M., Fox-Rushby, J., and Badia, X. (1998), A model of equivalence in the cultural adaptation of HRQoL instruments: The universalist approach, *Quality of Life Research*, 7, 323–335.

Herman, J. L. and Webb, N. M. (Eds.) (2007), Special issue of *Applied Measurement in Education: Alignment Issues*, 20, 1–135.

Hershfield, A. F., Rohling, N. G., Kerr, G. B., and Hursh-César, G. (1983), Fieldwork in rural areas, in M. Bulmer and D. P. Warwick (Eds.), *Social Research in Developing Countries*, pp. 241–252, London: Routledge.

Hill, D. and Willis, R. J. (2001), Reducing panel attrition: A search for effective policy instruments, *Journal of Human Resources*, 36, 416–438.

Ho, D. Y. F. (1976), On the concept of face, *American Journal of Sociology*, 81, 867–890.

Hofer, J. and Chasiotis, A. (2003), Congruence of life goals and implicit motives as predictors of life satisfaction: Cross-cultural implications of a study of Zambian male adolescents, *Motivation and Emotion*, 27, 251–272.

Hofer, J. and Chasiotis, A. (2004), Methodological considerations of applying a TAT-type picture-story-test in cross-cultural research: A comparison of German and Zambian adolescents, *Journal of Cross-Cultural Psychology*, 35, 224–241.

Hofer, J., Busch, H., Chasiotis, A., Kärtner, J., and Campos, D. (2008), Concern for generativity and its relation to implicit power motivation, generative goals, and satisfaction with life: A cross-cultural investigation, *Journal of Personality*, 76, 1–30.

Hofer, J., Chasiotis, A., and Campos, D. (2006), Congruence between social values and implicit motives: Effects on life satisfaction across three cultures, *European Journal of Personality*, 20, 305–324.

Hofer, J., Chasiotis, A., Friedlmeier, W., Busch, H., and Campos, D. (2005), The measurement of implicit motives in three cultures: Power and affiliation in Cameroon, Costa Rica, and Germany, *Journal of Cross-Cultural Psychology*, 36, 689–716.

Hoffmann, E. (2003), International Classification of Status in Employment, ICSE-93, in J. H. P. Hoffmeyer-Zlotnik and C. Wolf (Eds.), *Advances in Cross-National Comparison. A European Working Book for Demographic and Socio-Economic Variables*, pp. 125–136, New York: Kluwer Academic/Plenum Publishers.

Hoffmeyer-Zlotnik, J. H. P. (2008), Harmonisation of demographic and socio-economic variables in cross-national survey research, *Bulletin de Methodologie Sociologique*, 98, 5–24.

Hoffmeyer-Zlotnik, J. H. P. and Harkness, J. A. (Eds.) (2005), *ZUMA-Nachrichten Spezial Band 11, Methodological Aspects in Cross-National Research*, Mannheim, Germany: ZUMA. Retrieved from http://www.gesis.org/fileadmin/upload/forschung/ publikationen/zeitschriften/zuma_nachrichten_spezial/znspezial11.pdf.

Hofstede, G. (1980), *Culture's Consequences: International Differences in Work-Related Values*, Beverly Hills, CA: Sage.

Hofstede, G. (2001), *Culture's Consequences: Comparing Values, Behaviors, Institutions and Organizations Across Nations* (2nd ed.), Thousand Oaks, CA: Sage.

Holbrook, A. L., Green, M. C., and Krosnick, J. A. (2003), Telephone vs. face-to-face interviewing of national probability samples with long questionnaires: Comparisons of respondent satisficing and social desirability response bias, *Public Opinion Quarterly*, 67, 79–125.

Holsti, O. (1969), *Content Analysis for the Social Sciences and Humanities*, Reading, MA: Addison-Wesley.

Hönig, H. G. (1997), Position, power and practice: Functionalist approaches and translation quality assessment, *Current Issues in Language and Society,* 4(1), 6–34.

Hoodfar, H. (2008), Family law and family planning policy in pre- and post-revolutionary Iran, Chapter 4 in K. M. Yount and H. Rashad (Eds.), *Family in the Middle East: Ideational Change in Egypt, Iran, and Tunisia,* Oxford, UK: Routledge.

House, J. (1977/1981), *A Model for Translation Quality Assessment* (2nd ed.), Tübingen, Germany: Narr.

House, J. (1997), *Translation Quality Assessment: A Model Revisited,* Tübingen, Germany: Narr.

Howe, G. and McKay, A. (2007), Combining quantitative and qualitative methods in assessing chronic poverty: The case of Rwanda, *World Development,* 35, 197–211.

Howell, D. A. and Long Jusko, K. (2009), Methodological challenges, in H-D Klingemann (Ed.), *The Comparative Study of Electoral Systems* (pp. 50–84), New York: Oxford University Press.

Hox, J. (2002), *Multilevel Analysis: Techniques and Applications,* Mahwah, NJ: Erlbaum.

Hox, J. J. (2010), *Multilevel Analysis; Techniques and Applications* (2nd ed.), New York: The Psychology Press. (Forthcoming)

Hox, J. and de Leeuw, E. D. (2002), The influence of interviewers' attitude and behavior on household survey nonresponse: An international comparison, in R. M. Groves, D. Dillman, J. Eltinge, and R. Little (Eds.), *Survey Nonresponse,* pp. 103–120, New York: John Wiley & Sons.

Hox, J. J. and Maas, C. J. M. (2001), The accuracy of multilevel structural equation modeling with pseudobalanced groups and small samples, *Structural Equation Modeling,* 8(2), 157–174.

Hox, J. J., Maas, C. J. M., and Brinkhuis, M. (2009, submitted), The effect of estimation method and sample size in multilevel SEM. Paper, 6th International Multilevel Conference, Amsterdam, 2007.

Hoyle, L. and Wackerow, J. (2008), Exporting SAS datasets to DDI 3 XML Files: Data, metadata, and more metadata (http://www2.sas.com/proceedings/forum2008/137-2008.pdf).

Huddleston, R. and Pullum, G. K. (2005), *A Student's Introduction to English Grammar,* Cambridge: Cambridge University Press.

Hui, C. H. and Triandis, H. C. (1985), Measurement in cross-cultural psychology: A review and comparison of strategies, *Journal of Cross-Cultural Psychology,* 16, 131–152.

Hui, C. H. and Triandis, H. C. (1989), Effects of culture and response format on extreme response style, *Journal of Cross-Cultural Psychology,* 20, 296–309.

Hulin, C. L. (1987), A psychometric theory of evaluations of item and scale translations. Fidelity across languages, *Journal of Cross-Cultural Psychology,* 18, 115–142.

Hulin, C. L., Drasgow, F., and Parsons, C. K. (1983), *Item Response Theory: Application to Psychological Measurement,* Homewood, IL: Dow Jones–Irwin.

Hunter, J. and Landreth, A. (2006), Behavior coding analysis report: Evaluating bilingual versions of the non-response follow-up (NRFU) for the 2004 census test (Survey Methodology No. 2006-07), Washington, D.C.: U.S. Census Bureau. Retrieved from http://www.census.gov/srd/papers/pdf/rsm2006-07.pdf.

Iachan, R. and Dennis, M. L. (1993), A multiple frame approach to sampling the homeless and transient population, *Journal of Official Statistics,* 9(4), 747–764.

IDASA (Institute for Democracy in South Africa), and Center for Democracy and Development—Ghana, Michigan State University (2005–2006), *Afro-Barometer Survey Manual: Round 3 Surveys.*

IEA (International Association for the Evaluation of Educational Achievement), Guidelines for Translation and National Adaptations of the TIMSS 2007 Instruments.

Institute for Social and Economic Research (ISER) (2002), Weighting, imputation and sampling errors, Chapter 5, *British Household Panel Survey (BHPS) User Documentation*.

International IDEA (2008), Voter turnout. Accessed April 14, 2008, at http://www.idea.int/vt/index.cfm.

International Labour Office (ILO) (1990), International standard classification of occupations: ISCO-88, Genf: ILO.

International Standards Organization (2006), Standards for Marketing, Opinion, and Social Surveys, Standard 20252.

International Telecommunications Union (2007), *Measuring Village ICT in sub-Saharan Africa*, Geneva.

Inter-Parliamentary Union (2008), Parline database on national parliaments. Accessed April 14, 2008, at http://www.ipu.org/parline-e/parlinesearch.asp.

Iraq Family Health Survey Study Group (2008), Violence-related mortality in Iraq from 2002 to 2006, *New England Journal of Medicine*, 358(5), 484–493.

Iyengar, S. S., Lepper, M. R., and Ross, L. (1999), Independence from whom? Interdependence with whom? Cultural perspectives on ingroups versus outgroups, in D. Prentice and D. Miller (Eds.), *Cultural Divides*, pp. 273–301, New York: Sage.

Jabine, T. B., Straf, M. J., Tanur, J. M., and Tourangeau, R. (Eds.) (1984), *Cognitive Aspect of Survey Methodology: Building a Bridge between Disciplines,* Washington, DC: National Academy Press.

Jäckle, A., Roberts, C., and Lynn, P. (2006), Telephone versus face-to-face interviewing: Mode effects on data quality and likely causes, ISER Working Paper No. 2006-41, Colchester: University of Essex.

Jagodzinski, W. and Manabe, K. (2002), In search of Japanese religiosity, *Kwansei Gakuin University Social Sciences Review*, 7, 1–17.

Janis, I. L. (1951), *Air War and Emotional Stress: Psychological Studies of Bombing and Civilian Defense*, New York: McGraw-Hill.

Jansen, L. A. (2005), Local IRBs, multicenter trials, and the ethics of internal amendments, *IRB: Ethics and Human Research*, 27(4), 7–11.

Japec, L. (2005), Quality issues in interview surveys: Some contributions. PhD Thesis, Statistics Department, Stockholm University.

Jarvis, S. and Jenkins, S. P. (1998), How much income mobility is there in Britain? *Economic Journal,* 108(447), 428–443.

Javeline, D. (1999), Response effects in polite cultures: A test of acquiescence in Kazakhstan, *Public Opinion Quarterly*, 63, 1–28.

Ji, L., Schwarz, N., and Nisbett, R. E. (2000), Culture, autobiographical memory, and behavioral frequency reports: Measurement issues in cross-cultural studies, *Personality and Social Psychology Bulletin*, 26, 586–594.

Johnson, B. and Turner, L. A. (2003), Data collection strategies in mixed methods research, in A. Tashakkori and C. Teddlie (Eds.), *Handbook on Mixed Methods in the Behavioral and Social Sciences*, pp. 297–319, Thousand Oaks, CA: Sage Publications.

Johnson, D. R. and Creech, J. C. (1983), Ordinal measures in multiple indicator models: A simulation study of categorization error, *American Sociological Review*, 48, 398–407.

Johnson, T. P. (1998), Aproaches to establishing equivalence in cross-cultural and cross-national survey research, in J. Harkness (Ed.), *Cross-Cultural Survey Equivalence*, Mannheim, Germany: ZUMA-Nachrichten Spezial, 3, 1–40.

Johnson, T. P. and van de Vijver, F. (2003), Social desirability in cross-cultural research, in J. A. Harkness, F. van de Vijver, and P. Ph. Mohler (Eds.), *Cross-Cultural Survey Methods*, pp. 195–204, Hoboken, NJ: John Wiley & Sons.

Johnson, T. P., Cho, Y. I., Holbrook, A. L., O' Rourke, D., Warnecke, R. B, and Chávez, N. (2006), Cultural variability in the comprehension of health survey questions, *Annals of Epidemiology*, 16, 661–668.

Johnson, T. P., Kulesa, P., Cho, Y. I., and Shavitt, S. (2005), The relation between culture and response styles: Evidence from 19 countries, *Journal of Cross-Cultural Psychology*, 36, 264–277.

Johnson, T. P., O'Rourke, D., Burris, J., and Owens, L. (2002), Culture and survey nonresponse, in R. M. Groves, D. Dillman, J. L. Eltinge and R. J. A. Little (Eds.), *Survey Nonresponse*, pp. 55–69, Hoboken, NJ: John Wiley & Sons.

Johnson, T. P., O'Rourke, D., Chavez, N., Sudman, S., Warnecke, R., Lacey, L., et al. (1997), Social cognition and responses to survey questions among culturally diverse populations, in L. E. Lyberg, P. P. Biemer, M. Collin, E. D. de Leeuw, C. Dippo, N. Schwarz, et al. (Eds.), *Survey Measurement and Process Quality*, pp. 87–114, New York: John Wiley & Sons.

Joldersma, K. J. (2004), Cross-linguistic instrument comparability. Unpublished manuscript, Michigan State University.

Jöreskog, K. G. (1971), Simultaneous factor analysis in several populations, *Psychometrika*, 36, 409–426.

Jöreskog, K. G. (1990), New developments in LISREL: Analysis of ordinal variables using polychoric correlations and weighted least squares, *Quality and Quantity*, 24, 387–404.

Jöreskog, K.G. and Sörbom, D. (2001), LISREL 8.5 for Windows [computer software], Lincolnwood, IL: Scientific Software International, Inc.

Jowell, R. (1998), How comparative is comparative research? *American Behavioral Scientist*, 42(2), 168–177.

Jowell, R. (2008), Who needs attitude measurement at all and comparative measures in particular? PowerPoint presentation delivered on February 20, 2008. Accessed at http://www.natcen.ac.uk/natcen/pages/news_and_media_docs/RogerJ_slides_20_feb 08.pps#261,10,The ESS approach.

Jowell, R. and Eva, G. (2008), Cognitive evaluations as social indicators. Paper presented at the International Conference on Survey Methods in Multinational, Multiregional, and Multicultural Contexts (3MC), Berlin.

Jowell, R. and Eva, G. (2009), Happiness is not enough: Cognitive judgements as indicators of national wellbeing, *Social Indicators Research*, 91(3), 317–328.

Jowell, R., Kaase, M., Fitzgerald, R., and Eva, G. (2007a), The European Social Survey as a measurement model, in R. Jowell, C. Roberts, R. Fitzgerald, and G. Eva (Eds.), *Measuring Attitudes Cross-Nationally: Lessons from the European Social Survey*, pp. 1–31,Thousand Oaks, CA: Sage Publications.

Jowell, R., Roberts, C., Fitzgerald, R., and Eva, G. (Eds.) (2007b), *Measuring Attitudes Cross-Nationally: Lessons from the European Social Survey*, Thousand Oaks, CA: Sage Publications.

Juran, J. M. and Gryna, Jr, F. M. (1980), *Quality Planning and Analysis* (2nd ed.), New York: McGraw-Hill.

Jürges, H. (2005), Computing a Comparable Health Index, in A. Börsch-Supan et al. (Eds.), *Health, Ageing and Retirement in Europe. First Results from SHARE*, Mannheim: MEA, 357.

Jürges, H. (2007), Unemployment, life satisfaction and retrospective error, *Journal of the Royal Statistical Society: Series A (Statistics in Society)*, 170(1), 43–61.

Juster, F. T. and Suzman, R. (1995), An overview of the Health and Retirement Study, *Journal of Human Resources* 30, S7–S56.

Kalgraff Skjåk, K. and Harkness, J. A. (2003), Data collection methods, in J. A. Harkness, F. J. R. van de Vijver, and P. P. Mohler (Eds.), *Cross-Cultural Survey Methods*, pp. 179–193, Hoboken, NJ: John Wiley & Sons.

Kalsbeek, W. D. and Cross, A. R. (1982), Problems in sampling nomadic populations, *Proceedings of the American Statistical Association, Survey Research Methods Section*, pp. 398–402.

Kalton, G. (1977), Practical methods for estimating survey sampling errors, *Bulletin of the International Statistical Institute*, 47(3), 495–514.

Kalton, G., Lyberg, L., and Rempp, J.-M. (1998), Review of methodology, in U.S. Department of Education (1998), Adult Literacy in OECD Countries: Technical Report on the First International Adult Literacy Survey, Appendix A, NCES 98-053, S. Murray, I. Kirsch, and L. Jenkins (Eds.), Washington, DC.

Kalwij, A. and van Soest, A. (2005), Item non-response and alternative imputation procedures, in A. Börsch-Supan and H. Jürges (Eds.), *The Survey of Health, Ageing and Retirement in Europe—Methodology*, pp. 128–150, MEA: Mannheim.

Kane, E. W. and Macaulay, L. J. (1993), Interviewer gender and gender attitudes, *Public Opinion Quarterly*, 57(1), 1–28.

Kardam, F. (2005), *The Dynamics of Honor Killings in Turkey*, United Nations Development Program.

Karg, I. (2005), *Mythos PISA: Vermeintliche Vergleichbarkeit und die Wirklichkeit eines Vergleichs* [The myth PISA: Apparent comparability and the reality of a comparison], Gottingen: Vandenhoeck & Ruprecht.

Katz, D. and Hyman, H. (1954), Introduction, cross-national research: A case study, *Journal of Social Issues*, 10(4), 5–7.

Kazuo, K. (1995), Born-again Chinese religion, *Asian Folklore Studies*, 54, 127–130.

Keller, H. (2006), *Cultures of Infancy*, Mahwah, NJ: Erlbaum.

Kelley, J. and Evans, M. D. R. (1995), Class and class conflict in six western nations, *American Sociological Review*, 60, 157–178.

Kenny, D. (2009), Equivalence, in M. Baker and G. Saldanha (Eds.), *Routledge Encyclopedia of Translation Studies*, (2nd ed.), pp. 96–99, London: Routledge.

Kessler, R. C. and Üstün, T. B. (2008), Introduction, in R. C. Kessler and T. B. Üstün (Eds.), *Volume 1: Patterns of Mental Illness in the WMH Surveys*, New York: Cambridge University Press.

Kim, A. and Lim, B.-Y. (1999), How critical is back translation in cross-cultural adaptation of attitude measures? Paper presented at the American Educational Research Association, Montreal, Quebec, Canada.

King, G. and Wand, J. (2006), Comparing incomparable survey responses: Evaluating and selecting anchoring vignettes, *Political Analysis Advance Access*, 1093(10), 1–21.

King, G. and Wand, J. (2007), Comparing incomparable survey responses: Evaluating and selecting anchoring vignettes, *Political Analysis*, 15, 46–66.

King, G., Murray, C. J. L., Salomon, J. A., and Tandon, A. (2004), Enhancing the validity and cross-cultural comparability of measurement in survey research, *American Political Science Review*, 98(1), 198–207.

Kish, L. (1962), Studies of interviewer variance for attitudinal variables, *Journal of the American Statistical Association*, 57(297), 92–115.

Kish, L. (1965), *Survey Sampling*, New York: John Wiley & Sons.

Kish, L. (1994), Multipopulation survey designs: Five types with seven shared aspects, *International Statistical Review*, 62(2),167–186.

Kish, L., Groves, R. M., and Krotki, K. (1976), Sampling errors for fertility surveys, Occasional Paper No. 17, World Fertility Survey, Voorburg, Netherlands: International Statistical Institute.

Kitayama, S. and Cohen, D. (Eds.) (2007), *Handbook of Cultural Psychology*, New York: Guilford Press.

Kitayama, S. and Karasawa, M. (1997), Implicit self-esteem in Japan: Name letters and birthday numbers, *Personality and Social Psychology Bulletin*, 23, 736–742.

Kitayama, S. and Uchida, Y. (2003), Explicit self-criticism and implicit self-regard: Evaluating self and friend in two cultures, *Journal of Experimental Social Psychology*, 39, 476–482.

Kitayama, S., Duffy, S., and Uchida, Y. K. (2007), Self as cultural mode of being, in S. Kitayama and D. Cohen (Eds.), *The Handbook of Cultural Psychology,* pp. 136–174, New York: Guilford Press.

Klevmarken, A., Hesselius, P., and Swensson, B. (2005), The SHARE sampling procedures and calibrated design weights, in A. Börsch-Supan and H. Jürges (Eds.), *The Survey of Health, Ageing and Retirement in Europe—Methodology*, pp. 28–69, MEA: Mannheim,

Klungervik Greenall, A. (2009), Another way of saying the same thing: Gricean nondetachability and translation, in B. Frazer and K. Turner (Eds.), *Language in Life and a Life in Language: Jacob Mey—A festschrift*, pp. 121–130, Bingley, UK: Emerald.

Knudsen, K. (1997), Scandinavian neighbours with different character? Attitudes toward immigrants and national identity in Norway and Sweden, *Acta Sociologica*, 40, 223–243.

Knudsen, K. and Wærness, K. (1999), Reactions to global processes of change: Attitudes toward gender roles and marriage in modern nations, *Comparative Social Research. Family Change: Practices, Policies, and Values*, 18, 161–196.

Kohler, U. (1995), Individualisierung in der BRD 1953–1992. Ein empirischer Test der Individualisierungstheorie im Bereich des Wahlverhaltens, Mannheim: Master Thesis.

Kohn, M. L. (1987), Cross-national research as an analytical strategy, *American Sociological Review,* 52, 713–731.

Kok, F. K., Mellenbergh, G. J., and Van der Flier, H. (1985), Detecting experimentally induced item bias using iterative logit method, *Journal of Educational Measurement,* 22, 295–303.

Kolsrud, K., Skjåk, K. K., and Henrichsen, B. (2007), Free and immediate access to data, in R. Jowell et al. (Eds.), *Measuring Attitudes Cross-Nationally*, London: Sage Publications.

Kortmann, F. (1987), Problems in communication in transcultural psychiatry: The self-reporting questionnaire in Ethiopia, *Acta Psychiatrica Scandinavia*, 75, 563–570.

Krejci, J., Orten, H., Quandt, M., and Vavra, M. (2008), Strategy for collecting conversion keys for the infrastructure for data harmonisation, http://archiv.soc.cas.cz/download/640/WP9_T93report.doc.

Kreuter, F. and Kohler, U. (2007), Analyzing sequences of contacts. Presented at European Survey Research Association Conference, Prague.

Kreuter, F., Lemay, M., and Casas-Cordero, C. (2007), Using proxy measures of survey outcomes in post-survey adjustments: Examples from the European Social Survey (ESS), *Proceedings of the Survey Research Methods Section, American Statistical Association*, pp. 3142–3149.

Kreuter, F., Olson, K., Wagner, J., Yan, T., Ezzati-Rice, T. M., Casas-Cordero, C., et al. (2008), Using proxy measures and other correlates of survey outcomes to adjust for nonresponse: Examples from multiple surveys. Presented at the 19th International Workshop on Household Survey Nonresponse, Ljubljana.

Krippendorff, K. (1980), *Content Analysis: An Introduction to its Methodology,* Beverly Hills, CA: Sage.

Kroll, J. F. and De Groot, A. M. B. (Eds.) (2005), *Handbook of Bilingualism: Psycholinguistic Approaches*, New York: Oxford University Press.

Krosnick, J. A. (1991), Response strategies for coping with the cognitive demands of attitude measures in surveys, *Applied Cognitive Psychology*, 5(3), 213–236.

Krosnick, J. A. (1999), Survey research, *Annual Review of Psychology*, 50, 537–567.

Krosnick, J. A. and Fabrigar, L. (1997), Designing rating scales for effective measurement in surveys, in L. E. Lyberg, P. P. Biemer, M. Collins, L. Decker, E. deLeeuw, C. Dippo, et al. (Eds.), *Survey Measurement and Process Quality*, New York: John Wiley & Sons.

Krosnick, J. A., Narayan, S., and Smith, W. R. (1996a), Satisficing in surveys: Initial evidence, *New Directions for Program Evaluation*, 70, 29–44.

Krosnick, J. A., Narayan, S. S., and Smith, W. R. (1996b), Satisficing in surveys: Initial evidence, in M. T. Braverman and J. K. Slater (Eds.), *Advances in Survey Research*, pp. 29–44, San Francisco: Jossey-Bass.

Kuechler, M. (1987), The utility of surveys for cross-national research, *Social Science Research*, 16, 229–244.

Kuechler, M. (1998), The survey method: An indispensable tool for social science research everywhere? *American Behavioral Scientist*, 42(2), 178–200.

LaFromboise, T., Coleman, H. L. K., and Gerton, J. (1993), Psychological impact of biculturalism: Evidence and theory, *Psychological Bulletin*, 114, 395–412.

Lagos, M. (2008), International comparative surveys: Their purpose, content, and methodological challenges, in W. Donsbach and M. Traugott (Eds.), *Handbook of Public Opinion Research*, London: Sage.

Lalwani, A. K., Shavitt, S., and Johnson, T. (2006), What is the relation between cultural orientation and socially desirable responding? *Journal of Personality and Social Psychology*, 90, 165–178.

Landis, R. S., Beal, D. J., and Tesluk, P. E. (2000), A comparison of approaches to forming composite measures in structural equation models, *Organizational Research Methods*, 3, 186–207.

Lanham, R. A. (1974), *Style: An anti-textbook,* New Haven and London: Yale University Press.

Lazersfeld, P. F. (1950), The logic and foundation of latent structure analysis, in S. A. Stouffer, L. Guttman, E. A. Suchman, P. F. Lazersfeld, S. A. Star, and J. A. Clausen (Eds.), *Studies in Social Psychology in World War II*, Vol. IV, pp. 362–412, Princeton, NJ: Princeton University Press.

Lazarsfeld, P. F. (1952–53), The prognosis for international communications research, *Public Opinion Quarterly*, 16, 481–490.

Lazarsfeld, P. F. and Henry, N. W. (1968), *Latent Structure Analysis*, Boston: Houghton Mifflin.

Lee, E., Hu, M. Y., and Toh, R. S. (2004), Respondent non-cooperation in surveys and diaries: An analysis of item non-response and panel attrition, *International Journal of Market Research*, 46(3), 311–326.

Lee, J. W., Jones, P. S., Mineyama, Y., and Zhang, X. E. (2002), Cultural differences in responses to a Likert scale, *Research in Nursing and Health*, 25, 295–306.

Leech, N. L. and Onwuegbuzie, A. J. (2008), A typology of mixed methods research designs, *Quality and Quantity*, 2008 (available online).

Leitch, T. (2008), Adaptation studies at a crossroads, *Adaptation,* 1(1), 63–77.

Lensvelt-Mulders, G. (2008), Surveying sensitive topics, in E. D. de Leeuw, J. J. Hox, and D. A. Dillman (Eds.), *International Handbook of Survey Methodology,* New York: Psychology Press.

LeTendre, G. K. (2002), Advancements in conceptualizing and analyzing cultural effects in cross-national studies of educational achievement, in A. C. Porter and A. Gamoran

(Eds.), *Methodological Advances in Cross-National Surveys of Educational Achievement*, pp. 198–230, Washington, DC: National Academy Press.

Levin, J., Montag, I., and Comrey, A. L. (1983), Comparison of multitrait-multimethod, factor, and smallest space analysis on personality scale data, *Psychological Reports*, 53, 591–596.

Li, Q. (2003), The two distinct psychological spheres and opinion surveys in China, in *Shehuixue Yanjiu* (*Sociology Research*), pp. 40–44, Beijing: Chinese Academy of Social Sciences.

Lievesley, D. (2001), Making a difference: A role for the responsible international statistician? *The Statistician*, 50, 4, 367–406.

Lijtmaer, R. M. (1999), Language shift and bilinguals: Transference and counter-transference, *Journal of the American Academy of Psychoanalysis*, 27, 611–624.

Lincoln, Y. S. and Guba, E. G. (1985), *Naturalistic Inquiry*, Newbury Park, CA: Sage Publications.

Lincoln, Y. S. and Guba, E. G. (1986), But is it rigorous? Trustworthiness and authenticity in naturalistic evaluation, *New Directions for Program Evaluation*, 30, 73–84.

Linn, R. L. (2002), The measurement of student achievement in international studies, in A. C. Porter and A. Gamoran (Eds.), *Methodological Advances in Cross-National Surveys of Educational Achievement*, pp. 27–57, Washington, DC: National Academy Press.

Linshu, C. (1994), Studies on religions in modern China, *Numen*, 76–87.

Lipps, O. and Benson, G. (2005), Cross-national contact strategies, *Proceedings of the Survey Research Methods Section, American Statistical Association*, 3905–3914.

Lipset, S. M. (1986), Historical traditions and national characteristics—A comparative analysis of Canada and the United States, *Canadian Journal of Sociology*, 11(2), 113–155.

Little, R. J. A. (1988), Missing-data adjustments in large surveys, *Journal of Business and Economic Statistics*, 6, 287–296.

Little, R. J. A. and Rubin, D. B. (2002), *Statistical Analysis with Missing Data* (2nd ed.), Hoboken, NJ: John Wiley & Sons.

Little, R. J. A. and Su, H. L. (1989), Item non-response in panel surveys, in D. Kasprzyk, G. Duncan, and M. P. Singh (Eds.), *Panel Surveys*, pp. 400–425, New York: John Wiley & Sons.

Little, T. D. (2000), On the comparability of constructs in cross-cultural research: A critique of Cheung and Rensvold, *Journal of Cross-Cultural Psychology*, 31, 213–219.

Loosveldt, G., Carton, A., and Billiet, J. (2004), Assessment of survey data quality: A pragmatic approach focussed on interviewer tasks, *International Journal of Market Research*, 46, 65–82.

Loosveldt, G., Pickery, J., and Billiet, J. (2002), Item non-response as a predictor of unit non-response in a panel survey, *Journal of Official Statistics*, 18, 545–557.

Lowenthal, L. (1952), Introduction, special issue on international communications research, *Public Opinion Quarterly*, 16, v–x.

Lucas, R. E. and Diener, E. (2008), Can we learn about national differences in happiness from individual responses? A multilevel approach, in F. J. R. van de Vijver, D. A. Van Hemert, and Y. H. Poortinga (Eds.), *Individuals and Cultures in Multilevel Analysis*, pp. 221–246, Mahwah, NJ: Erlbaum.

Luz Guerrero, L. and Mangahas, M. (2004), The Philippines, in J. G. Geer (Ed.), *Public Opinion and Polling around the World: A Historical Encyclopedia*, Vol. 2, pp. 689–697, Santa Barbara: ABC-CLIO.

Lyberg, L. E. and Biemer, P. (2008), Quality assurance and quality control in surveys, in E. D. de Leeuw, J. J. Hox, and D. A. Dillman (Eds.), *International Handbook of Survey Methodology*, New York, London: Erlbaum, Taylor & Francis.

Lyberg, L. and Dean, P. (1992), Methods for reducing nonresponse rates: A review. Paper presented at the annual meeting of the American Association for Public Opinion Research, St. Petersburg, FL.

Lyberg, L., Biemer, P. P., Collins, M., de Leeuw, E., Dippo, C., Schwarz, N., et al. (Eds.) (1997) *Survey Measurement and Process Quality*, New York: John Wiley & Sons.

Lynn, P. (2001), Developing quality standards for cross-national survey research: Five approaches, *International Journal of Social Research Methodology*, 6(4), 323–337.

Lynn, P. (Ed.) (2009), *Methodology of Longitudinal Surveys,* Chichester, UK: John Wiley & Sons.

Lynn, P., Beerten, R., Laiho, J., and Martin, J. (2001), Recommended standard final outcome categories and standard definitions of response rate for social surveys, ISER Working Paper 2001-23, Colchester: University of Essex.

Lynn, P., Clarke, P., Martin, J., and Sturgis, P. (2002), The effects of extended interviewer efforts on nonresponse bias, in D. A. Dillman, J. D. Eltinge, R. M. Groves, and R. Little (Eds.), *Survey Nonresponse,* New York: John Wiley & Sons.

Lynn, P., Häder, S., Gabler, S., and Laaksonen, S. (2007), Methods for achieving equivalence of samples in cross-national surveys: The European Social Survey experience, *Journal of Official Statistics*, 23(1), 107–124.

Lynn, P., Japec, L., and Lyberg, L. (2006), What's so special about cross-national surveys? in J. A. Harkness (Ed.), *Conducting Cross-National and Cross-Cultural Surveys*, pp. 7–20, *ZUMA-Nachrichten Spezial*, 12, Mannheim: ZUMA.

Maas, C. J. M. and Hox, J. J. (2005), Sufficient sample sizes for multilevel modeling, *Methodology. European Journal of Research Methods for the Behavioral and Social Sciences*, 1, 85–91.

MacIsaac, D. (1976), *Strategic Bombing in World War Two: The Story of the United States Strategic Bombing Survey*, New York: Garland.

Maclure, M. and Willett, W. C. (1987), Misinterpretation and misuse of the kappa statistic, *American Journal of Epidemiology*, 126, 161–169.

Magidson, J. and Vermunt, J. K. (2004), Latent class models, in D. Kaplan (Ed.), *The Sage Handbook of Quantitative Methodology for the Social Sciences*, pp. 175–198, Thousand Oaks, CA: Sage.

Magidson, J. and Vermunt, J. K. (2005), Structural equation models: Mixture models, in B. Everitt and D. Howell (Eds.), *Encyclopedia of Statistics in Behavioral Science*, pp. 1922–1927, Chichester, UK: John Wiley & Sons.

Makaronidis, A. (2008), The EU harmonized index of consumer prices: Output harmonization for consumer price indices in the EU. Paper presented at the International Conference on Survey Methods in Multinational, Multiregional, and Multicultural Contexts (3MC), Berlin, June 25–28.

Mandelbaum, M. (1971), *History, Man, and Reason: A Study in Nineteenth-Century Thought*, Baltimore: The John Hopkins Press.

Marian, V. and Kaushanskaya, M. (2004), Self-construal and emotion in bicultural bilinguals, *Journal of Memory and Language*, 51, 190–201.

Marian, V. and Neisser, U. (2000), Language dependent recall of autobiographical memories, *Journal of Experimental Psychology: General*, 129, 361–368.

Marín, G. and Marín, B. V. (1991), Potential problems in interpreting data, in G. Marín and B. V. Marín (Eds.), *Research with Hispanic Populations*, pp. 101–123, Newbury Park, CA: Sage.

Marín, G., Gamba, R. J., and Marín, B. V. (1992), Extreme response style and acquiescence among Hispanics: The role of acculturation and education, *Journal of Cross-Cultural Psychology*, 23, 498–509.

Markus, H. R. and Kitayama, S. (1991), Culture and the self: Implications for cognition, emotion, and motivation, *Psychological Review*, 98, 224–253.

Marmot, M., Banks, J., Blundell, R., Lessof, C., and Nazroo, J. (Eds.) (2003), *Health, Wealth, and Lifestyles of the Older Population in England: The 2002 English Longitudinal Study of Ageing*, IFS: London.

Marshall, P. A. (2001), The relevance of culture for informed consent in US-funded international health research, Bethesda, MD: National Bioethics Advisory Commission.

Martin, E. A. (2004), Vignettes and respondent debriefing for questionnaire design and evaluation, in S. Presser et al. (Eds.), *Methods for Testing and Evaluating Survey Questionnaires*, pp. 149–172, Hoboken, NJ: John Wiley & Sons.

Martin, E. A. (2006), *Vignettes and Respondent Debriefings for Questionnaire Design and Evaluation* (Survey Methodology No. 2006-08). Washington, DC: U.S. Census Bureau. Retrieved from http://www.census.gov/srd/papers/pdf/rsm2006-08.pdf.

Martin, M. O., Mullis, I. V. S., and Chrostowski, S. J. (Eds.) (2004), TIMSS 2003 Technical Report, Chestnut Hill, MA: Boston College.

Martin, M. O., Mullis, I. V. S., Gonzalez, E. J., and Chrostowski, S. J. (2004), TIMSS 2003 International Science Report: Findings from IEA's Trends in International Mathematics and Science Study at the fourth and eighth grades, Chestnut Hill, MA: Boston College.

Martin, M. O., Mullis, I. V. S., and Kennedy, A. M. (Eds.), (2007), PIRLS 2006 Technical Report, Chestnut Hill, MA: Boston College.

Maxwell, B. (1996), Translation and cultural adaptation of the survey instruments, in M. O. Martin and D. L. Kelly (Eds.), *Third International Mathematics and Science Study (TIMSS) Technical Report, Volume I: Design and Development*, Chestnut Hill, MA: Boston College.

McAdams, D. P. (1982), Experiences of intimacy and power: Relationships between social motives and autobiographical memory, *Journal of Personality and Social Psychology*, 42, 292–302.

McCarty, C. (2003), Differences in response rates using most recent versus final dispositions in telephone surveys, *Public Opinion Quarterly*, 67, 396–406.

McClelland, D. C. (1987), *Human Motivation*, Cambridge: Cambridge University Press.

McClelland, D. C. and Pilon, D. A. (1983), Sources of adult motives in patterns of parent behavior in early childhood, *Journal of Personality and Social Psychology*, 44, 564–574.

McCutcheon, A. L. (1987), *Latent Class Analysis*, Thousand Oaks, CA: Sage.

McDonald, R. P. (1989), An index of goodness-of-fit based on non-centrality, *Journal of Classification*, 6, 97–103.

McGraw, K. O. and Wong, S. P. (1996), Forming inferences about some intraclass correlation coefficients, *Psychological Methods*, 1, 30–46.

McInnes, J. (1998), Analysing patriarchy capitalism and women's employment in Europe, *Innovation*, 11, 227–248.

McKay, R. B., Breslow, M. J., Sangster, R. J., Gabbard, S. M., Reynolds, R. W., Nakamoto, et al. (1996), Translating survey questionnaires: Lessons learned, *New Directions for Evaluation*, 70 (Summer), 93–105.

Mehta, P. D. and Neale, M.C. (2005), People are variables too: multilevel structural equations modeling, *Psychological Methods*, 10, 259–284.

Menon, G., Raghubir, P., and Schwarz, N. (1995), Behavioral frequency judgments: An accessibility-diagnosticity framework, *Journal of Consumer Research*, 22, 212–228.

Meredith, W. (1964), Notes on factorial invariance, *Psychometrika*, 29, 177–185.

Meredith, W. (1993), Measurement invariance, factor-analysis and factorial invariance, *Psychometrika*, 58, 525–543.

Messick, S. (1989), Validity, in R. L. Linn (Ed.), *Educational Measurement* (3rd ed.), pp. 13–104, London: Collier Macmillan Publishers.

Micklewright, J. and Schnepf, S. V. (2007), How reliable are income data collected with a single question? IZA Discussion Paper No. 3177, November, Bonn: IZA.

Mill, J. (1848), *The History of British India*, Vol. 1, London: James Madden.

Millar, J. ([1779] 1979), The origin of the distinction of ranks, in W. C. Lehmann (Ed.), *John Millar of Glasgow 1735–1801*, New York: Arno Press.

Miller, K. (2004), Implications of socio-cultural factors in the question response process, in P. Prufer, M. Rexroth, and F. J. J. Fowler (Eds.), *ZUMA-Nachrichten Spezial Band 9, QUEST 2003: Proceedings of the 4th Conference on Questionnaire Evaluation Standards*, pp. 172–189, Mannheim, Germany: ZUMA. Retrieved from http://www.gesis.org/fileadmin/upload/forschung/publikationen/zeitschriften/zuma_nachrichten_spezial/znspezial9.pdf.

Miller, K. (2005), Q-Bank: Development of a tested-question database, *Proceedings of the ASA Section on Government Statistics*, Alexandria, VA: American Statistical Association.

Miller, K., Eke, P. I., and Schoua-Glusberg, A. (2007), Cognitive evaluation of self-report questions for surveillance of periodontitis, *Journal of Periodontology*, 78, 1455–1462.

Miller, K., Fitzgerald, R., Caspar, R., Dimov, M., Gray, M., Nunes, C., et al. (2008), Design and analysis of cognitive interviews for cross-national testing. Paper presented at the International Conference on Survey Methods in Multinational, Multiregional, and Multicultural Contexts (3MC), Berlin, Germany, June 25–28, 2008.

Miller, K. E., Miller, O. P., Samad Quraishy, A., Quraishy, N., Nader Nasiry, M., Nasiry, S., et al. (2006), The Afghan Symptom Checklist: A culturally grounded approach to mental health assessment in a conflict zone, *American Journal of Orthopsychiatry*, 76, 423–433.

Miller, K., Mont, D., Maitland, A., Altman, B., and Madans, J. (2008), Implementation and results of a cross-national, structured-interview cognitive test. Paper presented at the International Conference on Survey Methods in Multinational, Multiregional and Multicultural Contexts, Berlin, Germany.

Miller, K., Willis, G., Eason, C., Moses, L., and Canfield, B. (2005a), Interpreting the results of cross-cultural cognitive interviews: A mixed-method approach, *ZUMA-Nachrichten Spezial, Issue* #10, 79–92.

Miller, K., Willis, G. B., Eason, C., Moses, L., & Canfield, B. (2005b), Interpreting the results of cross-cultural cognitive interviews: A mixed-method approach, in J. H. P. Hoffmeyer-Zlotnik and J. A. Harkness (Eds.), *ZUMA-Nachrichten Spezial Cross-National Research, Band 11*, pp. 79–92, Mannheim, Germany: ZUMA. Retrieved from http://www.gesis.org/fileadmin/upload/forschung/publikationen/zeit schriften/zuma_nachrichten_spezial/znspezial11.pdf.

Miller, M. J., Woehr, D. J., and Hudspeth, N. (2002), The meaning and measurement of work ethic: Construction and initial validation of a multidimensional inventory, *Journal of Vocational Behavior*, 60, 451–489.

Millsap, R. E. and Yun-Tein, J. (2004), Assessing factorial invariance in ordered-categorical measures, *Multivariate Behavioral Research*, 39, 479–515.

Minges, M. and Simkhada, P. (2002), A closer look at South Asia, *ITU News*. Retrieved from http://www.itu.int/itunews/issue/2002/10/ southasia.html.

Mohler, P. P. (2006), Sampling from a universe of items and the de-Machiavellization of questionnaire design, in M. Braun and P. P. Mohler (Eds.), *Beyond the Horizon of Measurement: Festschrift in Honor of Ingwer Bborg*, pp. 9–14, Mannheim: ZUMA. Retrieved from http://www.gesis.org/fileadmin/upload/forschung/publikationen/zeit schriften/zuma_nachrichten_spezial/znspezial10.pdf.

Mohler, P. (2007), What is being learned from the ESS? in R. Jowell et al. (Eds.), *Measuring Attitudes Cross-Nationally*, London: Sage Publications.

Mohler, P. (2008), Laudatio auf Bernhard von Rosenbladt, German Council for Social and Economic Data, Working Paper 35, Berlin: RatSWD.

Mohler, P. Ph. and Uher, R. (2003), Documenting comparative surveys for secondary analysis, in J. A. Harkness, F. J. R. van de Vijver and P. Ph. Mohler (Eds.), *Cross-Cultural Survey Methods*, pp. 311–327, Hoboken, NJ: John Wiley & Sons.

Mohler, P. Ph. and Züll, C. (2000), Observe! A Popperian critique of automatic content analysis, in Actes JADT, Paris: Lexicometrica.

Mohler, P. Ph. and Züll, C. (2001), Applied text theory: Quantitative analysis of answers to open-ended questions, in M. D. West (Ed.), *Applications of Computer Content Analysis*, S. 1–16, Westport, CT, Aplex Publishing.

Mohler, P. P., Pennell, B. E., and Hubbard, F. (2008), Survey documentation: Towards professional knowledge management in sample surveys, in E. D. de Leeuw, J. J. Hox and D. A. Dillman (Eds.), *International Handbook of Survey Methodology*, pp. 403–420, New York: Lawrence Erlbaum.

Mohler, P. Ph., Smith, T. W., and Harkness, J. A. (1998), Respondents' ratings of expressions from response scales: A two-country, two-language investigation on equivalence and translation, in J. A. Harkness (Ed.), *ZUMA-Nachrichten Spezial Band 3: Cross-cultural survey equivalence*, pp. 159–184, Mannheim, Germany: ZUMA. Retrieved from http://www.gesis.org/fileadmin/upload/forschung/publikationen/zeit schriften/zuma_nachrichten_spezial/znspezial3.pdf.

Mojab, S. and Abdo, N. (2004), *Violence in the Name of Honour: Theoretical and Political Challenges,* Istanbul: Istanbul Bilgi University Press.

Moors, G. (2003), Diagnosing response style behavior by means of a latent-class factor approach: Sociodemographic correlates of gender role attitudes and perceptions of ethnic discrimination reexamined, *Quality and Quantity*, 37(33), 277–302.

Moors, G. (2004), Facts and artifacts in the comparison of attitudes among ethnic minorities: A multigroup latent class structure model with adjustment for response style behavior, *European Sociological Review*, 20, 303–320.

Moors, G. (2008), Exploring the effect of a middle response category on response style in attitude measurement, *Quality and Quantity*, 42(6), 779–794.

Morganstein, D. and Marker, D. (1997), Continuous quality improvement in statistical agencies, in L. Lyberg, P. Biemer, M. Collins, E. de Leeuw, C. Dippo, N. Schwarz, and D. Trewin (Eds.), *Survey Measurement and Process Quality*, New York: John Wiley & Sons.

Morse, J. M. (2003), Principles of mixed methods and multimethod research design, in A. Tashakkori and C. Teddlie (Eds.), *Handbook on Mixed Methods in the Behavioral and Social Sciences*, pp. 189–208, Thousand Oaks, CA: Sage Publications.

Morton-Williams, J. (1993), *Interviewer Approaches*, Dartmouth: Aldershot.

MOT (1997), Approaches to instrument translation: Issues to consider, *Medical Outcomes Trust Bulletin* 5(4). http://www.outcomes-trust.org/bulletin/0797blltn.htm.

Mullis, I. V. S., Kennedy, A. M., Martin, M. O., and Sainsbury, M. (2006), PIRLS 2006 Assessment framework and specifications (2nd ed.), Chestnut Hill, MA: Boston College.

Mullis, I. V. S., Martin, M. O., Gonzalez, E. J., and Chrostowski, S. J. (2004), TIMSS 2003 International Mathematics Report: Findings from IEA's trends in international mathematics and science study at the fourth and eighth grades, Chestnut Hill, MA: Boston College.

Mullis, I. V. S., Martin, M. O., Kennedy, A. M., and Foy, P. (2007), PIRLS 2006 International Report: IEA's progress in International Reading Literacy Study in primary schools in 40 countries, Chestnut Hill, MA: Boston College.

Mullis, I. V. S., Martin, M. O., Ruddock, G. J., O'Sullivan, C. Y., Arora, A., and Erberber, E. (2005), TIMSS 2007 Assessment frameworks, Chestnut Hill, MA: Boston College.

Muthén, B. (1984), A general structural equation model with dichotomous, ordered categorical, and continuous latent variable indicators, *Psychometrika*, 49, 115–132.

Muthén, B. (1991), Multilevel factor analysis of class and student achievement components, *Journal of Educational Measurement*, 28, 338–354.

Muthén, B. and Asparouhov, T. (2002), Latent variable analysis with categorical outcomes: Multiple-group and growth modeling in MPlus. Mplus Web Notes.

Muthén, B. and Lehman, J. (1985), Multiple group IRT modeling: Applications to item bias analysis, *Journal of Educational Statistics*, 10, 133–142.

Muthén, L. K. and Muthén, B. O. (2007), *Mplus User's Guide* (5th ed.), Los Angeles, CA: Muthén & Muthén.

Nápoles-Springer, A. M., Santoyo-Olsson, J., O'Brien, H., Stewart, A. L. (2006), Using cognitive interviews to develop surveys in diverse populations, *Medical Care*, 44(11, suppl. 3), S21–S30.

Nastasi, B. K., Hitchcock, J. H., Burkholder, G., Varjas, K., Sarkar, S., and Jayasena, A. (2007), Assessing adolescents' understanding of and reactions to stress in different cultures: Results of a mixed-methods approach, *School Psychology International*, 28, 163–178.

National Bioethics Advisory Commission (2001), *Ethical and Policy Issues in International Research: Clinical Trials in Developing Countries, Volume II: Commissioned Papers and Staff Analysis*, Springfield, VA: U.S. Department of Commerce, NTIS.

National Information Standards Organization (2004), Understanding metadata, Bethesda, MD: NISO Press and http://www.niso.org/publications/press/UnderstandingMetadata.pdf.

Neuendorf, K. A. (2001), *The Content Analysis Guidebook*, Thousand Oaks, CA: Sage.

Nicholls, W. L., II and Kindel, K. K. K. (1993), Case management and communications for computer assisted personal interviewing, *Journal of Official Statistics*, 9(3), 623–639.

Nicoletti, C. and Buck, N. H. (2004), Explaining interviewee contact and co-operation in the British and German Household Panels, in M. Ehling and U. Rendtel (Eds.), *Harmonisation of Panel Surveys and Data Quality,* Wiesbaden: Federal Statistical Office.

Nicoletti, C. and Peracchi, F. (2006), The effects of income imputation on microanalyses: Evidence for the European Community Household Panel, *Journal of the Royal Statistical Society Series A*, 169(3), 625–646.

Nigerian Federal Ministry of Health (2007), National code of health research ethics, Abuja, Nigeria: Nigerian Federal Ministry of Health.

Nisbet, R. A. ([1969] 1975), *Social Change and History*, New York: Oxford University Press.

Nisbett, R. E. (2003), *The Geography of Thought: How Asians and Westerners Think Differently ...and Why,* New York: Free Press.

Nisbett, R. E. (2004), *The Geography of Thought*, New York: Simon & Schuster.

Nisbett, R. E. and Cohen, D. (1996), *Culture of Honor: The Psychology of Violence in the South,* Boulder, CO: Westview.

Norenzayan, A. and Schwarz, N. (1999), Telling what they want to know: Participants tailor causal attributions to researchers' interests, *European Journal of Social Psychology*, 29, 1011–1020.

Nunnally, J. C. and Bernstein, I. H. (1994), *Psychometric Theory* (3rd ed.), New York: McGraw-Hill.

Oberski, D., Saris, W. E., and Kuipers, S. (2007), SQP: Survey Quality Predictor [computer program]. Available at http://www.sqp.nl/.

Oberski, D., Saris, W. E., and Hagenaars, J. (2007), Why are there differences in measurement quality across countries? in G. Loosveldt, M. Swyngedouw, and B. Cambré (Eds.), *Measuring Meaningful Data in Social Research*, Leuven: Acco.

Ochs, E. (1979), Transcription as theory, in E. Ochs and B. Schieffelin (Eds.), *Developmental Pragmatics*, pp. 43–72, New York: Academic Press.

OECD (Organisation for Economic Co-Operation and Development) (1999), Classifying educational programmes, Manual for ISCED-97 Implementation in OECD Countries. Paris: http://www.oecd.org/dataoecd/41/42/1841854.pdf (08/01/21).

Olsson, U. (1979), On the robustness of factor analysis against crude classification of the observations, *Multivariate Behavioral Research*, 14, 485–500.

O'Muircheartaigh, C. (1984), *The Magnitude and Pattern of Response Variance in the Lesotho Fertility Survey*, WFS Scientific Report No. 70, The Hague: International Statistical Institute.

O'Muircheartaigh, C. (2007), Sampling, in W. Donsbach and M. W. Traugott (Eds.), *Sage Handbook of Public Opinion Research*, Chapter 27, Thousand Oaks, CA: Sage Publications Ltd.

Ongena, Y. P., Dijkstra, W. and Smit, J. H. (2008), Training and monitoring interviewers in administering CAPI event history calendar instruments. Paper presented at the 63rd conference of the American Association for Public Opinion Research, New Orleans, Louisiana, May 15–18.

ORC Macro (2005), *Malaria Indicator Survey: Guidelines for the Malaria Indicator Survey Interviewer Training*, Calverton, MD.

O'Shea, R., Bryson, C., and Jowell, R. (2001), Comparative Attitudinal Research in Europe. Retrieved from http://www.europeansocialsurvey.org.

Oshinsky, D. M. (2005), *Polio: An American Story*, Oxford: Oxford University Press.

Ostini, R. and Nering, M. L. (2006), *Polytomous Item Response Theory Models*, Thousand Oaks, CA: Sage Publications.

Owens, L., Johnson, T. P., and O'Rourke, D. (1999), Culture and item nonresponse in health surveys, in M. L. Cynamon and R. A. Kulka (Eds.), *Seventh Conference on Health Survey Research Methods*, DHHS Publication No. [PHS] 01-1013, pp. 69–74, Hyattsville, MD: National Center for Health Statistics, 2001.

Øyen, E. (1990), *Comparative Methodology: Theory and Practice in International Social Research*, London: Sage Publications.

Oyserman, D. (1993), The lens of personhood, *Journal of Personality and Social Psychology*, 65, 993–1009.

Oyserman, D. and Lee, S. W. S. (2007), Priming 'culture': Culture as situated cognition, in S. Kitayama and D. Cohen (Eds.), *Handbook of Cultural Psychology*, pp. 252–279, New York: Guilford Press.

Oyserman, D. and Lee, S. W. S. (2008a), Does culture influence what and how we think? Effects of priming individualism and collectivism, *Psychological Bulletin*, 134, 311–342.

Oyserman, D. and Lee, S. W. S. (2008b), A situated cognition perspective on culture: Effects of priming cultural syndromes on cognition and motivation, in R. Sorrentino and S. Yamaguchi (Eds.), *Handbook of Motivation and Cognition across Cultures*, pp. 237–265, New York: Elsevier.

Oyserman, D. and Sorensen, N. (2009), Understanding cultural syndrome effects on what and how we think: A situated cognition model, in R. Wyer, C-Y Chiu, and Y-Y Hong (Eds.), *Understanding Culture: Theory, Research and Application*, New York: Psychology Press.

Oyserman, D., Coon, H., and Kemmelmeier, M. (2002), Rethinking individualism and collectivism: Evaluation of theoretical assumptions and meta-analyses, *Psychological Bulletin*, 128, 3–73.

Oyserman, D., Kemmelmeier, M., and Coon, H. (2002), Cultural psychology, a new look, *Psychological Bulletin*, 128, 110–117.

Oyserman, D., Sorensen, N., Reber, R., and Chen, S. X. (2009), Connecting and separating mind-sets: Culture as situated cognition, *Journal of Personality and Social Psychology*, 97(2), 217–235.

Ozgur, S. and Sunar, D. (1982), Social psychological patterns of homicide in Turkey: A comparison of male and female convicted murderers, in C. Kagitcibasi (Ed.) *Sex Roles, Family and Community in Turkey*, Bloomington: Indiana University Press.

Paik, C. and Michael, W. B. (1999), A construct validity investigation of scores on a Japanese version of an academic self-concept scale for secondary school students, *Educational and Psychological Measurement*, 59, 98–110.

Paltridge, B. (2006), *Discourse analysis: An introduction*, London, New York: Continuum.

Palutis, B. (2008), Rebuilding the Nielsen Sample in New Orleans after Katrina, *Public Opinion Quarterly*, 72(3), E1–E38.

Pan, Y. (2004), Cognitive interviews in languages other than English: Methodological and research issues. Paper presented at the American Association for Public Opinion Research conference, May 13–16, Phoenix, AZ.

Pan, Y. (2008), Cross-cultural communication norms and survey interviews, in H. Sun and D. Kádár (Eds.), *It's the Dragon's Turn—Chinese Institutional Discourse(s)*, pp. 17–76, Berne: Peter Lang.

Pan, Y. and de la Puente, M. (2005), Census Bureau guidelines for the translation of data collection instruments and supporting materials: Documentation on how the guideline was developed, Research Report Series #2005-06, Statistical Research Division, Washington, DC: U.S. Bureau of the Census.

Pan, Y., Craig, B, and Scollon, S. (2005), Results from Chinese cognitive interviews on the Census 2000 Long Form: Language, literacy, and cultural issues, in Statistical Research Division's Research Report Series (Survey Methodology #2005-09), Washington, DC: U.S. Bureau of the Census.

Pan, Y., Hinsdale, M., Park, H., and Schoua-Glusberg, A. (2006), Cognitive testing of translations of ACS CAPI materials in multiple languages, in *Statistical Research Division's Research Report Series* (RSM#2006/09), Washington, DC: U.S. Census Bureau.

Park, A. and Jowell, R. (1997a), *Consistencies and Differences in a Cross-National Survey*, London: NATCEN. http://www.gesis.org/en/data_service/issp/data/1995_National_ Identity.htm.

Park, A. and Jowell, R. (1997b), *Consistencies and Differences in a Cross-National Survey*, London: SCPR, *International Social Science Journal*, 17(4), 665–685.

Patel, V. (1995), Explanatory models of mental illness in sub-Saharan Africa, *Social Science & Medicine*, 40, 1291–1298.

Payne, S. (1951), *Art of Asking Questions*, Princeton: Princeton University Press.

Peabody, D. (1962), Two components in bipolar scales: Direction and intensity, *Psychological Review*, 69, 65–73.

Pennell, B.-E., Mneimneh, Z., Bowers, A., Chardoul, S., Wells, J. E., Viana, M. C., et al. (2008), Implementation of the World Mental Health Survey initiative, in R. C. Kessler and T. B. Üstün (Eds.), *Volume 1: Patterns of Mental Illness in the WMH Surveys*, New York: Cambridge University Press.

Peristiany, J. G. (Ed.) (1965), *Honour and Shame: The Values of Mediterranean Society*, London: Weidenfeld and Nicolson.

Pervizat, L. (1998), Honor killings. Paper presented at the 54th session of the United Nation Commission on Human Rights.

Peterson, M. F. and Quintanilla, S. A. R. (2003), Using emics and etics in cross-cultural organizational studies: Universal and local, tacit and explicit, in D. Tjosvold and K. Leung (Eds.), *Cross-Cultural Management: Foundations and Future*, pp. 73–102, Surrey, United Kingdom: Ashgate Publishing.

Philippen, M., Loosveldt, G., Stoop, I., and Billiet, J. (2003), Noncontact rates and interviewer calling strategies in the ESS. Presented at International Workshop on Household Nonresponse, Leuven, Belgium.

Phillips, L. (1996), *Lieder Line by Line, and Word for Word*, Oxford: Oxford University Press.

Pilmis, O. (2006), Does public opinion finally exist? The use of ESS to test Pierre Bourdieu's theory, *Quantitative Methods in the Social Sciences: Research Methods Festival Session 2006*, Oxford.

PISA (Programme for International Student Assessment) (2005), *Main Study National Project Manager's Manual*, Paris: OECD.

Pitt-Rivers, J. (1965), Honor and social status, in J. G. Peristiany (Ed.), *Honor and Shame: The Values of Mediterranean Society*, pp. 18–77, London: Weidenfeld and Nicolson.

Pitt-Rivers, J. (1977), *The fate of Shechem, or the Politics of Sex: Essays in the Anthropology of the Mediterranean*, Cambridge, UK: Cambridge University Press.

Pleis, J. R., Lethbridge-Cejku, M. (2006), Summary health statistics for U.S. adults: National Health Interview Survey, 2005, *Vital Health Stat Series* 10, 232:1–153.

Podsakoff, P. M., MacKenzie, S. B., Lee, J. Y., and Podsakoff, N.P. (2003), Common method biases in behavioral research: A critical review of the literature and recommended remedies, *Journal of Applied Psychology*, 88, 879–903.

Popper, K. R. (1935), Induktionslogik "und" Hypothesenwahrscheinlichkeit, *Erkenntnis*, 5, Bd., 170–172.

Popper, K. R. (1972), *Objective Knowledge—an Evolutionary Approach*, Oxford: Oxford University Press.

Porter, A. C. and Gamoran, A. (Eds.), (2002). *Methodological Advances in Large-Scale Cross-National Education Surveys*, Washington, DC: National Research Council.

Potaka, L. and Cochrane, S. (2004), Developing bilingual questionnaires: Experiences from New Zealand in the development of the 2001 Maori language survey, *Journal of Official Statistics*, 20(2), 289–300.

Presser, S. and Blair, J. (1994), Survey pretesting: Do different methods produce different results? in P. V. Marsden (Ed.), *Sociological Methodology*, Vol. 24, pp. 73–104, Washington, DC: American Sociological Association.

Presser, S., Rothgeb, J. M., Couper, M. P., Lessler, J. T., Martin, E. A., Martin, J., et al. (Eds.) (2004), *Methods for Testing and Evaluating Survey Questionnaires*, Hoboken, NJ: John Wiley & Sons.

Przeworski, A. and Teune, H. (1970), *The Logic of Comparative Social Inquiry*, New York: Wiley Interscience.

Puglisi, J. T. (2000), IRB knowledge of local research context. http://www.hhs.gov/ohrp/humansubjects/guidance/local.htm.

Rabe-Hesketh, S. and Skrondal, A. (2008), *Multilevel and Longitudinal Modeling Using Stata* (2nd ed.), College Station TX: Stata Press.

Rabe-Hesketh, S., Skrondal, A. and Pickles, A. (2004), *GLLAMM Manual*. www.gllamm.org/.

Radio Free Europe (1968), The impact of industrialization on the family structure in Austria, Czechoslovakia and Hungary, Audience and Public Opinion Research Department. February.

Radio Free Europe-Radio Liberty (1979), Three measurements of East European attitudes to the Helsinki conference, 1973–1978, East European Area Audience and Opinion Research. July.

Radio Free Europe-Radio Liberty (1985), Czechoslovak, Hungarian and Polish expectations concerning domestic developments. East European Area Audience and Opinion Research. October.

Rajagopalan, K. (2009), Pragmatics today: From a component of linguistics to a perspective of language, in B. Frazer and K. Turner (Eds.), *Language in Life and a Life in Language: Jacob Mey—A Festschrift*, pp. 335–342, Bingley, UK: Emerald.

Raju, N. S. (1999), DFITPS6: A Fortran program for calculating polytomous DIF/DTF [computer program], Chicago: Illinois Institute of Technology.

Raju, N. S. and Ellis, B. B. (2002), Differential item and test functioning, in F. Drasgow and N. Schmitt (Eds.), *Measuring and Analyzing Behavior in Organizations: Advances in Measurement and Data Analysis*, pp. 156–188, San Francisco: Jossey-Bass.

Raju, N. S., van der Linden, W., and Fleer, P. (1995), An IRT-based internal measure of test bias with implications for differential item functioning, *Applied Psychological Measurement*, 19, 353–368.

Ralis, M., Suchman, E.A., and Goldsen, R. K. (1958), Applicability of survey techniques in northern India, *Public Opinion Quarterly*, 22(3), 245–250.

Ramos-Sanchez, L. (2007), Language switching and Mexican Americans' emotional expression, *Journal of Multicultural Counseling and Development*, 35(3), 154–169.

Rässler, S. and Riphahn, R. T. (2006), Survey item nonresponse and its treatment, *Allgemeines Statistisches Archiv*, 90(1), 213–228.

Raudenbush, S. W. and Bryk, A. S. (2002), *Hierarchical Linear Models*, Thousand Oaks: Sage.

Reckase, M. D. (1979), Unifactor latent trait models applied to multifactor tests: Results and implications, *Journal of Educational Statistics*, 4, 207–230.

Reif, K. and Melich, A. (1991), *Euro-Barometer 30: Immigrants and Out-Groups in Western Europe*, October-November 1988 File, Ann Arbor, MI: Inter-University Consortium for Political and Social Research.

Richard, M. O. and Troffoli, R. (2009), Language influence in responses to questionnaires by bilingual respondents: A test of the Whorfian hypothesis, *Journal of Business Research*, 62(10), 987–994.

Riphahn, R. T. and Serfling, O. (2005), Item non-response on income and wealth questions, *Empirical Economics*, 30(2), 521–538.

Roberts, R. P. (2002), Translation, in R. B. Kaplan (Ed.), *The Oxford Handbook of Applied Linguistics*, pp. 429–442, Oxford: Oxford University Press.

Rodriguez Mosquera, P. M. R., Fischer, A. H., Manstead, A. S. R., and Zaalberg, R. (2008), Attack, disapproval, or withdrawal? The role of honor in anger and shame responses to being insulted, *Cognition & Emotion*, 22(8), 1471–1498.

Rodriguez Mosquera, P. M. R., Manstead, A.S.R., and Fischer, A.H. (2000), The role of honor-related values in the elicitation, experience, and communication of pride, shame, and anger: Spain and the Netherlands compared, *Personality and Social Psychology Bulletin*, 26, 833–844.

Rodriguez Mosquera, P. M. R., Manstead, A. S. R., and Fischer, A. H. (2002), The role of honor concerns in emotional reactions to offences, *Cognition & Emotion*, 16, 143–163.

Rokkan, S. (1969), Cross-national survey research: Historical, analytical and substantive contexts, in S. Rokkan, S. Verba, J. Viet, and E. Almasy (Eds.), *Comparative Survey Analysis*, pp. 3–55, The Hague: Mouton.

Romney, A. K., Batchelder, W. H., and Weller, S. C. (1987), Recent applications of cultural consensus theory, *American Behavioral Scientist*, 31, 163–177.

Ross, C. E. and Mirowsky, J. (1984), Socially-desirable response and acquiescence in a cross-cultural survey of mental health, *Journal of Health and Social Behavior*, 25, 189–197.

Ross, M., Xun, W. Q. E., and Wilson, A. (2002), Language and the bicultural self, *Personality and Social Psychology Bulletin*, 28, 1040–1050.

Rost, J. and Langeheine, R. (1997), A guide through latent structure models for categorical data, in J. Rost and R. Langeheine (Eds.), *Applications of Latent Trait and Latent Class Models in the Social Sciences*, pp. 13–37, New York: Waxmann.

Rost, J., Carstensen, C., von Davier, M. (1997), Applying the mixed Rasch model to personality questionnaires, in J. Rost and R .Langeheine (Eds.), *Applications of Latent Trait and Latent Class Models in the Social Sciences*, pp. 324–332, Münster, Germany: Waxmann Verlag GmbH. Retrieved 03/28/2009 from http://www.ipn.uni-kiel.de/aktuell/buecher/rostbuch/c31.pdf.

Rubin, D. B. (1976), Inference with missing data, *Biometrika*, 63, 581–592.

Rubin, D. B. (1987), *Multiple Imputation for Nonresponse in Surveys*, New York: John Wiley & Sons.

Rudmin, F. W. and Ahmadzadeh, V. (2001), Psychometric critique of acculturation psychology: The case of Iranian migrants in Norway, *Scandinavian Journal of Psychology*, 42, 41–56.

Ryan, T. P. (2000), *Statistical Methods for Quality Improvement*, New York: John Wiley & Sons.

Sadana, R., Mathers, C. D., Lopez, A. D., Murray, C. J. L., and Moesgaard-Iburg, K. (2002), Comparative analysis of more than 50 household surveys of health status, in C. J. L. Murray, J. A. Salomon, C. D. Mathers, and A. D. Lopez (Eds), *Summary Measures of Population Health: Concepts, Ethics, Measurement and Applications*, Geneva: World Health Organization.

Saldanha, G. (2009), Linguistic approaches, in M. Baker and G. Saldanha (Eds.), *Routledge Encyclopedia of Translation Studies* (2nd ed.), pp. 148–151, London: Routledge.

Sale, J. E. M. and Brazil, K. (2004), A strategy to identify critical appraisal criteria for primary mixed-method studies, *Quality & Quantity*, 38, 351–365.

Sam, D. L. and Berry, J. W. (Eds.) (2006), *The Cambridge Handbook of Acculturation Psychology*, Cambridge, UK, Cambridge University Press.

Samejima, F. (1969), Estimation of latent ability using a pattern of graded scores, *Psychometrika* Monograph Supplement No. 17.

Sandelowski, M. (1995), Focus on qualitative research. Sample size in qualitative research, *Research in Nursing and Health*, 18, 170–183.

Sanderson, S. K. (1990), *Social Evolutionism. A Critical History*, Oxford: Basil Blackwell.

Saris, W. E. (1997), Comparability across mode and country, in W. E. Saris and M. Kaase (Eds.), *Eurobaramoter Measurement for Opinions in Europe*, pp. 125–139, ZUMA-Nachrichten Spezial 2, Mannheim: ZUMA.

Saris, W. E. (2004), Comparability across countries of responses in the ESS. Paper presented at the Sixth International Conference on Social Science Methodology, Amsterdam.

Saris, W. (2008), Alternative measurement procedures and models for political efficacy, Barcelona: ESADE.

Saris, W. E. and Gallhofer, I. N. (2002), Report on the MTMM experiments in the pilot studies and a proposal for round 2 of the ESS. Report for the ESS.

Saris, W. and Gallhofer, I. (2007a), Can questions travel successfully? in R. Jowell et al. (Eds.), *Measuring Attitudes Cross-Nationally*, London: Sage Publications.

Saris, W. E. and Gallhofer, I. (2007b), *Design, Evaluation, and Analysis of Questionnaires for Survey Research*, Wiley Series in Survey Methodology, Hoboken, NJ: John Wiley & Sons.

Saris, W. E. and Gallhofer, I. (2007c), Estimation of the effects of measurement characteristics on the quality of survey questions, *Survey Research Methods*, 1(1), 29–43.

Saris, W. E. and Kaase, M. (Eds.) (1997), *ZUMA-Nachrichten Spezial Band 2, Eurobarometer: Measurement Instruments for Opinions in Europe*, Mannheim, Germany: ZUMA. Retrieved from http://www.gesis.org/fileadmin/upload/forschung/publikationen/zeitschriften/zuma_nachrichten_spezial/znspezial2.pdf.

Saris, W., Krosnick, J., and Shaeffer E. (2005), Comparing questions with agree/disagree response options to questions with construct-specific response options. Unpublished.

Saris, W. E., Satorra, A., and Coenders, G. (2004), A new approach to evaluating the quality of measurement instruments: The split-ballot MTMM design, *Sociological Methodology*, 34, 311–347.

Saris, W. E., Satorra, A., and Sorbom, D. (1987), The detection and correction of specification errors in structural equation models, *Sociological Methodology*, 17, 105–129.

Saris, W. E., Satorra, A., and van der Veld, W. (2009), Testing structural equation models or detection of misspecifications? *Structural Equation Modeling: An Interdisciplinary Journal*, 16(4), 561–582.

Satorra, A. (1990), Robustness issues in structural equation modeling: A review of recent developments, *Quality and Quantity*, 24, 367–386.

Saville-Troike, M. (1989), *The Ethnography of Communication,* Cambridge, MA: Blackwell.

Schäler, R. (2009), Localization, in M. Baker and G. Saldanha (Eds.), *Routledge Encyclopedia of Translation Studies* (2nd ed.), pp. 157–161, London: Routledge.

Schegloff, E. (1972), Sequencing in conversational openings, in J. Gumperz and D. Hymes (Eds.), *Directions in Sociolinguistics*, pp. 346–380, New York: Holt, Rinehart and Winston.

Scheuch, E. K. (1968), The cross-cultural use of sample surveys: Problems of comparability, in S. Rokkan (Ed.), *Comparative Research Across Cultures and Nations*, pp. 176–209, Paris: Mouton.

Scheuch, E. K. (1973), Entwicklungsrichtlinien bei der Analyse Sozialwissenschaftlicher Daten, in R. König (Ed.), *Handbuch der Empirischen Sozialforschung*, Vol. 1, pp. 161–226, Stuttgart: Enke.

Scheuch, E. K. (1989), Theoretical implications of comparative survey research: Why the wheel of cross-cultural methodology keeps being reinvented, *International Sociology*, 4(2), 147–167.

Scheuch, E. K. (1993), The cross-cultural use of sample surveys: Problems of comparability, *Historical Social Research*, 18(2), 104–138.

Scheuren, F. (2001), Macro and micro paradata for survey assessment. Unpublished manuscript. Washington: Urban Institute.

Schmidt, S. and Bullinger, M. (2003), Current issues in cross-cultural quality of life instrument development, *Archives of Physical Medicine Rehabilitation*, 84, S29–34.

Schmitt, N. and Stults, D. M. (1986), Methodology review: Analysis of multitrait–multimethod matrices, *Applied Psychological Measurement*, 10, 1–22.

Schneider, S. L. (2008), Nominal comparability is not enough: Evaluating cross-national measures of educational attainment using ISEI scores, Department of Sociology, University of Oxford: Sociology Working Papers; Paper Number: 2008-04. http://www.sociology.ox.ac.uk/documents/working-papers/2008/2008-04.pdf.

Schnell, R. and Kohler, U. (1995), Empirische Untersuchungen einer Individualisierungshypothese am Beispiel der Parteipräferenz von 1953–1992, *Kölner Zeitschrift für Soziologie und Sozialpsychologie*, 47, 634–658.

Schnell, R. and Kreuter, F. (2005), Separating interviewer and sampling-point effects, *Journal of Official Statistics*, 21(3), 389–410.

Schober, M. F. and Conrad, F. G. (1997), Does conversational interviewing reduce survey measurement error? *Public Opinion Quarterly*, 61, 576–602.

Scholtz, E., Harkness, J., and Faaß, T. (2008), ISSP study monitoring 2005. Accessed April 14, 2008, from http://www.gesis.org/Publikationen/Berichte/GESIS_Berichte/gesis_mb_08_04.pdf.

Scholz, E. (2005), Harmonization of survey data in the International Social Survey Programme (ISSP), *ZUMA Nachrichten Spezial–Methodological Aspects in Cross-National Research*, 11, 183–200.

Schoua-Glusberg, A. (1992), Report on the translation of the questionnaire for the National Treatment Improvement, Evaluation Study, Chicago: National Opinion Research Center.

Schräpler, J. P. (2004), Respondent behavior in panel studies—A case study for income-nonresponse by means of the German Socio-Economic Panel (SOEP), *Sociological Methods & Research*, 33(1), 118–156.

Schuman, H. and Converse, J. M. (1971), The effects of black and white interviewers on black responses in 1968, *Public Opinion Quarterly*, 35(1), 44–68.

Schuman, H. and Presser, S. (1981), *Questions and Answers in Attitude Surveys: Experiments in Question Form, Wording, and Context*, New York: Academic Press.

Schwartz, S. H. (2004), Mapping and interpreting cultural differences around the world, in H. Vinken, J. Soeters, and P. Ester (Eds.), *Comparing Cultures: Dimensions of Culture in a Comparative Perspective,* pp. 43–73, Leiden, The Netherlands: Brill.

Schwartz, S. H. and Sagiv, L. (1995), Identifying culture-specifics in the content and structure of values, *Journal of Cross-Cultural Psychology*, 26, 92–116.

Schwarz, N. (1994), Judgment in a social context: Biases, shortcomings and the logic of conversation, in M. Zanna (Ed.), *Advances in Experimental Social Psychology*, 26, pp. 123–126, San Diego, CA: Academic Press.

Schwarz, N. (1996), *Cognition and Communication: Judgmental Biases, Research Methods, and the Logic of Conversation*, Hillsdale, NJ: Erlbaum.

Schwarz, N. (1999), Self-reports: How the questions shape the answers, *American Psychologist*, 54, 93–105.

Schwarz, N. (2003a), Culture-sensitive context effects: A challenge for cross-cultural surveys, in J. Harkness, F. J. R. van de Vijver, and P. Mohler (Eds.), *Cross-Cultural Survey Methods*, pp. 93–100, Hoboken, NJ: John Wiley & Sons.

Schwarz, N. (2003b), Self-reports in consumer research: The challenge of comparing cohorts and cultures, *Journal of Consumer Research*, 29, 588–594.

Schwarz, N. and Bless, H. (1992), Scandals and the public's trust in politicians: Assimilation and contrast effects, *Personality and Social Psychology Bulletin*, 18, 574–579.

Schwarz, N. and Bless, H. (2007), Mental construal processes: The inclusion/exclusion model, in D. A. Stapel and J. Suls (Eds.), *Assimilation and Contrast in Social Psychology*, Philadelphia, PA: Psychology Press.

Schwarz, N., Hippler, H. J., Deutsch, B., and Strack, F. (1985), Response scales: Effects of category range on reported behavior and subsequent judgments, *Public Opinion Quarterly*, 49, 388–395.

Schwarz, N., Knäuper, B., Hippler, H. J., Noelle-Neumann, E., and Clark, F. (1991), Rating scales: Numeric values may change the meaning of scale labels, *Public Opinion Quarterly*, 55, 570–582.

Schwarz, N., Strack, F., and Mai, H. P. (1991), Assimilation and contrast effects in part-whole question sequences: A conversational logic analysis, *Public Opinion Quarterly*, 55, 3–23.

Scollon, R. and Scollon, S. W. (2001), *Intercultural Communication: A Discourse Approach* (2nd ed.), Oxford: Basil Blackwell.

Seale, C., Charteris-Black, J., Dumelow, C., Locock, L., and Ziebland, S. (2008), The effect of joint interviewing on the performance of gender, *Field Methods*, 20(2), 107–128.

Sechrest, L., Fay, T. L., and Zaidi, S. M. H. (1972), Problems of translation in cross-cultural research, *Journal of Cross-Cultural Psychology*, 3(1), 41–56.

Serpell, R. (1990), Audience, culture and psychological explanation: A reformulation of the emic-etic problem in cross-cultural psychology, *The Quarterly Newsletter of the Laboratory of Comparative Human Cognition*, 12(3), 99–132.

Sev'er, A. and Yurdakul, G. (2001), Culture of honor, culture of change: A feminist analysis of honor killings in rural Turkey, *Violence Against Women*, 7, 964–998.

Shaules, J. (2007), *Deep Culture: The Hidden Challenges of Global Living*, Clevedon: Multilingual Matters Ltd.

Shin, H. B. and Bruno, R. (2003), Language use and English-speaking ability: 2000, Census 2000 Brief, U. S. Department of Commerce, Economics and Statistics Administration, U.S. Bureau of the Census. Retrieved from http://www.census.gov/prod/2003pubs/c2kbr-29.pdf.

Shiomi, K. and Loo, R. (1999),Cross-cultural response styles on the Kirton Adaptation-Innovation Inventory, *Social Behavior and Personality*, 27, 413–420.

Shishido, K., Iwai, N., and Yasuda, T. (2006), Study of response categories in the Cross-National Surveys, Experience of JGSS team with South Korea, Taiwan and China, *Proceedings of the 79th Conference of the Japan Sociological Society*, 106.

Shively, W. P. (2005), Democratic design: The comparative study of electoral systems project, *Public Opinion Pros,* retrieved from http://www.publicopinionpros.com/from_field/2005/oct/shively.asp.

Shrout, P. E. and Fleiss, J. L. (1979), Intraclass correlations: Uses in assessing reliability, *Psychological Bulletin*, 86, 420–428.

Shuy, R. (1993), *Language Crimes: The Use and Abuse of Language Evidence in the Courtroom*, Cambridge, MA: Blackwell.

Singelis, T. M., Yamada, A. M., Barrio, C., Laney, J. H., Her, P., Ruiz-Anaya, A. R., et al. (2006), Metric equivalence of the Bidimensional Acculturation Scale, the Satisfaction with Life Scale, and the Self-Construal Scale across Spanish and English language versions, *Hispanic Journal of Behavioral Sciences*, 28, 231–244.

Singer, E. (2006), Nonresponse bias in household surveys, *Public Opinion Quarterly*, 70(5), 637–645.

Singer, E. (2008), Ethical issues in surveys, in E. D. deLeeuw, J. J. Hox, and D. A. Dillman. (Eds.), *International Handbook of Survey Methodology*, pp. 78–96, New York/London: Lawrence Erlbaum Associates/Taylor & Francis Group.

Singer, E., van Hoewyk, J., Gebler, N., Raghunathan, T., and McGonagle, K. (1999), The effect of incentives on response rates in interviewer-mediated surveys, *Journal of Official Statistics*, 15(2), 217–230.

Sinha, J. B. P. (2004), Emic and etic approaches to Asian management research, in K. Leung and S. White (Eds.), *Handbook of Asian Management*, pp. 19–52, Norwell, MA: Kluwer Academic Publishers.

Sirken, M., Hermann, D., Schechter, S., Schwarz, N., Tanur, J., and Tourangeau, R. (Eds.) (1999), *Cognition and Survey Research*, New York: John Wiley & Sons.

Skevington, S. M. (2002), Advancing cross-cultural research on quality of life: Observations drawn from the WHOQOL development, *Quality of Life Research*, 11, 135–144.

Skevington, S. M. and Tucker, C. (1999), Designing response scales for cross-cultural use in health care: Data from the development of the UK WHOQOL, *British Journal of Medical Psychology*, 72, 51–61.

Skevington, S. M., Sartorius, N., Amir, M., and the WHOQOL group (2004), Developing methods for assessing quality of life in different cultural settings: The history of the WHOQOL instruments, *Social Psychiatry & Psychiatric Epidemiology*, 39(1), 1–8.

Skinner, C. J., Holt, D., and Smith, T. M. F. (1989), *Analysis of Complex Surveys*, New York: John Wiley & Sons.

Skrondal, A. and Rabe-Hesketh, S. (2004), *Generalized Latent Variable Modeling: Multilevel, Longitudinal, and Structural Equation Models*, Boca Raton FL: Chapman Hall/CRC Press.

Smelser, N. J. (1973), The methodology of comparative analysis, in D. P. Warwick and S. Osherson (Eds.), *Comparative Research Methods*, pp. 42–88, Englewood Cliffs, NJ: Prentice-Hall.

Smith, A. D. (1973), *The Concept of Social Change*, London: Routledge & Kegan Paul.

Smith, D. M., Schwarz, N., Roberts, T. R., and Ubel, P. A. (2006), Why are you calling me? How study introductions change response patterns, *Quality of Life Research*, 15, 621–630.

Smith, E. E. (1995), Concepts and categorization, in E. E. Smith and D. N. Osherson (Eds.), *Thinking (An Invitation to Cognitive Science)*, Vol. 3, pp. 3–34, Cambridge, MA: MIT Press.

Smith, P. B. (2004), Acquiescent response bias as an aspect of cultural communication style, *Journal of Cross-Cultural Psychology*, 35(1), 50–61.

Smith, P. B. and Fischer, R. (2008), Acquiescence, extreme response bias, and culture: A multilevel analysis, in F. van de Vijver, D. A. Van Hemert, and Y. H. Poortinga (Eds.), *Multilevel Analysis of Individuals and Cultures*, pp. 285–314, New York: Lawrence Erlbaum Associates.

Smith, T. W. (1978), In search of house effects: A comparison of responses to various questions by different survey organizations, *Public Opinion Quarterly*, 42, 443–463.

Smith, T. W. (1982), House effects and the reproducibility of survey measurements: A comparison of the 1980 GSS and the 1980 American National Election Study, *Public Opinion Quarterly*, 46, 54–68.

Smith, T. W. (1995), Little things matter: A sampler of how differences in questionnaire format can affect survey responses. Paper presented at the Annual Conference of the American Statistical Association, Ft. Lauderdale, FL, 1046–1051. Retrieved from http://www.amstat.org/sections/srms/Proceedings/papers/1995_182.pdf.

Smith, T. W. (2003), Developing comparable questions in cross-national surveys, in J. A. Harkness, F. J. R. van de Vijver, and P. Mohler, *Cross-Cultural Survey Methods*, pp. 69–92, Hoboken, NJ: John Wiley & Sons.

Smith, T. W. (2004a), Crossnational survey research: The challenge and the promise, *ICPSR Bulletin*, 14, 3–12.

Smith, T. W. (2004b), Developing and evaluating cross-national survey instruments, in S. Presser et al. (Eds.), *Methods for Testing And Evaluating Survey Questionnaires*, pp. 431–452, Hoboken, NJ: John Wiley & Sons.

Smith, T. W. (2005), Laws of studying societal change, *Survey Research*, 2, 1–5. Retrieved from http://www.srl.uic.edu/Publist/Newsletter/2005/05v36n2.pdf.

Smith, T. W. (2007), Formulating the laws for studying societal change. Paper presented to the FCSM Research Conference, Arlington, VA.

Smith, T. W. (2008a), The 2008 ISSP bibliography: A report. Retrieved from http://www.issp.org.

Smith, T. W. (2008b), Codes of ethics and standards in survey research, in W. Donsbach and M. Traugott (Eds.), *Handbook of Public Opinion Research*, London: Sage.

Smith, T. (forthcoming), Surveying across nations and cultures, in J. D. Wright and P. V. Marsden (Eds.), *Handbook of Survey Research* (2nd ed.), San Diego: Elsevier.

Smith, T. W., Kim, J., Koch, A., and Park, A. (2005), Social-science research and the General Social Surveys, *ZUMA-Nachrichten*, 56, 68–7.

Smith, T. W., Mohler, P. P., Harkness, J. A., and Onodera, N. (2009), Methods for assessing and calibrating response scales across countries and languages, in M. Sasaki (Ed.), *New Frontiers in Comparative Sociology*, pp. 45–96, Leiden; Boston: Brill.

Snell Hornby, M. (2006), *The Turns of Translation Studies. New Paradigms or Shifting Viewpoints?* Amsterdam: Benjamins.

Snijders, T. A. B. and Bosker, R. J. (1999), *Multilevel Analysis. An Introduction to Basic and Advanced Multilevel Modelling*, London: Sage.

Spiess, M. and Goebel, J. (2003), Evaluation of the ECHP imputation rules. CHINTEX Working Paper No. 17, Wiesbaden: Statistisches Bundesamt. http://www.destatis.de/chintex/proj_des/wp_7.htm/.

Starick, R. (2005), Imputation in longitudinal surveys: The case of HILDA. Research Paper of the Australian Bureau of Statistics. ABS Catalogue no. 1352.0.55.075.

Starick, R. and Watson, N. (2007), Evaluation of alternative income imputation methods for the HILDA Survey, HILDA Project Technical Paper Series No. 1/07, Melbourne Institute of Applied Economic and Social Research, University of Melbourne.

Steenkamp, J. B. E. M. and Baumgartner, H. (1998), Assessing measurement invariance in cross national consumer research, *Journal of Consumer Research*, 25, 78–90.

Steiger, J. H. (1990), Structural model evaluation and modification: An interval estimation approach, *Multivariate Behavioral Research*, 25, 173–180.

Stening, B. W. and Everett, J. E. (1984), Response styles in a cross-cultural managerial study, *Journal of Social Psychology*, 122, 151–156.

Stern, E. (1948), The universe, translation, and timing, *Public Opinion Quarterly*, 12, 711–715.

Stevenson, H. W. and Stigler, J. W. (1992), *The Learning Gap*, New York: Simon & Schuster.

Stewart, F. (1994), *Honor*, Chicago, University of Chicago Press.

Stocking, G. W. Jr. (1968), *Race, Culture, and Evolution*, New York: The Free Press.

Stocking, G. W. Jr. (1987), *Victorian Anthropology*, New York: The Free Press.

Stocking, M. L. and Lord, F. M. (1983), Developing a common metric in item response theory, *Applied Psychological Measurement*, 7, 201–210.

Stoop, I. (2007), If it bleeds, it leads: The impact of media reported events, in R. Jowell et al. (Eds.), *Measuring Attitudes Cross-Nationally*, London: Sage Publications.

Stoop, I., Devacht, S., Loosveldt, G., Billiet, J., and Philippens, M. (2003), Developing a uniform contact description form. Paper presented at the 14th International Workshop on Household Survey Non-Response, Sept 22–24, Leuven, Belgium.

Storti, C. (1999), *Figuring Foreigners Out: A Practical Guide*, Yarmouth, ME: Intercultural Press, Inc.

Strack, F. and Martin, L. (1987), Thinking, judging, and communicating: A process account of context effects in attitude surveys, in H. J. Hippler, N. Schwarz, and S. Sudman (Eds.), *Social Information Processing and Survey Methodology*, pp. 123–148, New York: Springer Verlag.

Strotgen, R. and Uher, R. (2000), ISSP data wizard—Computer assisted merging and archiving of distributed international comparative data, IASSIST Quarterly, Winter, 13–18.

Struwig, J. and Roberts, B. (2006), Data collection methodology in South Africa, Sixth ZUMA Symposium of Cross-Cultural Survey Methodology, 1–5 December, Mannheim, Germany. http://www.hsrc.ac.za/Research_Publication-6459.phtml.

Suchman, L. and Jordan, B. (1990), Interactional troubles in face-to-face interviews, *Journal of the American Statistical Association*, 85, 232–241.

Sudman, S. and Bradburn, N. M. (1974), *Response Effects in Surveys*, Chicago: Aldine.

Sudman, S., Bradburn, N. M., and Schwarz, N. (1996), *Thinking about Answers: The Application of Cognitive Processes to Survey Methodology*, San Francisco: Jossey-Bass.

Sugarman, J., Popkin, B., Fortney, J., and Rivera, R. (2001), International perspectives on protecting human research subjects, Bethesda, MD: National Bioethics Advisory Commission.

Sun, H. (2008), Participant roles and discursive actions: Chinese transactional telephone interactions, in H. Sun and D. Kádár (Eds.), *It's the Dragon's Turn: Chinese Institutional Discourses*, pp. 77–126, Bern: Peter Lang.

Suzuki, N. and Yamagishi, T. (2004), An experimental study of self-effacement and self-enhancement among the Japanese, *Japanese Journal of Social Psychology*, 20, 17–25.

Suzuki, L. A., Ponterotto, J. G., and Meller, P. J. (2008), *Handbook of Multicultural Assessment: Clinical, Psychological, and Educational Applications*, San Francisco, CA: Jossey Bass.

Swann, W., Pelham, B., Krull, D. (1989), Agreeable fancy or disagreeable truth? Reconciling self-enhancement with self-verification, *Journal of Personality and Social Psychology*, 57, 782–791.

Swires-Hennessy, E. and Drake, M. (1992), The optimum time at which to conduct interviews, *Journal of the Market Research Society*, 34, 61–72.

Symons, K., Matsuo, H., Beullens, K., and Billiet, J. (2008), Response based quality assessment in the ESS—Round 3: An update for 19 countries, London: Centre for Comparative Social Surveys, City University. Retrieved 23rd September 2008 from http://ess.nsd.uib.no/index.jsp?year=2007&country=&module=documentation.

Szabo, S., Orley, J., and Saxena, S. (1997), An approach to response scale development for cross-cultural questionnaires, *European Psychologist*, 2, 270–276.

Tafforeau, J., Lopez Cobo, M., Tolonen, H., Scheidt-Nave, C., and Tinto, A. (2005), Guidelines for the Development and Criteria for the Adoption of Health Survey Instruments, Luxembourg: European Communities.

Tan, C.-B. (1983), Chinese religion in Malaysia: A general view, *Asian Folklore Studies*, 42, 217–25.

Tannen, D. and Wallat, C. (1993), Interactive frames and knowledge schemas in interaction: Example from a medical examination/interview, in D. Tannen (Ed.), *Framing in Discourse*, pp. 57–76, New York and Oxford: Oxford University Press.

Tashakkori, A. and Teddlie, C. (Eds.) (2003), *Handbook on Mixed Methods in the Behavioral and Social Sciences*, Thousand Oaks, CA: Sage Publications.

Taylor, M. F., Brice, J., Buck, N., Prentice-Lane, E. (Eds.) (2005), *British Household Panel Survey User Manual Volume A: Introduction, Technical Report and Appendices*, Colchester: University of Essex.

Teddlie, C. and Tashakkori, A. (2003), Major issues and controversies in the use of mixed methods in the social and behavioral sciences, in A. Tashakkori and C. Teddlie (Eds.), *Handbook on Mixed Methods in the Behavioral and Social Sciences*, pp. 3–50, Thousand Oaks, CA: Sage Publications.

Teiser, S. (1995), Popular religion, *The Journal of Asian Studies*, 378–395.

Thissen, D. (2003), Multilog 7.03: A computer program for multiple, categorical item analysis and test scoring utilising item response theory, Chicago: Scientific Software, Inc.

Thomassen, J., Rosenstone, S. J., Klingemann, H. D., and Curtice, J. (1994), The comparative study of electoral systems. Retrieved January 14, 2008, from http://www.cses.org/plancom/module1/stimulus.htm.

Thornton, A. (2001), The developmental paradigm, reading history sideways, and family change, *Demography*, 38(4), 449–465.

Thornton, A. (2005), *Reading History Sideways: The Fallacy and Enduring Impact of the Developmental Paradigm on Family Life*, Chicago: University of Chicago Press.

Thornton, A., Ghimire, D. J., and Mitchell, C. (2005), The measurement and prevalence of developmental thinking about the family: Evidence from Nepal. Presentation at the Population Association of America Annual Meeting, Philadelphia, March 31–April 2.

Tourangeau, R. (1984), Cognitive science and survey methods: A cognitive perspective, in T. Jabine, M. Straf, J. Tanur, and R. Tourangeau (Eds.), *Cognitive Aspects of Survey Methodology: Building a Bridge Between Disciplines*, pp. 73–100, Washington, DC: National Academy Press.

Tourangeau, R. and Smith, T. W. (1996), Asking sensitive questions: The impact of data collection mode, question format, and question context, *Public Opinion Quarterly*, 60(2), 275–304.

Tourangeau, R., Couper, M. P., and Conrad, F. (2007), Color, labels, and interpretive heuristics for response scales, *Public Opinion Quarterly*, 71, 91–112.

Tourangeau, R., Rips, L. J., and Rasinski, K. (2000), *The Psychology of Survey Response*, New York: Cambridge University Press.

Trafimow, D., Silverman, E. S., Fan, R. M. T., and Law, J. S. F. (1997), The effect of language and priming on the relative accessibility of the private self and the collective self, *Journal of Cross-Cultural Psychology*, 28, 107–123.

Triandis, H. C. (1976), Approaches toward minimizing translation, in R. W. Brislin (Ed.), *Translation: Applications and Research*, pp. 229–243, New York: Gardner Press.

Triandis, H. C. (1989), The self and social behavior in differing cultural contexts, *Psychological Review*, 96, 506–520.

Triandis, H. C. (1995), *Individualism and Collectivism*, Boulder, CO: Westview Press.

Triandis, H. C. and Marin, G. (1983), Etic plus emic versus pseudoetic: A test of a basic assumption of contemporary cross-cultural psychology, *Journal of Cross-Cultural Psychology*, 14, 489–500.

Triandis, H. C., Davis, E. E., Vassiliou, V., and Nassiakou, M. (1965), Some methodological problems concerning research on negotiations between monolinguals, Technical Report No. 28, Urbana: Department of Psychology, University of Illinois.

Tucker, C. (1983), Interviewer effects in telephone surveys, *Public Opinion Quarterly*, 47(1), 84–95.

Tulving, E. and Thompson, D. (1973), Encoding specificity and retrieval processes in episodic memory, *Psychological Review*, 80, 352–373.

Twumasi, P. A. (2001), *Social Research in Rural Communities*, Accra, Ghana: Ghana Universities Press.

Tylor, E. B. (1871), *Primitive Culture*, Vol. I, London: John Murray, Albermarle Street.

Ueno, K. and Nakatani, I. (2003), Japan, in J. Georgas, L. G. Weiss, F. J. R. van de Vijver, and D. H. Saklofske (Eds.), *Culture and Children's Intelligence: Cross-Cultural Analysis of the WISC-III*, pp. 215–225, Amsterdam, Boston: Academic Press.

UNDP (United Nations Development Programme) (2003), Human development report 2003: Millennium development goals: A compact among nations to end human poverty, New York: Oxford Press.

UNESCO (1999), *Operational Manual for ISCED-1997*, Montreal: UNESCO Institute for Statistics.

UNESCO (2003), International Standard Classification of Education, ISCED 1997, in J. H. P. Hoffmeyer-Zlotnik and C. Wolf (Eds.), *Advances in Cross-National Comparison. A European Working Book for Demographic and Socio-Economic Variables*, pp. 195–220, New York: Kluwer Academic/Plenum Publishers.

UNESCO Institute for Statistics (UIS) (2004), Literacy Assessment and Monitoring Programme (LAMP), Montreal: UNESCO Institute for Statistics.

United Nations (2005), Household sample surveys in developing and transition countries, *Studies in Methods*, Series F., No. 96, New York: United Nations.

United Nations Economic Commission for Europe Population Activities Unit (2008), Harmonisation and dissemination of the data of the Generations and Gender Surveys http://www.unece.org/pau/_docs/ggp/2008/GGP_2008_IWG006_BgDocHarmDissRpr t.pdf.

United States Bureau of the Census (1965), Atlantida: a case study in household sample surveys, International Statistics Programs Office, Washington, DC.

United States Bureau of the Census (2004), Census Bureau guideline: Language translation of data collection instruments and supporting materials. U.S. Bureau of the Census, U.S. Department of Commerce, Washington, D.C. Retrieved from http://www.census. gov/cac/www/pdf/Language%20Translation%20 Guidelines.pdf.

United States Bureau of the Census (2006), America Speaks: A demographic profile of foreign-language speakers for the United States: 2000. Retrieved from http://www.census.gov/population/www/socdemo/hh-fam/AmSpks.html.

United States Bureau of the Census (2007a), Table S1601, Language Spoken at Home (2006 American Community Survey). Accessed on April 21, 2009, http://factfinder. census.gov/servlet/STTable?_bm=y&-geo_id=01000US&-qr_name=ACS_2007_ 1YR_ G00_S1601&-ds_name=ACS_2007_1YR_G00_&-_lang=en&-redoLog=false&-CON TEXT=st.

United States Bureau of the Census (2007b), Table R1603, Percent of People 5 Years and Over Who Speak English Less Than "Very Well." Accessed on April 21, 2009, http://factfinder.census.gov/servlet/GRTTable?_bm=y&-_box_head_nbr=R1603&- ds_ name=ACS_2006_EST_G00_&-format=US-30.

United States Bureau of the Census (2007c), Table S1602, Linguistic Isolation (2007 American Community Survey), Accessed on April 21, 2009, http://factfinder.census. gov/servlet/STTable?_bm=y&-geo_id=01000US&-qr_name=ACS_2006_EST_G00_ S1602&-ds_name=ACS_2007_EST_G00_&-_lang=en&-redoLog=false&-CONTEXT =st.

United States Department of Education (1998), Adult Literacy in OECD Countries: Technical Report on the First International Adult Literacy Survey, NCES 98-053, S. Murray, I. Kirsch, and L. Jenkins (Eds.), Washington, DC.

United States Department of Health and Human Services (2001), Mental health: Culture, race, and ethnicity—A supplement to mental health: A report of the Surgeon General. Rockville, MD: U.S. Department of Health and Human Services, Substance Abuse and Mental Health Services Administration, Center for Mental Health Services. Retrieved from http://www.surgeongeneral.gov/library/mentalhealth/cre/sma-01-3613.pdf.

United States Strategic Bombing Survey (1947a), The effects of strategic bombing on German morale, Volume 1, Washington, DC: U. S. Government Printing Office.

United States Strategic Bombing Survey (1947b), The effects of strategic bombing on Japanese morale, Volume 1, Washington, DC: U. S. Government Printing Office.

Uskul, A. K., Oyserman, D., Schwarz, N., Lee, S. W., and Xu, A. J. (2008), Culture and question format: Rating scale use in cultures of honor, modesty, and enhancement. Manuscript submitted for publication.

Van Belle, G. (2008), Statistical Rules of Thumb (2nd ed.), Hoboken: John Wiley & Sons.

Vandello, J. A. and Cohen, D. (2003), Male honor and female fidelity: Implicit cultural scripts that perpetuate domestic violence, Journal of Personality and Social Psychology, 84, 997–1010.

Vandenberg, R. J. and Lance, C. E. (2000), A review and synthesis of the measurement invariance literature: Suggestions, practices and recommendations for organizational research, Organizational Research Methods, 3, 4–69.

van Deth, J. W. (1998), Equivalence in comparative political research, in J. W. Van Deth (Ed.), Comparative Politics, The Problem of Equivalence, pp. 1–19, London: Routledge.

van Deth, J. W. (2003), Using published survey data, in J. A. Harkness, F. J. R. van de Vijver and P. Ph. Mohler (Eds.), *Cross-Cultural Survey Methods*, pp. 291–309, Hoboken, NJ: John Wiley & Sons.

van de Vijver, F. J. R. (2003), Bias and equivalence: Cross-cultural perspectives, in J. A. Harkness, F. J. R. van de Vijver, and P. Mohler (Eds.), *Cross-Cultural Survey Methods*, pp. 143–155, Hoboken, NJ: John Wiley & Sons.

van de Vijver, F. J. R., and Leung, K. (1997), *Methods and Data Analysis for Cross-Cultural Research*, Thousand Oaks, CA: Sage.

van de Vijver, F. J. R. and Poortinga, Y. (1997), Towards an integrated analysis of bias in cross-cultural assessment, *European Journal of Psychological Assessment*, 13, 29–37.

van de Vijver, F. J. R. and Poortinga, Y. H. (2002), Structural equivalance in multilevel research, *Journal of Cross-Cultural Psychology*, 33(2), 141–156

van de Vijver, F. J. R., Hofer, J., and Chasiotis, A. (2010), Methodological aspects of cross-cultural developmental studies, in M. H. Bornstein (Ed.), *Handbook of Cultural Developmental Science* (pp. 21-37). Florence, KY: Taylor & Francis.

van de Vijver, F. J. R., Van Hemert, D. A., and Poortinga, Y. H. (Eds.) (2008a), *Individuals and Cultures in Multilevel Analysis*, Mahwah, NJ: Erlbaum.

van de Vijver, F. J. R., Van Hemert, D. A., and Poortinga, Y. H. (2008b), *Multilevel Analysis of Individuals and Cultures*, New York: Lawrence Erlbaum.

van Dijk, J. J. M., Manchin, R., van Kesteren, J., Nevala, S., and Hideg, G. (2005), The burden of crime in the EU, Research Report: A comparative analysis of the European Crime and Safety Survey (EU ICS) 2005.

Van Hemert, D. A., van de Vijver, F. J. R., Poortinga, Y. H., and Georgas, J. (2002), Structural and functional equivalence of the Eysenck Personality Questionnaire within and between countries, *Personality and Individual Differences*, 33, 1229–1249.

Van Herk, H., Poortinga, Y. H., and Verhallen, T. M. M. (2004), Response styles in rating scales: Evidence of method bias in data from six EU countries, *Journal of Cross-Cultural Psychology*, 35, 346–360.

van Widenfelt, B. M., Treffers, P. D. A., de Beurs, E., Siebelink, B. M., and Koudijs, E. (2005), Translation and cross-cultural adaptation of assessment instruments used in psychological research with children and families, *Clinical Child and Family Psychology Review*, 8(2), 135–147.

Varughese, G. (2007). Practical challenges: Conducting survey research in Afghanistan, *Public Opinion Pros,* retrieved from http://www.publicopinionpros.com/from_field/2007/apr/varughese_printable.asp.

Vehovar, V., Batagelj, Z., Manfreda, K. L., and Zaletel, M. (2002), Nonresponse in web surveys, in R. M. Groves, D. A. Dillman, J. L. Eltinge, and R. J. A. Little (Eds.), *Survey Nonresponse*, pp. 229–242, New York: John Wiley & Sons.

Venuti, L. (1995), *The Translator's Invisibility: A History of Translation*, London: Routledge.

Verba, S. (1969), The uses of survey research in the study of comparative politics: Issues and strategies, in S. Rokkan, S. Verba, J. Viet, and E. Almasy (Eds.), *Comparative Survey Analysis*, pp. 56–106, The Hague: Mouton.

Verba, S. (1971), Cross-national survey research: The problem of credibility, in I. Vallier (Ed.), *Comparative Methods in Sociology: Essays on Trends and Applications*, pp. 309–356, Berkeley: University of California Press.

Verba, S. (1980), On revisiting the Civic Culture: A personal postscript, in G. A. Almond, and S. Verba (Eds.) *The Civic Culture Revisited*, pp. 394–410, Boston: Little, Brown and Company.

Verba, S., Nie, N., and Kim, J.-O. (1978), *Participation and Political Equality: A Seven Nation Comparison*, Chicago: University of Chicago Press.

Verma, V. (2002), Comparability in multi-country survey programmes, *Journal of Statistical Planning and Inference*, 102(1), 189–210.

Verma, V., Scott, C., and O'Muircheartaigh, C. (1980), Sample designs and sampling errors for the World Fertility Survey, *Journal of the Royal Statistical Society*, Series A (General), 143(4), 431–473.

Villar, A. (2006), Agreement answer scales and their impact on response styles across cultures. Paper presented at the Annual Conference of the Midwest Association for Public Opinion Research (MAPOR), Chicago, IL.

Villar, A. (2008), What's behind cultural differences in response styles? Paper presented at the Conference on Multinational, Multicultural and Multiregional Survey Methods (3MC) Berlin, Germany. http://www.csdiworkshop.org/pdf/3mc2008_proceedings/session_51/Villar.pdf.

Villar, A. (2009), Agreement answer scale design for multilingual surveys: Effects of translation-related changes in verbal labels on response styles and response distributions. Unpublished doctoral dissertation, University of Nebraska, Lincoln, Nebraska.

von Elm, E., Altman, D. G., Egger, M., Pocock, S. J., Gøtzsche, P. C., and Vandenbroucke, J. P. (2008), The Strengthening the Reporting of Observational Studies in Epidemiology (STROBE) statement: Guidelines for reporting observational studies, *Journal of Clinical Epidemiology*, 61(4), 344–349.

Voogd, L. (2008), Panel on survey agencies: TNS. Paper presented at the International Conference on Survey Methods in Multinational, Multiregional, and Multicultural Contexts (3MC), Berlin, Germany. http://www.csdiworkshop.org/pdf/3mc2008_proceedings/session_05/de_Voogd.pdf.

Wagner, A. K., Gandek, B., Aaronson, N. K., Acquadro, C., Alonso, J., Apolone, G., et al. (1998), Cross-cultural comparisons of the content of SF-36 translations across 10 countries: Results from the IQOLA Project, International Quality of Life Assessment, *Journal of Clinical Epidemiology*, 51, 925–932.

Wagner, G. G., Frick, J. R., and Schupp, J. (2007), The German Socio-Economic Panel Study (SOEP)—Evolution, scope and enhancements, *Schmoller's Jahrbuch—Journal of Applied Social Science Studies*, 127, 139–169.

Wagner, J. R. (2008), Adaptive survey design to reduce nonresponse bias. Unpublished dissertation, University of Michigan, Ann Arbor, MI.

Wake, V. Y. (2006), The triad in prenatal genetic counseling discourse. Ph.D. Dissertation. Washington, DC: Georgetown University.

Wang, Q. and Ross, M. (2007), Culture and memory, in S. Kitayama and D. Cohen (Eds.), *Handbook of Cultural Psychology*, pp. 645–667, New York: Guilford Press.

Ward, C., Bochner, S., and Furnham, A. (2001), *The Psychology of Culture Shock*, London: Routledge.

Ware, J. E., Kosinski, M., and Keller, S. D. (1994), SF-36 physical and mental health summary scales: A user's manual, Boston, MA: The Health Institute.

Warner, U. and Hoffmeyer-Zlotnik, J. H. P. (2005), Measuring income in comparative social survey research,. *ZUMA Nachrichten Spezial—Methodological Aspects in Cross-National Research*, 11, 203–222.

Warwick, D. P. and Osherson, S. (1973), *Comparative Research Methods: An Overview*, Englewood Cliffs, NY: Prentice-Hall.

Watkins, D. (1989), The role of confirmatory factor analysis in cross-cultural research, *International Journal of Psychology*, 24, 685–701.

Watkins, D. and Cheung, S. (1995), Culture, gender, and response bias: An analysis of responses to the Self-Description Questionnaire, *Journal of Cross-Cultural Psychology*, 26, 490–504.

Watson, I. (2005), The earnings of casual employees: The problem of unobservables. Paper presented at the 2005 HILDA Survey Research Conference, The University of Melbourne, 29 September 2005.

Watson, N. (Ed.) (2005), *HILDA User Manual—Release 3.0*, Melbourne Institute of Applied Economic and Social Research, University of Melbourne.

Watson, N. and Wooden, M. (2009), Identifying factors affecting longitudinal survey response, in P. Lynn (Ed.), *Methodology of Longitudinal Surveys*, London: John Wiley & Sons.

Weber, R. P. (1985), *Basic Content Analysis*, London: Sage.

Weeks, M. F., Kulka, R. A., and Pierson, S. A. (1987), Optimal call scheduling for a telephone survey, *Public Opinion Quarterly*, 44, 101–114.

Weintraub, K. J. (1978), *The Value of the Individual: Self and Circumstance in Autobiography*, Chicago, IL: University of Chicago Press.

Weijters, B., Schillewaert, N., and Geuens, M. (2008), Assessing response styles across modes of data collection, *Journal of Academy of Marketing Science*, 36, 409–422.

Welkenhuysen-Gybels, J., Billiet, J., and Cambré, B. (2003), Adjustment for acquiescence in the assessment of the construct equivalence of Likert-type score items, *Journal of Cross-Cultural Psychology*, 34, 702–722.

Werner, O. and Campbell, D. T. (1970), Translating, working through interpreters, and the problem of decentering, in R. Naroll and R. Cohen, (Eds.), *A Handbook of Method in Cultural Anthropology*, pp. 398–420, New York, American Museum of Natural History.

White, T. (2008), An international conference examines stigma research, American Sociological Association Footnotes. Retrieved from: http://www.asanet.org/footnotes/mar08_R/fn5.html.

Widaman, K. F. (1985), Hierarchically nested covariance structure models for multitrait-multimethod data, *Applied Psychological Measurement*, 9, 1–26.

Willis, G. B. (2004), Overview of methods for developing equivalent measures across multiple cultural groups, in S. B. Cohen and J. M. Lepkowski (Eds.), Eighth Conference on Health Survey Research Methods (DHHS No. PHS-04-1013), pp. 91–96, Hyattsville, MD: US Department of Health and Human Services.

Willis, G. B. (2005), *Cognitive Interviewing: A Tool for Improving Questionnaire Design*, Thousand Oaks, CA: Sage.

Willis, G. (2009), Developing a model to conceptualize the results of comparative pretesting. Paper presented at the International Workshop on Comparative Survey Design and Implementation, Ann Arbor, MI.

Willis, G. B. and Zahnd, E. (2006), Response errors in cross-cultural surveys. Paper presented at the American Association for Public Opinion Research, Montreal, Quebec, Canada, May 18–21.

Willis, G. B. and Zahnd, E. (2007), Questionnaire design from a cross-cultural perspective: An empirical investigation of Koreans and non-Koreans, *Journal of Health Care for the Poor and Underserved*, 18(4 Suppl.), 197–217.

Willis, G., Lawrence, D., Hartman, A., Kudela, M., Levin, K., and Forsyth, B. (2008), Translation of a tobacco survey into Spanish and Asian languages: The Tobacco Use Supplement to the Current Population Survey, *Nicotine and Tobacco Research*, 10(6), 1075–1084.

Wilss, W. (1998), Decision making in translation, in Baker, M. (Ed.), *Encyclopedia of Translation Studies*, pp. 57–60, London: Routledge.

Woehr, D. J., Arciniega, L., and Lim, D. (2007), Examining work ethic across populations: A comparison of the multidimensional work ethic profile across three diverse cultures, *Educational and Psychological Measurement*, 67, 154–168.

World Bank (2003), Sustainable Development in a Dynamic World: Transforming Institutions, Growth, and Quality of Life, New York: Oxford University Press.

World Bank (2006), Information and communications for development: Global trends and policies, Washington, DC: World Bank.

World Medical Association. (1964), World Medical Association Declaration of Helsinki, 2007, retrieved from http://www.wma.net/e/policy/pdf/17c.pdf.

Wothke, W. and Browne, M. W. (1990), The direct product model for the MTMM matrix parameterized as a second order factor analysis model, *Psychometrika*, 55, 255–262.

Wyer, R. S. (1969), The effects of general response style on measurement of own attitude and the interpretation of attitude-relevant messages, *British Journal of Social & Clinical Psychology*, 8, 104–115.

Yamagishi, T. and Yamagishi, M. (1994), Trust and commitment in the United States and Japan, *Motivation and Emotion*, 18, 129–166.

Yamaguchi, S. (1994), Collectivism among the Japanese: A perspective from the self, in U. Kim, H. C. Triandis, C. Kagitcibasi, S-C. Choi, and G. Yoon (Eds.), *Individualism and Collectivism: Theory, Method, and Applications*, pp. 175–188, London: Sage.

Yang, C. K. (1962), Religion in Chinese society, A study of contemporary social functions of religion and some of their historical factors, *American Sociological Review*, 27, 439–440.

Yang, Y., Chin, T. Z., Harkness, J. A., and Villar, A. (2008), Response styles and comparative research. Paper presented at The International Conference on Survey Methods in Multinational, Multiregional, and Multicultural Contexts, Berlin, Germany. http://www.csdiworkshop.org/pdf/3mc2008_proceedings/session_50/Yang.pdf.

Yirmibesoglu, V. (1997), Compilation of newspaper articles on honor killings. Paper presented to Women for Women's Human Rights, Istanbul. WWHR Archive.

Yount, K. M. and Rashad, H. (Eds.) (2008), *Family in the Middle East: Ideational Change in Egypt, Iran, and Tunisia*, Oxford, UK: Routledge.

Yu, D. S. F., Lee, D. T. F., and Woo, J. (2004), Issues and challenges of instrument translation, *Western Journal of Nursing Research*, 26, 307–320.

Zax, M. and Takahashi, S. (1967), Cultural influences on response style: Comparisons of Japanese and American college students, *Journal of Social Psychology*, 71, 3–10.

Zhang, Z. X. (2007), Obstacles to harmony in organization: Mismatch between leadership philosophy and employee value, *PKU Business Review*, 1, 24–29. [In Chinese]

Zheng, X. and Rabe-Hesketh, S. (2007), Estimating parameters of dichotomous and ordinal item response models with GLLAMM, *The Stata Journal*, 7, 313–333.

Zhou, B. and McClendon, M. J. (1999), Cognitive ability and acquiescence, *Proceedings of the Survey Research Methods Section*, American Statistical Association, 1003–1012.

Zucha, V. (2005), The level of equivalence in the ISSP 1999 and its implications on further analysis, in J. H. P. Hoffmeyer-Zlotnik and J. A. Harkness (Eds.), *ZUMA— Nachrichten Spezial No. 11, Methodological Aspects in Cross-National Research*, Mannnheim: ZUMA.

INDEX

WILEY SERIES IN SURVEY METHODOLOGY
Established in Part by Walter A. Shewhart and Samuel S. Wilks

Editors: *Mick P. Couper, Graham Kalton, J. N. K. Rao, Norbert Schwarz, Christopher Skinner*
Editor Emeritus: *Robert M. Groves*

The *Wiley Series in Survey Methodology* covers topics of current research and practical interests in survey methodology and sampling. While the emphasis is on application, theoretical discussion is encouraged when it supports a broader understanding of the subject matter.

The authors are leading academics and researchers in survey methodology and sampling. The readership includes professionals in, and students of, the fields of applied statistics, biostatistics, public policy, and government and corporate enterprises.

ALWIN · Margins of Error: A Study of Reliability in Survey Measurement
BETHLEHEM · Applied Survey Methods: A Statistical Perspective
*BIEMER, GROVES, LYBERG, MATHIOWETZ, and SUDMAN · Measurement Errors in Surveys
BIEMER and LYBERG · Introduction to Survey Quality
BRADBURN, SUDMAN, and WANSINK ·Asking Questions: The Definitive Guide to Questionnaire Design—For Market Research, Political Polls, and Social Health Questionnaires, *Revised Edition*
BRAVERMAN and SLATER · Advances in Survey Research: New Directions for Evaluation, No. 70
CHAMBERS and SKINNER (editors) · Analysis of Survey Data
COCHRAN · Sampling Techniques, *Third Edition*
CONRAD and SCHOBER · Envisioning the Survey Interview of the Future
COUPER, BAKER, BETHLEHEM, CLARK, MARTIN, NICHOLLS, and O'REILLY (editors) · Computer Assisted Survey Information Collection
COX, BINDER, CHINNAPPA, CHRISTIANSON, COLLEDGE, and KOTT (editors) · Business Survey Methods
*DEMING · Sample Design in Business Research
DILLMAN · Mail and Internet Surveys: The Tailored Design Method
FULLER · Sampling Statistics
GROVES and COUPER · Nonresponse in Household Interview Surveys
GROVES · Survey Errors and Survey Costs
GROVES, DILLMAN, ELTINGE, and LITTLE · Survey Nonresponse
GROVES, BIEMER, LYBERG, MASSEY, NICHOLLS, and WAKSBERG · Telephone Survey Methodology
GROVES, FOWLER, COUPER, LEPKOWSKI, SINGER, and TOURANGEAU · Survey Methodology, *Second Edition*
*HANSEN, HURWITZ, and MADOW · Sample Survey Methods and Theory, Volume 1: Methods and Applications
*HANSEN, HURWITZ, and MADOW · Sample Survey Methods and Theory, Volume II: Theory
HARKNESS, BRAUN, EDWARDS, JOHNSON, LYBERG, MOHLER, PENNELL, and SMITH (editors) · Survey Methods in Multinational, Multiregional, and Multicultural Contexts
HARKNESS, van de VIJVER, and MOHLER (editors) · Cross-Cultural Survey Methods
KALTON and HEERINGA · Leslie Kish Selected Papers
KISH · Statistical Design for Research
*KISH · Survey Sampling

*Now available in a lower priced paperback edition in the Wiley Classics Library.

KORN and GRAUBARD · Analysis of Health Surveys

LEPKOWSKI, TUCKER, BRICK, DE LEEUW, JAPEC, LAVRAKAS, LINK, and SANGSTER (editors) · Advances in Telephone Survey Methodology

LESSLER and KALSBEEK · Nonsampling Error in Surveys

LEVY and LEMESHOW · Sampling of Populations: Methods and Applications, *Fourth Edition*

LUMLEY · Complex Surveys: A Guide to Analysis Using R

LYBERG, BIEMER, COLLINS, de LEEUW, DIPPO, SCHWARZ, TREWIN (editors) · Survey Measurement and Process Quality

MAYNARD, HOUTKOOP-STEENSTRA, SCHAEFFER, VAN DER ZOUWEN · Standardization and Tacit Knowledge: Interaction and Practice in the Survey Interview

PORTER (editor) · Overcoming Survey Research Problems: New Directions for Institutional Research, No. 121

PRESSER, ROTHGEB, COUPER, LESSLER, MARTIN, MARTIN, and SINGER (editors) · Methods for Testing and Evaluating Survey Questionnaires

RAO · Small Area Estimation

REA and PARKER · Designing and Conducting Survey Research: A Comprehensive Guide, *Third Edition*

SARIS and GALLHOFER · Design, Evaluation, and Analysis of Questionnaires for Survey Research

SÄRNDAL and LUNDSTRÖM · Estimation in Surveys with Nonresponse

SCHWARZ and SUDMAN (editors) · Answering Questions: Methodology for Determining Cognitive and Communicative Processes in Survey Research

SIRKEN, HERRMANN, SCHECHTER, SCHWARZ, TANUR, and TOURANGEAU (editors) · Cognition and Survey Research

SUDMAN, BRADBURN, and SCHWARZ · Thinking about Answers: The Application of Cognitive Processes to Survey Methodology

UMBACH (editor) · Survey Research Emerging Issues: New Directions for Institutional Research No. 127

VALLIANT, DORFMAN, and ROYALL · Finite Population Sampling and Inference: A Prediction Approach